D. Hamilton · **Diagnostic Nuclear Medicine**

Springer

Berlin
Heidelberg
New York
Hong Kong
London
Milan
Paris
Tokyo

David Hamilton

Diagnostic Nuclear Medicine

A Physics Perspective

With Contribution by Peter J. Riley

With 145 Figures and 34 Tables

Springer

David Hamilton, PhD FIPEM
Riyadh Al Kharj Hospital Programme
X990 Military Hospital
PO Box 7897
Riyadh 11159
Saudi Arabia

Contributor

Peter J. Riley, MSc
Riyadh Al Kharj Hospital Programme
X990 Military Hospital
PO Box 7897
Riyadh 11159
Saudi Arabia

ISBN 3-540-00690-7 Springer-Verlag Berlin Heidelberg New York

Library of Congress Cataloging-in-Publication Data
Hamilton, David (David I.), 1951– Diagnostic nuclear medicine : the physics perspective / David Hamilton, with contributions by Peter J. Riley. p. ; cm. Includes bibliographical references and index.
ISBN 3-540-00690-7 (hard : alk. paper)
1. Medical physics. 2. Nuclear medicine. 3. Radioisotope scanning. I. Riley, Peter J., MSC. II. Title.
[DNLM: 1. Health Physics, 2. Radionuclide Imaging. 3. Nuclear Medicine–methods. 4. Quality Assurance, Health Care. 5. Radiation Protection. WN 203 H217 2004
R895.H355 2004 616.07'575–dc22 2004041650

Springer-Verlag is a part of Springer Science+Business Media
springeronline.com

Cover design: Erich Kirchner, Heidelberg
Typesetter: K. Detzner, Speyer

Printed on acid-free paper 21/3150hs 5 4 3 2 1 0

To Tricia, Patrick, Edward, Suzanne and Simon

Acknowledgements

I should like to thank Dr UJ Miola, and all the staff and students in nuclear medicine at RKH for their inspiration; Nicola Kay Morgan, of the Medical Illustration Department at RKH, for her imagination in producing the illustrations; Dr Robert Shields at Manchester Royal Infirmary, Dr Sharok Kimiaei at King Faisal Specialist Hospital and Michael Northrop at RKH for their reviews of the manuscript; and Dr Ute Heilmann at Springer-Verlag for encouragement of the project.

Physics is about seeing further, better, deeper

Stephen Hawking

Preface

In the development of many medical technologies the beginning is characterised by an emphasis on the basic scientific principles of the technology and the optimisation of the functional aspects of the technology. As a technology matures there is a tendency for the underlying principles to be forgotten as the clinical applications begin to develop and the focus moves to an understanding of the clinical application. This maturity brings with it new challenges for those involved in the use of the technology. An acceptance of the methodology may lead to a scaling back of the basic training of staff into the fundamentals of the techniques and lead to a lack of questioning as to those issues which lead to the optimisation in clinical applications. This lack of basic training may ultimately lead to a stifling of research and development of the technology as a whole as trained staff becomes a scarce commodity.

Nuclear medicine is no exception to this development cycle. As a medical specialty the discipline has matured. The basic imaging technology has become more reliable in everyday use requiring less input from scientific staff. Clinical procedures have become protocols which are often followed without due understanding of the basic principles underlying the imaging procedure. This is clearly demonstrated when new radiopharmaceuticals are introduced into the market place. The optimisation of the imaging process becomes the major obstacle to the clinical utility of the product and is exacerbated by the lack of skilled staff and basic learning material. The introduction of new radiopharmaceuticals is the life blood of nuclear medicine in all its facets and each new product will bring its own challenges in its application and optimisation. The radiopharmaceutical component must also be seen alongside the recent development of positron emission tomography as a clinical tool and the process of development turns full circle with the introduction and rapid development of a new technology in the combination of positron emission tomography with X-ray computed tomographic imaging.

The process of optimisation of a clinical procedure requires an in-depth knowledge of all aspects of the diagnostic or therapeutic process. The clinical question to be investigated must be clearly defined and the components of the process which impact upon the outcome of the patient procedure optimised in the light of this clinical question. This process is encompassed in the concept of 'clinical quality control'.

This textbook provides a resource developed from a clinical quality control viewpoint. As such it is aimed at all the professional groups which combine to deliver nuclear medicine services. From an initial discussion of the fundamentals of radiation science, radiation biology and radiation protection the authors move through the fundamentals of detection systems, both imaging and non-imaging to sections on clinical procedure optimisation and quality control. The clinical procedures section provides detailed guidance on the principles of optimisation in the clinical context and will equip the reader with an understanding which should permit the introduction of new procedures. The book should prove an invaluable resource for clinicians, scientists and technologists. Its concentration on quality control in the clinical context is timely and essential in a matured discipline. However, this in no way detracts from usefulness of the book into the future as this emphasis on the combination of basic principles and clinical application must be the foundation of future developments.

The authors have written the book from their considerable experience in the delivery of clinical nuclear medicine services as well as the regular provision of national and international training courses to a range of professional staffs. An active interest in the research and development of the subject is clearly reflected in the presentation of the subject material. Whilst the authors come from a scientific and technological background the material has been presented so as to be accessible to all those practicing nuclear medicine or undertaking academic courses at the outset of their professional careers.

Professor Peter Jarritt
Chief Executive, Northern Ireland Regional Medical Physics Agency

Contents

Introduction

Abbreviations

DPET dual photon emission tomography
FDG fluorodeoxyglucose
SPET single photon emission tomography

The charm of diagnostic nuclear medicine derives from its ability to provide functional information, as a number, graph or image; by assessing the regional accumulation of administered radioactive materials. Although images reflect anatomical constraints, the discipline is complementary to other imaging modalities, which are essentially morphological.

Nuclear medicine only became a recognised medical specialty in the 1960s, but its history dates back to the 1896 discovery of natural radioactivity. Early use was exclusively therapeutic, with biological tracer studies only beginning in the 1920s, using radium. The discipline then came to be dominated, for three decades, by one radioactive element, iodine, which was used for both the investigation and treatment of thyroid disease. Initially, in the 1930s, it was the naturally occurring ^{130}Iodine that was used. Such naturally occurring radionuclides are not suitable for medical use, however, delivering high radiation doses to the patient because of their long physical half-lives and their biologically damaging radioactive emissions. This problem was alleviated by the commercial introduction of artificial radionuclides in 1946, following the development of the cyclotron charged particle accelerator in 1930 and the nuclear reactor in 1942. Reactor products were the ones first introduced into medicine, with ^{131}Iodine replacing ^{130}I and maintaining the dominance of the element into the 1950s, during which time it was likely to be the only radionuclide in use for the clinical work of a nuclear medicine department. Radionuclides continued to be almost exclusively produced by reactor until the 1960s, when products from the first medical cyclotron, built in 1953 at Hammersmith Hospital and opened in 1955 (McCready 2000), and that built at Oak Ridge National Laboratory, in 1956, were made available for clinical work. In the 1960s, most imaging using these types of radionuclide, was restricted to centres in close proximity to a cyclotron because detection of brain tumours was a priority, and clearer images could be obtained by detecting the annihilation photons from the short half-life β^+ emitting radionuclides produced in these machines. With the development of better imaging equipment, good images could eventually be obtained using non-annihilation photons. The real growth in nuclear medicine dates from 1962, when the single photon emitting metastable radionuclide ^{99}Technetium^m (^{99}Tc^m) was introduced. This has taken over the role of iodine as the dominant radionuclide. Because of its suitability for diagnostic imaging, by 1972, it was being used in approximately half of all clinical nuclear medicine investigations undertaken in the US, increasing to around 90% at the present time. A similar dependence on one radionuclide has arisen in dual photon imaging. Here, the major development occurred in 1978 with the production of fluorodeoxyglucose (^{18}FDG) which is now used in around 98% of clinical investigations.

In the 1930s and 1940s, the external mapping of in-vivo radionuclide distributions involved hand held gas ionisation detectors (Wagner 1998), a technique which continued even into the 1960s (McCready 2000). The number of photons available for such detection is severely limited, however, because of the limitation on the amount of radioactive material that can be administered to a patient. For such work, the detection sensitivity of these instruments is not adequate and the alternative of scintillation detection is more appropriate. These were first used clinically, for brain imaging, in 1952. The scintillant of choice in all current commercial nuclear medicine, non-annihilation photon, imaging systems, the thallium activated sodium iodide crystal NaI(Tl), was introduced in 1954. It has now replaced gas ionisation as the detector of choice in other nuclear medicine counting equipment, such as sample and contamination counters. Gas detectors remain useful mainly for high activity measurements, for instance as activity calibrators or exposure meters. Even with improved sensitivity, the use of hand held detectors for distribution mapping is an extremely tedious experience for both patients and staff, and provides only a very crude mapping of the distribution of radionuclide. To improve on this, automated methods of obtaining repeat measurements were developed; employing mechanisms for moving the detector across the patient, uniformly and continuously. This was implemented as the rectilinear scanner, which was first used in 1955. An example is shown in Fig. 1. The main limitation of the rectilinear scanner is its long acquisition duration, which is a consequence of its sequential data acquisition. Direct imaging, by allowing a large area of scintillation crystal to be exposed simultaneously, became possible in the early 1960s with the introduction of the γ camera (Anger 1964), see Fig. 2.

Great work was undertaken using the early γ cameras but, since their introduction, remarkable improvements have been made, particularly in terms of spatial linearity

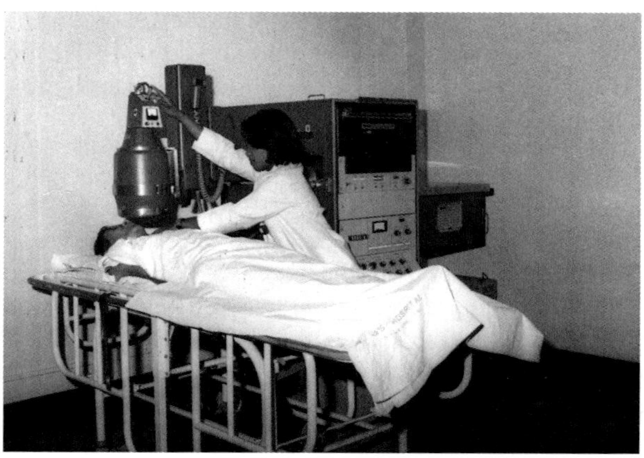

Fig. 1. Rectilinear scanner

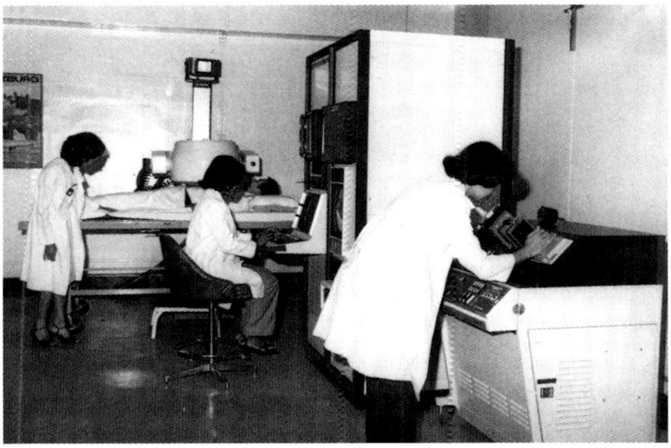

Fig. 2. Early gamma camera

and resolution, and uniformity and energy resolution, which have resulted in vastly improved images. Initially intended for investigating static radionuclide distributions, their potential for performing dynamic studies was first recognised in the late 1960s and made available for routine work in smaller hospitals during the middle 1970s. The interfacing of dedicated computers allowed the processing of images and the generation of numerical and graphical representations of the data. This application, of a dedicated clinical data acquisition and processing computer system, was one of the first in medical imaging. During this period also, the field of view of the cameras was increased which, in addition to improving static and dynamic imaging, enhanced the quality of whole body imaging investigations. Image contrast and spatial appreciation, in certain investigations, were improved by the widespread availability of single photon tomographic (SPET) facilities in the 1980s. The fields of view were changed from circular to rectangular in the middle 1990s. Also at this time, multiple camera heads were introduced to allow radiation to be detected from different orientations, simultaneously, thereby addressing the fundamental restriction to detection sensitivity imposed by the low photon emission rate from the body. The recent significant advances in digital electronics and computing has initiated a rapid further advancement in image quality and information extraction. Sophisticated acquisition and analysis packages are now available for routine use and fusion with images from other modalities is increasingly utilised. Correction techniques have moved from the research arena into routine practice, including scatter, resolution and transmission attenuation compensation methods. This period has also seen the clinical implementation of positron or dual photon emission tomography which had started as, and for a long time remained, a research tool. Dedicated systems have proliferated but also being evaluated are coincidence gamma cameras, capable of both single and dual photon (DPET) detection (Wagner 1998).

These developments have led to the present status, which is one of a mature, versatile, accomplished, but surprisingly complicated imaging modality. With all of the enormous improvements in quality, the question is still asked: "Why is it that with techniques that have been around for decades there is a failure to agree on acquisition, processing and display protocols?" (Coakley 2000). There is a need for consistency in practicing the discipline to ensure the overall quality of the investigations (Bergmann et al. 1995) and it is suggested that it may be prudent to standardise acquisition and processing techniques so that inter-centre variation is reduced as much as possible (Tindale and Barber 2000).

Aim of the Book

Diagnostic nuclear medicine is unique in its reliance on, and benefit from, the application of a range of advanced physical principles directly to the clinical situation. It is the integration of these principles with a knowledge of physiology and pathology that is essential for effectively producing and interpreting the diagnostic information. The aim of this book is to describe the practice of diagnostic nuclear medicine from the physics perspective, in order to help the practitioner maximise the diagnostic information available and ensure that the available resources are used in the best way by optimising the performance of the investigations. This is done by documenting the principles which determine the radiations, radioactive materials, production methods and safety restraints that must be used, the design parameters and the operating characteristics of the equipment, the techniques for optimising the data, and the methods for ensuring maximal information quality. In this way, the potential for and the constraints on both the discipline and the diagnostic information are explored.

Structure of the Book

The book is organised into four sections: Radiation, Detection Systems, Clinical Procedures and Quality Assurance.

Radiation

This section describes the characteristics required of the radioactivity and radiopharmaceuticals. It discusses the dangers involved in their use and the techniques necessary to protect both the environment and the patient.

Detection Systems

This section describes the basic principles of radiation detection, including a consideration of measurement statistics. It describes the principles of various types of equipment, including: non-imaging systems, gamma cameras, tomographic systems and computers, emphasising their performance characteristics, potentials and limitations.

Clinical Procedures

This section describes how the equipment must be used in order to optimise the various clinical investigations: non-imaging, single photon planar and tomographic imaging, and dual photon tomographic imaging.

Quality Assurance

This section describes the principles of quality assurance and the methods of testing the radiopharmaceuticals and the various types of equipment: non-imaging and probe detectors, gamma cameras, coincidence gamma cameras and dedicated dual photon imaging systems, displays and computers.

References

Anger HO (1964) Scintillation camera with multichannel collimators. J Nucl Med 5:515–531
Bergmann H, Busemann-Sokole E, Horton PW (1995) Quality assurance and harmonisation of nuclear medicine investigations in Europe. Eur J Nucl Med 22:477–480
Coakley AJ (2000) Editor's comment. Nucl Med Commun 21:220
McCready VR (2000) Milestones in nuclear medicine. Eur J Nucl Med 27(Suppl):S49–79
Tindale WB, Barber DC (2000) Diagnostic impact: a challenge for quantitative nuclear medicine. Nucl Med Commun 21:217–219
Wagner HN (1998) A brief history of positron emission tomography (PET). Semin Nucl Med XXVIII:213–220

PART I

Radiation

Radioactivity

Abbreviations

EC electron capture
SI System Internationale

The radiations required to practice diagnostic nuclear medicine are mainly medium energy photons (γ and x), with the discrete energy of γ photons being optimal for detection (NCRP 1983). An indication whether unwanted and hazardous particulate radiation (e^-, β^-, Auger e^-), low energy photon radiation (characteristic x), or the precursor of photon radiation (β^+) will also be produced, can be obtained by consideration of the decay mechanism of the radioactive atom. All radionuclides used in diagnostic nuclear medicine are produced artificially and are chosen on the basis of their emissions and to have half-lives commensurate with the duration of the investigations. Some useful radionuclides do, however, have half-lives which are inconveniently short for logistical purposes. In such circumstances, a system of a parent decaying to a daughter has to be used. Radionuclide production is by nuclear reactor or charged particle accelerator; the method determining which decay mechanism operates and which radiations will be produced.

Radionuclide Characteristics Required for Diagnosis

The half-life of the radionuclide should be sufficiently long to allow measurement throughout an investigation but must not be so excessive as to impose an unduly high radiation burden on either the patient or close contacts. It must also allow for logistical considerations such as quality control testing, transport and storage before use. The main characteristics of radionuclides in use in clinical nuclear medicine are summarised in Appendix V.1. The requirements vary slightly among the different categories of investigation.

Imaging

An ideal radionuclide for diagnostic imaging is one which disintegrates with the emission of photon radiation of high enough energy to escape the body, without experiencing too much attenuation, but of low enough energy to avoid degrading the sensitivity of the detector. γ photons are preferred over particulate radiation, because they interact less with tissue thus causing less damage and being more likely to escape the body. The probability of interaction does increase with reducing energy, however, and the preferred energy range is >100 keV. Energies below this can be tolerated but the numbers of photons available for external detection are increasingly reduced by interaction within tissues over-lying the source. x and bremsstrahlung radiation can also be used to create images but, being low energy, suffer the same disadvantages as low energy γ emissions. High energy emissions reduce the sensitivity of the detection system, and therefore the preferred upper limit, for single photon detection systems, is ~200 keV. Energies above this can be used but the probability of interaction with the detection system will be reduced and the sensitivity for detecting emissions decreased. Collimation of the higher energy photons also causes degradation in the image. In dual photon systems, the detected radiation has an energy of 511 keV, but extremely good images are produced because of the specialised imaging technique used.

Non-Imaging, External Probe Detection

These single photon systems can operate at a much wider energy range than imaging systems. They work best with photon emissions of sufficiently high energy to avoid excessive attenuation in over-lying tissues but, because they do not produce an image, can better accommodate the reduced photon numbers occasioned by lower emission energies. Also, because their radiation stopping properties are better than the imaging systems, they do not experience as severe a reduction in detection sensitivity at higher emission energies, as the imaging systems.

Non-Imaging, Sample Measurement

The energy characteristics required for diagnostic in-vitro studies are even less critical than for probe detection. Photon radiation is still preferred to particulate because of lower tissue damage, except for a particular type of sample detection system which utilises β^- radiation. A much wider photon energy range can be accommodated. Lower energies can be tolerated because tissue attenuation is not a compromising factor and exclusion of scattered radiation is not important. Higher energies are acceptable because, as in probe systems, radiation stopping properties are good. Also measurement duration can be increased, without inconveniencing patients and staff, to compensate for any reduction in detection sensitivity.

Radioactive Decay Mechanisms

The emission characteristics of radionuclides can be predicted to a certain extent by considering their relationship to stable nuclides. A chart of such stable nuclides (see Fig. I.1.1) shows a curvilinear relationship between the number of neutrons, N, and the number of protons, atomic number = Z, in the nucleus. For light nuclides, $Z <\sim 20$, the stable ratio is the line of identity, $N \approx Z$; but as Z increases, stability tends towards a higher neutron to proton ratio, $N \approx 1.5 \times Z$. The line of stability ends at Bismuth, mass number, A = 209; all heavier nuclides being unstable.

Nuclides situated away from the line of stability are radioactive and will alter the nucleon ratio in order to achieve stability. In general, the further they are away from the line of stability and the greater the imbalance of the neutron to proton ratio, the more incentive there is to reorganise this ratio and the shorter their half-lives will be. Radioactive decay results in the release of nuclear energy, most of which is imparted to emitted particles and photons with a small, usually insignificant fraction, being imparted to the nucleus in terms of recoil energy. The source of this energy is a conversion of some of the mass of the nucleus. Each radionuclide has a unique set of disintegration characteristics, including: mode of decay, emissions and average lifetime of the nucleus.

Radionuclides that are neutron rich will decay in order to reduce the neutron to proton ratio. Those of concern in nuclear medicine decay by β^- emission. Radionuclides that are proton rich will decay in order to increase the neutron to proton ratio, by electron capture or by β^+ emission. Exceptions to these rules are radionuclides which have odd numbers of both protons and neutrons and lie near the line of

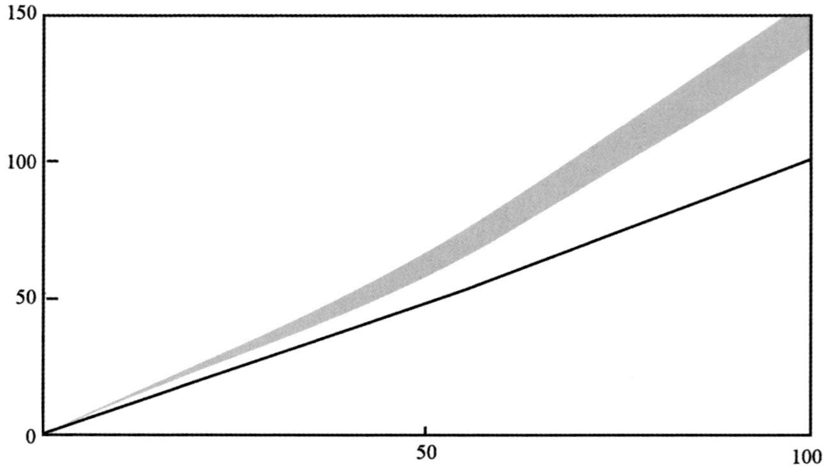

Fig. I.1.1. Chart of nuclide stability, showing line of identity and approximate stability zone. The ordinate represents the number of neutrons, N, and the abscissa the number of protons, atomic number = Z

stability. The instability of these radionuclides originates from this oddness and they can decay in either direction away from the line of stability. Secondary emissions are associated with these events, comprising: γ, characteristic x-ray and e^-.

β^- Emission, Neutron Rich

For neutron rich radionuclides, the decay to stability is achieved by expelling an electron from the nucleus which, because it originates in the nucleus, is called a β^- particle. Essentially, a neutron is transformed into a proton, represented by:

$${}^{A}_{Z}P \xrightarrow{\beta^-} {}^{A}_{Z+1}D$$

The parent radionuclide (P) and daughter product (D) represent different chemical elements because (Z) increases by one, thus the process results in a transmutation of elements. (A) does not change, that is the total number of nucleons remains the same, and this is therefore an isobaric decay mode, the parent and daughter being isobars. The energy released is characteristic of the decay process and is shared between the β^- particle and a neutrino (ν):

$$n \rightarrow p^+ + \beta^- + \nu + \text{kinetic energy}$$

The neutrino has no mass and no electrical charge. It undergoes virtually no interactions with matter and is, therefore, essentially undetectable. Its only practical consequence is that it carries away some of the energy released in the decay process. The energy sharing is, more or less, random and the β^- particles are emitted with a continuous distribution of energies up to a maximum. They usually receive less than half and only rarely all of the energy. There is a characteristic average for each radionuclide which is ~1/3 of the maximum. When the β^- particle leaves the atom, it interacts with the orbital electrons of other atoms and loses velocity; the energy released appearing as x-rays, labelled bremsstrahlung or braking radiation.

e^- Capture (EC) and β^+ Emission, Proton Rich

For proton rich radionuclides, the decay to stability is achieved either by the nucleus capturing an e^- (EC) or by expelling a β^+; essentially resulting in the transformation of a proton into a neutron. EC occurs more frequently in the atoms with higher (Z) because the inner orbital electrons tend to be closer to the nucleus and are more easily captured. In lower (Z) atoms, EC gives way to β^+ emission. However, for the β^+ decay to occur, the difference in energy between the parent and the daughter nuclei must be at least 1.022 MeV, a figure which is explained in the following section describing β^+ decay. If this is not the case, then only EC can occur. Some radionuclides can decay by either mode, and for each there is a fixed proportion of disintegrations which take each route.

e^- Capture

Nuclear capture of an orbital e^-, usually from an inner orbital, can be represented by:

$$^{A}_{Z}P \xrightarrow{EC} {}^{A}_{Z-1}D$$

Again, this is an isobaric decay with a transmutation of elements and, again, a neutrino is emitted from the nucleus and carries away some of the transition energy.

$$p^+ + e^- \rightarrow n + \nu + \text{kinetic energy}$$

The orbital electron vacancy left in the daughter atom will be filled by a more peripheral orbital e^-, repeated in cascade fashion, until the outermost vacancy is filled by a free e^- (see Fig. I.1.2). The energy released by electrons cascading to inner orbitals will appear as photons which are called x-rays, because their origin is in the elec-

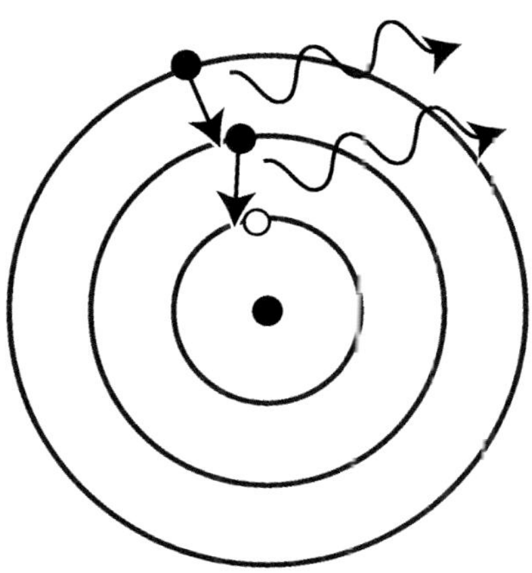

Fig. I.1.2. Cascade of orbital e^- filling a vacancy and causing emission of characteristic x-rays

tron orbitals. They are characteristic of the initial radionuclide, the parent, not of the daughter and are called characteristic x-rays.

β^+ *Decay*

Positron (positively charged electron) emission, from the nucleus, can be represented by:

$$^{A}_{Z}P \xrightarrow{\beta^+} {}^{A}_{Z-1}D$$

Again, this is an isobaric decay with a transmutation of elements and the β^+ emission occurs with an energy spectrum similar to that of β^-.

$$p^+ \rightarrow n + \beta^+ + \nu + \text{energy}$$

The minimum transition energy requirement of 1.022 MeV occurs because of the need to conserve the energy equivalence of two electrons; the β^+ itself and an orbital e^-, which is lost due to the decrease in the Z. Therefore the maximum kinetic energy, carried away by the β^+ particle, is the transition energy minus this 1.022 MeV. Being positively charged and of the same mass as an electron, the β^+ particle is extremely reactive. Once outside the atom, it loses its kinetic energy in collisions with the orbital electrons of other atoms and approaches rest after a tortuous path, usually within a few mm and $\sim 10^{-9}$ s. It will then interact with a free e^- during which their masses are converted into energy causing the emission of two 0.511 MeV annihilation photons which leave the site of the annihilation in almost opposite directions.

γ Emission

When the nucleus has decayed by any one of the above processes, it may be left in an excited state. In which case being unstable, it will adjust its energy levels, to achieve a ground state. If the energy adjustment is $>\sim 100$ keV, a photon will be emitted which, because of its origin in the nucleus, is called a γ photon. If the excited state has a relatively long life-time ($>\sim 10^{-12}$ s) it is called metastable or isomeric. γ emission does not result in transmutation and the process may be represented by:

$$^{A}_{Z+1}D^* \xrightarrow{\gamma} {}^{A}_{Z+1}D$$

More than one γ photon may be emitted before the excited radionuclide achieves the ground state. The relative proportions of these are determined by probability values characteristic of the metastable nucleus. Unlike β^- particles, γ photons are emitted with discrete energies which are characteristic of the daughter nucleus. These are the most appropriate emissions for use in diagnostic nuclear medicine because they contribute to the diagnostic information and involve a low radiation burden to the patient. However, a γ photon may not escape the atom cleanly, but may be internally converted, in which case there is reduced contribution to the diagnostic information and there is an increased radiation dose to the patient because of the resulting emissions.

e^- and x Photon Emission from Internal Conversion

The γ photon may interact with the orbital, usually inner, electrons of its own atom. Such an interaction may reduce the energy of the photon or may cause its total elimination in what is called internal conversion. In the latter case, the orbital electron with which interaction occurred will be expelled from the atom as a fast e^-, called a conversion e^- (see Fig. I.1.3). The orbital electron vacancy will be filled by the cascade process described for e^- capture.

The energy of the γ photon, in excess of the binding energy of the ejected orbital e^-, is converted to kinetic energy of the conversion e^-. This highlights the important difference between this process and β^- decay in that conversion electrons are emitted with discrete, not continuous, energies. In the decay of metastable atoms, some conversion electrons will always be emitted and thus metastable radionuclides cannot really be considered to be "pure" γ emitters. Whether a γ photon or a conversion electron is emitted is a matter of probabilities, which are characteristic to each radionuclide. The probability of internal conversion increases as Z increases because, with higher Z, the inner e^- shell will be closer to the nucleus and there will be a greater chance of interaction. The probability also increases with reducing photon energy because it will be more comparable with the binding energies of the electrons. For instance, 9% of γ photons emitted from $^{99}Tc^m$ are internally converted whereas from ^{125}I, with a much lower photon energy, the proportion is 93%.

Auger e^-

A characteristic x photon emitted by the cascade process described above may, itself, not escape the confines of the atom. It may interact with an outer orbital e^-. In this situation it is likely to be eliminated and the orbital e^- expelled from the atom as a very low energy Auger e^- (see Fig. I.1.4). The energy is low because it is the difference

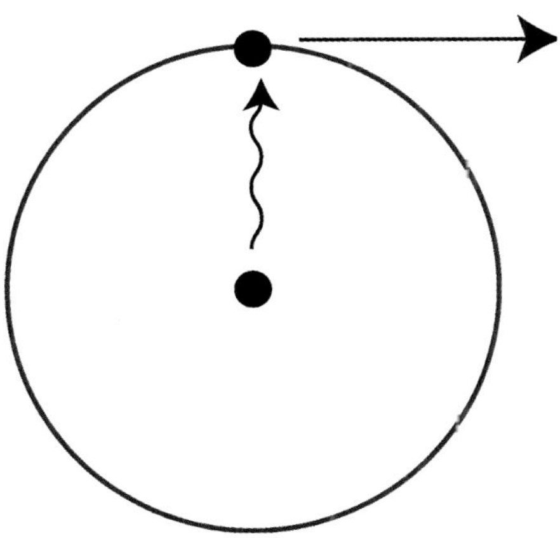

Fig. I.1.3. Internal conversion of a γ photon resulting in the emission of a conversion e^-

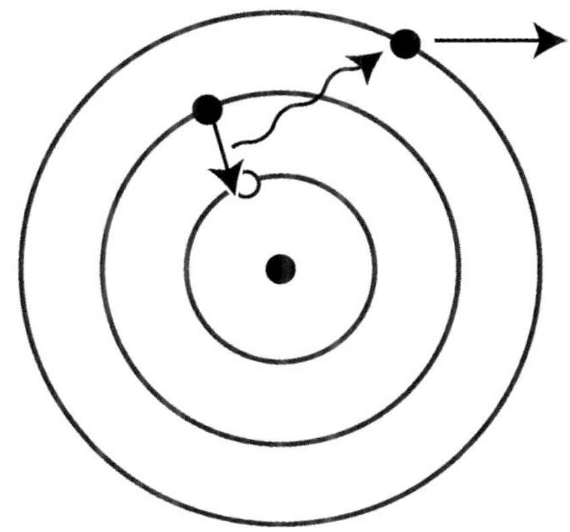

Fig. I.1.4. Interaction of characteristic x photon and outer orbital e^- causing emission of Auger e^-

between the energy of the preceding e^- re-arrangement and the binding energy of the ejected e^-. Following such a process, the orbital vacancy, which contained the Auger e^-, will be filled by electrons from more peripheral orbitals, continuing the cascade process and resulting in the emission of additional characteristic x-rays or Auger e^-. The probability that an orbital vacancy will result in the emission of a characteristic x-ray or an Auger e^- is termed the fluorescent yield (ω). Heavier atoms, $Z > \sim 35$, have a high ω and are more likely to emit characteristic x-ray photons, whereas lighter atoms, with lower ω, are more likely to emit Auger e^-. In a particular type of Auger process, Coster-Kronig e^- are emitted; when transitions are restricted to those between e^- sub-shells of the same major e^- shell.

Radioactive Decay

The decay of an unstable nucleus is not affected, to any significant extent, by events occurring outside the nucleus. It is spontaneous or random, in that the exact moment at which a given nucleus will decay cannot be predicted, but the behaviour of a large number of nuclei can be described by equations, which are presented in the following sections. Comprehensive derivations can be found in texts such as Parker et al. (1984).

Decay Constant

For each radionuclide, there is a unique probability that a nucleus will decay in a particular period of time, determined by the energy levels of the nucleus. For example, a nucleus of ^{131}I has a 1 in 10^6 chance of decaying s^{-1}, which is called the decay constant, λ. A $\lambda = 0.01\ s^{-1}$ indicates that, on average, 1% of the radioactive atoms present will decay each second, but there is a statistical variation in the number of atoms decaying in each second. In radionuclides that have several different decay paths, each path (1 ... n) will have a particular λ with the total decay constant being: $\lambda = \lambda_1 + \lambda_2 + \lambda_3 + \ldots + \lambda_n$.

Activity

Activity, A = total number of nuclei decaying $s^{-1} = \lambda N$, where N is the number of radioactive atoms present in the sample. Because λ is a constant, A decreases as N decreases. The SI unit of activity is the Becquerel, Bq, which represents 1 radioactive disintegration s^{-1}; its relationship to the older Curie (Ci) unit is documented in Appendix V.2. The range of activities used in diagnostic nuclear medicine is large; activities of 100 GBq may be handled, 1–1000 MBq administered and as little as 10 Bq assayed.

Decay

Because a constant fraction of the total number of radioactive atoms decays per time interval, the decay describes an exponential pattern (see Fig. I.1.5), i.e. from ΔN atoms decaying in a short period of time Δt:

$$\Delta N = -\lambda N \Delta t$$

the result is:

$$N = N_0 e^{-\lambda t} \qquad (I.1.1)$$

where N represents the number of radioactive atoms present at time t and N_0 represents the number at time, t = 0. The decay factor $e^{-\lambda t}$ is the fraction of radioactive atoms remaining after time t.

Since $A = \lambda N$,

$$A = A_0 e^{-\lambda t} \qquad (I.1.2)$$

On a semi-log plot this will appear as a straight line. It will never reach zero, but will only tend towards it.

Half-Life

The half-life, $T_{1/2}$, of a radionuclide is the time required for its activity to decay to half of its initial level. At $T_{1/2}$,

$$A = 1/2\, A_0$$

and therefore:

$$0.693 = \lambda T_{1/2}$$

Once administered, the radionuclide is removed from the body physiologically with a biological half-life. Since the physical and biological eliminations effectively form two decay paths, the effective decay constant is:

$$\lambda_e = \lambda_{ph} + \lambda_b$$

and the effective half-life, T_e, is:

$$1/T_e = 1/T_{ph} + 1/T_b \qquad (I.1.3)$$

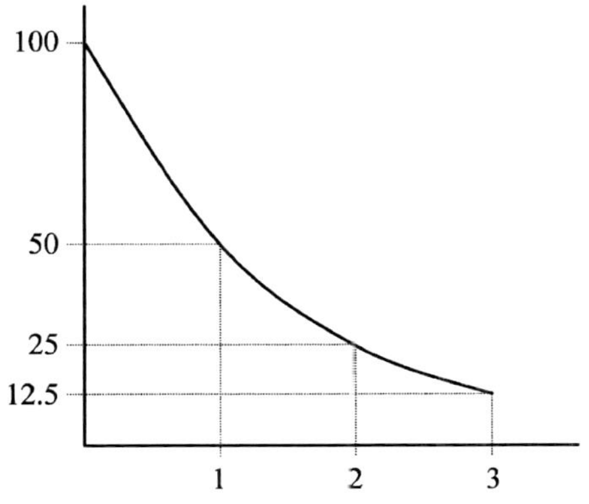

Fig. I.1.5. Exponential decay of radioactive atoms. The ordinate represents the percentage of the initial radioactive atoms remaining and the abscissa represents multiples of the half-life

The optimum physical half-life for radionuclides used in diagnostic nuclear medicine, is one which is similar to the half-time of the physiological process being investigated (NCRP 1983), and a rule of thumb is:

$$T_{ph} = 0.693\, T_i$$

where T_i is the time interval between the administration and the end of the investigation, which is usually determined by the physiological process.

Mean-Life

The average lifetime, τ, of a radionuclide is important in radiation dosimetry calculations, described in Chapter I.3, and is given by $\tau = 1/\lambda$.

$$\text{ie. } \tau = 1.44\, T_{1/2} \qquad (I.1.4)$$

Mixed Radionuclides

In a mixture of unrelated radionuclides, the total activity is the sum of the individual activities. With decay over time, the activity-time curve assumes that of the radionuclide with the longest $T_{1/2}$.

Parent-Daughter Decay

When the daughter nuclide is also radioactive, the activity equation describing the behaviour of the daughter is complicated because these atoms are decaying as they are being formed:

$$dN_p/dt = -\lambda_p N_p$$

$$dN_d/dt = -\lambda_d + \lambda_p N_p$$

the solution of these is the Bateman equation:

$$N_d = N_p(0)\, \lambda_p/(\lambda_d - \lambda_p)\, (e^{-\lambda p\, t} - e^{-\lambda d\, t}) + N_d(0)\, e^{-\lambda d\, t} \qquad (I.1.5)$$

$N_d(0)e^{-\lambda d\, t}$ describes the behaviour of the daughter atoms which were already present at time, t=0.

From above,

$$A_d = A_p(0)\, \lambda_d/(\lambda_d - \lambda_p)\, (e^{-\lambda p\, t} - e^{-\lambda d\, t}) + A_d(0)\, e^{-\lambda d\, t} \qquad (I.1.6)$$

This can be used to describe three particular circumstances. In the description, $A_d(0)e^{-\lambda d\, t}$ is assumed to be 0.

Parent Half-Life $\gg$ Daughter Half-Life (Secular Equilibrium)

When the half-life of the parent = 100–1000 times that of the daughter, the parent activity remains approximately constant, ie $\lambda_p \approx 0$:

$$A_d \approx A_p(0)\, (1 - e^{-\lambda d\, t})$$

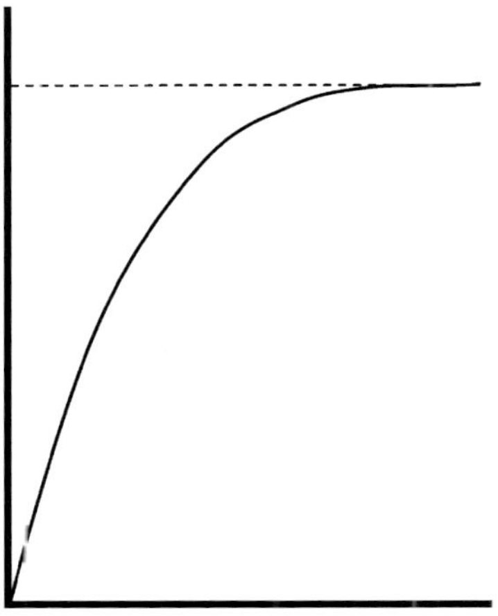

Fig. I.1.6. Secular equilibrium showing the build up of the daughter activity until it equals that of the parent. The ordinate represents activity and the abscissa represents time

The activity of the daughter builds up until it equals that of the parent, and the parent and daughter achieve secular equilibrium (see Fig. I.1.6):

$$A_d = A_p \qquad (I.1.7)$$

An example is the ^{81}Rb-$^{81}Kr^m$ generator.

Parent Half-Life > Daughter Half-Life (Transient Equilibrium)

When the half-life of the parent = 10–100 times that of the daughter, there is a decrease in the activity of the parent and $\lambda_p \neq 0$. In this case, the daughter activity builds up to exceed that of the parent, and then decreases in parallel with the parent; the parent and daughter achieving transient equilibrium (see Fig. I.1.7):

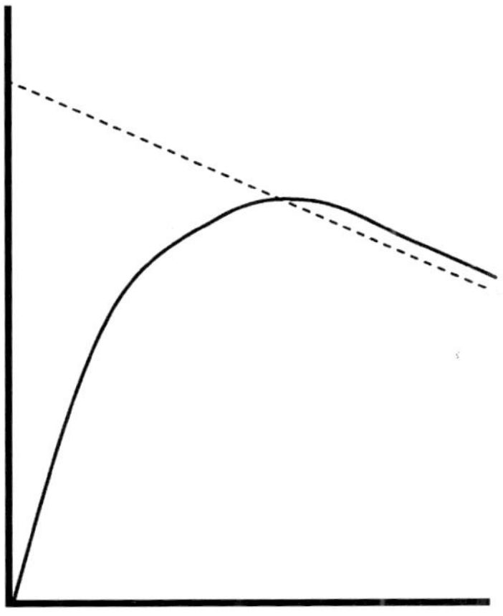

Fig. I.1.7. Transient equilibrium showing the daughter activity building up to exceed that of the parent, and then decreasing in parallel with the parent. The ordinate represents activity and the abscissa represents time

$$A_d = A_p T_p/(T_p - T_d) = A_p \lambda_d/(\lambda_d - \lambda_p) \qquad (I.1.8)$$

An example is the ^{99}Mo-^{99}Tc^m generator.

Parent Half-Life < Daughter Half-Life

When the half-life of the daughter is longer than that of the parent, equilibrium is not achieved. The daughter activity increases, until it reaches a maximum and then decreases. Eventually, when the activity of the parent is essentially zero, the remaining daughter activity decays with its own half-life.

Time to Maximum Daughter Activity

The time required for maximum daughter activity to be achieved can be calculated from,

$$\begin{aligned} A_d &= A_p(0)\,\lambda_d/(\lambda_d - \lambda_p)\,(e^{-\lambda_p t} - e^{-\lambda_d t}) \\ t_{max} &= 1.44 \ln(T_p/T_d)\, T_p T_d/(T_p - T_d) = \ln(\lambda_d/\lambda_p)/(\lambda_d - \lambda_p) \end{aligned} \qquad (I.1.9)$$

Production of Radionuclides

Two methods of radionuclide production are the nuclear reactor and the charged particle accelerator.

Reactor – Neutron Rich

The earliest method of producing commercial artificial radionuclides involved immersing a target material in the neutron flux of a nuclear reactor. Since neutrons (n) do not carry a net electrical charge, they are not repelled by the target nuclei, can interact with it and convert it into a radioactive nucleus in a process called neutron activation. Two reactions can occur. The most common is the (n, γ) reaction in which the product nucleus is formed in an excited state and immediately undergoes de-excitation with the emission of a γ photon:

$${}^{A}{}_{Z}X\,(n,\gamma)\,{}^{A+1}{}_{Z}X$$

The target and product nuclei are different isotopes of the same element, cannot be chemically separated and hence the product is not target- or carrier-free. Even in intense neutron fluxes, only a small fraction of the target nuclei are activated and thus the product is likely to have a very low specific activity because of a large amount of unactivated target material. Carrier-free products can be obtained using this reaction, however, when a short-lived intermediate product is activated:

$${}^{130}Te\,(n,\gamma)\,{}^{131}Te \xrightarrow{\beta^-} {}^{131}I$$

Also the activated nucleus, although still the same element, may become a different chemical species because of the breaking of chemical bonds by the large amount of energy involved in the activation. Such nuclei can be separated chemically from the non-activated carrier. Less common types of reaction result in the ejection of a proton (p^+) or an alpha particle, (α^{2+} ie $2p^-+2n$)

$$^{A}{}_{Z}X\,(n,p^+)\,^{A}{}_{Z-1}Y$$

$$^{A}{}_{Z}X\,(n,\alpha^{2+})\,^{A-3}{}_{Z-2}Y$$

In these reactions, the target and product nuclei do not represent the same element and can be separated chemically. An example is the ^{99}Mo – $^{99}Tc^m$ generator, in which the parent ^{99}Mo is obtained from a nuclear reactor, either by an (n,γ) reaction with ^{98}Mo as the target or by fission of uranium. Fission is being used more frequently, especially for higher activity generators; the advantage being that a higher specific activity is possible and therefore generator columns can be smaller allowing lower elution volumes, which may be required for some labelling procedures. A disadvantage is the higher level of radionuclide impurities and the greater processing required.

Accelerator – Proton Rich

The later method of commercial radionuclide production involves the charged particle accelerator, the cyclotron. Only positively charged particles, such as deuteron (d, ie p^++n) and α^{2+}, are used, since it is not practical to accelerate e^- in such equipment, due to large relativistic mass increases experienced even at relatively low energies. Usually, the product is a different element from the target because the addition of a positive charge changes the atomic number of the nucleus. Thus, cyclotron products can usually be separated, chemically, from the target and are therefore carrier-free. Cyclotrons generally produce smaller quantities of radioactivity than can be obtained from nuclear reactors, and their products tend to be more expensive. They are, however, attractive for diagnostic nuclear medicine studies because of the high photon to particle emission ratios that are obtained in EC decay. The activation reactions encountered in cyclotrons are:

$$^{A}{}_{Z}X\,(d^+,n)\,^{A+1}{}_{Z+1}Y$$

$$^{A}{}_{Z}X\,(\alpha^{2+},np^+)\,^{A+2}{}_{Z+1}Y$$

$$^{A}{}_{Z}X\,(\alpha^{2+},2n)\,^{A+2}{}_{Z+2}Y$$

$$^{A}{}_{Z}X\,(p^+,n)\,^{A}{}_{Z+1}Y$$

Impurities

In cyclotron production contamination, with other radionuclides created in the target, can occur. If administered, these will result in a higher radiation dose to the patient and, if of higher energy, will also degrade the image (NCRP 1983). The simple situation is when the contaminating radionuclide has a shorter half-life. Storing, for an appropriate time, allows the impurity to decay to an acceptable level before the product is used. An example is the production of ^{67}Ga by α^{2+} bombardment of copper. Natural copper is composed of 31% ^{65}Cu and 69% ^{63}Cu and the required reaction is $^{65}Cu(\alpha^{2+},2n)^{67}Ga$ but contamination from the reaction $^{63}Cu(\alpha^{2+},n)^{66}Ga$ occurs. ^{66}Ga, however, has a half-life of 9.3 h compared with the 78 h half-life of ^{67}Ga and so can be allowed to decay to acceptable levels, by storage for 1–2 days. A more difficult problem is when the impurity has a longer half-life. An example is the contamination of ^{123}I by ^{124}I and ^{125}I. ^{123}I is produced by the reaction $^{121}Sb(\alpha^{2+},2n)^{123}I$, ^{124}I by $^{121}Sb(\alpha^{2+},n)^{124}I$ and ^{125}I by $^{123}Sb(\alpha^{2+},2n)^{125}I$. Two solutions are feasible. An enriched target containing a higher percentage of ^{121}Sb can be used; this is expensive and will only minimise the ^{125}I, however. The energy of the cyclotron beam can be increased; pure ^{123}I being obtained, with a high yield, using high energy p^+ of ~200 MeV:

$$^{127}I\,(p^+,5n)^{123}Xe \xrightarrow{EC} {}^{123}I$$

If impure ^{123}I is used, an expiry time is usually placed on the product to avoid a build-up of the contaminants to unacceptable levels (Parker et al. 1984).

References

NCRP (1983) National Council on Radiation Protection and Measurements report 73. Protection in nuclear medicine and ultrasound diagnostic procedures in children. NCRP, Bethesda, Maryland

Parker RP, Smith PHS, Taylor DM (1984) Basic science of nuclear medicine. Churchill Livingstone, Edinburgh

Radiopharmaceuticals

Abbreviations

FDG fluorodeoxyglucose
w/v weight per volume

Some radionuclides can be administered directly to the patient and provide useful information. These follow the behaviour of the equivalent non-radioactive element in the body, such as iodine. Many, however, need to be combined with a body fluid sample or pharmaceutical in order to accumulate in the required target tissue. The latter is a radiopharmaceutical; a medicinal product, which contains a radionuclide as an integral part of the main ingredient and in which the mass of the principal ingredient is generally too low to produce a pharmacological response (NCRP 1983). The practice of using radiopharmaceuticals has the advantage that many different investigations can be undertaken using the same radionuclide, allowing the detection to be undertaken with the same detection parameters. Diagnostic nuclear medicine would not be as widespread and cost efficient, as it is, were it not for the parent – daughter radionuclide technique, which is utilised in the form of radionuclide generators. These allow the use of short half-life radionuclides, remote from the production facilities. Many radiopharmaceuticals are administered intravenously. Their handling, therefore, requires particular care from the biological as well as the radiological protection aspect, especially for those that are manufactured in the imaging unit, itself.

Required Characteristics

The radiopharmaceutical must achieve the required distribution in the body rapidly and should remain there for a time sufficient to allow the investigation to be undertaken completely. Ideally after study, the radiopharmaceutical should be excreted rapidly and the tissues involved in the excretory process should not receive excessive radiation exposure. It is preferable that tissues, other than the target tissue, not accumulate the radiopharmaceutical, even if this does not interfere directly with the test, because the administered activity will then be determined by the radiation absorbed dose to this tissue rather than to the target tissue.

Radionuclide

The choice of radionuclide is made on the basis of providing the maximum diagnostic information for the minimum radiation dose to the patient; the information content per unit radiation dose being referred to as the Figure of Merit (NCRP 1983).

Specific Activity

Stable isotopes of the radionuclide may be present; these are called carrier. When these are not present, the material is carrier-free. The ratio of radioactive to total mass of the element is the specific activity, which has units of Bq g^{-1}. In most circumstances, a high specific activity is desirable because a moderate amount of activity contains only a very small mass of the element and can be administered to a patient without causing a pharmacologic response, which is an essential requirement of a “tracer” study.

Radioactive Concentration

This is the activity per unit volume (Bq l^{-1}).

Pharmaceutical

The pharmaceutical is stable before labelling with the radionuclide and can usually be stored for long periods in accordance with good pharmaceutical practice. It must be non-toxic in the amounts that need to be administered, it must localise in the target tissue or compartment in a reasonable time, and the uptake in pathology should demonstrate a difference from normal. There are two types of pharmaceutical, depending on whether the distribution changes during the investigation (dynamic) or not (static). For static processes, the distribution should remain unchanged over the complete investigation. For dynamic processes, the distribution should change significantly within a reasonable time, not too short to preclude the acquisition of good diagnostic data and not too long to make the examination inconvenient.

Radiopharmaceuticals

Radiopharmaceuticals can be divided into two categories; those obtained from outside the hospital ready for use; and those that have to be reconstituted in the hospital before they are suitable for use.

For radionuclides and radiopharmaceuticals acquired from outside the hospital; on receipt, the package documentation must be compared with the order details and the container labels, and the package must be confirmed to be in good condition (UKRG 2001). The activity must be measured and the product must be securely stored, taking into account considerations of both good pharmaceutical practice and radiation protection. Radiopharmaceuticals that are manufactured in the hospital, are divided into two categories based on how they are produced: from radionuclides and pharmaceutical kits or, more unusually, from raw materials.

Stability describes how well the radionuclide and pharmaceutical remain associated, and good stability is essential for ensuring accurate diagnostic data. The in vitro stability determines the time that a radiopharmaceutical can be stored without significant deterioration. The in vivo stability determines how closely the distribution of the radionuclide resembles that of the pharmaceutical; they should be similar, at least for the investigation duration.

$^{99}Tc^m$ Radiopharmaceuticals

Because of its very attractive physical characteristics, $^{99}Tc^m$ is the most widely used radionuclide in diagnostic nuclear medicine. It decays, with a half-life of six hours, to ^{99}Tc, which is itself radioactive and decays with the emission of β^- particles and γ photons to ^{99}Ru. However, as the half-life of ^{99}Tc is 2.1×10^5 years, decay of 3 GBq of $^{99}Tc^m$ will produce only ~10 Bq of ^{99}Tc, which does not present a radiological problem to the patient. Because the half-life of $^{99}Tc^m$ is so short, it has to be obtained on the day of the investigation. It is usually used as a radiopharmaceutical by combining the radionuclide, in the form of sterile sodium pertechnetate $Na^+(^{99}Tc^mO_4)^-$, with a non-active labelling compound in a kit containing pre-sterilised pre-dispensed freeze dried components enclosed in a vial. The radiopharmaceutical is produced by introducing some $^{99}Tc^m$ into the kit vial and shaking, boiling or incubating the mixture. As pertechnetate $(^{99}Tc^mO_4)^-$, the technetium is in its most stable oxidation state of 7+ and is chemically unreactive. To enable labelling, the kit must force the reduction of the $^{99}Tc^m$(VII) to $^{99}Tc^m$(IV) using a stannous (Sn^{2+}) chloride or fluoride, usually complexed to the labelling compound:

$$\begin{aligned} 3Sn^{2+} &\rightarrow 3Sn^{4+} + 6e^- \\ 2(^{99}Tc^mO_4)^- + 16H^+ + 6e^- &\rightarrow 2(^{99}Tc^m)^{4+} + 8H_2O \end{aligned} \quad (I.2.1)$$

This reduced state is very reactive and will combine with a variety of chelating compounds. The activity and volume of sodium pertechnetate introduced depends on the kit requirements, the number of patients and the injection times. The kit contains other additives which act as stabilisers, buffers, antioxidants or bactericides and further dilution must not be undertaken after reconstitution. Because they do not contain radioactivity, the kits are often referred to as 'cold' kits and, although such kits can be stored for long periods without deteriorating, they should not be used beyond their expiry date, because of sterility concerns, possible ineffective labelling and chemical decomposition of the ligand. Sterility, half-life and radiochemical stability mean that the $^{99}Tc^{m}$ complexes are generally only stable for a maximum of 8 hours.

Impurities

Although Sn^{2+} is present much in excess of the $^{99}Tc^{m}$, there are a few competing reactions that result in impurities being produced within the kit. Free pertechnetate $(^{99}Tc^{m}O_4)^-$ may persist due the failure of the reducing agent. Also, $(^{99}Tc^{m})^{4+}$ can become re-oxidised back to $(^{99}Tc^{m}O_4)^-$, which may then not re-label. Such a re-oxidation is prevented by the presence of a nitrogen atmosphere within the kit vial and for this reason air must not be introduced into the vial before the reaction has finished. Sn^{2+} can hydrolyse when saline is introduced; producing tin hydroxides, which compete with the pharmaceutical for the $(^{99}Tc^{m})^{4+}$, resulting in the production of reduced hydrolysed $^{99}Tc^{m}$-tin colloid impurities. $(^{99}Tc^{m})^{4+}$ can itself undergo hydrolysis in aqueous solutions to form various $^{99}Tc^{m}$-oxide impurities.

Generators

Whilst the physical characteristics of $^{99}Tc^{m}$ make it an almost ideal radionuclide for many procedures, it would never have attained such widespread acceptability without being available from a generator. The principle, of this, is that a long half-life parent radionuclide decays to produce a relatively short half-life daughter radionuclide, whose chemical characteristics are different from that of the parent and allow chemical separation. Examples of radionuclide generators are listed in Table I.2.1. The generator activity is normally described in terms of the parent activity, and the daughter activity available for use depends on the time required for the system to reach equilibrium, which is determined by the relative half-lives of the parent and daughter.

Table I.2.1. Examples of radionuclide generators, showing the half-lives of the parent and daughter

parent	daughter	decay product	final product
^{99}Mo 67 h	$^{99}Tc^{m}$ 6 h	^{99}Tc	^{99}Ru
^{81}Rb 4.7 h	$^{81}Kr^{m}$ 13 s	^{81}Kr	
^{113}Sn 115 days	$^{113}In^{m}$ 1.67 h	^{113}In	

^{99}Mo–$^{99}Tc^m$

For $^{99}Tc^m$, the parent radionuclide is ^{99}Mo, which is in the form of ammonium molybdate $(NH_4)^{2+}(MoO_4)^{2-}$ adsorbed on positively charged alumina (Al_2O_3) contained in a glass or plastic column (see Fig. I.2.1). It decays to its daughter in the form of pertechnetate, $(^{99}Tc^mO_4)^-$.

Elution

$^{99}Tc^m$ is eluted from the column, as sodium pertechnetate, $Na^+(^{99}Tc^mO_4)^-$, by drawing 0.9% w/v sterile sodium chloride (Na^+Cl^-) through the column. The ^{99}Mo remains on the column, during elution, because the 2^- charge of the $(^{99}MoO_4)^{2-}$ ion binds more strongly to the alumina than the $(^{99}Tc^mO_4)^-$ ion, which has only a single negative charge. Almost all of the activity is removed in the first 5 ml, extra being merely diluent (Marengo et al. 1999). The eluate can be administered intravenously without modification. However, it is often further diluted, with more 0.9% w/v NaCl, before reconstitution of a kit to fulfil the requirements of the kit and for convenience during administration.

Available Activity

Not all the activity on the column can be eluted from it; the proportion that can be, is determined by the elution efficiency, a typical value being between 80–90%. One of the reasons for this is radiolysis due to the radiation arising from the activity on the column. Free radicals are formed which are extremely reactive and can cause reduction of the $^{99}Tc^mO_4^-$ to a lower oxidation state, which will not elute from the column. Oxidising agents may be added, either to the eluent or to the column itself to improve the yield. The severity of the problem depends on the type of column, of which there are two: the wet column and the dry column. The former leaves some saline on the column, after elution, which allows radiolysis. The latter draws air over the column, following elution, to avoid radiolysis. In both types, to avoid accidental radioactive leakage in transit, it is important that the column be kept dry during transport.

Fig. I.2.1. ^{99}Mo–$^{99}Tc^m$ generator column showing passage of NaCl during elution

Activity Profile

The build-up of $^{99}Tc^m$ activity following each elution is characterised by the Bateman equation for transient equilibrium (see equation I.1.8 in Chapter I.1), where the $^{99}Tc^m$ activity should eventually exceed that of the ^{99}Mo and then appear to decay with the half-life of the ^{99}Mo. However, 12.4% of the ^{99}Mo decays directly to ^{99}Tc without passing through the $^{99}Tc^m$ metastable state and therefore only 87.6% of the ^{99}Mo decays to $^{99}Tc^m$. This makes the factor by which the activity of $^{99}Tc^m$ eventually exceeds that of ^{99}Mo at equilibrium:

$$(0.876) * \lambda_d/(\lambda_d - \lambda_p) \quad \text{(I.2.2)}$$

which simplifies to 0.876 * 1.1. Therefore, the actual $^{99}Tc^m$ activity present at equilibrium is 0.964 times the ^{99}Mo activity. Conveniently, equilibrium is established ~23 hours after each elution. Therefore, the maximum single elution activity can be obtained once every day, although a greater total activity can be produced by eluting smaller activities more frequently. The variation in $^{99}Tc^m$ activity on the generator column after elution and during growback is shown in Fig. I.2.2.

The $^{99}Tc^m$ in the eluate is essentially carrier-free, as are all radionuclides derived from generator systems. However, the concentration of ^{99}Tc relative to $^{99}Tc^m$ will increase on storage both in the generator and in the eluate. This may adversely affect the labelling efficiency of certain kits, particularly those with low Sn^{2+} ion content, since the 2 technetium species will compete chemically. The measurement of the ^{99}Tc concentration is not easy, but the amount of $^{99}Tc^m$ as a fraction of both $^{99}Tc^m$ and ^{99}Tc can be estimated using the radioactive decay equation for N_d, see equation I.1.5 in Chapter I.1. The increase of this with time is shown in Fig. I.2.3.

Thus, if the time between elutions exceeds 10 hours, there will be more ^{99}Tc than $^{99}Tc^m$. To prevent the presence of too much carrier ^{99}Tc, a $^{99}Tc^m$ eluate should be obtained from relatively recently, <24 hour, eluted generator. This means that the first elution after generator delivery should be discarded, if possible. The generator activity should be chosen to suit the investigational needs of the hospital and this depends on both the reference and delivery dates. Since the activity of the eluate will vary during the week with the decay of ^{99}Mo, the work load has to be adjusted to this, although the concentration of the eluate can be adjusted to keep the radioactive concentration constant. For larger workloads, it is sometimes more appropriate to take delivery of two lower activity generators per week, on different days, rather than one very high activ-

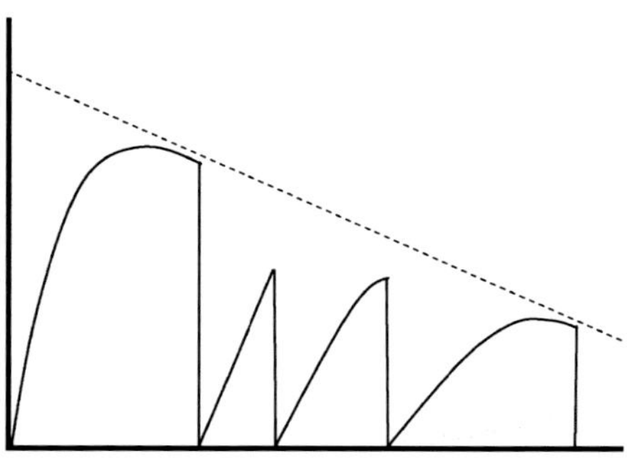

Fig. I.2.2. Variation in $^{99}Tc^m$ activity on the generator column after elution and during growback. The ordinate is the $^{99}Tc^m$ activity on the generator column and the abscissa is time. The variation in activity of ^{99}Mo on the column is shown superimposed

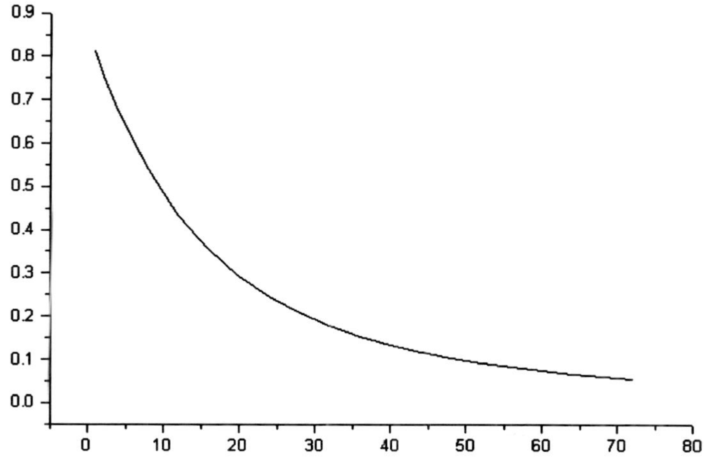

Fig. I.2.3. The amount of $^{99}Tc^m$ as a fraction of both $^{99}Tc^m$ and ^{99}Tc i.e. total Tc, showing the decrease in this fraction with time. The ordinate represents the amount of $^{99}Tc^m$ as a fraction of both $^{99}Tc^m$ and ^{99}Tc and the abscissa represents time in hours

ity generator. This avoids having to deal with very high eluate activities and radioactive concentrations, and also reduces the eluate activity variation during the week.

Other Generators

^{81}Rb–$^{81}Kr^m$

This is an example of a generator which is continuously eluted, the $^{81}Kr^m$ being removed by passing humidified air over the column and delivered to the patient via a non-rebreathing face mask (see Fig. III.2.10 in Chapter III.2). The activity of $^{81}Kr^m$ formed per second, $A_d = \lambda_d A_p$. The main disadvantages of this generator are the high cost and short useful life of 12 hours.

^{113}Sn–$^{113}In^m$

This is particularly suitable when logistics are poor because it can be used for 6–12 months. The main disadvantage is the high, 393 keV, photon energy of the $^{113}In^m$.

Dual Photon Radiopharmaceuticals

Almost all, i.e. ~95%, of dual photon imaging investigations use ^{18}F, as fluorodeoxyglucose, FDG; almost exclusively in the fields of neurology, cardiology and oncology (Delbeke and Sandler 2000). One of its advantages is its reasonably long half-life of 110 minutes which allows transport to sites 1–2 hour distant from the cyclotron. The other, most frequently used, radionuclides are ^{15}O, ^{13}N and ^{11}C. These have very short half-lives and therefore can only be used with a cyclotron on site. Other radionuclides can be produced using generator systems, e.g. ^{68}Ge–^{68}Ga, ^{82}Sr–^{82}Rb, ^{62}Zn–^{62}Cu, but these are not widely used at present.

Handling of Radiopharmaceuticals

Because most radiopharmaceuticals are administered by intravenous injection; whenever they are produced, an extremely high standard of cleanliness must be maintained by good pharmaceutical practice. The demands of this, often, reinforce those required for protection against radioactive contamination when solutions of radioactive materials are manipulated.

Even closed procedures, such as the elution of generators, the preparation of radiopharmaceuticals using kits, and manipulations in which the ingredients are not exposed to the environment, should take place in an environment where the risks of microbial and particulate contamination are minimised. This is usually achieved using a workstation, itself situated in a clean room. It also involves the operator being adequately trained and wearing suitable clothing to maintain the clean environment and to protect against radioactive contamination.

Facilities

In order to maintain a clean environment, it may be necessary to access the working area via a changing room and to introduce materials via a hatch. The clean area may require two work stations, one for manipulations and one for the storage and elution of generators. Open procedures including blood handling, require more severe restrictions which may include making the working area an aseptic area. To maintain such aseptic conditions, access via a changing room and a clean area may be required.

Workstations

Workstations can protect the radiopharmaceutical, the operator or both. The first type should not be used in nuclear medicine because it blows sterile air across the radiopharmaceutical and out onto the operator, raising the possibility of radioactive contamination.

The operator can be protected using fume hoods, with total exhaust, which draw room air across the radiopharmaceutical (see Fig. I.2.4). They can be used, in clean areas, for work with gases and oral preparations of volatile radiopharmaceuticals. The exhaust must be situated carefully outside the building to avoid ingress of radioactive contamination into occupied areas. In these units, the airflow across the opening to the working area should be within manufacturer's specification but excessively high airflow rates should be avoided to minimise turbulence. Airflow is usually controlled by the extent of opening of the working aperture and maximum working limits for the opening should be marked.

Both the operator and the radiopharmaceutical can be protected using work stations (see Fig. I.2.5). Air is passed through a high efficiency particle filter to produce a laminar flow of sterile air over the working area. This then passes through grilles or holes in the work surface. This is either totally recirculated with no demand for extra

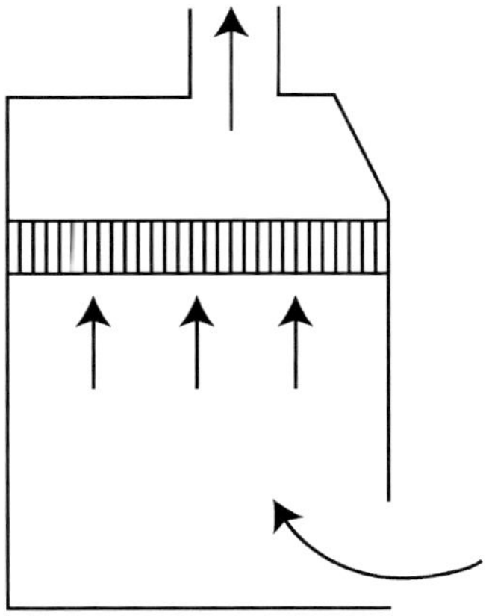

Fig. I.2.4. Fume hood showing movement of air from the room across the work area and then to total exhaust

room air, which is not recommended for working with radiopharmaceuticals because of the potential for build up of radioactive contamination, or totally or partially (>20%) exhausted to the outside environment; in which case, the external venting must be carefully situated. Partial exhausting can be back into the laboratory if it is via a high efficiency filter. Partially exhausting systems are suitable for most applications, with totally exhausting systems mainly being used for gases, volatile radionuclides and blood handling. The protection factor for operator protection is the contamination which would be experienced outside the workstation to that inside the work station and should comply with the class standard for the type of equipment.

The interior surfaces should be smooth, impervious and durable. Radiological shielding can be incorporated using integral lead sheeting under the work surface and lead glass at the front panel. Any extra shielding should be as compact as possible to avoid impairing the protection factor. Monitoring should be undertaken regularly to ensure that any filters do not accumulate sufficient radioactive material to become a radiological hazard.

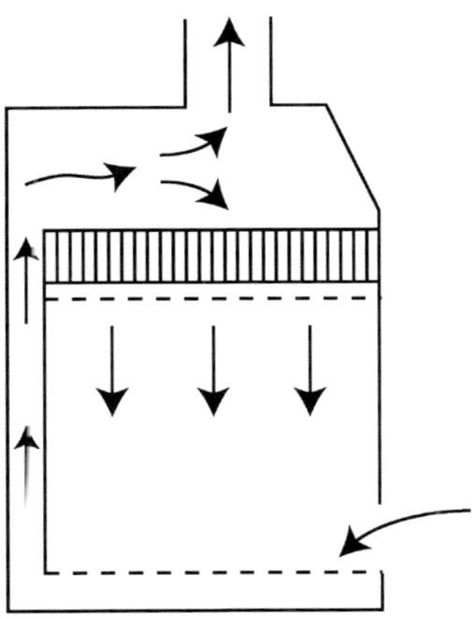

Fig. I.2.5. Partially exhausting workstation showing air moving through a high efficiency particle filter to produce a laminar flow of sterile air over the working area and then to partial exhaust and partial recirculation

References

Delbeke D, Sandler MP (2000) The role of hybrid cameras in oncology. Semin Nucl Med XXX: 268–280

Marengo M, Aprile C, Bagnara C, Bolzati C, Bonada C, Candini G, Casati R, Civollani S, Colombo FR, Compagnone G, del Dottore F, Di Guglielmo E, Ferretti PP, Lazzari S, Minoia C, Pancaldi D, Ronchi A, Sanita di Toppi G, Saponaro R, Torregiani T, Uccelli L, Vecchi F, Piffanelli A (1999) Quality control of $^{99}Mo/^{99}Tc^{m}$ generators: results of a survey of the Radiopharmacy Working Group of the Italian Association of Nuclear Medicine (AIMN). Nucl Med Commun 20:1077–1084

NCRP (1983) National Council on Radiation Protection and Measurements report 73. Protection in nuclear medicine and ultrasound diagnostic procedures in children. NCRP, Bethesda, Maryland

UKRG (2001) Report of a joint working party: the UK Radiopharmacy Group and the NHS Pharmaceutical Quality Control Committee. Quality assurance of radiopharmaceuticals. Nucl Med Commun 22:909–916

Biological Effects of Radiation

Abbreviations

BEIR	United States National Academy of Sciences Biological Effects of Ionizing Radiation committee
BSS	basic safety standards
DNA	deoxyribose nucleic acid
HVT	half-value thickness
IACRS	Inter-Agency Committee on Radiation Safety
IAEA	International Atomic Energy Agency
ICRP	International Commission on Radiological Protection
ICRU	International Commission on Radiation Units and Measurements
LET	linear energy transfer
MIRD	Medical Internal Radiation Dose committee
RBE	relative biological effectivenesses
SI	specific ionisation
TVT	tenth-value thickness
WHO	World Health Organization

The radiation that has to be used in diagnostic nuclear medicine is, unfortunately, of sufficiently high energy to be ionising. Thus, although there is a benefit to the patient, there is also a risk to both the patient and the community as a whole. Recommendations on controlling this risk are made by a number of organisations, based on an evaluation of the detriment to the individual caused by the absorbed radiation dose. Of prime importance, in this, is the estimation of the radiation dose and hence the likely biological consequences, resulting from the interaction of the radiation with matter. It has been stated that: a key factor in sustaining the growth in nuclear medicine while retaining the confidence of the referring clinician, the patient and the public, is for practitioners to maintain an up to date knowledge of the radiation risks associated with the procedures, an understanding of the methodology used to assess these risks and an appreciation of the associated limitations (Mountford 1997).

Radiation Protection

Nuclear medicine demands one of the most comprehensive radiation protection regimens of all the diagnostic radiation services. This is because a wide variety of radionuclides are used, they are unsealed and they are administered to patients who become sources and who also excrete radioactive materials. It is difficult to compare absorbed doses from nuclear medicine procedures with those received in other modalities. In diagnostic radiology, the dose is almost entirely limited to the area being examined, the whole body radiation dose due to scattered radiation being low; whilst there is usually a significant whole body dose from radionuclides. Another major difference is that an organ is uniformly irradiated in radionuclide procedures but the entrance and exit doses can differ by a factor of 100 in x-ray procedures. The ultimate comparison of different diagnostic procedures involves the calculation of the actual risk of cancer mortality and genetic damage.

Origin of Advice

The non-governmental scientific organisation with responsibility for making recommendations and providing guidance on minimising the detriment, caused by the use of ionising radiation, is the International Commission on Radiological Protection (ICRP). This was formed, in 1923, at the 2nd International Congress of Radiology as the International x-ray and Radium Protection Commission. It was reorganised and renamed, in 1950, to its present title. Pressure for the creation of the organisation came from the radiological professions of a number of countries who had, by the early 1920s, adopted detailed recommendations on the proper management of x-ray and radioactive sources. The Commission works closely with its sister body, the International Commission on Radiation Units and Measurements (ICRU) and has official links with the World Health Organization (WHO) and the International Atomic Energy Agency (IAEA). The commission issues regular information reports and fundamental recommendations appear every 10–15 years, the latest being Publication 26 (1977) and Publication 60 (1991). This last publication has been followed by a user's edition (ICRP 1992) and an application document whose purpose is to clarify how the recommended system of radiological protection, described in ICRP 60, should be applied in medicine (ICRP 1996; Harding and Thomson 1997).

The Inter-Agency Committee on Radiation Safety (IACRS) was formed by the IAEA, in 1990, to help achieve consistency in the practical application of radiation protection. It has produced the "International basic safety standards for protection against ionising radiation and for the safety of radiation sources" (the basic safety standards – BSS) (IAEA 1996), which is itself based on the recommendations of the ICRP. These reports are intended help regulatory and advisory bodies to provide guidance regarding the fundamental principles on which appropriate radiological protection can be based. Because of the differing conditions that exist in various countries, the reports are not intended to be regulatory texts and national bodies have to develop their own structures of legislation, regulation, authorisations, licences, codes of practice and guidance material, from the guidance given. It is these national regulations which actually implement the ICRP recommendations.

Risk of Biological Consequences

There is some discussion as to whether exposure to low levels of radiation is beneficial, presumed to be due to stimulation of immunological or other reparative processes, and there is some experimental evidence of this radiation hormesis (Overbeek et al. 1994; van Wyngaarden and Pauwels 1995; Sont et al. 2001; Johansson 2003). However, this cannot be anticipated by extrapolating from the effects observed at high doses, and has not been proven in humans. Therefore, all exposures to ionising radiations are regarded as harmful. The biological consequences originate at the cellular level. If damage occurs and is not adequately corrected by repair mechanisms, it may prevent the cell from surviving or reproducing, or it may result in a viable but modified cell.

Deterministic – Cell Death

Most organs and tissues are unaffected by the death of even substantial numbers of cells but, if this number is large enough, there will be observable harm. These injuries include extensive cellular death, leading to total or partial loss of organ function, e.g. bone marrow depression, as well as subsequent tissue changes such as impairment of blood supply and fibrosis. The probability of causing such harm will be zero at small doses but above some threshold the severity of the harm will increase with increasing dose. This type of effect, previously called "non-stochastic", is now called "deterministic" by the ICRP (ICRP 1991), meaning "causally determined by preceding events". The terminology was changed because cell killing by irradiation is itself a stochastic process and the term "non-stochastic" was considered an inappropriate descriptor. These effects do not occur in routine work and are the result of accidents, hence radiation protection is aimed at preventing deterministic effects.

Stochastic – Cancer

The only effects possible at low radiation doses are stochastic effects. If the irradiated cell is not killed but is modified and remains viable, after a prolonged and variable latency period, a cancer may develop. Hence radiation protection is aimed at reducing the chances of stochastic effects occurring.

Hereditary

If a cell, whose function is to transmit genetic information to later generations, is damaged then resulting effects may be expressed in the children of the exposed person. This type of stochastic effect is called 'hereditary'. Although convincing statistical evidence for hereditary change in humans has not been demonstrated, the allowance is made for mutagenesis in all calculations of risk.

Foetus

The foetus is very sensitive to radiation damage, The effects are dose related and are also related to the time in the pregnancy. They include spontaneous abortion, malformation, developmental retardation and possibly an increased risk of cancer in later life.

Models of Biological Effects

A knowledge of biological effects helps to assess the risks associated with radiation exposure. The dose of radiation absorbed by an individual is documented as the Gray (Gy). The number, D, of Gy absorbed do not change the severity of the cancer but do alter the chance of developing it. The link between D and the effect, e, of the radiation at low doses, <0.5 Gy, has only been evaluated using extrapolations from high doses and dose-rates. This makes the nature of the link uncertain and various models have been proposed. The United States National Academy of Sciences Biological Effects of Ionizing Radiation committee (BEIR) adopted a linear quadratic model, in 1980. In this, a linear component adds to a quadratic component, with the latter becoming dominant in the high-dose region.

$$e = aD + bD^2 \qquad (I.3.1)$$

where

e is the effect of radiation

D is the dose

a, b is the constants

At ~1 Gy, the linear and quadratic components are believed to be of ~equal magnitude. At lower dose rates, however, the value of b is smaller and the quadratic term becomes of diminishing importance. Hence, ICRP Publication 26 (ICRP 1977) recommended, for nuclear medicine, a linear relationship without threshold, i.e. even small doses of radiation might cause cancer.

Detriment – Effective Dose

The ICRP uses "detriment" to assess the overall harm caused by exposure to ionising radiation. This concept is quantified by effective dose, E, which is a uniform whole body dose giving the same risk as a non-uniform dose actually experienced; thus enabling comparison among many situations. It has units of Sievert (Sv) and does not take age or sex into account. The concept of detriment was introduced in 1977 when it was quantified by the effective dose equivalent (H_E) but this name was felt to be unnecessarily cumbersome. E is estimated by calculating the equivalent dose (H_T) to individual organs or tissue (T), weighting them by a factor (W_T) to take account of the relative sensitivity of each type of tissue to radiation and then summing the products. E is thus the sum of the weighted equivalent doses in all organs and tissues of the body:

$$E = \sum_T W_T H_T \qquad (I.3.2)$$

Because of revisions to H_T there are small differences in the values of E and H_E but these differences are generally small for photon radiation, being <10% for most monoenergetic photon energies and common radiation distributions.

Tissue Weighting Factor

W_T represents the relative contribution of an organ or tissue to the total detriment and was introduced because the relationship between the probability of stochastic effects and H_T is found to depend on the organ or type of tissue irradiated. This is because some cell populations in the body are more radiosensitive than others, related to the rate of cellular replication and the stage in the cell cycle. The most sensitive organs: red bone marrow, lung, thyroid, bone, gonads and the female breast; are generally those with the highest rate of cellular replication. The values of W_T are chosen to be independent of the type and energy of the radiation. They were revised in 1991 but these revisions do not contribute to the differences between E and H_E.

High relative sensitivity tissues with $W_T = 0.20$ are gonads. Medium relative sensitivity tissues with $W_T = 0.12$ are lung, red bone marrow, stomach, colon. Medium/low relative sensitivity tissues with $W_T = 0.05$ are thyroid, liver, oesophagus, breast, bladder. Low relative sensitivity tissues with $W_T = 0.01$ are skin, bone surfaces. The remaining tissues are allocated a medium / low relative sensitivity (ICRP 1991). The sum of these W_T values are one, so that a uniform H_T over the whole body gives a value of E equal to the uniform H_T. The values of W_T are for the occupational age range and should, roughly, be increased by a factor of 2 in paediatrics (Gadd et al. 1999) and decreased by a factor of 0.2 in geriatrics. W_T only concerns stochastic effects, and therefore E is not applicable for therapeutic administrations since, in this case, the absorbed dose is so large that deterministic effects occur.

Equivalent Dose

H_T is introduced because the probability of stochastic effects occurring depends, not only on the absorbed dose, but also on the type and energy of the radiation. To account for the varying biological effects of different types of radiation, the absorbed dose is weighted by a factor related to the quality of the radiation.

$$H_T = \sum_R W_R D_{T,R} \tag{I.3.3}$$

Where W_R is the radiation weighting factor and $D_{T,R}$ is the absorbed dose averaged over the tissue or organ T, due to radiation R. This is where the unit Sv is introduced into the equation.

Radiation Weighting Factor

W_R is introduced because different types and energies of radiation have different relative biological effectivenesses (RBE) of inducing stochastic effects. This is determined solely by their linear energy transfer (LET), a measure of the density of ionisation along the track of an ionising particle. Thus, the values of W_R depend only on the type and energy of the radiation and not on the organ or tissue. The current value of $W_{R,}$ for photons and electrons, is one.

Dose Rate

The rate at which radiation is absorbed is important in determining the severity of the biological damage (NCRP 1983). Provision was made for a weighting factor, which would have accommodated any effect of the dose rate. This has not been attempted because of limited data and, in practice, the weighting factor has always been ascribed a value of one. Therefore, the ICRP has decided to exclude it.

Dosimetry

The most widely used and biologically meaningful expression of the radiation dose is the absorbed dose, D. This is the radiation energy deposited in a mass of absorber such that 1 Gy = 1 Joule energy deposited in one kg (Zanzonico 2000). It can be calculated for both external and internal exposure, and a reference source has been produced detailing these for many radionuclides (Delacroix et al. 1998).

External

The radiation doses absorbed by tissue can be calculated for photons or β^- particles.

γ *or x Photons*

For an external photon source, only the radiation energy absorbed per kg of tissue need be calculated and, to achieve this, only an estimate of radiation exposure is required. The latter refers to the amount of ionisation of air caused by γ or x-ray photons and is presented in units of Coulomb kg^{-1}. If the exposure is known at a certain location, the absorbed dose that would be delivered to a person, at that location, can be estimated by means of the f factor:

$$D = f \cdot \text{exposure} \tag{I.3.4}$$

For soft tissues, $f \approx 0.026$. For bone and low energy photons, <100 keV, the f factor may be higher than this because energy absorption in bone is greater than energy absorption by air at these energies, due to photoelectric absorption by the heavier elements (Ca and P) in bone. Radiation exposure from γ and x-ray emitters can be estimated using the exposure rate constant, Γ (C kg^{-1} h^{-1} MBq^{-1}); the calculation of which is based on the number of emissions from the radionuclide, their energies and the absorption coefficient of air at these energies. This represents the exposure rate (C kg^{-1} h^{-1}), at 1 m, from an unshielded 1 MBq source. Γ can be measured experimentally or calculated. For practical purposes, only energies above ~20 keV should be included in the calculation (Γ_{20}) because the penetration of lower energy photons is too low (Sorensen and Phelps 1987).

$$\text{exposure rate} = \Gamma_{20}\, n e^{-\mu x} d^{-2} \tag{I.3.5}$$

where
n is the number of MBq in the source
d is the distance (m) of the point at which the exposure is required
μ is the linear attenuation coefficient of the shielding material (m^{-1})
x is the thickness of the shielding material (m)

Such exposure rate constants and absorbed dose rates have been calculated, for a number of radionuclides (Groenewald and Wasserman 1990; Delacroix et al. 1998).

β^- *Particles from Contamination of the Skin*

β^- particles are completely absorbed within the first few mm of tissue, e.g. 1 MeV particles can penetrate ~5 mm. For dosimetry, the penetration is assumed to be equal to their average range, which is ~1/3 of their maximum range. Additionally, in the case of skin contamination, it is assumed that half of the total β^- energy will be absorbed in tissue and the other half will travel away from the skin.

$$\text{absorbed dose rate} = 0.5 \times 3.6\ 10^6\ A\ E_\beta / V \quad \text{Gy h}^{-1} \tag{I.3.6}$$

where
A is the activity of contaminant (Bq)
E_β is the energy emitted per Bq (J s^{-1})
V is the volume irradiated = area contaminated * penetration depth (cm)

and

$$E_\beta = E_m \, 1.602 \, 10^{-13}$$

where E_m is the mean energy in MeV. This is calculated as $E_m = \sum_i n_i E_i$ when there are i different β^- particles, each of mean energy E_i and emitted in proportion n_i.

Values for both uniform deposits and droplets have been calculated for many radionuclides (Delacroix et al. 1998). Examples of the absorbed dose rates from β^- particles, relevant to routine work, involve skin contamination either by pure β^- emitting radionuclides or by $^{99}Tc^m$.

Pure β^- Emitting Radionuclides

The absorbed dose rate from 1 kBq of ^{32}P, which emits β^- particles of E_m ~0.6 MeV and average range in tissue of ~0.27 cm, spread over an area of skin 1 cm in diameter is ~8.6×10^{-4} Gy h^{-1}.

$^{99}Tc^m$

In this radionuclide, electron emission occurs in 12% of disintegrations, mainly because of internal conversion, and has a mean energy of 0.123 MeV and average range of 0.02 cm. Thus, $E_m = \sum_i n_i E_i = 0.0148$ MeV. For the same contamination activity and area as in the previous example, the absorbed dose rate ~2.7×10^{-4} Gy h^{-1}.

Internal – MIRD

A knowledge of the radiation dose received by different organs in the body is essential to an evaluation of the risks and benefits of any procedure (Stabin et al. 1999). Absorbed doses may be calculated, for ingested radioactivity, by considering one organ as a target and other organs as sources and using an absorbed fraction technique. The most usual scheme is MIRD, developed by the Medical Internal Radiation Dose committee, established in 1964 by the American Society of Nuclear Medicine with the object of providing 'the best possible estimate of the absorbed dose to patients resulting from the diagnostic or therapeutic use of internally administered radiopharmaceuticals' (Loevinger et al. 1988, 1991; Watson et al. 1993; Toohey et al. 2000). MIRD have published many pamphlets, which provide information, and reports, that detail dose estimates for various radiopharmaceuticals (Howell et al. 1999; Stabin et al. 1999). In this scheme, the source and target can be the same and the method involves calculation of:

1. activity in the source organ or organs
2. total radiation energy emitted by (1)
3. fraction of (2) absorbed by the target organ

Activity in the Source Organ

The requirement, in this section, is to estimate the cumulated activity Ã in the source organ (Bq s). The variation in activity with time can be complex due to physical decay, and biological uptake and excretion. This makes analysis difficult unless simplifying assumptions can be made concerning the biological behaviour. For instance, if the effective removal of radionuclide is exponential (Simpkin 1999), in which case:

$$\tilde{A} = \text{mean-life} \times \text{initial activity}$$

Radionuclide imaging techniques including planar scintigraphy, rectilinear scanning, single-photon emission computed tomography and positron emission tomography have all been used to provide data from which this information can be obtained (Thomas et al. 1988; Mountford 1996; Ott 1996; Stabin et al. 1999). Significant errors may be introduced into the calculations at this stage because of variations in metabolism and the distribution of radionuclides among patients, especially in different disease states. Also within an organ, the distribution of the radionuclide may be inhomogeneous rather than the assumed homogeneous.

Total Radiation Energy Emitted by the Source

This is calculated as:

$$\tilde{A}\,(\Delta_1 + \Delta_2 + \Delta_3 + \ldots) \quad \text{kg Gy}$$

where Δ is the equilibrium absorbed dose constant, which is the energy emitted per unit of $\tilde{A}$. This must be calculated for each type of emission from the radionuclide:

$$\Delta_i = 1.602\ 10^{-13}\ n_i\,E_i \quad \text{J} \tag{I.3.7}$$

where E_i is the average energy (MeV) of the ith emission and n_i is the relative frequency of that emission, i.e. the number emitted per disintegration.

Fraction of Energy Absorbed by the Target Organ

The absorbed fraction, φ, is determined by the type and energy of the radiation and on the anatomical relationship of the source and target. For charged particles and low energy, <11 keV, photons; all of the energy is absorbed within the source tissue, which is therefore also the target tissue, with $\varphi = 1$. When the source tissue is not the target tissue, $\varphi = 0$. For higher energy photons, φ must be determined for each type of emission and for each source–target pair. For the ith emission:

$$\varphi_i\,(r_t \leftarrow r_s)$$

where r_t is the target organ, r_s is the source organ

The specific absorbed fraction:

$$\Phi = \varphi / m_t$$

takes into account, m_t, the mass of the target organ. For a given organ pair, Φ is the same regardless of which organ is the source and which is the target. Significant errors can be introduced into the calculation at this stage, because of variations in patient anatomy.

Final Calculation of Absorbed Dose

The average absorbed dose, $\bar{D}$ (Gy), is:

$$\bar{D}\,(r_t \leftarrow r_s) = \tilde{A}\,\Sigma_i\,\Phi_i\,(r_t \leftarrow r_s)\,\Delta_i \quad \text{Gy} \tag{I.3.8}$$

This calculation has been simplified by the calculation of S factors, the mean dose in the target organ per unit cumulated activity in the source organ:

$$\bar{D}(r_t \leftarrow r_s) = \tilde{A}\,\bar{S}(r_t \leftarrow r_s)\,\Delta_i \quad \text{Gy} \tag{I.3.9}$$

Such S factors have been published for different source–target organs and radionuclides, for a "standard" human, by MIRD. Above 100 keV there is little variation in φ and it is acceptable to combine all photon emissions into a mean value. Each photon below 100 keV should be treated separately, however. The total dose to the target tissue is obtained by summing the doses from all of the source tissues in the body.

Implementation

The MIRDOSE3 program is currently the main tool available to estimate absorbed doses in nuclear medicine (Stabin 1996). It implements the MIRD scheme, which originally assumed a 70 kg hermaphrodite and in which no account was taken of variations in adult body size or shape, or the effect of disease. However, factors are now included for children of different ages and for pregnant females at different stages of gestation. Radiation dose estimates have also been published, for children of different ages for many different nuclear medicine procedures (Stabin and Gelfand 1998). There are several situations where the above calculations are modified (Bardies and Myers 1996). To develop more patient-specific dosimetry, different mathematical models for adults of varying height, based on the MIRD model design, have also been developed (Clairand et al. 1999, 2000; Stabin et al. 1999; Zanzonico 2000)

Limitations

The MIRD scheme assumes a uniform radionuclide concentration in each source organ. When the accumulation is non-uniform, as is particularly likely to occur in disease states, the calculated dose to a target organ will be an average which may misrepresent the risk to the organ (Simpkin 1999). This is because large radionuclide distribution non-uniformities can result in absorbed doses, adjacent to the distribution, being one or more orders of magnitude different to the average estimates. Similar discrepancies are experienced with charged particles (NCRP 1983). Auger electrons emitted from nuclei bound to the DNA present a special problem because it is not realistic to average the absorbed dose over the whole mass of DNA as would be required by the present definition of H_T (Thierens et al. 2001).

Biological Effects

The appearance of harmful effects is the last, and slowest, of 4 stages in the induction of biological damage and is the functional expression of molecular changes which result primarily from radiation-induced chemical changes in the complex molecules of the cells. The first 3 stages cause the molecular damage and are very rapid (Parker et al. 1984):

Physical Interaction

The initial, physical, interaction is the shortest lasting $\sim 10^{-17}$ to 10^{-15} s. During this phase, energy is deposited in the cell causing ionisation and excitation. The absorption of energy by a water molecule results in the ejection of an electron.

$$H_2O \rightarrow H_2O^+ + e^- \tag{I.3.10}$$

Physico-Chemical Interaction

In the second, physico-chemical, interaction which lasts $\sim 10^{-14}$ to 10^{-3} s, the positive ion dissociates to yield a hydrogen ion, H^+, and a hydroxyl free radical, OH^*.

$$H_2O^+ \rightarrow H^+ + OH^* \tag{I.3.11}$$

The electron reacts with other water molecules which then dissociate to form hydrogen free radicals H^* and hydroxyl ions OH^-.

$$H_2O + e^- \rightarrow H_2O^-$$

$$H_2O^- \rightarrow H^* + OH^- \tag{I.3.12}$$

The H^+ and OH^- are normally present in living cells and do not contribute to the radiation damage. The OH^* and H^*, which contain unpaired electrons, are very reactive, being electron donors, i.e. reducing agents, or electron acceptors, i.e. oxidising agents, and may undergo many reactions. For example, two OH^* radicals may combine to form hydrogen peroxide H_2O_2, which is a powerful oxidising agent.

$$OH^* + OH^* \rightarrow H_2O_2 \tag{I.3.13}$$

Similar types of reaction may occur with other cellular components, producing excited ions and other types of free radicals.

Chemical Phase

In the third, chemical, phase lasting a few seconds to hours, the products of earlier reactions, especially the free radicals, react with the organic molecules within the cell.

Linear Energy Transfer

The ionisation described in the 'physical interaction' above, tends to occur in clusters such that the average energy deposition per ion cluster is ~ 100 eV. These clusters may occupy a volume of $1\text{–}5 \times 10^{-6}$ m^3. Thus at the subcellular level the deposition of energy is highly non-uniform with amounts of energy, very large by biochemical standards, being deposited in very small volumes. This can result in dramatic biological consequences, e.g. in DNA. For a given D, the frequency or severity of the biological effects is generally higher for densely, rather than sparsely, ionising radiations (Zanzonico 2000). This means that it is greater for high rather than low linear energy transfer (LET) radiation, where LET is the rate at which energy is transferred to the atoms of the medium through which the radiation is passing, or the energy deposited per unit length of track (keV μm^{-1}).

The LET is a measure of local deposition of energy along the track and does not include radiation losses since these result in the production of bremsstrahlung radiation whose energy may be deposited some distance away from the track. This component, however, is very small in nuclear medicine. Electrons, in the energy range 10 keV to 10 MeV, have an average LET of 0.2–2 keV μm^{-1} in soft tissue. Lower energy electrons have higher values of LET, however. The products of photon interactions are secondary photons and high energy electrons and it is the high energy electrons that are ultimately responsible for the deposition of energy in matter. Because of this, the average LET of photons is the same as for electrons of similar energy.

Specific Ionisation

Specific ionisation (SI) refers to the total number of primary and secondary ionisations, per unit track length, along a charged particle track.

$$\mathrm{SI} \propto \mathrm{LET}$$

For electrons, SI increases as energy decreases because energy loss rates increase as the electrons slow down. Below ~0.1 keV, the kinetic energy is inadequate to ionise efficiently and SI decreases rapidly to zero. Because SI increases as the particle slows down, there is a marked increase in ionisation density near the end of its track. This is termed the Bragg ionisation peak. For electrons, this peak occurs when the energy has been reduced to less than ~1 keV and accounts for only a small fraction of the total energy deposited.

Interaction of Radiation with Matter

The ionisation causing the biological effects, is a result of electrical interactions between charged particles and the orbital electrons during close encounters with atoms in the absorbing medium, although the final dissipation of the transferred energy is as heat. The particles lose energy and slow down as a result of these interactions which continue until their total kinetic energy is expended.

Ionisation

Ionisation occurs when an orbital e^- is ejected as a secondary e^-, leaving a positively charged ion (see Fig. I.3.1). The average energy required to form such an ion pair is ~34 eV, the incident particle losing this energy to overcome the binding energy of the orbital electron and to give kinetic energy to the secondary electron. Most interactions involve outer shell orbital electrons but some do involve inner shell electrons and these result in emission of characteristic x-rays and Auger or Coster-Kronig electrons. The secondary electron may be sufficiently energetic to cause secondary ionisations and excitations. In a few cases, a distinct separate track, a delta (δ) ray, can be formed. Interactions involving, generally, less energy may result in the orbital electron not

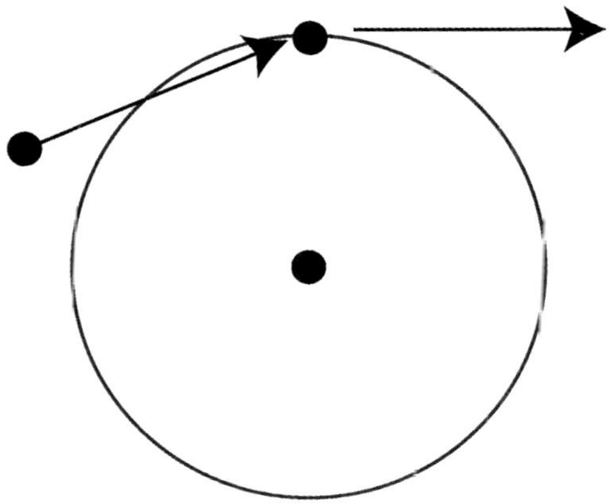

Fig. I.3.1. The process of ionisation, showing ejection of an orbital electron

being ejected but being raised to an excited state. The energy transferred to an atom in such an excitation interaction is dissipated as heat or as the emission of electromagnetic radiation in the infrared, visible or ultraviolet part of the spectrum.

The energy losses in these interactions are termed collisional losses. In tissue, the energy expended by an electron in collision processes is roughly equally divided between ionisation and excitation. Collisional loss rates decrease with increasing electron energy because the speed of the electron increases, it spends less time in the vicinity of the atom and has a reduced likelihood of interacting. They also decrease with increasing atomic number of the absorbing medium because, in atoms of higher atomic number, inner shell electrons are "screened" from the incident electron by layers of outer shell electrons, making interactions with inner shell electrons less likely. Therefore, gram for gram, lighter elements are better absorbers of electron energy than heavier atoms.

Nuclear Capture

Should the charged particle actually penetrate the orbital electron cloud and interact with the nucleus, nuclear capture is a possibility

Bremsstrahlung Radiation

A result, more likely than nuclear capture, is deflection of the incident particle which results in a rapid deceleration in which the change in kinetic energy appears as bremsstrahlung radiation. The energy of bremsstrahlung photons appears as a continuous spectrum which can vary from nearly zero, when the incident particle is only slightly deflected, up to a maximum equal to the full kinetic energy of the incident particle, lost in one interaction. Most interactions, however, result in emissions in the low energy range. These energy losses are termed radiation losses.

Relative Importance

Radiation losses are usually only a few % of total losses and increase with decreasing incident particle mass, increasing incident particle energy and increasing atomic number of the target atom. Therefore, in the energy range encountered in nuclear medicine, collisional losses dominate.

β^- Particles

β^- particles can lose a large fraction of their kinetic energy in an individual interaction and can undergo large deflections which, together with the higher probability of nuclear deflections, make the tracks tortuous. So, although the track lengths of monoenergetic electrons are similar, the penetration in a given direction varies enormously. For β^- particles, in which the kinetic energy varies on emission, the maximum range is applicable only to those β^- particles with the maximum energy and the average penetration is much less, normally about a fifth of the maximum range.

$$\text{range, R} \propto 1/\varrho \qquad \text{(I.3.14)}$$

where ϱ is the density of the absorbing material

To normalise for these density effects, R is expressed in g cm^{-2} of absorber. When plotted against absorber thickness on semi-log axes, there is an almost straight line decline, because even thin absorbers can remove a few electrons, which gradually merges with a long relatively flat tail representing, not electron transmission, but the detection of relatively penetrating bremsstrahlung photons.

β^+ Particles

The interactions of energetic β^+ with matter is similar to that of β^- and the range is equivalent to that of a β^- with the same energy. At the end of their path they annihilate with an electron and two 511 keV photons are created.

Photons

Photons are removed from a beam, and charged particles are created, by two effects: absorption, when the photon is entirely eliminated, and scatter, when the photon is deflected. At the energies encountered in nuclear medicine, there are two mechanisms of absorption, the photoelectric effect and pair production, and one of inelastic scatter, the Compton effect.

Attenuation Coefficient

All three mechanisms contribute to beam attenuation and the total attenuation coefficient is the sum of the separate attenuation coefficients, which indicate the fractional decrease in beam intensity due to that process (see Fig. I.3.2):

$$\text{total } \mu = \text{photoelectric } \mu^{pe} + \text{pair production } \mu^{pp} + \text{Compton } \mu^{c} \qquad \text{(I.3.15)}$$

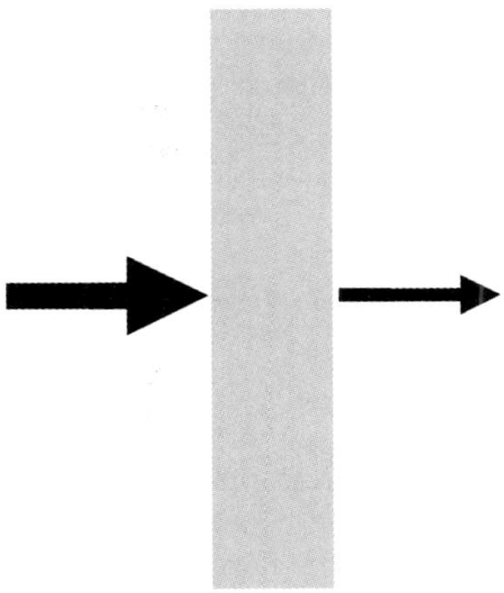

Fig. I.3.2. The effect of attenuation, the number of photons in a beam are reduced by an attenuating medium

For a thin absorber, of thickness Δx such that the incident beam intensity, I, is reduced by only a small amount ΔI, i.e. <10%, the fractional decrease in beam intensity,

$$\Delta I/I = -\mu_l \Delta x \quad cm^{-1} \tag{I.3.16}$$

where μ_l is the linear attenuation coefficient of the absorber, which increases linearly with absorber density, ϱ.

$$\mu_m = \mu_l/\varrho \quad cm^2\,g^{-1} \tag{I.3.17}$$

is the mass attenuation coefficient which is independent of ϱ and depends only on the atomic number, Z, of the absorber and the energy, E, of the incident photon.

Photoelectric Effect

In the photoelectric effect, the mass attenuation coefficient falls rapidly with increasing energy and obeys a cube law over the part of the energy range important in nuclear medicine

$$\mu^{Pe}{}_m \propto 1/E_\gamma^3 \tag{I.3.18}$$

The variation with atomic number also, in general, obeys a cube law

$$\mu^{Pe}{}_m \propto Z^3 \tag{I.3.19}$$

Pair Production

In pair production, the mass attenuation coefficient is ~proportional to the energy

$$\mu^{PP}{}_m \propto E_\gamma - 1.02 \tag{I.3.20}$$

where E_γ is in MeV

The electron–positron pair are created in the nuclear field and the coefficient is proportional to Z

$$\mu^{PP}{}_m \propto Z \tag{I.3.21}$$

Compton Scattering

In Compton scattering, the mass attenuation coefficient is independent of Z and decreases with increasing photon energy

$$\mu^{C}{}_{m} \propto 1/E_{\gamma} \tag{I.3.22}$$

Relative Importance

In low Z materials, such as body tissue, and the range of energies normally encountered in nuclear medicine (50 keV to several MeV), the Compton effect is the most important interaction type accounting for >99.7% of all interactions at 511 keV (Zaidi and Hasegawa 2003). However, after one or more Compton interactions, the photon will have a low enough energy that a photoelectric interaction will then become the most probable. In high Z materials such as lead, the photoelectric effect is important up to 500 keV. It is the combination of the high density and high atomic number which makes lead such an excellent material for protection against photon radiation.

Attenuation

In high Z materials there is, however, an enormous range in the attenuation coefficients, due primarily to the cubic dependence on atomic number and energy; this latter explaining why lead tenth-value thicknesses increase so rapidly with increasing γ energy. For example, it is 0.9 mm for 150 keV photons but 20 mm for 511 keV annihilation photons.

Narrow-Beam Geometry

When an absorber is thick enough that the probability of photon interaction is not small, i.e. >10%, the amount of transmission detected depends on whether scattered photons are recorded as part of the transmitted beam. In narrow-beam geometry, the contribution of scattered photons is minimised. The intensity, of a monoenergetic photon beam, measured through an absorber thickness of (x) is

$$I(x) = I(0)\, e^{-\mu l x} \tag{I.3.23}$$

where I(0) is the intensity of the incident beam and $e^{-\mu l x} = tf$, is the transmission factor. This demonstrates that photons do not have a definite maximum range, there is always some finite probability that a photon will penetrate even the thickest absorber. Analogous to the concept of half-life, the thickness of an absorber that decreases incident beam intensity to half is the half-value thickness (HVT)

$$HVT = 0.693/\mu_l \tag{I.3.24}$$

Similarly for the tenth-value thickness (TVT)

$$TVT = 2.30/\mu_l \tag{I.3.25}$$

The mean free path of photons in an absorber is the average distance travelled by a photon in the absorber before experiencing an interaction

$$X_m = 1/\mu_l \tag{I.3.26}$$

Broad-Beam Geometry

Broad-beam geometries are experienced in practical situations and involve a considerable amount of scattering. The build-up factor, B, is the amount by which transmission is increased relative to narrow beam conditions and this depends on both photon energy and the transmission factor, which now becomes, $tf = Be^{-\mu lx}$. B is 1 for narrow-beam geometries, but will increase with depth for the broad beam geometries of nuclear medicine imaging, until a plateau is reached (Zaidi and Hasegawa 2003).

Beam Hardening

The transmission curve for a polyenergetic photon beam is a sum of exponentials, one for each of the energies. The curve initially declines steeply as the lower energy, i.e. softer, components are absorbed. It then declines less steeply, reflecting the greater penetration of the higher energy, i.e. harder, components. In this way, the average energy of the photons in the beam increases with increasing absorber thickness, a process known as beam hardening.

Photoelectric Effect

The photoelectric effect results in an ionised atom. In this process, the total energy of the photon is imparted to an orbital electron which is ejected from the atom as a photoelectron (see Fig. I.3.3). Photoelectron ranges in soft tissue are very small, 0.023–4.4 mm for the energy range 34–1240 keV. They tend to travel in a forward direction, especially at high photon energies, with kinetic energy equal to the energy of the incident photon minus its binding energy. Consequently, photoelectrons cannot be ejected unless the energy of the incident photon exceeds the binding energy of an orbital electron. Because of this, the contribution to attenuation demonstrates abrupt increases, as much as 5–6 times, at energies corresponding to the binding energies, called absorption edges. The orbital electron most likely to interact, will be the one with binding energy closest to, but less than, the energy of the incident photon. Thus, electrons in the more inner shells are more likely to be ejected, as the energy of the in-

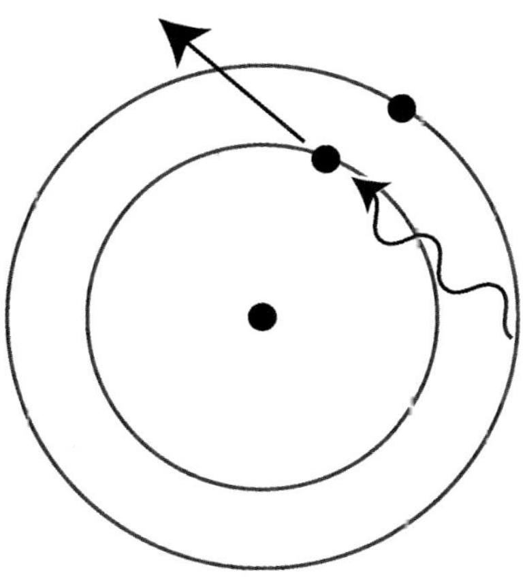

Fig. I.3.3. The photoelectric effect showing ejection of an orbital electron with total loss of a photon from outside the atom

cident photon increases. The resulting orbital vacancies will lead to the emission of characteristic x photons or Auger electrons.

Pair Production

At γ energies >1.022 MeV, pair production is possible. This occurs when a photon interacts with the electric field of a charged particle, usually an atomic nucleus but occasionally an electron, and the total energy of the photon is used to create a positive–negative electron pair. The minimum energy is required to provide the rest mass of the pair, any excess being imparted to the electrons as kinetic energy, shared randomly between them usually within the range 20%–80%.

Compton Scattering

Compton scattering involves interaction with a weakly bound outer orbital electron but without the total disappearance of the incident photon (see Fig. I.3.4). Some energy is transferred to the orbital electron which is ejected and is called a recoil electron. Because interaction is with an outer orbital electron, the amount of energy transferred does not depend on density, atomic number or any other property of the absorbing material. Thus for photons of the same energy, gram for gram, all materials will cause a similar amount of Compton scattering. The incident photon is effectively deflected through a scattering angle which is determined by the amount of the energy transferred:

$$E_{SC} = E_I / [1 + (E_I/511)(1 - \cos\theta)] \qquad (I.3.27)$$

where
E_I, E_{SC} are the energies of the incident and scattered photons (keV)
θ is the scatter angle

This equation indicates that very small energy transfers result in very small θ. In the case of the maximum energy transfer, which is still not total, θ=180°, i.e. the incident photon is backscattered. The equation also indicates that, as the energy of the incident

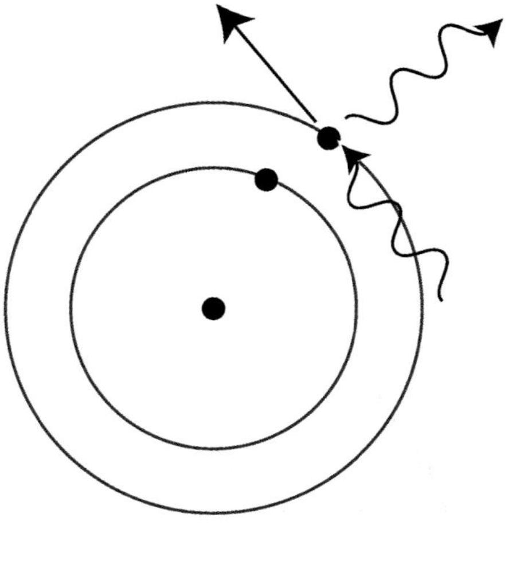

Fig. I.3.4. Compton scattering of a photon from outside the atom and, again, ejection of an orbital electron

photon decreases, the proportion that can be transferred to the recoil electron decreases. Also, at relatively low, i.e. 10–100 keV energies, the proportion transferred tends to be either a minimum or a maximum with less in between, and for higher energy incident photons, i.e. >500 keV, the proportion transferred tends to be a minimum. For body tissues, where Z<20, this process is dominant over most of the energy range applicable in nuclear medicine.

The angle through which scattering occurs is random, with high-energy photons tending to scatter in the forward direction. For low-energy photons, the probabilities for forward and backward scattering are approximately equal and approximately twice that for side scattering (Simpkin 1999).

References

Bardies M, Myers MJ (1996) Computational methods in radionuclide dosimetry. Phys Med Biol 41: 1941–1955

Clairand I, Ricard M, Gouriou J, di Paola M, Aubert B (1999) DOSE3D: EGS4 Monte Carlo code-based software for internal radionuclide dosimetry. J Nucl Med 40:1517–1523

Clairand I, Bouchet LG, Ricard M, Durigon M, di Paola M, Aubert B (2000) Improvement of internal dose calculations using mathematical models of different adult heights. Phys Med Biol 45: 2771–2785

Delacroix D, Guerre JP, Leblanc P, Hickman C (1998) Radionuclide and radiation protection data handbook. Rad Prot Dos 76:1–2

Gadd R, Mountford PJ, Oxtoby JW (1999) Effective dose to children and adolescents from radiopharmaceuticals. Nucl Med Commun 20:569–573

Groenewald W, Wasserman H (1990) Constants for calculating ambient and directional dose equivalents from radioactive point sources. Health Phys 58:655–658

Harding K, Thomson WH (1997) Radiological protection and safety in medicine – ICRP 73. Eur J Nucl Med 24:1207–1209

Howell RW, Wessels BW, Loevinger R in collaboration with the Medical Internal Radiation Dose Committee Watson EE, Bolch WE, Brill AB, Charkes ND, Fisher DR, Hays MT, Howell RW, Robertson JS, Siegel JA, Thomas SR, Wessels BW (1999) The MIRD perspective 1999. J Nucl Med 40: 3S–10S

IAEA (1996) International Atomic Energy Agency international basic safety standards for protection against ionizing radiation and for the safety of radiation sources: a safety standard. IAEA, Vienna (International Atomic Energy Agency safety series 115)

ICRP (1977) International Commission on Radiological Protection publication 26; 1977 recommendations of the International Commission on Radiological Protection. Pergamon, Oxford

ICRP (1991) International Commission on Radiological Protection publication 60; 1990 recommendations of the International Commission on Radiological Protection. Pergamon, Oxford

ICRP (1992) International Commission on Radiological Protection publication 60 – user's edn; 1990 recommendations of the International Commission on Radiological Protection. Pergamon, Oxford

ICRP (1996) International Commission on Radiological Protection publication 73 radiological protection and safety in medicine. Pergamon, Oxford

Johansson L (2003) Hormesis, an update of the present position. Eur J Nucl Med 30:921–933

Loevinger R, Budinger TF, Watson EE (eds) (1988) MIRD primer for absorbed dose calculations. Society of Nuclear Medicine, New York

Loevinger R, Budinger TF, Watson EE (eds) (1991) MIRD primer for absorbed dose calculations, revised. Society of Nuclear Medicine, New York

Mountford PJ (1996) Internal dosimetry: developments and limitations. Eur J Nucl Med 23:491–493

Mountford PJ (1997) Risk assessment of the nuclear medicine patient. Br J Radiol 70:671–684

NCRP (1983) National Council on Radiation Protection and Measurements report 73. Protection in nuclear medicine and ultrasound diagnostic procedures in children. NCRP, Bethesda, Maryland

Ott RJ (1996) Imaging technologies for radionuclide dosimetry. Phys Med Biol 41:1885–1894

Overbeek F, Pauwels EKJ, Broerse JJ (1994) Carcinogenic risk in diagnostic nuclear medicine: biological and epidemiological considerations. Eur J Nucl Med 21:997–1012

Parker RP, Smith PHS, Taylor DM (1984) Basic science of nuclear medicine. Churchill Livingstone, Edinburgh

Simpkin DJ (1999) Radiation interactions and internal dosimetry in nuclear medicine. Radiographics 19:155–167

Sont WN, Zielinski JM, Ashmore JP, Jiang H, Krewski D, Fair ME, Band PR, Letourneau EG (2001) First analysis of cancer incidence and occupational radiation exposure based on the national dose registry of Canada. Am J Epidemiol 153:309–317

Sorenson JA, Phelps ME (1987) Physics in nuclear medicine, 2nd edn. Grune and Stratton, New York

Stabin MG (1996) MIRDOSE: personal computer software for internal dose assessment in nuclear medicine. J Nucl Med 37:538–546

Stabin MG, Gelfand MJ (1998) Dosimetry of pediatric nuclear medicine procedures. Q J Nucl Med 42:93–112

Stabin MG, Tagesson M, Thomas SR, Ljungberg M, Strand SE (1999) Radiation dosimetry in nuclear medicine. Appl Radiat Isot 50:73–87

Thierens HM, Monsieurs MA, Brans B, van Driessche T, Christiaens I, Dierckx RA (2001) Dosimetry from organ to cellular dimensions. Comput Med Imaging Graph 25:187–193

Thomas SR, Maxon HR III, Kereiakes JG (1988) Techniques for quantitation of in vivo radioactivity. In: Gelfand MJ, Thomas SR (eds) Effective use of computers in nuclear medicine. McGraw-Hill, New York, pp 348–383

Toohey RE, Stabin MG, Watson EE (2000) The AAPM/RSNA physics tutorial for residents: internal radiation dosimetry: principles and applications. Radiographics 20:533–546

Van Wyngaarden KE, Pauwels EKJ (1995) Hormesis: are low doses of radiation harmful or beneficial? Eur J Nucl Med 22:481–486

Watson EE, Stabin MG, Siegel JA (1993) MIRD formulation. Med Phys 20:511–514

Zaidi H, Hasegawa B (2003) Determination of the attenuation map in emission tomography. J Nucl Med 44:291–315

Zanzonico PB (2000) Internal radionuclide radiation dosimetry: a review of basic concepts and recent developments. J Nucl Med 41:297–308

Protection of the Community

Abbreviations

ALARA	as low as reasonably achievable
ALI	annual limit on intake
BEIR	United States National Academy of Sciences Biological Effects of Ionizing Radiation committee
FDG	fluorodeoxyglucose
IAEA	International Atomic Energy Agency
ICRP	International Commission on Radiological Protection
LET	linear energy transfer
TLD	thermoluminescent dosimeter
UNSCEAR	United Nations Scientific Committee for the Effects of the Atomic Radiation

Although diagnostic nuclear medicine provides a benefit to the patient; to the community, it produces mainly a risk. These risks have to be kept to reasonable levels otherwise more damage will be done than justifies the benefits achieved. The control of risk is exercised on the amount of radiation that individuals absorb by managing the work that is undertaken and the method in which it is undertaken. Ultimate control is in the form of limits which should not be exceeded. The basis on which these are determined is the risk of cancer. These form the principles of radiation protection. From them, it is possible to design a practical implementation. Particular risks to certain groups of individuals can be identified. Those working in nuclear medicine departments are at risk from both sources and patients who have been administered. The rest of the community can be protected from physical sources but they remain at risk from the radioactive patient.

Effective Dose Limits

The ICRP recommends limits on effective dose for three categories of people: occupationally exposed persons, members of the public and pregnant workers (ICRP 1991a).

Occupationally Exposed Persons

To protect occupationally exposed persons against stochastic effects, the recommended limit on effective dose is 20 mSv per year, averaged over 5 years, i.e. 100 mSv in 5 years, with a further provision that the effective dose should not exceed 50 mSv in any single year.

Annual Limits on Intake (ALI)

For internal exposure, annual limits on intake (ALIs) are listed in publication 61 (ICRP 1991b) and, including the respiratory tract, in publication 68 (ICRP 1995), based on a committed effective dose, E(τ), of 20 mSv. The estimated intakes are allowed to be averaged over 5 years to provide some flexibility. For ingestion, values vary markedly being 0.8 MBq for ^{131}I and 1000 MBq for $^{99}Tc^{m}$.

Committed Effective Dose

E(τ), is estimated by calculation of the committed equivalent dose, $H_T(\tau)$. This is the H_T that will be absorbed over τ years following an intake. If τ is not specified, it is 50 years for adults and from intake to age 70 years for children.

Deterministic Effects

Even if radiation absorbed doses are at their limit for long periods, the restrictions are adequate to avoid deterministic effects except in the lens of the eye, which makes no contribution to the effective dose, and the skin, for which the effective dose limits provide adequate protection against stochastic effects, but which requires an additional limit to prevent deterministic effects in localised exposures.

Lens of the Eye

For the lens of the eye, the presently recommended annual equivalent dose limit is 150 mSv, which is the same as the previous 1977 limit, and is based on the estimated threshold for cataract formation, of >0.15 Sv.

Skin

The additional recommended annual limit, for the skin, is 500 mSv, averaged over any 1 cm^2 regardless of the area exposed. Applied to the skin of the face, this limit will also provide protection, for the lens of the eye, against localised exposure to poorly penetrating radiation such as β^- particles. The same limit can be applied to all the tissues in the hands and feet.

Members of the Public

For members of the public, the recommended annual limit on effective dose is 1 mSv, based on the level of the natural background. In special circumstances, a higher value can be allowed in a single year, provided that the average over 5 years does not exceed 1 mSv per year.

Lens of the Eye, and the Skin

As for workers, additional limits are also needed for the lens of the eye and localised areas of skin. For these, an arbitrary reduction factor of 10 has been adopted, leading to annual limits of 15 mSv for the lens and 50 mSv averaged over any 1 cm^2 area of skin regardless of the area exposed.

Pregnant Workers

For pregnant workers, the ICRP considers that a foetus should have the same protection as members of the general public, but that the recommended dose limits for occupational exposure should achieve this, so that no special dose limits are recommended for women of reproductive capacity. However, once a woman has declared pregnancy and for the remainder of the pregnancy, in order to ensure the member of public limit of 1 mSv to the foetus, a supplementary equivalent dose limit of 2 mSv is applied to the surface of the woman's abdomen and intake of radionuclides is limited to ~1/20 of the ALI. A more severe restriction, of 1.3 mSv, is suggested as being necessary for technologists and nursing staff involved in nuclear medicine (Mountford and Steele 1995; Mountford and O'Doherty 1999).

Previous Recommendations

Setting limits on the risks to which an individual may be subjected is difficult. For the above dose limits, the ICRP adopted a more comprehensive approach than in its previous recommendations (ICRP 26), published in 1977. In these, the recommended annual effective dose equivalent limit for workers was set at 50 mSv, with an over-riding condition that any single organ or tissue should be limited to 500 mSv, except the lens of the eye which should be limited to 150 mSv. Committed doses were regulated by Annual Limits on Intake (ALI) published in ICRP 30 (ICRP 1979, 1980, 1982). There were also subsidiary limits for trainees, for the abdomen for women of reproductive capacity and for those known to be pregnant. The limit during the declared term of pregnancy was 10 mSv.

These recommendations were made assuming that an annual occupational death probability of about 10^{-3} to the most exposed individuals would be at the border of being unacceptable and that the corresponding extra imposed annual death probability for members of the public at the annual limit of 1 mSv (ICRP 1985) would be about 10^{-5}. For the occupational dose limits, the ICRP attempted to use a comparison with the rates of accidental death in industries not associated with radiation. These comparisons are not altogether satisfactory for a number of reasons. For example, standards of industrial safety are neither constant nor uniform world-wide; the mortality data relate to averages over whole industries whereas dose limits apply to individuals; the quantitative comparisons were limited to mortality data, although the inclusion of non-fatal conditions on both sides of the comparison would have led to less restrictive dose limits; and, finally, there are few grounds for believing that society expects the same standard of safety across a wide range of industries (ICRP 1991a).

Basis of Present Recommendations

The aim in the 1991a publication, was to establish, for a set of practices, a level of dose above which the consequences for the individual would be widely regarded as unacceptable. A dose limit was defined as representing a selected boundary in the region between 'unacceptable' and 'tolerable' for the situation to which the dose limit was to apply. It was admitted that this approach was inevitably subjective, but that it made it possible to consider a wide range of inter-related factors, more properly called attributes. The attributes associated with mortality were given as:

- the lifetime attributable probability of death
- the time lost if the attributable death occurs
- the reduction of life expectancy (a combination of the first 2 attributes)
- the annual distribution of the attributable probability of death
- the increase in the age specific mortality rate, i.e. the probability of dying in a year at any age, conditional on reaching that age

The ICRP also decided to allow for morbidity due to non-fatal cancer and hereditary disorders by using a number of non-fatal conditions weighted for severity and for the period of life lost or impaired.

Cancer Risk

The probability of a fatal cancer was estimated by relying mainly on studies of the Japanese survivors of the atomic bombs and their assessment by bodies such as UNSCEAR and BEIR. UNSCEAR is the United Nations Scientific Committee for the Effects of the Atomic Radiation; it is a committee of the United Nations General Assembly and was established in 1955. With the recalculation of dosimetry from the atomic bomb survivors of Hiroshima and Nagasaki and increasing public concern about ionising radiations, the various international committees re-estimated the effect of radiation. The review has led to a revision of the risk estimate for radiation induced fatal cancers from the previously accepted value of 2.5% (i.e. 1 in 40) per Sv, to 3–10%. Consequently, the effect of radiation could be taken to be 4 times the previous risk, except that these values have been estimated for high doses, i.e. bomb data. The dose reduction factor which modifies the risk for the effect at low doses and low dose-rates, i.e. occupational exposure, was also reviewed. In September 1987, the ICRP met in Como, Italy and subsequently issued a preliminary statement with their advice on the possible outcome of the re-assessment of cancer risk estimates resulting from the revised radiation dosimetry of the Japanese survivors. The final recommendations were published in Publication 60 (ICRP 1991a).

Over the years, starting in about 1972, the UNSCEAR and BEIR Committees have made major risk evaluations resulting in estimates of the risk associated with 1 Gy of acute low LET uniform whole body irradiation, using two models: the additive, which predicts a constant excess of induced cancer throughout life unrelated to the spontaneous rate of cancer; and the multiplicative, which predicts that the excess of induced cancers will increase with time as a constant multiple of the spontaneous or natural rate of cancer and consequently will increase with age in that population. Estimates based on these models have come closer with time. The estimates based on the multiplicative model have changed the least. The previous preference for the additive model is part of the reason that the risk estimates appear to have increased by a factor of ~3–4 in comparison to those used in ICRP 26 (ICRP 1977).

Effective Dose Limits

The ICRP's conclusion was that a regular annual effective dose of 50 mSv, corresponding to a lifetime value of 2.4 Sv, was probably too high. In particular, the reduction of life expectancy at this level, of 1.1 years, and the fact that there would be a probability exceeding 8% that the radiation hazards in a worker's occupation would be the cause of his death, albeit at a late age, would be widely seen as excessive for a group of occupations, many of which are of recent origins and should therefore be setting an example. The ICRP decided that its dose limits should be set in such a way and at such a level that the total effective dose received in a full working life would be prevented from exceeding about 1 Sv, received moderately uniformly year by year and that the application of its system of radiological protection should be such that this figure would only rarely be approached. The idea of monitoring the effective dose over a 5 year period was introduced because it was thought that the previous system, of maintaining a rigid control period of one year with no credit taken for any earlier years of low effective dose, was too inflexible.

Principles of Radiation Protection

The latest system of dose limitation is based on the following general principles of radiological protection, stated by the ICRP.

Justification

A person should be exposed to ionising radiation only if the exposure produces sufficient benefit to justify the detriment.

Optimisation

All exposures should be kept as low as reasonably achievable (ALARA), economic and social factors being taken into account. Procedures should be optimised to produce the best combination of maximum benefit and minimum exposure.

Limitation

The ultimate level of protection is provided by establishing: dose constraints, which should be imposed to limit doses to individuals; and risk constraints, which should be imposed to reduce the risk of accidental exposures.

As late as the 1950s, there was a tendency to regard compliance with individual dose limits as being a satisfactory system of protection. The advice that all exposures should be kept as low as possible was noted but not often applied consciously. Since the 1950s, however, much more emphasis has been placed on keeping all exposures 'as low as reasonably achievable, economic and social factors being taken into account' (ALARA). This has decreased individual doses substantially (Wrixon 1994), has greatly reduced the number of situations in which the limits play a major role in the overall system of protection and has also changed the purpose of the limits. Initially, their main function was to avoid directly observable, non-malignant, effects but subsequently they were also intended to limit the incidence of cancer and hereditary effects. This has resulted in the annual limit for occupational whole body exposure being reduced substantially.

Practical Implementation of Recommendations

The practical implementation of the above recommendations must be very straightforward, very practical, clearly documented and regularly reviewed. It should be controlled by a Radiation Safety Committee.

Local Rules

Local rules should document the practical implementation of the radiation protection regimen and should address legal obligations, radionuclides and activities in use, controlled areas, storage procedures, work practices, radiation safety personnel, monitoring procedures, contingency and decontamination procedures, waste disposal procedures, record keeping and transport arrangements.

Control of Sources

One of the most important responsibilities is the control of radiation sources. These should be received in a specified area and their receipt recorded. Care must be taken in their unpacking, with protective gloves being worn and any damaged packages being wipe tested for contamination. Any radioactive gases and volatile materials, e.g. iodine, should only be unpacked in a well ventilated area, such as a fume cupboard. The package contents must be confirmed against the transport documentation and the packaging discarded only after contamination monitoring and label removal. Solid sources, e.g. check sources, should be wipe tested for contamination on receipt and then at regular intervals. The radionuclides must be transferred immediately to a designated, clearly labelled, well ventilated and secure store.

Designation of Working Areas

Source control is emphasised by formally designating the work areas, in which they are used. The ICRP uses two descriptions, controlled and supervised (ICRP 1991a). A controlled area is one in which procedures and practices, to control radiation exposures, are followed. A supervised area is one in which the working conditions are kept under review but special procedures are not normally required. Outside these areas, workers should not need to be regarded as occupationally exposed, because the doses received will be below the limits for public exposure. The distinction between controlled and supervised areas has usually been determined on the basis that the doses to workers in supervised areas receive less than 3/10 of the occupational dose limits. The ICRP now regards this definition as being too arbitrary and recommends that the designation of controlled and supervised areas be decided on operational experience, taking into account the expected level and the likely variations of the doses and intakes and the potential for accidents. The recommendations have been expanded, in ICRP 73, to describe a controlled area as one where possible occurrence of minor mishaps requires workers to have well-established procedures and practices aimed at controlling radiation exposures (ICRP 1996; Harding and Thomson 1997). Once classified, the areas should be clearly labelled and the boundaries well defined.

Facilities

Minimisation of ambient radiation levels and reduction of the spread of contamination can be achieved by good facility design. The situation of the radiopharmacy, dispensary, injection and patient waiting areas and waste storage areas should be considered in relation to measurement and imaging areas, in which it is necessary to maintain very low radiation levels. Movement of patients and radioactive sources past such sensitive areas should be minimised. These areas should be shielded to a level appropriate to the use of adjacent areas taking into account that transient very high radiation levels may be experienced, particular where a ^{99}Mo–^{99}Tc^{m} generator is stored and used. The layout of rooms should be such that movement of radioactive materials is minimised and can be undertaken in a safe manner. Separate toilet facilities for ambulatory patients as well as facilities for emptying and decontaminating containers of human excreta for bed or stretcher patients, must be available. Walls, floors, benches and furniture, such as chairs, should have smooth, impervious and durable finishes to allow for repeated cleaning and ease of decontamination. Covings should be used where walls meet floors and benches and bench tops should have a lip at the front to contain any spillages. All joints must be sealed by impervious sealant, to avoid the build-up of contamination. The ventilation systems of rooms in which volatile materials or gases are used should be specially vented in such a way as to prevent contamination of the air supply of any building.

Classification of Workers

In previous recommendations, the ICRP defined two types of worker based on the expected level of individual annual dose. This was originally intended to help in the choice of workers to be subject to individual monitoring and medical surveillance. In recent years, this has come to be regarded as inappropriate and the ICRP no longer recommends such a classification.

Minimisation Techniques

There are four main factors to be considered for radiation protection when dealing with unsealed radioactive materials. These are time, distance and shielding for protection against external exposure and avoidance of the spread of contamination for protection against internal exposure. Sometimes the last factor can be compromised by the first three and a balance must be achieved.

Time

Because the absorbed dose is proportional to the exposure duration, it is extremely important to spend as little time as possible near x or γ photon emitting sources. All work must be undertaken as swiftly as possible without compromising safety, particularly with regard to contamination. Because patients are also sources of radiation, the time spent in their vicinity should be kept to a minimum, subject to clinical demands. Waste and contaminated items should be dealt with as quickly as possible.

Distance

The separation between the source and the operator must be as large as practical. A discrete distance should be maintained from patients, again subject to clinical demands. The dose rate will vary approximately inversely as the separation up to around 3 m, if the radionuclide is distributed homogeneously, because it can be approximated by a line (Siegel et al. 2002). The dose rate from syringes and vials will vary inversely as the square of the separation, however, because they can be approximated by point sources. Syringes one size larger than necessary should be used, so that they are never more than half-full, and should be handled from the unfilled end to increase the source to finger distance. This precaution also reduces the risk of the plunger becoming dislodged. Remote handling tools and carrying boxes should be used wherever appropriate, but not at the expense of increasing the manipulation duration to such an extent that any benefit is lost.

Shielding

Sources must always be adequately shielded, using lead or tungsten for photons and Perspex for charged particles. Transport shielding, particularly for generators, is not sufficient protection during storage. Supplementary shielding should be used to reduce the dose rate, and this should be situated as close to the source as possible, to reduce the quantity required. The amount required to reduce exposure to an acceptable level depends on the activity and the energy of the photons. For example, the tenth value thickness of lead, is 0.9 mm for 150 keV photons whereas it is 20 mm for 511 keV annihilation photons. Syringe shields are recommended, during manufacturing, dispensing or administration, in view of the ALARA principal. Lead aprons are not very effective in nuclear medicine, however, and are not usually recommended, therefore. Although vials are manipulated using vial shields, extra shielding, for instance lead glass face and body shields and safety glasses to reduce the eye dose, may be required during use. Distance and shielding factors, for protection against annihilation radiation, at 511 keV, are significantly more severe than for single photon radiation.

Contamination Control

The possibility of contamination is highest during radiopharmaceutical dispensing procedures, ventilation scanning (Mountford 1991) and administration. The causes of the latter are lack of experience by the individual performing the injection, followed closely by radioactive syringe disposal problems, and injection technique (Mosman et al. 1999).

Contamination needs to be considered under two categories: one is its ingestion, possible by three routes: penetration through skin, ingestion and inhalation, and the other its spread. To avoid the former, restrictions on smoking, drinking and eating are applied in controlled and supervised areas, and cuts must be covered by waterproof plasters. Protective clothing must always be worn during wet operations Gloves must be worn to avoid skin contamination and possible ingestion. It is important to take precautions against aerosols, which are often created when a needle is removed from a vial, such as wearing gloves over the sleeves of a protective coat. To reduce this possibility, over-pressure of closed vials must be reduced by drawing gas back into the syringe, which must not then be expelled into the atmosphere. Contamination by inhalation poses a problem where radioactive gases are used or radioiodine, particularly ^{131}I and ^{125}I, is handled. These should be used in rooms with air extracts and work with radioiodine should be undertaken in a fume hood.

To minimise the risk of contamination spread, in controlled and supervised areas, taps must not cause splashing and paperwork must be undertaken only in specific areas. All wet handling operations should be undertaken over non-absorbent and cleanable drip trays, and in specific areas. All equipment and clothing used during a wet handling procedure, including vial shields, must be regarded as potentially contaminated. Regular changing of gloves during wet procedures and removal by turning them inside out helps to restrict contamination spread.

Monitoring

Exposure

Exposure monitoring should be undertaken with a calibrated survey meter, at regular intervals, in areas where radioactive materials are used and, particularly, stored to ensure that unexpected high exposure areas have not arisen.

Contamination

Monitoring for radioactive contamination, of both personnel and area, should be undertaken after each wet handling procedure and at regular intervals. This will usually be undertaken with a calibrated contamination monitor (MacDonald and Gibson 2002). However, monitoring in areas of high background activities or monitoring solid check sources necessitates wipe testing with a moist swab, and assuming that 10% of the contamination is absorbed by the swab. To monitor environmental contamination of radioaerosols or gases, air concentration can be measured by drawing air continuously through a detector or through a filter, e.g. activated charcoal when radioiodine is used, which can then be measured. Aerosols, however, will eventually settle on the floor and surface contamination can also be undertaken.

Personal

Personal monitoring should be used for those working in controlled areas (ICRP 1991a, 1996; Harding and Thomson 1997) where it is appropriate, adequate and feasible (IAEA 1996). It should be considered for those working regularly in supervised areas (ICRP 1991a, 1996; Harding and Thomson 1997). A supplementary function of individual monitoring is in confirming the classification of workplaces and detecting aberrations in environmental conditions. The results should be displayed for review by all involved.

External

Monitors should be worn on the most exposed part of the body, normally anteriorly, at breast height. Those manipulating radioactive materials should also wear an extremity monitor, under gloves to avoid contamination, on the hand most likely to receive the highest exposure. Pregnant workers should wear a separate monitor on the abdomen, to monitor exposure of the foetus, and this should be assessed more frequently than the usual monitor. Two methods are used for external dosimetry: active and passive monitoring. An active monitor is usually a Geiger-Mueller or semiconductor detector which gives an immediate assessment. A passive monitor is usually a photographic film or thermoluminescent dosimeter (TLD), which is changed at regular intervals, but not longer than 3 months.

Internal

Sometimes, it is necessary to assess internal contamination. Routinely, this should only be for workers in controlled areas, designated specifically on the basis of the contamination control and in which significant intakes may be expected. It is usually when therapeutic activities of ^{131}I are handled and involves external monitoring of the activity accumulated by the thyroid, either 1–3 days following a procedure or at regular intervals.

Reference Levels

These are levels which trigger a response (Wrixon 1994; ICRP 1996; Harding and Thomson 1997). They include: recording levels, above which a monitoring value should be recorded; investigation levels, above which the cause should be determined; and intervention levels, above which remedial action may be necessary.

Accidents

The probability of accidents and the severity of any that occur are reduced by contingency planning incorporating risk assessment, decontamination training (Mountford 1991) and provision of a portable decontamination kit. In the event of an accident, priority must be given to medical necessities. Access to the area can then be restricted, and decontamination of personnel undertaken, followed by decontamination of the room and equipment. Spills can be dealt with using absorbents or by closing the room, particularly when radioactive gases or aerosols are involved, followed by monitoring and decontamination. Finally, any persons or equipment involved in the decontamination should be monitored for contamination. For decontamination of personnel, the skin should be washed in warm water, rather than hot water which opens the capillaries.

Transport

Within a hospital, the radioactive material should be doubly contained in a suitably shielded and labelled container, the outer one containing absorbent material.

For transport outside a hospital, the IAEA defines allowable types of packaging and internationally accepted labelling (IAEA 1985). Type A is used for the lower activities, encountered in hospital work, and is designed to withstand most commonly encountered minor accidents. The labelling is according to dose categories shown in Table I.4.1. The IAEA defines an 'excepted package' for low activities where the external dose rate is minimal, i.e. <5 μSv h^{-1}. An exact indication of the radiation arising from the package is provided by the transport index, which is 100 times the maximum dose rate, in millisieverts per hour, at a distance of 1 m from the surface of the package.

Waste

Restrictions are imposed on the storage and disposal of radioactive waste and comprehensive records must be kept. All radioactive warnings must be erased before removal from the department. Spent generators are usually returned to the manufacturer, since disposal within the hospital is not recommended as dismantling is required. This route is also useful for long half-life solid sources. Shorter half-life, i.e. <60 days, solid waste may be stored, usually for 10 half-lives, until it has decayed to levels easily dealt with. This is aided by separating radionuclides by half-life, to minimise storage durations for the shorter half-life materials. The decayed material may then be disposed with ordinary solid waste, if its activity is less than twice the background activity; or incinerated, if its activity is higher than this but lower than nationally authorised limits. Depending on national regulations, liquid waste, including that from patients, may be dispersed into environment, via drains, as long as there is sufficient dilution to avoid environmental impact and there is no possibility of blockage and irradiation hazard from activity pooling in the drains themselves. Some used radioactive gases are disposed directly to the atmosphere, again with sufficient dilution to avoid environmental impact.

Table I.4.1. IAEA package labelling categories, figures are given as millisieverts per hour

Label category	Maximum dose at surface	Maximum dose at 1 m
I (white)	0.005	
II (yellow)	0.5	0.01
III (yellow)	2	0.1

Hazards to Nuclear Medicine Staff

The potential radiological hazards to staff can be characterised in terms of the type of work. The main group of staff who are likely to receive relatively higher personal doses are those involved in the preparation, dispensing and administration of radiopharmaceuticals. On the surface of a syringe or vial containing a single patient dose of bone imaging agent the dose rate may be 1000 mSv h^{-1}. In these cases, the radiation dose to the hands is the most crucial. However using syringe shields, forceps for remote handling and lead containers for shielding vials, the radiation dose to the pulp of the finger can be reduced (Harding et al. 1985a).

Radiopharmacy, Dispensary, Injection Area

Radiation doses to the fingers can be significant, with the fingertips usually receiving the highest. These can be monitored using TLD tapes (Williams et al. 1987; Dhanse et al. 2000). Even measuring syringe activities, in activity meters, can double the finger doses (Hastings et al. 1997). Doses vary substantially among institutions, related to work load and practices. Dominant hand extremity doses, in a month, are reported to be as low as 1.6 mSv (Mackenzie 1997) and as high as 12 mSv without syringe shields (Harding et al. 1985a), with an annual figure of up to 300 mSv (Batchelor et al. 1991; Jansen et al. 1994). These figures may be worse in large central radiopharmacies, but may actually be as low as 83 mSv. Per 10 GBq of $^{99}Tc^{m}$ handled, the dose is reported as 0.18–1.2 mSv, with all lead glass syringe shields and improved dispensing procedures being credited for a fall from 0.40 to 0.18 mSv in one institution (Dhanse et al. 2000). These figures will also be worse with β^{+} emitting radionuclides (Heller 1996) since, unshielded, a 200 MBq ^{18}FDG dose gives about 2 $^{1}/_{2}$ times the dose rate compared to a 600 MBq $^{99}Tc^{m}$ bone dose.

Conventional syringe shields are constructed of lead or tungsten. They reduce the dose rates less than anticipated, however, by a factor of 8 for the dominant hand and only marginally for the non-dominant hand (Harding et al. 1985a). These may also present a problem in patients with difficult veins and lighter, and thinner bismuth shields are becoming popular. Dose rates can be reduced using short cannula rather than direct needle administrations (Batchelor 1986; Batchelor et al. 1991).

Although extremity doses are dominant, whole body doses can also be significant. The whole body absorbed dose rate during radionuclide manipulation ~2 μSv h^{-1} (Lundberg et al. 2002). Whole-body doses are generally below 0.2 mSv per month (Harding et al. 1990b). This depends on work patterns. For example, when $^{99}Tc^{m}$-sestamibi replaced ^{201}Tl for stress myocardial perfusion imaging, the mean whole-body film badge readings increased from 240 to 560 μSv $month^{-1}$ for radiopharmacy technologists (Culver and Dworkin 1993). To reduce whole body doses when injecting ^{18}FDG, administrations should take place behind a body shield.

Imaging Rooms

In the imaging rooms, it is the whole body dose, rather than the extremity dose, that is important and nuclear medicine staff tend to have the highest doses in health care.

The main factor, in this, is proximity to patients (Boutcher and Haas 1985; Harding et al. 1990b). During imaging, therefore, staff should, if possible, remain at least 1 m from the patient. The whole body absorbed dose rate is 0.2–2 μSv h^{-1}, during imaging procedures (Lundberg et al. 2002); the average whole body exposure to the technologist from a typical $^{99}Tc^{m}$ imaging procedure being 0.2–5.5 μSv (Harding et al. 1985b, 1990b) with the average approximating 1.5 μSv (Clarke et al. 1992; Harding and Thomson 1997) and resulting in 1–4 mSv $year^{-1}$. Again, this depends on work patterns. For example, when $^{99}Tc^{m}$-sestamibi replaced ^{201}Tl for stress myocardial perfusion imaging, the mean whole-body film badge readings increased from 100 to 450 μSv $month^{-1}$ for nuclear medicine technologists (Culver and Dworkin 1993). Whole-body doses have recently been measured to be 0.2–0.4 μSv. Excepting equilibrium angiocardioscintigraphy, at 1.0±0.5 μSv, and $^{99}Tc^{m}$-sestamibi SPET, at 1.7±1.0 μSv; these are slightly lower than values published in the last 20 years. This is a consequence of more favourable working conditions, resulting from technological improvements, e.g. dual-head γ cameras capable of whole-body studies, safer shielding and increased distance from patients (Chiesa et al. 1997).

A particular concern, in imaging rooms, relates to ventilation studies. In this, patient compliance is an important factor in minimising doses to staff; clear instructions and practice being a vital part of the procedure (Greaves et al. 1995). With $^{81}Kr^{m}$, although the gas is exhaled into the room, the radiation dose results almost entirely from external exposure, with the internal dose accounting for <1%. For a single study, the dose ~0.04 μSv compared to ~0.2 μSv for a perfusion study. The doses with ^{133}Xe will be higher. When radioaerosol systems are used, a major hazard arises from the administration apparatus, which retains up to 97% of the loaded activity. The best way of dealing with this is to leave it to decay for 24 hours before handling. A reduction in the risk of contamination during administration can be obtained by using a mouthpiece and nose clip, rather than a face mask (Harding et al. 1990b).

Flood sources present a significant hazard; the surface dose rate at the edge of a ^{57}Co source can ~500 μSv h^{-1} and, if handles are not part of the construction, should be fitted by the user (Harding et al. 1990a). Alternatively, a vacuum grip or attenuating gloves can be used (Mountford 1997). A large portion of the radiation dose received, during use of this source, is due to scattered radiation and x-rays produced by γ interactions within the camera. This dose can be reduced significantly if quality control checks are performed with the flood source lying directly on the inverted gamma camera head rather than placing the flood source on an imaging table under the gamma camera (Rudin and Johnson 2000).

The dose rate from ^{18}FDG is much higher than from single photon radionuclides, the mean dose rates measured at 0.1, 0.5, 1.0 and 2.0 m from the mid-thorax on patients immediately after injection with ^{18}FDG were 391.7, 127.0, 45.3 and 17.1 μSv h^{-1} (Benatar et al. 2000). The separate contributions, from each phase of the procedure, were: 0.11±0.04 μSv for daily quality assurance, 0.3±0.1 μSv for syringe preparation, 2.8±1.8 μSv for injection, and 1.7±1.5 μSv for a whole-body scan. These values, high in comparison with conventional procedures, are attributable to the higher specific γ constant of ^{18}F, and the longer time required for accurate positioning (Chiesa et al. 1997). However, with correct planning, emerging oncological techniques such as PET, high dose ^{67}Ga, and high dose ^{201}Tl should not represent a significantly greater occupational radiation hazard than conventional nuclear medicine procedures (White et al. 2000).

Pregnancy

Foetal doses have been measured to be 1.12–1.17 μSv for three types of $^{99}Tc^{m}$ scan, and 6.7–9.0 μSv for whole-body ^{131}I scans (Mountford and Steele 1995). It is considered that a nuclear medicine technologist could perform up to 8 imaging studies per day during the declared term and still remain within the ICRP limits for pregnancy (Clarke et al. 1992). However, the suggested, more restrictive dose limit of 1.3 mSv to the maternal abdominal surface, corresponds to six $^{99}Tc^{m}$ studies or one diagnostic ^{131}I investigation per day (Mountford and Steele 1995), during the declared term of pregnancy. Activities where it is probably not wise to involve a pregnant technologist include: dealing with radioactive spills, using radioaerosols or unshielded krypton generators, imaging very ill patients (Harding and Mountford 1993), and installing technetium generators (Harding and Thomson 1997).

Hazards from the Radioactive Patient

A radioactive patient is an important source of radiation exposure in nuclear medicine, up to 7 μSv h^{-1} at 1 m (Mountford et al. 1991), and twice this for patients administered ^{18}FDG, on discharge from the department (Cronin et al. 1999). Through penetrating radiations, x or γ photons, the patient can irradiate persons at a distance. Through physical contact or excretion and exhalation into the environment, the patient becomes a source of internal contamination. Within the hospital; staff, other patients and visitors are protected by restricting the area around the radioactive patient. Patients should only be discharged from hospital when their activity is below nationally allowed limits and the implications for the community must be considered. Following oral administration, the patient should be observed for a short period in case of vomiting. Outside the hospital; fellow travellers, family members, work colleagues and social contacts are protected by modifying the behaviour of the radioactive patient. However, for diagnostic studies there is little need for restrictions (BIR 1999), even for ^{18}FDG (Cronin et al. 1999), except in the situation of a prolonged journey home by public transport following administration with high levels of $^{99}Tc^{m}$ and an outpatient, administered ^{111}In, who has to look after a fretful infant (Mountford et al. 1991; Thomson and Harding 1995).

Accommodation

In hospital, care should be exercised regarding patient accommodation and toilet facilities. For instance, the area where radioactive patients wait for their investigations should be separated, by at least 3 m, from areas occupied by staff, or shielding should be used between the two areas.

Family

Due to the short effective half-life of most diagnostic radiopharmaceuticals and their sporadic use, there is usually very little radiation hazard to the patient's family. For instance, for $^{99}Tc^{m}$, the approximate absorbed dose rate close to a patient immediately after administration is 27–36 $\times 10^{-9}$ Gy h^{-1} MBq^{-1} and after 2 hours it is 13–36. For ^{18}F, the maximum surface doses ~5 $\times 10^{-2}$ Gy (Shukla et al. 1994). The radiation dose to the partner of patients receiving $^{99}Tc^{m}$ is unlikely to exceed 0.12 mSv (Thomson et al. 1993) and sleeping restrictions are unnecessary (Thomson and Harding 1995). Nevertheless, even the very small doses that might be received could be avoided by minimising prolonged contact between a patient and members of the family during the first few hours after administration (Mountford 1997).

Children

Close contact doses have been estimated (Mountford and Coakley 1986, 1989; Mountford et al. 1991) by multiplying the dose rate near the surface of a patient by an effective exposure time, which accounts for the intermittency of close contact (Greaves et al. 1996) and for dose rate decrease with radionuclide decay (Mountford 1987). The close contact dose to a young infant from a radioactive patient who has undergone a diagnostic investigation is not likely to exceed the member of public limit of 1 mSv, if the usual administration activities are used; except for $^{99}Tc^{m}$ myocardial perfusion investigations (Greaves and Tindale 1999), and when >20 MBq of ^{111}In-leucocytes (Mountford and Coakley 1989; Mountford et al. 1991; Thomson and Harding 1995) or >112 MBq of ^{131}I is administered (Mountford 1987).

Parents

The critical group will be parents of young infants who require extended periods of close contact. A parent could receive a dose of 1 mSv only if the infant had been administered as much as 500 MBq of $^{99}Tc^{m}$. This would be very unlikely because an infant young enough to necessitate the required duration of close contact would be too young to receive such a high administration activity (Mountford 1997).

Other Patients

The additional absorbed dose to other patients waiting in the same room as a patient administered with radiopharmaceutical for a diagnostic procedure will be <0.2% of that from their own administration (Harding et al. 1990a, 1994).

Members of the Public

Scenarios postulated for exposure of co-workers and family members yielded doses between 7–20 µSv for ^{111}In, ^{201}Tl and $^{99}Tc^{m}$ radiopharmaceuticals (Benedetto et al. 1989). The total absorbed dose to a relative or friend accompanying an injected patient in the nuclear medicine waiting room ~13 µSv, which is not affected by whether a separate "injected" waiting room is available. This is <2% of the annual dose limit and indicates that 2 waiting rooms for patients, before and after injection, are not necessary (Harding et al. 1990a, 1994). A possible area of concern involves ^{18}FDG, however. Depending on the number of patients in a waiting area at any one time, accompanying persons may approach the limits set for members of the public (Benatar et al. 2000).

Ultrasound

Because of the close contact, ultrasound operators may receive a radiation dose of 25 μSv from a radioactive patient. Considering the repeated likelihood of this situation, ultrasound scans should be carried out before radiopharmaceutical administration or on another day (Harding et al. 1990b; Mountford 1997).

Nursing

The total absorbed dose to a nurse accompanying an injected patient in the nuclear medicine waiting room ~3 μSv, which is ~0.3% of the annual dose limit, and is not affected by whether a separate "injected" waiting room is available (Harding et al. 1990a, 1994). For diagnostic investigations, a ward nurse may receive ~20 μSv day^{-1} (Harding et al. 1985b), but this could be higher when caring for a high dependency adult or paediatric patient classified as totally helpless and if the patient had undergone a high activity $^{99}Tc^{m}$ procedure (Mountford et al. 1991) or some ^{111}In and ^{201}Tl investigations (Greaves and Tindale 1999). With totally helpless patients the nurse might spend 50% of the working shift with the patient and the dose from a bone scan patient may ~99 μSv. The nursing activities which contributed most to this figure were dealing with incontinence and catheter care, treating pressure areas and feeding (Harding et al. 1990b). In the case of a severely ill patient, therefore, care should be shared by staff rotation. Ambulatory patients should be encouraged to use the toilet rather than a bedpan or bottle, and to flush twice, although normal hygiene procedures are usually adequate. If the patient is incontinent, they should be catheterised before administration. The catheter bag should be emptied regularly but stored, for 24 hours in the case of $^{99}Tc^{m}$ and longer for longer half-life radionuclides, before disposal. Should the patient vomit after an oral administration or become incontinent, normal hygienic procedures should be adopted. Contaminated clothing should be stored, for 24 hours in the case of $^{99}Tc^{m}$ and longer for longer half-life radionuclides, before being sent to the laundry.

Special nursing considerations are not required for patients receiving ^{18}FDG (Cronin et al. 1999).

Pregnancy

Doses to the foetus of pregnant nursing staff caring for an adult patient, administered $^{99}Tc^{m}$, were 0.9–1.6 μSv (Mountford and Steele 1995).

Portering

Assuming that ~1/2 of nuclear medicine patients are in-patients who are accompanied from their ward in a wheelchair or bed by a porter and the porter is usually 0.5 m from the patient, the monthly dose to a single porter accompanying all nuclear medicine patients would be 0.1 mSv (Harding et al. 1990b; Mountford 1997).

Pathology

The dose to clinical chemistry department staff over one month, from blood and urine specimens taken from nuclear medicine patients, was 0.002 mSv and 0.02 mSv, respectively (Harding et al. 1990b). Blood, urine and faeces and any other sample should be labelled as radioactive since they may interfere with laboratory measurements. If possible $^{99}Tc^{m}$ samples should be retained for 24 hours before being sent for analysis.

Theatre and Mortuary

Restrictions are not usually necessary with respect to whole-body external radiation hazard to staff during surgery, post-mortem or preparation for burial or cremation. Normal surgical and post-mortem room procedures are sufficient to minimise the risk of ingestion and skin contamination (Mountford 1997). Precautions may need to be considered when ^{131}I is involved, however. Because there is little data describing the risk to theatre staff from ^{131}I patients, appropriate precautions will have to be derived on an individual basis.

In sentinel node localisation, no additional procedures are required for the protection of staff, provided that the usual procedures for biohazards are in place; theatre swabs may need to be stored temporarily before disposal, however. Injecting and imaging on the day before surgery is preferred, compared with injecting and imaging before surgery on the same day, since this gives lower radiation doses to staff, lower activity in excised specimens and waste, and provides a higher count rate giving better image quality (Morton et al. 2003). The average dose rate at working distances following administration of 950 MBq of $^{99}Tc^{m}$ for intraoperative radioimmunoscintimetry was 6.4 μSv h^{-1} (Bares et al. 1992). For surgeons: the radiation dose to the hands has been estimated to be 5–94 μSv per patient (Alazraki et al. 2002); after 100 procedures of sentinel lymph node biopsy and radioguided occult lesion localisation, involving the injection of $^{99}Tc^{m}$ close to or into the lesion, the mean absorbed dose to the hands was 0.45 mGy and the mean effective dose 0.09 mSv. Absorbed doses to all hospital personnel involved in the procedures were very low compared to the annual limits recommended by the ICRP (Cremonesi et al. 1999). Radiation doses to pathology personnel who handle the radioactive sentinel node and primary tumour specimen for a limited period would be no greater than that received by the surgeon (Alazraki et al. 2002).

References

Alazraki N, Glass EC, Castronovo F, Valdes Olmos RA, Podoloff D (2002) Procedure guideline for lymphoscintigraphy and the use of intraoperative gamma probe for sentinel lymph node localization in Melanoma of intermediate thickness 1.0. J Nucl Med 43:1414–1418

Bares R, Muller B, Fass J, Buell U, Schumpelick V (1992) The radiation dose to surgical personnel during intraoperative radioimmunoscintigraphy. Eur J Nucl Med 19:110–112

Batchelor S (1986) The value of syringe shields in a nuclear medicine department. Nucl Med Commun 7:76

Batchelor S, Penfold A, Aric I, Huggins R (1991) Radiation dose to the hands in nuclear medicine. Nucl Med Commun 12:439–444

Benatar NA, Cronin BF, O'Doherty MJ (2000) Radiation dose rates from patients undergoing PET: implications for technologists and waiting areas. Eur J Nucl Med 27:583–589

Benedetto AR, Dziuk TW, Nusynowitz ML (1989) Population exposure from nuclear medicine procedures: measurement data. Health Phys 57:725–731

BIR (1999) Working Party of the Radiation Protection Committee of the British Institute of Radiology: patients leaving hospital after administration of radioactive substances. Br J Radiol 72: 121–125

Boutcher S, Haas T (1985) External radiation doses to nuclear medicine technologists from procedures using ^{99m}Tc radiopharmaceuticals. Can J Radiogr Radiother Nucl Med 16:161–165

Chiesa C, De Sanctis V, Crippa F, Schiavini M, Fraigola CE, Bogni A, Pascali C, Decise D, Marchesini R, Bombardieri E (1997) Radiation dose to technicians per nuclear medicine procedure: comparison between technetium-99m, gallium-67, and iodine-131 radiotracers and fluorine-18 fluorodeoxyglucose. Eur J Nucl Med 24:1380–1389

Clarke EA, Thomson WH, Notghi A, Harding LK (1992) Radiation doses from nuclear medicine patients to an imaging technologist: relation to ICRP recommendations for pregnant workers. Nucl Med Commun 13:795–798

Cremonesi M, Ferrari M, Sacco E, Rossi A, de Cicco C, Leonardi L, Chinol M, Luini A, Galimberti V, Tosi G, Veronesi U, Paganelli G (1999) Radiation protection in radioguided surgery of breast cancer. Nucl Med Commun 20:919–924

Cronin B, Marsden PK, O'Doherty MJ (1999) Are restrictions to behaviour of patients required following fluorine-18 fluorodeoxyglucose positron emission tomographic studies? Eur J Nucl Med 26:121–128

Culver CM, Dworkin HJ (1993) Comparison of personnel radiation dosimetry from myocardial perfusion scintigraphy: technetium-99m-sestamibi versus thallium-201. J Nucl Med 34: 1210–1213

Dhanse S, Martin CJ, Hilditch TE, Elliott AT (2000) A study of doses to the hands during dispensing of radiopharmaceuticals. Nucl Med Commun 21:511–519

Greaves CD, Tindale WB (1999) Dose rate measurements from radiopharmaceuticals: implications for nuclear medicine staff and for children with radioactive parents. Nucl Med Commun 20: 179–187

Greaves CD, Sanderson R, Tindale WB (1995) Air contamination following aerosol ventilation in the gamma camera room. Nucl Med Commun 16:901–904

Greaves CD, Tindale WB, Flynn PJ (1996) A survey of close contact regimes between patients undergoing diagnostic radioisotope procedures and children. Nucl Med Commun 17:554–561

Harding K, Thomson WH (1997) Radiological protection and safety in medicine – ICRP 73. Eur J Nucl Med 24:1207–1209

Harding LK, Mountford PJ (1993) Pregnant employees in a nuclear medicine department. Nucl Med Commun 14:345–346

Harding LK, Hesslewood S, Ghose SK, Thomson WH (1985a) The value of syringe shields in a nuclear medicine department. Nucl Med Commun 6:449–454

Harding LK, Mostafa AB, Roden L, Williams N (1985b) Dose rates from patients having nuclear medicine investigations. Nucl Med Commun 6:191–194

Harding LK, Harding NJ, Warren H, Mills A, Thomson WH (1990a) The radiation dose to accompanying nurses, relatives and other patients in a nuclear medicine department waiting room. Nucl Med Commun 11:17–22

Harding LK, Mostafa AB, Thomson WH (1990b) Staff radiation doses associated with nuclear medicine procedures – a review of some recent measurements. Nucl Med Commun 11:271–277

Harding LK, Bossuyt A, Pellet S, Reiners C, Talbot J (1994) Radiation doses to those accompanying nuclear medicine department patients: a waiting room survey. EANM Task Group Explaining Risks. European Association of Nuclear Medicine. Eur J Nucl Med 21:1223–1226

Hastings DL, Hillel PG, Jeans SP, Waller ML (1997) An assessment of finger doses received by staff while preparing and injecting radiopharmaceuticals. Nucl Med Commun 18:785–790

Heller SL (1996) Radiation safety in the central radiopharmacy. Semin Nucl Med XXVI:107–118

IAEA (1985) International Atomic Energy Agency regulations for the safe transport of radioactive materials. IAEA, Vienna (International Atomic Energy Agency safety series 6)

IAEA (1996) International Atomic Energy Agency international basic safety standards for protection against ionizing radiation and for the safety of radiation sources: a safety standard. IAEA, Vienna (International Atomic Energy Agency safety series 115)

ICRP (1977) International Commission on Radiological Protection publication 26, 1977 recommendations of the International Commission on Radiological Protection. Pergamon, Oxford

ICRP (1979) International Commission on Radiological Protection publication 30 limits for intakes of radionuclides by workers, part 1. Pergamon, Oxford

ICRP (1980) International Commission on Radiological Protection publication 30 limits for intakes of radionuclides by workers, part 2. Pergamon, Oxford

ICRP (1982) International Commission on Radiological Protection publication 30 limits for intakes of radionuclides by workers, part 3. Pergamon, Oxford

ICRP (1985) International Commission on Radiological Protection statement from the Paris meeting of the ICRP. Pergamon, Oxford
ICRP (1991a) International Commission on Radiological Protection publication 60, 1990 recommendations of the International Commission on Radiological Protection. Pergamon, Oxford
ICRP (1991b) International Commission on Radiological Protection publication 61, annual limits on intake of radionuclides by workers based on the 1990 recommendations. Pergamon, Oxford
ICRP (1995) International Commission on Radiological Protection publication 68, dose coefficients for intakes by workers. Pergamon, Oxford
ICRP (1996) International Commission on Radiological Protection publication 73, radiological protection and safety in medicine. Pergamon, Oxford
Jansen SE, van Aswegen A, Lotter MG, Herbst CP, Otto AC (1994) Staff radiation doses during eight years in a nuclear medicine radiopharmacy. Nucl Med Commun 15:114–118
Lundberg TM, Gray PJ, Bartlett MI (2002) Measuring and minimizing the radiation dose to nuclear medicine technologists. J Nucl Med Technol 30:25–30
MacDonald J, Gibson CJ (2002) Calibration of radioactive contamination monitors for practical use in hospitals. Nucl Med Commun 23:573–580
Mackenzie A (1997) Reduction of extremity dose in the radiopharmacy. Nucl Med Commun 18: 578–581
Morton R, Horton PW, Peet DJ, Kissin MW (2003) Quantitative assessment of the radiation hazards and risks in sentinel node procedures. Br J Radiol 76:117–122
Mosman EA, Peterson LJ, Hung JC, Gibbons RJ (1999) Practical methods for reducing radioactive contamination incidents in the nuclear cardiology laboratory. J Nucl Med Technol 27:287–289
Mountford PJ (1987) Estimates of close contact doses to young infants from surface dose rates on radioactive adults. Nucl Med Commun 8:857–863
Mountford PJ (1991) Techniques for radioactive decontamination in nuclear medicine. Semin Nucl Med 21:82–89
Mountford PJ (1997) Risk assessment of the nuclear medicine patient. Br J Radiol 70:671–684
Mountford PJ, Coakley AJ (1986) The radiation dose to an infant following maternal radiopharmaceutical administration. Br J Radiol 59:957–958
Mountford PJ, Coakley AJ (1989) Body surface dosimetry following re-injection of ^{111}In-leucocytes. Nucl Med Commun 10:497–501
Mountford PJ, O'Doherty MJ (1999) Exposure of critical groups to nuclear medicine patients. Appl Radiat Isot 50:89–111
Mountford PJ, Steele HR (1995) Fetal dose estimates and the ICRP abdominal dose limit for occupational exposure of pregnant staff to technetium-99m and iodine-131 patients. Eur J Nucl Med 22:1173–1179
Mountford PJ, O'Doherty MJ, Forge NI, Jeffries A, Coakley AJ (1991) Radiation doses from adult patients undergoing nuclear medicine investigations. Nucl Med Commun 12:767–777
Rudin MJ, Johnson WH (2000) The influence of flood source placement on radiation exposure during quality control testing. J Nucl Med Technol 28:88–93
Shukla AK, Ito K, Nishino M, Ishigaki T, Kato T, Ikeda M, Ota T, Aoyama Y, Yamashita M, Makino N (1994) Radiation surface doses to patients and staff during positron emission tomography. Radiat Med 12:89–92
Siegel JA, Marcus CS, Sparks RB (2002) Calculating the absorbed dose from radioactive patients: the line-source versus point-source model. J Nucl Med 43:1241–1244
Thomson WH, Harding LK (1995) Radiation protection issues associated with nuclear medicine out-patients. Nucl Med Commun 16:879–892
Thomson WH, Mills AP, Smith NB, Mostafa AB, Notghi A, Harding LK (1993) Day and night radiation doses to patients' relatives: implications of ICRP 60. Nucl Med Commun 14:275
White S, Binns D, Johnston VV, Fawcett M, Greer B, Ciavarella F, Hicks R (2000) Occupational exposure in nuclear medicine and PET. Clin Positron Imaging 3:127–129
Williams ED, Laird EE, Forster E (1987) Monitoring radiation dose to the hands in nuclear medicine: location of dosemeters. Nucl Med Commun 8:499–503
Wrixon AD (1994) Current NRPB recommendations on optimisation of protection of workers. J Radiol Prot 14:219–227

CHAPTER I.5

Protection of the Patient

Abbreviations

ALARA	as low as reasonably achievable
BSA	body surface area
DPET	dual photon emission tomography
EANM	European Association of Nuclear Medicine
ED	effective dose
EDE	effective dose equivalent
ICRP	International Commission on Radiological Protection
MIRD	Medical Internal Radiation Dose committee
SPET	single photon emission tomography
WHO	World Health Organization

Although, in a properly undertaken investigation, the patient receives a benefit, there is also a risk, which must be assessed and controlled (Overbeek et al. 1994). The level of this risk can be appreciated by comparison with other investigations and risks and it can be controlled, most directly, by limiting the administration activity. This must not be so severe as to reduce the efficacy of the investigation, however. Of particular concern are the nursing child and the foetus of an irradiated patient, because they are particularly sensitive to radiation. They also derive no direct benefit from the investigation, although the indirect benefit is the care of the mother and the possibility that delaying an essential procedure may present a greater risk to the foetus (Sharp et al. 1998). It is the responsibility of the nuclear medicine department to keep radiation absorbed doses to patients ALARA and to provide comprehensive information and advice. Guidance is provided in ICRP reports: Publication 52 (1987) "Protection of the Patient in Nuclear Medicine", and Publication 53 (1994) ad 80 (1999) "Radiation Dose to Patients from Radiopharmaceuticals", the last using ICRP 60 (1991) dosimetry. These reports supercede Publication 17 (1971) "Protection of the Patient in Radionuclide Investigations" because, in the time between the reports, the number of radiopharmaceuticals increased, knowledge of their pharmacokinetics improved, and acquisition and processing techniques became more complex.

Risk to the Patient

The ionising radiation involved with the administration of unsealed radioactive materials presents a hazard to patients which must be balanced against the potential benefit. This balance has been evaluated and has resulted in limitation on the activity that can be administered; which varies by radionuclide, pharmaceutical, investigation and patient status. In the case of diagnosis, a compromise has to be achieved between administering too low an activity, in which case non-diagnostic images or sample information may be obtained (Piepsz et al. 1991); and administering too high an activity, in which case the patient will be exposed to an unacceptable risk. This is achieved with the aid of the effective dose (ED), which was developed to aid in the protection of workers but has recently been used in patients. The values of ED are lower, by an average of 11%, than the old effective dose equivalent, EDE. For $^{99}Tc^{m}$ radiopharmaceuticals, the average reduction is 20% but for ^{125}I, ^{123}I and ^{131}I radiopharmaceuticals, there is an average increase of 3%, mainly because of a higher tissue weighting factor for the thyroid.

The risk to a patient is from stochastic effects. Apart from accidents, deterministic effects do not occur in diagnostic nuclear medicine because the absorbed doses are well below the threshold levels for these effects. The EDs to adult patients range up to a few tens of millisieverts (ICRP 1994, 1999), and one method of appreciating the risk is to compare these with examples from diagnostic radiology. For instance, a chest x-ray ~0.04 mSv whereas a barium enema ~8.7 mSv (NRPB 1993a). Another method of comparison is with natural background radiation, which is taken to be 1 mSv $year^{-1}$.

Table I.5.1. Statement of risks Sv^{-1} for the whole population (ICRP 1994)

	Sv^{-1}	risk per mSv
Non-fatal cancer	0.010	1 in 100,000
Severe hereditary effects	0.013	1 in 77,000
Fatal cancer	0.050	1 in 20,000
Total	0.073	

To estimate the probability of radiation damage and to provide a comparison with natural occurrence, the risk is expressed in absolute terms, by taking the ICRP statement of risks Sv^{-1} for the whole population shown in Table I.5.1. For a $^{99}Tc^{m}$ diagnostic study with an ED of 5 mSv, the risk of fatal cancer would be 1 in 4000 (ICRP 1994; Mountford 1997). These figures can then be compared to other risks, such as the cumulative risk for fatal childhood cancer up to the age of 15 years, which is 1 in 1300 (ICRP 1991).

Control of Risk

Optimisation of Procedure

The second component of the ICRP system of dose limitation is optimisation (ICRP 1991), which requires that all exposures be kept ALARA. The performance of each diagnostic procedure must be optimised. This involves ensuring that: radiopharmaceutical quality is adequate, every part of the imaging chain is functioning optimally and the system is sufficiently reliable to minimise the possibility of data loss, and acquisition and processing are undertaken to produce the maximum diagnostic information. Therefore, the ED received by the patient should not be higher than required to produce the necessary diagnostic information. The optimal activity is a subject of discussion. Below a minimum activity, deterioration in diagnostic information will be occasioned (ICRP 1996; Harding and Thomson 1997). Above this threshold, it will increase sharply with increasing activity but then no further improvement will be achieved with increasing activity. A substantial reduction in absorbed dose can also be achieved by increasing fluid intake and by administering blocking agents. The latter can reduce the dose to the thyroid substantially when using radioisotopes of iodine.

Control of Administrations

Administration of radiopharmaceuticals has generally been subject to greater control than x-ray exposure and responsibility, for keeping within specified limits, rests with the department. This can be difficult to assess in certain cases, e.g. inhalation imaging using radioactive gases or aerosols (Thomas 2001).

Misadministration involves errors in radiopharmaceutical type, activity and access route. To reduce the potential, all radiopharmaceuticals must be adequately labelled and, immediately before administration, the activity must be measured in an activity meter (Hastings et al. 1997). Most radiopharmaceuticals are administered intravenously, other routes of administration include oral, inhalation, intrathecal, intraperitoneal and subcutaneous. The error in activity, which constitutes a misadministration, depends on local regulation. Because diagnostic investigations generally result in absorbed doses to individual organs of less than a few tens of milligray and misadministration, therefore, carries a relatively small risk of tissue damage, the misadministration level may be 50% of prescribed activity (NCRP 1983). The commonest problem is extravasation of an intravenous injection, a 'tissued injection'. It is good practice to use a cannula rather than direct needle administration for longer half-life radionuclides and to monitor the site of any suspect administration. Apart from the loss of diagnostic information, the most serious consequence is that a high radiation dose may be absorbed by the tissue around the injection site. With the majority of $^{99}Tc^{m}$ radiopharmaceuticals, such doses are low. With longer half-life radionuclides, however, the absorbed dose can be severe and intervention may be required (NCRP 1983).

A recent additional contribution to patient doses has been introduced by transmission attenuation measurements in tomographic acquisitions. Therefore, the evaluation of the radiation doses received by patients needs to include the contribution of transmission procedures. However, measurements show that the dose received by a patient during a transmission scan adds little to the typical dose received in a routine nuclear medicine procedure. Radiation dose, therefore, does not represent a limit to the generalised use of transmission attenuation correction (Almeida et al. 1998). A summary of EDE estimates for different transmission scanning sources and geometries has been compiled. This reveals a discrepancy between reported results, even when using the same experimental set-up, scanning source geometries and dosimeters (Zaidi and Hasegawa 2003). In brain SPET using ^{201}Tl, ^{153}Gd, and ^{99}Tc^{m} line sources, EDEs of 18.6, 11.9, and 26.3 μSv were measured for a typical 20-minute investigation (van Laere et al. 2000). The maximum ED from a typical myocardial perfusion study using ^{153}Gd line source transmission measurement, was 1.3 μSv for male and 1.9 μSv for female patients (Perisinakis et al. 2002). ED values, from transmission radiation in μSv, were estimated as: 110±10 for cardiac DPET, 37±3 for brain DPET, 11±2 for cardiac SPET, 5.8±2.6 for brain SPET (Almeida et al. 1998).

Dose Limits

If the procedure is medically justified and optimised, the dose to the patient will be as low as is compatible with the procedure. Additional application of dose limits might cause detriment to the patient. The ICRP therefore recommends that dose limits should not be applied to medical exposures. For similar reasons, it is not appropriate to include the doses incurred by patients, in the course of diagnostic examinations, when considering compliance with dose limits applied to occupational or public exposures (ICRP 1991).

Research Investigations

When considering research studies, the risk involved should be associated to one of the three categories defined by the World Health Organization (WHO) on the basis of the ED (mSv) accumulated in one year shown in Table I.5.2.

Risk to Children

The effects of ionising radiation are considered to be more important in younger than in older patients.

Table I.5.2. WHO research categories

I	Within the variation of natural background
II	Within the limits for members of the public
III	Within the limits for occupationally exposed persons

Control of Risk

The administered activity is reduced, to optimise ED and data quality. The aim is to produce a count rate density approximately equal to that of an adult (Mountford 1991; Piepsz et al. 1991). The reduction is generally in proportion to age or to either an adult weight of 70 kg or surface area of 1.73 m^2 (Mountford 1991). Both assume that the radiopharmaceutical distribution is similar in both children and adults, which is not always the case (NCRP 1983; Ruotsalainen et al. 1996). The reduction by weight results in lower activities than the body surface area, BSA, approach and in some circumstances has been found to be unsatisfactory for some imaging studies and more appropriate for non-imaging studies. The most logical index to use is the BSA since the growth of many organs follows it closely. The calculation of BSA is:

$$BSA = 0.007184 \times W^{0.425} \times H^{0.725} \tag{I.5.1}$$

The European Association of Nuclear Medicine, EANM, paediatric task group has compiled a list of minimum activities, below which the acquired data may be inadequate (Piepsz et al. 1990, 1991).

Breast Feeding

Because many radionuclides are excreted in milk, a nursing infant may be exposed to this risk. There is really no acceptable level of risk because there is no direct benefit to the child from the investigation, and an analysis of the dosimetric implications of breast feeding has been undertaken (Mountford and Coakley 1989).

Control of Risk

Consideration must be given to interrupting breast feeding (Thomson and Harding 1995), during which time the milk should be expressed and discarded (Mountford 1997). The period of interruption must be long enough to ensure that, when feeding is resumed, the activity ingested by the child will result in a radiation dose less than the member of the public limit of 1 mSv. On this basis and because of the difficulties involved in resuming breast feeding after lengthy interruptions, radiopharmaceuticals have been grouped into 3 categories of interruption recommendations listed in Table I.5.3. A fourth category has been introduced because of the lack of biological data, in which case the dosimetric calculations mentioned above are undertaken, category III: must be interrupted for a short calculated period with measurement of activity in milk samples (Mountford and Coakley 1989). Lists of radiopharmaceuticals in each category have been compiled by a number of authors (Mountford and Coakley 1989; Rubow et al. 1994; Mountford 1997; Toohey et al. 2000).

Table I.5.3. Three categories of breast feeding interruption

I	Need not be interrupted
II	Must be interrupted for a short fixed period (<48 hours)
IV	Must be discontinued

Risk to the Foetus

The foetus will be at risk from activity in adjacent maternal organs and from any radioactive material passing across the placenta (Mountford 1991). For example, in the latter stages of pregnancy, iodine and $^{99}Tc^{m}$ may cause damage to the foetal thyroid. A list of uterine absorbed doses has been compiled for various investigations, which are mainly <10 mGy (Smith and Warner 1976; ICRP 1994; Zanzonico 2000). The accuracy of estimation is limited by the absence of activity data for all organs in the pregnant woman and the lack of data on placental radiopharmaceutical transfer. In the early stages of pregnancy, it can be assumed that the foetal dose is the same as the dose to the uterus. The latest version of the MIRD dosimetry software (MIRDOSE3) contains factors for the pregnant female at different stages of pregnancy, allowing an estimate of the foetal dose from adjacent organs (Mountford 1997; Toohey et al. 2000). This has been used, with data for placental transfer of radiopharmaceuticals, to produce estimates of foetal absorbed doses throughout pregnancy (Mountford 1999; ICRP 2001). Foetal dose tables have been published for all common radiopharmaceuticals (Russell et al. 1997a,b). In almost every case, if the investigation is medically indicated, the diagnostic benefit to the mother outweighs the risk to the foetus (Toohey et al. 2000). A particular concern is ^{131}I, because this easily crosses the placenta and the foetal thyroid begins concentrating iodine at ~10 weeks, and this can deliver 500–1100 mGy MBq^{-1} to the foetal thyroid and 0.05–0.28 mGy MBq^{-1} to the foetal whole body (ICRP 2000). It is therefore critical to avoid administration of even diagnostic quantities of ^{131}I (Zanzonico 2000) in later pregnancy.

Deterministic Effects

The principal deterministic effects are death, growth malformations and severe mental retardation, but the thresholds for these are much higher than experienced in diagnostic investigations. A possible no-threshold deterministic effect is a decrease in intelligence quotient but, at diagnostic levels, this would be undetectable (ICRP 1991; Mountford 1991, 1997, 1999;Mountford and Harding 1993; ACOG 1995; Sharp et al. 1998). Prenatal doses from most properly undertaken diagnostic investigations present no measurably increased risk of these effects (ICRP 2000).

Stochastic Effects

Possible stochastic effects to the foetus are heritable disease and cancer, and the risks up to the age of 15 years are (NRPB 1993b; Sharp et al. 1998) are shown in Table I.5.4. The embryo and foetus are very radiosensitive during prenatal development, the nature and sensitivity of induced biological effects depending upon dose and developmental stage at irradiation (ICRP 2003). The risks are most significant during organogenesis and the early foetal period, less in the 2nd trimester and least in the 3rd trimester but throughout most of pregnancy, the embryo/foetus is assumed to be at about the same risk for potential carcinogenic effects of radiation as are children (ICRP 2000).

Although other investigations will produce higher absorbed doses (Goldstein et al. 1988), one of the more common investigations during pregnancy is ventilation – perfu-

Table I.5.4. Risks of heritable disease and cancer up to the age of 15 years (NRPB 1993b; Sharp et al. 1998)

	Gy^{-1}	Risk per mGy
Heritable disease	0.024	1 in 42,000
Excess fatal cancer	0.030	1 in 33,000
Excess fatal and non-fatal cancer	0.060	1 in 17,000

sion imaging (Mountford and Harding 1993; ACOG 1995). In the worst case for this, the uterine absorbed dose will be ~0.7 mGy (Marcus et al. 1985; Mountford and Harding 1993) and, for this, the corresponding risks are given in Table I.5.5 (Mountford 1997).

Control of Risk

Activity in the urinary bladder is a significant source of radiation to the foetus and, in investigations likely to produce such accumulation, administration should be undertaken when the bladder is partially full of urine. Following the study, frequent voiding should be encouraged.

Females of Reproductive Capacity

Local regulations determine restrictions on administrations to females of reproductive capacity. The most common techniques are the 10 day rule and the 28-day rule (Mountford and Harding 1993), supplemented, if necessary, by pregnancy test. Often, for female patients of reproductive capacity but not known to be pregnant, no special administration limitation is implemented except when a foetus would receive an absorbed dose of tens of milligray, because such a dose will approximately double the natural total cancer risk of the unborn child (NRPB 1993). If the limit is taken as 20 mGy (Mountford 1997), the only diagnostic procedure which will result in such a foetal dose is ^{131}I-iodide scintigraphy in thyroid cancer. This is often restricted to the first 10 days of the menstrual cycle (NRPB 1993; Sharp et al. 1998).

If pregnancy is discovered after a procedure, the foetal dose must be estimated (Russell et al. 1997b) and provided to the clinical director of the investigation. However, except in the case of ^{131}I administration, termination is not a consideration in diagnostic nuclear medicine because foetal whole body doses below 100 mGy should not be considered a reason for this (ICRP 2000).

Preconception

Local regulations will provide guidance on the necessity of delaying pregnancy following a diagnostic investigation, which in the case of ^{131}I may be because of non-radia-

Table I.5.5. Risks from a uterine absorbed dose of ~0.7 mGy (Mountford 1997)

Heritable disease	1 in 60,000
Excess fatal cancer	1 in 48,000

tion related medical concerns. However, most radionuclide distributions experienced in contemporary routine diagnostic investigations have insufficiently long effective half-lives to expose a subsequent embryo significantly, and pre-conception irradiation of either parent's gonads has not been shown to result in increased cancer or malformations in the children (ICRP 2000). Concern that pre-conception exposure of men may increase the cancer risk to the foetus is not proven (Elliott and Sumner 1991; Mountford and Harding 1993; Doll et al. 1994), and long term data for females treated with ^{131}I have shown no abnormalities in their children (Schlumberger et al. 1996; Mountford 1997, 1999).

References

ACOG (1995) American College of Obstetricians and Gynecologists guidelines for diagnostic imaging during pregnancy. Int J Gynaecol Obstet 51:288–291

Almeida P, Bendriem B, de Dreuille O, Peltier A, Perrot C, Brulon V (1998) Dosimetry of transmission measurements in nuclear medicine: a study using anthropomorphic phantoms and thermoluminescent dosimeters. Eur J Nucl Med 25:1435–1441

Doll R, Evans HJ, Darby SC (1994) Paternal exposure not to blame. Nature 367:678–680

Elliott AT, Sumner DJ (1991) The Gardner Report (editorial). Nucl Med Commun 12:1–2

Goldstein HA, Ziessman HA, Fahey FH, Collea JV, Alijani MR, Helfrich GB (1988) Renal scans in pregnant transplant patients. J Nucl Med 29:1364–1367

Harding K, Thomson WH (1997) Radiological protection and safety in medicine – ICRP 73. Eur J Nucl Med 24:1207–1209

Hastings DL, Hillel PG, Jeans SP, Waller ML (1997) An assessment of finger doses received by staff while preparing and injecting radiopharmaceuticals. Nucl Med Commun 18:785–790

ICRP (1971) International Commission on Radiological Protection publication 17. Protection of the patient in radionuclide investigations. Pergamon, Oxford

ICRP (1987) International Commission on Radiological Protection publication 52. Protection of the patient in nuclear medicine. Pergamon, Oxford

ICRP (1991) International Commission on Radiological Protection publication 60. 1990 recommendations of the International Commission on Radiological Protection. Pergamon, Oxford

ICRP (1994) International Commission on Radiological Protection publication 53, 2nd edn. Radiation dose to patients from radiopharmaceuticals (including addendum 1). Pergamon, Oxford

ICRP (1996) International Commission on Radiological Protection publication 73. Radiological protection and safety in medicine. Pergamon, Oxford

ICRP (1999) International Commission on Radiological Protection publication 80. Radiation dose to patients from radiopharmaceuticals. Pergamon, Oxford

ICRP (2000) International Commission on Radiological Protection publication 84. Pregnancy and medical radiation. Pergamon, Oxford

ICRP (2001) International Commission on Radiological Protection publication 88. Doses to the embryo and fetus from intakes of radionuclides by the mother. Pergamon, Oxford

ICRP (2003) International Commission on Radiological Protection publication 90. Biological effects after prenatal irradiation (embryo and fetus). Pergamon, Oxford

Marcus CS, Mason GR, Kuperus JH, Mena I (1985) Pulmonary imaging in pregnancy. Maternal risk and fetal dosimetry. Clin Nucl Med 10:1–4 Mountford PJ (1991) Radiation protection for the parent and child in diagnostic nuclear medicine. Eur J Nucl Med 18:940–943

Mountford PJ (1997) Risk assessment of the nuclear medicine patient. Br J Radiol 70:671–684

Mountford PJ (1999) Radiation, conception and pregnancy. Nucl Med Commun 20:979–981

Mountford PJ, Coakley AJ (1989) A review of the secretion of radioactivity in human breast milk: data quantitative analysis and recommendations. Nucl Med Commun 10:15–27

Mountford PJ, Harding LK (1993) Nuclear medicine and the pregnant patient. Nucl Med Commun 14:625–627

NCRP (1983) National Council on Radiation Protection and Measurements report 73. Protection in nuclear medicine and ultrasound diagnostic procedures in children. NCRP, Bethesda, Maryland

NRPB (1993a) National Radiological Protection Board document 4(2), occupational, public and medical exposure. HMSO, London

NRPB (1993b) National Radiological Protection Board document 4(4), statement on diagnostic medical exposures to ionising radiation during pregnancy. HMSO, London

Overbeek F, Pauwels EKJ, Broerse JJ (1994) Carcinogenic risk in diagnostic nuclear medicine: bio-

logical and epidemiological considerations. Eur J Nucl Med 21:997–1012
Perisinakis K, Theocharopoulos N, Karkavitsas N, Damilakis J (2002) Patient effective radiation dose and associated risk from transmission scans using ^{153}Gd line sources in cardiac spect studies. Health Phys 83:66–74
Piepsz A, Hahn K, Roca I, Ciofetta G, Toth G, Gordon I, Kolinska J, Gwidlet J (1990) Paediatric Task Group of the European Association of Nuclear Medicine. A radiopharmaceutical schedule for imaging in paediatrics. Eur J Nucl Med 17:127–129
Piepsz A, Gordon I, Hahn K (1991) Paediatric nuclear medicine. Eur J Nucl Med 18:41–66
Rubow S, Klopper J, Wasserman H, Baard B, van Niekerk M (1994) The excretion of radiopharmaceuticals in human breast milk: additional data and dosimetry. Eur J Nucl Med 21:144–153
Ruotsalainen U, Suhonen-Polvi H, Eronen E, Kinnala A, Bergman J, Haaparanta M, Teras M, Solin O, Wegelius U (1996) Estimated radiation dose to the newborn in FDG-PET studies. J Nucl Med 37:387–393
Russell JR, Stabin MG, Sparks RB (1997a) Placental transfer of radiopharmaceuticals and dosimetry in pregnancy. Health Phys 73:747–755
Russell JR, Stabin MG, Sparks RB, Watson E (1997b) Radiation absorbed dose to the embryo/fetus from radiopharmaceuticals. Health Phys 73:756–769
Schlumberger M, de Vathaire F, Ceccarelli C, Delisle M-J, Francese C, Couette J-E, Pinchera A, Parmentier C (1996) Exposure to radioactive iodine-131 for scintigraphy or therapy does not preclude pregnancy in thyroid cancer patients. J Nucl Med 37:606–612
Sharp C, Shrimpton JA, Bury RF (1998) National Radiological Protection Board document, diagnostic medical exposures – advice on exposure to ionising radiation during pregnancy. HMSO, London
Smith EM, Warner GG (1976) Estimates of radiation dose to the embryo from nuclear medicine procedures. J Nucl Med 17:836–839
Thomas SR (2001) MIRD Pamphlet 18 administered cumulated activity for ventilation studies. J Nucl Med 42:520–526
Thomson WH, Harding LK (1995) Radiation protection issues associated with nuclear medicine out-patients. Nucl Med Commun 16:879–892
Toohey RE, Stabin MG, Watson EE (2000) The AAPM/RSNA physics tutorial for residents: internal radiation dosimetry: principles and applications. Radiographics 20:533–546
Van Laere K, Koole M, Kauppinen T, Monsieurs M, Bouwens L, Dierck R (2000) Nonuniform transmission in brain SPECT using ^{201}Tl, ^{153}Gd, and ^{99m}Tc static line sources: anthropomorphic dosimetry studies and influence on brain quantification. J Nucl Med 41:2051–2062
Zaidi H, Hasegawa B (2003) Determination of the attenuation map in emission tomography. J Nucl Med 44:291–315
Zanzonico PB (2000) Internal radionuclide radiation dosimetry: a review of basic concepts and recent developments. J Nucl Med 41:297–308

PART II

Detection Systems

Radiation Detection

Abbreviations

FWHM	full width at half maximum
MDA	minimum detectable activity
PMT	photomultiplier tube
APD	avalanche photodiode

In addition to chemical radiation detection such as film, passive detectors include thermoluminescent and photoluminescent detectors. There are many active radiation detector types, but these can be generally characterised into those that use ionisation and those that use scintillation effects. They are each appropriate to different applications in diagnostic nuclear medicine and, to ensure accurate results, it is important that the operational characteristics and limitations are appreciated. In determining the significance of the measurements obtained, the influence of the statistics of the radioactive decay process itself must always be considered.

Ionisation Detectors

These detect a current pulse, created by applying a potential difference across the detecting medium, in which charged ions originate from the ionisation, of a gas or a semiconductor, caused by the radiation (see Fig. II.1.1). Because the charge created by the ionisation is small, a multiplicative effect has to be introduced.

Gas-Filled Detectors

These comprise ionisation chambers, proportional counters and Geiger-Mueller detectors and, in them, the ionising radiation interacts with the walls of the detector or the gas molecules, producing ion pairs. Their detection characteristics are determined purely by the magnitude of the applied potential difference (V).

When $V=0$, the ion pairs produced by the ionisation recombine and no current flows between the electrodes.

When $V>0$, the applied electric field causes some ion pairs to reach the electrodes and a current pulse is produced. The magnitude of the current depends on: V, the electrode separation; and the gas used and its geometry, volume, pressure and temperature. Of these, the most important is V; as it increases, the current pulse increases. The response characteristic is however quite complex, resulting in 5 distinct regions of the curve.

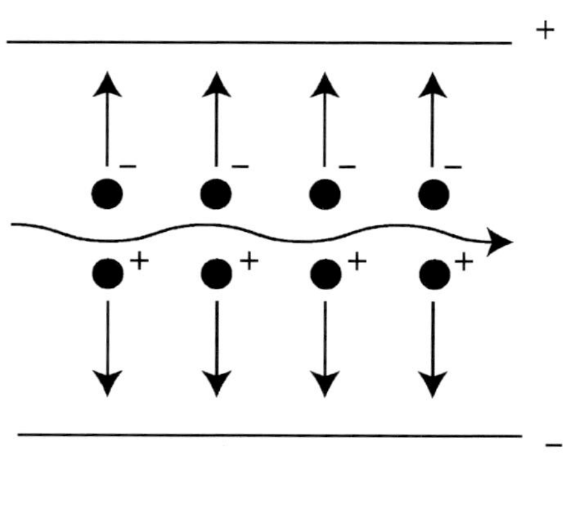

Fig II.1.1. Ionisation detection showing charged ions, originating from ionisation of the gas atoms, creating a current pulse under the influence of an electric field

Region 1 Recombination

V is low enough that some ion pairs recombine, thus producing an incomplete collection. As V increases, more ion pairs are collected and the current pulse increases.

Region 2 Ionisation Plateau

V is high enough to collect most of the ion pairs and there is little recombination. In this region, the current pulse does not increase appreciably, with increasing V, because all the ion pairs are collected. The current pulse produced by a single photon or particle is too small to be detected and the operation of the instrument relies on summing a number of pulses. The detectors operating in this region are ionisation chambers.

Region 3 Proportional Region

V is high enough to impart enough energy to the ion pairs, so that some of them produce secondary ion pairs through collisions with neutral gas molecules. The number, of such secondary ion pairs, increases with increasing V, resulting in a larger current pulse; proportional to the energy of the radiation and large enough for the detection of a single photon or particle. The detectors operating in this region are proportional chambers.

Region 4 GM Plateau

V is high enough to produce a large number of secondary ion pairs and excitations, which then strike the metallic electrodes and more neutral gas molecules to produce more ionisations and excitations. The de-excitation of the molecules produces ultraviolet light, which produces more ionisation. The result is that the total gas volume is in a temporary state of discharge and the amount of current produced is more or less independent of both V and the energy of the radiation. The detectors operating in this region are Geiger-Mueller counters.

Region 5 Continuous Discharge

V is so high that radiation is not necessary to produce discharge, the electrons are pulled out from the atomic shells, the atoms become ionised and a discharge may be established without radiation initiation.

Semiconductor Detectors

An incident photon will produce ionisation in the depletion region, this is when electrons are excited into a free state called a conduction band, where they can flow almost as freely as in a metal. A current pulse will be created by charges moving through the semiconductor under the influence of an electric field, which sweeps an area in the middle free of charge carriers (see Fig. II.1.2). In the absence of an ionising effect, the electric field will maintain a very low current through the crystal. This is called a leakage current, however, it is swamped during an ionising event (Hoffman et al. 1999).

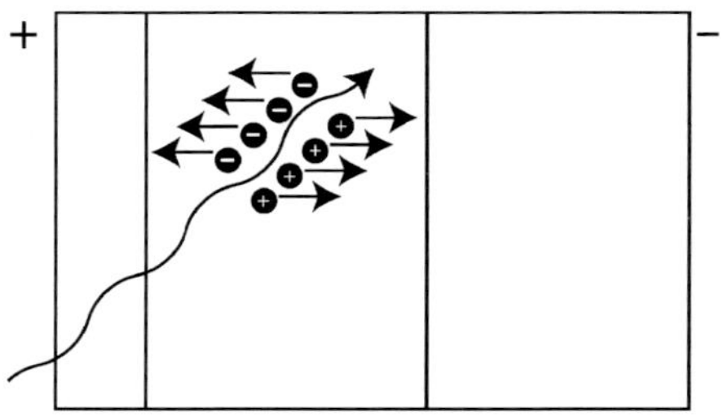

Fig II.1.2. Semiconductor detection showing charges, originating from ionisation created in the depletion layer, creating a current under the influence of an electric field

The two most common semiconductor radiation detectors are germanium and silicon crystals doped with lithium, Ge(Li) and Si(Li). The former being used with photons and the latter primarily with charged particles and low energy photons. The attraction of Ge(Li) detectors is their good energy resolution, which is ~1% at 500 keV and is a consequence of the high number of electrons produced per keV of incident photon energy. However, they exhibit a low detection sensitivity, they have to be used at a very low temperature and large crystals cannot be produced (Zanzonico and Heller 2000). CdTe, and the improved CdZnTe, detectors are ideal for low energy photons but can be used with the relatively high energy photons from $^{99}Tc^{m}$ (Hoffman et al. 1999).

Luminescent Detectors

Luminescence is the emission of light following electronic excitation. Thermoluminescent and photoluminescent detectors are used as integrators of radiation exposure. Absorbed radiation energy is stored by electrons moving into higher energy levels. This energy is released, as light, following stimulation, either by heat or by laser light, which allows the electrons to move back to their original energy levels.

Scintillation Detectors

This is the type of detector almost exclusively used in diagnostic investigations. A radiation pulse in the visible spectrum is produced immediately after the interaction of the ionising radiation in a scintillation medium (see Fig. II.1.3). The amount of light created is small and a multiplicative effect is achieved by conversion to an electronic pulse and amplification. An advantage of such a conversion is that the energy of the incident γ photon can be determined. The main characteristics of a scintillator are: the

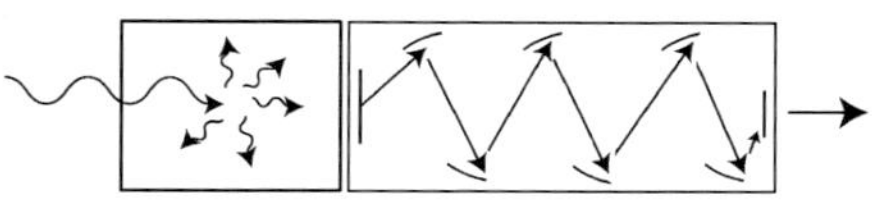

Fig II.1.3. Scintillation detection showing photon interaction in a scintillation crystal generating light photons. These create photoelectrons in a photomultiplier tube and their numbers are increased by high voltage amplification

intrinsic detection efficiency, the light conversion efficiency and the phosphorescent decay time. The intrinsic detection efficiency depends on the linear attenuation coefficient of the scintillator. Light conversion efficiency, which partly determines the energy resolution, is the amount of light produced per keV of absorbed radiation. Phosphorescent decay time, which partly determines the maximum counting rate of the detector, is the time required for the light to be emitted. A photomultiplier tube (PMT) is traditionally used to convert the light to a voltage pulse, the amplitude of which is directly related to the number of light photons impinging on the PMT. Consequently, the voltage of the pulse produced is directly proportional to the energy of the incident radiation. Avalanche photodiodes (APD) are now also being used. Scintillators are used in liquid or crystal form.

Liquid

Liquid scintillants are used to detect β^- particles in samples of body substances, particularly when the low energy emitters ^3H and ^{14}C are used. The samples are introduced directly into the scintillant, to overcome the short ranges of the charged particles in solids and liquids, which would lead to their absorption in the sample itself or in the walls or window of the detector. Most effective scintillators are solids and have to be dissolved in a solvent, to form as colourless a solution as possible, which will allow a homogeneous mixture of the scintillator and radioactive sample. The scintillator molecules thus act as the radiation detector, however most of the interactions are with the solvent molecules rather than directly with the scintillator molecules. Homogeneous mixing of the radioactive sample and detector thus has two advantages. The first is that, because each radioactive atom of the sample is practically surrounded from all directions by scintillator molecules, the geometric efficiency is close to 100%. The second is that there is little material between the radiation source and the scintillator to cause the loss of β^- particles.

Crystal

For in vivo investigations, it would not be practical to use a liquid detector and a crystal scintillator has to be used. The one used almost exclusively in diagnostic nuclear medicine, except for dedicated dual photon imaging systems, is sodium iodide. Pure NaI will only scintillate at low temperatures. However, incorporation of a thallium impurity (doping) at a concentration of ~0.2% produces crystal imperfections, known as luminescent centres, which can be excited by ionising radiation at room temperature (Fleming et al. 2002). Doping at 1 part in 10^6 increases the light output by >10 times. The light conversion efficiency, to a certain extent, varies with temperature and, therefore detector efficiency changes slightly with room temperature. The effectiveness of NaI(Tl) for stopping photons depends on the linear attenuation coefficient, which determines its intrinsic detector efficiency, and on the thickness of the crystal. For the most usual combination of 140 keV photons ($^{99}Tc^m$) and 9.5 mm thick crystal, this effectiveness is high at 84%. Also its light conversion efficiency is high at 13% (Madsen 1996) and its phosphorescent decay time is adequate for the activities encountered in diagnostic work (Zanzonico and Heller 2000). NaI(Tl) is, however, hygroscopic and must be hermetically sealed to prevent ingress of moisture, usually using thin sheets of aluminium or steel on the front and side faces, which only slightly attenuate the incident photons. Such ingress produces a yellow discolouration, which attenuates the light output. Even a sealed crystal eventually undergoes yellowing and this ultimately limits the lifetime of a detector to 10–15 years (Fleming et al. 2002). The rear surface, of the crystal, has a transparent covering so that the emitted light can be detected by the PMTs. Light is directed to these by coating the front and side faces with a reflecting material such as magnesium oxide. These crystals are very susceptible to mechanical damage, causing cracking, by abrupt temperature changes. A rule of thumb is that a temperature change >5°C per hour may cause breakage (Zanzonico and Heller 2000).

Light Production

A pulse of light is produced when the scintillator atoms return to ground state, following excitation by energy absorption from an incident photon. The light is produced in a very small volume, <1 mm, which is mainly determined by the range of photoelectrons and Compton recoil electrons. Even though only a small fraction, typically 10%, of the photon energy absorbed is converted into light photons, the number produced is proportional to the energy of the incident photon. Per keV, this is ~30 for NaI(Tl), ~10 for plastic scintillators (Hoffman et al. 1999) and ~5 for liquid scintillators (Fleming et al. 2002). To avoid loss of efficiency, the PMT light sensitivity is chosen to match the wavelength of the emitted light.

Photomultiplier Tube

PMTs are light-sensitive devices which convert the pulse of scintillation light into an electronic pulse and then amplify it. It comprises an evacuated glass tube with a flat window at one end, through which the light enters and which is coupled to the crystal. This coupling is critical, because of potential light loss that would reduce detection efficiency, and is accomplished using optical grease or direct bonding. The crystal and the PMT are surrounded by a light-tight enclosure which prevents external light from reaching the PMT, which would otherwise impose noise on the output signal. The internal construction of PMTs consists of a photocathode facing the window, several electrodes, termed dynodes, at increasing positive potentials and an anode. Light photons reaching the photocathode, release low energy, 0.1–1 keV, electrons by photoelectric interaction, with ~1 electron being produced per keV of incident γ photon energy (Zanzonico and Heller 2000). Amplification of the electronic pulse is achieved by accelerating the electrons through the dynode chain such that at the last dynode the multiplicative factor is 10^5–10^8.This is highly dependent on the voltage applied to the PMT, a change of 1% altering the gain by as much as 10%.

Pulse Spectrum

With a stable PMT the amplitude of the output pulse, v, is linearly dependent on the energy deposited, E_d, in the crystal by the incident γ photon. This is the basis of the energy discrimination abilities of a scintillator, which are discussed in 'Pulse height analysis' below. However, because the interaction may comprise a combination of Compton scattering and photoelectric effect, E_d can vary even when the incident photons are monoenergetic, E_γ. In the case of photoelectric interaction, all of the γ photon energy is deposited in the crystal, and $E_d = E_\gamma$. In the case of Compton scattering, however, $E_d < E_\gamma$, and there is no simple relationship between E_d and v. The consequence is a spectrum which comprises two regions: a photopeak, and a broad continuum at lower energies, termed the Compton plateau (see Fig. II.1.4).

The photopeak, also called the total absorption peak, represents interactions in which there is practically complete transfer of E_γ to the crystal, primarily by photoelectric interaction. A few events in this region may occur as a result of single or multiple Compton interactions, followed by photoelectric interaction. Multiple Compton interactions are, however, much less probable than single photoelectric or Compton interactions. The Compton plateau represents interactions in which there is an incomplete transfer of E_γ to the crystal, as a result of Compton interaction; the maximum E_d, for this, occurring when the incident γ photon is backscattered and corresponding to the Compton edge. Often, additional but much smaller peaks can be recognised in a spectrum.

K Escape Peak

This peak is barely visible in the spectrum, but is mostly observed when Eγ ~50–150 keV and occurs ~28 keV below the photopeak. In the 50–150 keV energy range, most photons interact, by the photoelectric effect, with the K shell electrons of iodine atoms within a few mm of the surface of the crystal. Because of the interaction being close to the crystal surface, the characteristic K x-ray, at ~28 keV, has a good chance of escaping the crystal without depositing its energy and $E_d \sim E\gamma - 28$.

Summation Peak

Energy peaks, corresponding to the sum of individual Eγ, can be encountered when multiple, usually two, photons interact with the crystal, simultaneously or almost simultaneously, i.e. within $\sim 0.25 \times 10^{-6}$ s. This may happen when the radionuclide emits several γ photons almost simultaneously or when high activities are present, in which case pile-up pulses occur. Pulse pile-up is when PMT pulses cannot be separated temporally and are summed.

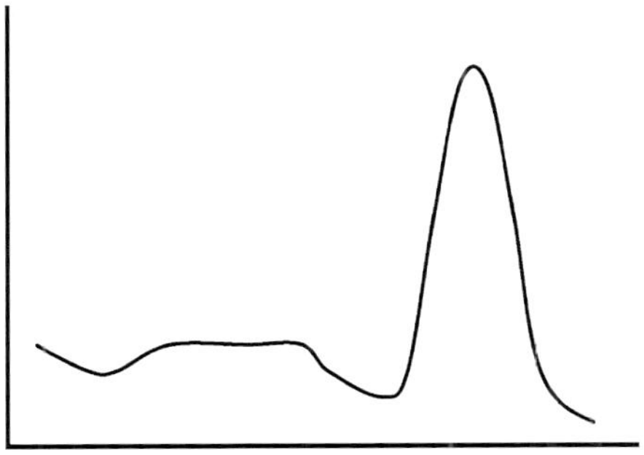

Fig II.1.4. Energy spectrum obtained in scintillation detection. The ordinate represents the number of photons detected at each energy and the abscissa represents the energy of the photons. As well as the photopeak, there is a broad continuum at lower energies

Back-Scattered Peak and Lead X-Ray Peak

Photons will interact with any surrounding lead shields, and will produce both scattered photons and characteristic lead K x rays of ~80 keV. These can then interact with the crystal resulting in a low energy photopeak.

Pulse Height Analysis

Only the photoelectric events should be recorded and the Compton scattering events should be excluded. This can be achieved by pulse height analysis as the Compton plateau and the photopeak are generally separated by a valley. Only events within an energy range, defined by energy level and window, are then accepted (see Fig. II.1.5).

Energy Resolution

The effectiveness of photopeak and Compton plateau separation, depends on the width of the photopeak which is determined by the energy resolution of the system. Production of light photons in a crystal and electron multiplication in a PMT are statistical processes, and a NaI(Tl) detector does not produce pulses of constant height, v, even for $E_d = E_\gamma$. Instead, the pulse heights are distributed as a Gaussian distribution around the photopeak, and the full width at half maximum, FWHM (see Fig. II.1.6), of the distribution is used to indicate the energy resolution (Zanzonico and Heller 2000):

$$\%\ \text{energy resolution} = 100 \cdot \text{FWHM}/E_\gamma \qquad \text{(II.1.1)}$$

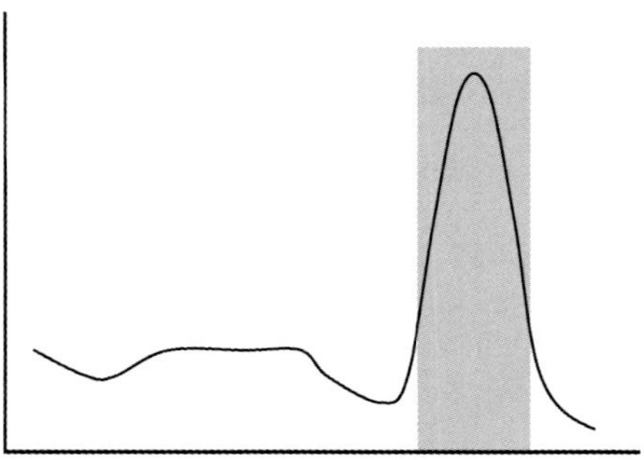

Fig II.1.5. Acceptance window imposed on the energy spectrum by the pulse height analyser. Only photons detected with energies in the shaded area are accepted. The upper and lower limits would usually be positioned symmetrically around the photopeak

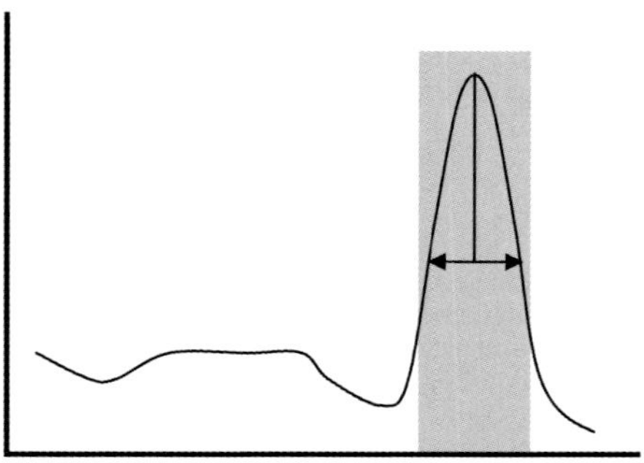

Fig II.1.6. Definition of the full width at half maximum of the photopeak

Energy resolution improves non-linearly as Eγ increases, because increasing numbers of light photons are produced in the interactions thereby reducing statistical uncertainty. For NaI(Tl), at the 140 keV of $^{99}Tc^{m}$ it is typically ~10% (Zanzonico and Heller 2000) and at the 662 keV of ^{137}Cs it is ~7% (Fleming et al. 2002). As the energy resolution improves, it is easier to distinguish photons of different energies. The importance of this is in the rejection of Compton scattering events and, as energy resolution improves, more events originating from scattered photons may be eliminated whilst discarding fewer events originating from unscattered photons (Zanzonico and Heller 2000) by using a narrower PHA window. A rule of thumb is to set a PHA window width to be about twice the energy resolution of the detector at the Eγ of interest. Periodic measurement of energy resolution is also used to diagnose slow deterioration of the crystal due to moisture leaks and PMT degradation.

Simultaneous Measurement of Different γ Energies

For two radionuclides, $E\gamma_1 < E\gamma_2$, the relationship of the energies determines the accuracy with which their measurements can be separated. The photopeak of γ_2 is isolated and γ_1 causes no interference. However, the Compton plateau of γ_2 lies over the photopeak of γ_1 and contributes to the measurement (see Fig. II.1.7). Correction is undertaken by determining the number of events from γ_2 which appear in the photopeak window of γ_1, as a proportion of the counts in the photopeak window of γ_2. This has to be done in the absence of γ_1. These are then subtracted from subsequent mixed measurements in the photopeak window of γ_1.

Measurement Statistics

Because of the random errors, the number of counts recorded in successive measurements form a frequency distribution around the true value. This is the basis of determining the probable error in any measurement involving radioactivity (Sorensen and Phelps 1987). An indication of the likely accuracy of any single measurement is obtained from the width of the frequency distribution (see Fig. II.1.8). This width is defined using the variance, σ^2, whose magnitude is such that 68.3% of measurements fall within $\pm\,\sigma$ of the true value; 95.4% within $\pm\,2\sigma$ and 99.7% within $\pm\,3\sigma$. Where σ is the standard deviation.

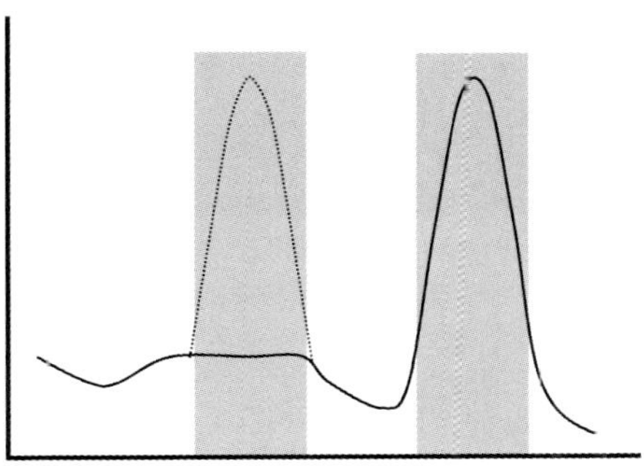

ig II.1.7. Energy spectrum for photons of different energies. The photopeak at the lower energy is contaminated by higher energy photons which have undergone Compton scatter

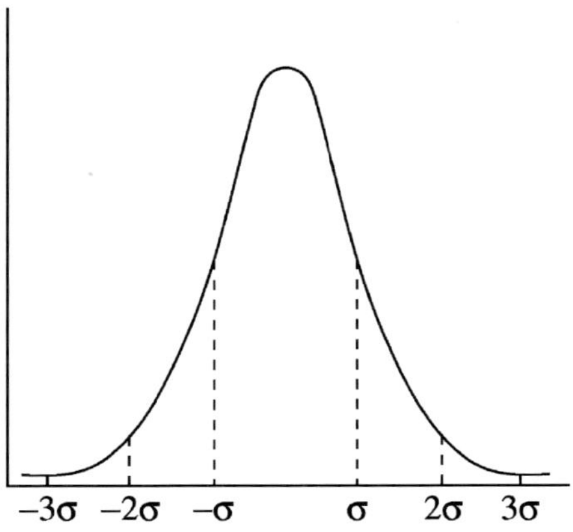

Fig II.1.8. Frequency distribution quantifying the probability of error in any measurement involving radioactivity. The vertical lines on either side of the mean represent multiples of the standard deviation, σ

Poisson

Counting statistics follow the Poisson frequency distribution, in which the probability of obtaining a certain count, n, when the true count is m is:

$$P(n) = e^{-m}\, m^n / n! \qquad (II.1.2)$$

This equation indicates that the mathematical description of the Poisson distribution is complicated. For a low value of m, the Poisson distribution is asymmetric, because it is not possible to have a negative count and the Poisson distribution is only defined for non-negative integers (see Fig. II.1.9). However, an important feature of this distribution is that the standard deviation is always:

$$\sigma = \sqrt{m} \qquad (II.1.3)$$

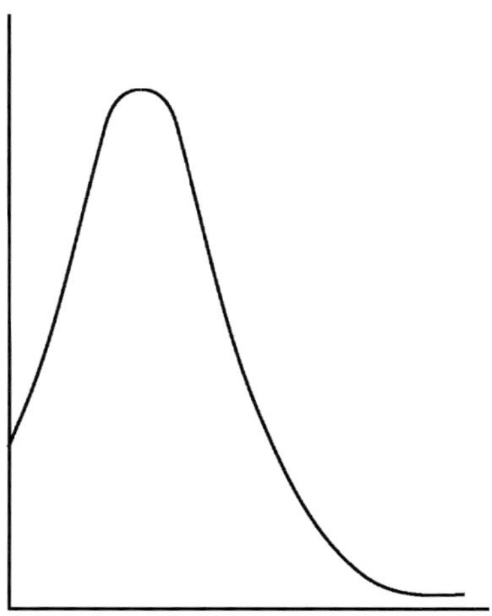

Fig II.1.9. Asymmetric Poisson frequency distribution showing that the function can only be defined for positive numbers

Gaussian

For $m = 10$, the Poisson distribution is very similar to the Gaussian distribution, and for $m \geq 20$, the distributions are virtually indistinguishable. Also known as the normal distribution, the Gaussian distribution is a symmetric 'bell-shaped' distribution that is much easier to describe than the Poisson distribution:

$$P(n) = [1/\sqrt{(2\pi\sigma^2)}]\, e^{-x} \tag{II.1.4}$$

where $x = (n-m)^2/2\sigma^2$

Standard Deviation

However, σ is not fixed in the Gaussian distribution, as it is in the Poisson distribution. The distribution is defined for any value of n and not just non-negative values. Therefore, in counting statistics, the Gaussian distribution with $\sigma = \sqrt{m}$ is a useful approximation. When a single measurement n is taken, it is assumed that $n \approx m$ and so $\sigma = \sqrt{n}$. If the measurement is multiplied by a constant k:

$$\sigma = k\sqrt{n} \tag{II.1.5}$$

If more than one measurement is made, and they are then summed or subtracted:

$$\sigma_{12\ldots} = \sqrt{(n_1 + n_2 \ldots)} \tag{II.1.6}$$

If n_2 is a background component, this equation indicates that the background will increase the statistical uncertainty in the results. On the other hand, if a number of measurements, m, are used to compute an average, n_{av}; this is a more reliable estimate of the true value than any one of the individual measurements, and the uncertainty is smaller:

$$\sigma_{av} = \sqrt{(n_{av}/m)} \tag{II.1.7}$$

The number of counts needed to be kσ% confident that the true count is within *p*% of the measured counts is given by:

$$C = [k/(p/100\%)]^2 \tag{II.1.8}$$

where $k = 1$ for 68.3% confidence etc.

For example, 10,000 counts are required to be within 95% confident that the true count is within 2% of the measured value. The 95% confidence interval on 10,000 counts is ±200 counts and this 200 counts represents 2% of 10,000 counts.

The difference between two measurements n_1 and n_2 may be real or may be due to statistical variation. Differences, i.e. $n_1 - n_2$, of less than 2σ, i.e. $2\sqrt{(n_1 + n_2)}$, are not considered statistically significant because there is >5% probability that the difference is caused by random error. Differences >3σ are considered significant because the probability of them being caused by random error is <1%.

% Standard Deviation

A index of statistical error, which is more useful than σ, is $\%\sigma$ i.e. coefficient of variation:

$$\%\sigma = 100\,\sigma/n = 100\sqrt{n}/n = 100/\sqrt{n} \tag{II.1.9}$$

This demonstrates that the error is reduced by acquiring more counts because, although σ increases, $\%\sigma$ decreases. Thus, there is less relative variability in large count values. If more than one measurement is made, and they are then summed:

$$\%\sigma = 100/\sqrt{(n_1 + n_2)} \tag{II.1.10}$$

or subtracted, e.g. background subtraction:

$$\%\sigma = 100/\sqrt{(n_1 + n_2)}/(n_1 - n_2) \tag{II.1.11}$$

or multiplied or divided, e.g. area normalisation:

$$\%\sigma = \sqrt{(\%\sigma_1{}^2 + \%\sigma_2{}^2)} \tag{II.1.12}$$

When the measurement is multiplied by a constant, $\%\sigma = 100/\sqrt{n}$ i.e. there is no improvement is counting statistics.

Counting Rates

The counting rate, $r = n/t$, when n counts are recorded during a measurement period t.

$$\sigma_r = \sigma_n/t = (1/t)\sqrt{n} = \sqrt{[n/t^2]} = \sqrt{[r/t]} \tag{II.1.13}$$

Because the error in the measurement of t is usually small, the error in r mainly results from the error in n. The uncertainty in the difference of two counting rates, r_1 and r_2, determined using counting times t_1 and t_2:

$$\sigma_{12} = \sqrt{(r_1/t_1 + r_2/t_2)} \tag{II.1.14}$$

This indicates that uncertainties are relatively large when there are only small differences between high counting rates and when background counting rates are high. To reduce the error of a count rate r, the averaging time t has to be increased, i.e. longer counting times produce smaller uncertainties in estimated counting rates. The $\%\sigma_r$ of a count rate:

$$\%\sigma_r = 100\,\sigma_r/r = (100/r)\sqrt{(r/t)} = 100/\sqrt{(rt)} = 100/\sqrt{n} \tag{II.1.15}$$

Minimum Detectable Activity (MDA)

For each radionuclide, there is a minimum activity that can be detected on a particular counting system. This is the activity which would give a count of n_{MDA} over a duration, t.

Where

$n_{MDA} = 3\sigma$

$\sigma = \sqrt{n_b}$

n_b is the background count in the same duration

The counting rate corresponding to the MDA is $3\sqrt{(r_b/t)}$, where r_b is the background rate.

References

Fleming J, Williams N, Skrypniuk J, Thorley P, Rose M, Driver I (2002) Report 85 radioactive sample counting – principles and practice. Institute of Physics and Engineering in Medicine, York

Hoffman EJ, Tornai MP, Janecek M, Patt BE, Iwanczyk JS (1999) Intraoperative probes and imaging probes. Eur J Nucl Med 26:913–935

Madsen MT (1996) Scintillation detectors and scintillation detector counting systems. In: Henkin RE, Boles MA, Dillehay GL, Halama JR, Karesh SM, Wagner RH, Zimmer AM (eds) Nuclear medicine. Mosby, St Louis, pp 74–84

Sorenson JA, Phelps ME (1987) Physics in nuclear medicine, 2nd edn. Grune and Stratton, New York

Zanzonico P, Heller S (2000) The intraoperative gamma probe: basic principles and choices available. Semin Nucl Med XXX:33–48

Non-Imaging

Abbreviations

FWHM	full width at half maximum
GM	Geiger-Mueller
PHA	pulse height analyser
PMT	photomultiplier tube
TLD	thermoluminescent dosimeter

The efficiency of the detection process is an important consideration when dealing with measurements on patients, since the activity that can be administered is limited. The total detection efficiency is affected by both detector efficiency and detection geometry. Many investigations in diagnostic nuclear medicine require the representation of radiation levels as numbers rather than as images. For the safety of the patient and for consistency in investigation, it is essential to measure the activity before administration to the patient. This is usually undertaken using ionisation chambers. Values of physiological parameters can be calculated by following the distribution of radioactivity in the body by measuring the activity in body fluid samples. This is undertaken using either crystal or liquid scintillation counters. The distribution of activity within the body can be determined using external measurements with probe systems, based on scintillation or semiconductor crystals. The safe use of radionuclides demands conscientious monitoring of the environment and staff, which is undertaken using a variety of detectors.

Detection Efficiency

Detection of radiation is an inefficient process. In the case of a scintillation detector presenting a flat surface to a source of radiation, the overall efficiency (E) is given as:

$$E = E_g * E_i \tag{II.2.1}$$

where
E_g is the geometric efficiency
E_i is the intrinsic efficiency of the detector

E_i = detected events/incident events, and depends primarily on the linear attenuation coefficient, μ_l, and the thickness of the detector. In a scintillation detector where the counts are obtained only from the photopeak, the intrinsic efficiency is known as the photopeak efficiency. Geometric efficiency is evaluated by assuming that there will be no preferred direction for the emission of radiation from a radionuclide; the radiations from a source being emitted with equal probability in all directions, and only a fraction being incident on the sensitive volume of the detector, as shown in Fig. II.2.1.

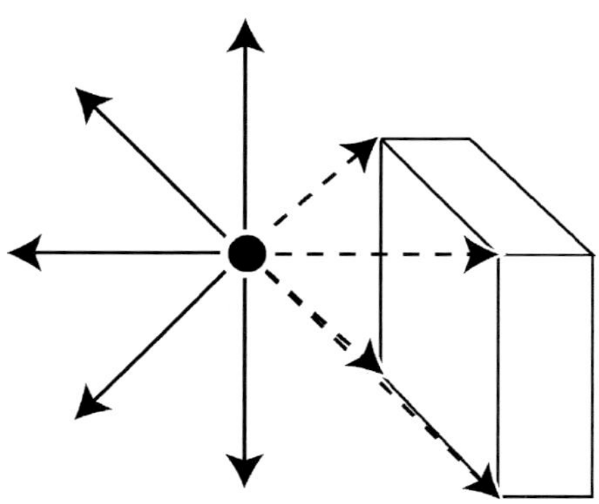

Fig. II.2.1. Geometric efficiency of a flat radiation detector showing the dependence on the solid angle seen by the source

E_g = events incident on detector/total radiation emissions, and depends on two factors: the distance, x, between the source and the detector and the radius, r, of the cross-sectional area of the detector.

$$\text{when} \quad x \gg r, \; E_g \propto 1/(4\pi x^2) \quad \text{and} \quad \propto r^2 \qquad \text{(II.2.2)}$$

E_g is a maximum, ~50%, when x = 0 i.e. when the source is adjacent to the face of the detector. For E_g to be >50%, the sample must be surrounded by the detector.

Administration Activity

Before administration to a patient, the activity must be accurately measured, usually in an ionisation chamber. These gas-filled detectors operate in region 2 of the voltage characteristic curve, the ionisation plateau which is described in Chapter II.1. Because the current pulse does not change appreciably with applied voltage, these detectors are very stable and reliable. A rare gas, such as Argon, is often used at high pressure, ~20 atmospheres, to increase the density of the detecting medium and hence the intrinsic detection efficiency. To increase the geometrical efficiency and decrease geometrical variation, a 4π counting geometry is used. In this, the detector is formed around a well into which the activity can be introduced and thus positioned close to the centre of the detector. The well is protected, from contamination, by a removable plastic liner and the activity is introduced in a plastic 'dipper' which helps to achieve geometric consistency (see Fig. II.2.2). The outer walls are often lead shielded to reduce any effects of extraneous sources and to provide operator protection from activity in the well.

The ionisation current, which is proportional to activity, is amplified and measured by an electrometer, on which activity is displayed directly as Bq. Each radionuclide requires its own amplification, or calibration, factor as the current depends on the types, energies and abundances of all the emitted radiations, as well as primarily on the amount of photon energy. The accuracy of the measurement is dependent on the measurement geometry, and the error diminishes as the source is moved nearer to the centre of the detector. Correction factors may therefore need to be applied when this is not the case. Such factors vary with radionuclide and depend on whether the source is

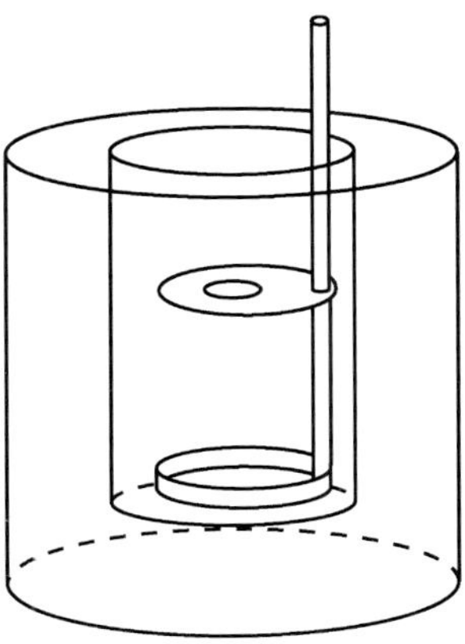

Fig. II.2.2. Activity calibrator showing the position of the 'dipper' which holds the source, in relation to the centre of the detector

contained in a syringe or vial; on container construction, size and shape; and on the volume of the source (Tyler et al. 2002).

Application of a particular calibration factor does not discriminate against other radionuclides as would, for instance, selecting an energy window in scintillation counting. An activity will be displayed for any radionuclide, no matter what factor is set. Thus, these detectors cannot measure the separate activities of more than one radionuclide in a mixture, except in the particular circumstance of measuring ^{99}Mo breakthrough in a ^{99}Mo–^{99}Tc^{m} mixture, which is described in Chapter IV.2.

Calibration for an ionisation chamber–electrometer pair is established at construction, with any intended shielding in place, and the units are not interchangeable with others. Although the initial accuracy may degrade with time because of changes in gas pressure and electrometer drift, these systems are accurate over a wide range of activities, from less than one MBq to several hundred GBq. At very high activities, low readings may occur because of ionisation chamber saturation, the thresholds being different for different radionuclides. Practically, however, the only radionuclide for which this may be a problem is ^{99}Tc^{m}, as this is the only one used at such high activities

Operation

The appropriate calibration factor should be applied and, before any activity is moved into the vicinity of the detector, the background activity should be established. If this is high, the well liner or dipper should be decontaminated, extraneous sources removed or background subtraction applied. The source to be measured should then be introduced into the well using the dipper, great care being taken to maintain the correct geometry. Usually, a vial is placed on the bottom of the dipper and a syringe is suspended from the higher shelf. It is particularly important that the shelf geometry should not change over time and that, if a number of dippers are in use, they should all have the same geometry. Sufficient time should be allowed for the activity reading to stabilise, this will be longer for lower activities.

Activity in Patient Samples

In many procedures it is necessary to determine the amount of radioactivity in a body sample, which is often compared with a standard activity. This is undertaken using a scintillation detector. A review of these techniques is given by Fleming et al. (2002) and Todd et al. (1980). The choice of detector and geometry is primarily dictated by the detection sensitivity.

Crystal Scintillation

These are used with photon emitting radionuclides and are well-type detectors with 4π counting geometry. They can be used manually, where samples are measured individually, or automatically, where a number of samples are loaded and measured in sequence. The well in the crystal is a small cylindrical hole, of diameter ~19 mm and depth ~38 mm, which allows the sample to be positioned very close to the centre of the crystal (see Fig. II.2.3). E_g increases with decreasing sample volume, being ~85% at 4 ml and approaching 95% for samples <1 ml. It is preferable, therefore, for the sample volume to be <2 ml but, at acceptance, the optimal volume should be determined and consistently used. E_i increases as the crystal size increases. For medium energies, a diameter of 45 mm and depth of 50 mm is adequate, but for high energies, >500 keV, 76×76 mm is better. The dead time of the system is a consequence of the resolving time, and limits the maximum activity that can be measured without significant, <5%, count loss. The true count rate can be obtained from the observed count rate by:

$$C_t = C_o/(1 - C_o t) \qquad \text{(II.2.3)}$$

where
C_t is the true count rate
C_o is the observed count rate
t is the resolving time

This correction should be used when counting losses >1%, i.e. when:

$$100\, C_o t > 1 \qquad \text{(II.2.4)}$$

A rule of thumb, to avoid counting losses due to dead time, is to keep the total (not just the photopeak) count rate $<10^4$ counts s^{-1}, which corresponds to ~40 kBq of the usually used radionuclides. Energy resolution depends to a slight extent on the size and shape of the crystal, being poorer for a well-type than a cylindrical type, and is usually defined as the %FWHM of the 662 keV photopeak of ^{137}Cs.

Operation

Because of the detection sensitivity dependence on sample geometry, all samples should be the optimal volume. For routine use, the PHA window width should include virtually all of the photopeak. Each acquisition should be undertaken for a set time or number of counts, such that the number of counts collected achieves a certain %error, calculated on the basis explained in 'Measurement statistics' in Chapter II.1.

Fig. II.2.3. Well type scintillation detector showing position of the source relative to the centre of the crystal, which is on top of its photomultiplier tube and completely encased in a light tight shield

Liquid Scintillation

Liquid scintillation counting is the preferred method for β^- particle detection, and for some low energy x-ray and γ photons. The sample detector vial and the PMT are enclosed in a light tight compartment to exclude external light from the PMT. The sample detector vial is often viewed by two opposed PMTs rather than one (see Fig. II.2.4). In this arrangement, electronic noise can be reduced significantly using coincidence circuitry, thus enhancing the sensitivity of low-energy β^- particles that would otherwise be lost in the PMT noise.

Activity Distribution in a Patient

Once radioactivity is administered to a patient, its distribution can be assessed by external monitoring; using a probe system consisting of a crystal and PMT system. One or more of these may be used in a complete system.

Probes

The components are similar to those incorporated in in-vitro crystal scintillation systems, however, a cylindrical NaI(Tl) crystal without a well is used. For medium energies, a thickness ~25 mm is adequate but thicker crystals, of ~50 mm, are required for higher energies. Three factors need to be considered in the in-vivo assessment of radionuclide distributions: collimation, scattering and attenuation.

Collimation

To localise the activity distribution, it has to be assessed over a well defined area or volume, and any photons originating outside this should not be able to interact with the detector. This is achieved by protecting the detector using shielding and by defining the field of view by attaching a collimator to the front of the crystal. These are usually both made of lead. The field of view is determined by the length and radius of the collimator; becoming smaller as the length increases and the radius decreases, and larger as the length decreases and the radius increases (Fig. II.2.5). The field of view also increases with distance from the collimator. A smaller field of view excludes more scattered radiation from the detector, a larger the field of view improves sensitivity.

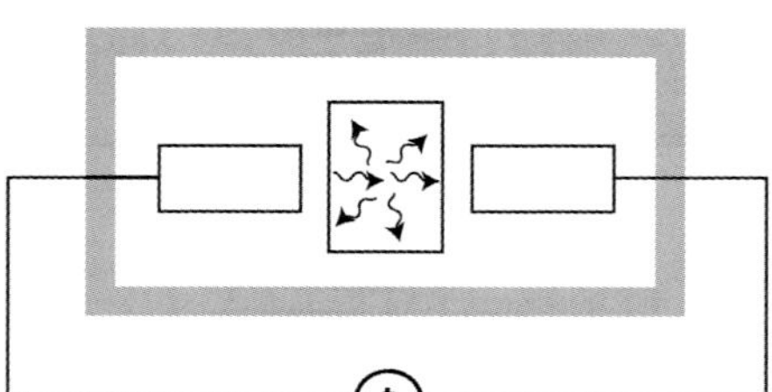

Fig. II.2.4. Liquid scintillation detection showing dual detectors connected by coincidence circuitry and totally encased in a light tight shield

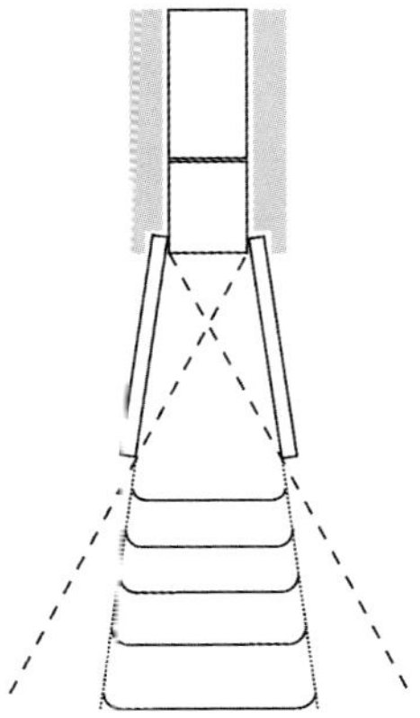

Fig. II.2.5. Detector collimator showing the field of view and the profiles describing detection sensitivity. Sensitivity decreases with increasing distance from the detector but the sensitivity at a particular distance is reasonably constant across the field of view

Scattering

Even with collimation, a large proportion of the radiation events detected represent photons whose initial direction is altered by Compton scattering (see Fig. II.2.6). These degrade the information content but can be partially eliminated by energy discrimination, the effectiveness of which depends on the energy resolution of the detector which determines the width of the PHA window that can be used practically.

The window width determines the proportion of scattered to unscattered events that are recorded. The narrower the window the more scattered radiation is excluded, but this reduces sensitivity as more of the unscattered radiation is also rejected. There are two better ways to compensate for scattered radiation. One way is the simultaneous use of one or more windows at an energy lower than the photopeak; with the counts, or a fraction of the counts, in the lower energy windows being subtracted from the counts in the photopeak. Another way is the simultaneous use of two windows that are only a few keV in width and set just below and above the photopeak (see Fig. II.2.7), usually appropriate if scattering from higher energy photons is involved. Here the average counts from these two windows is subtracted from the window set at the photo-

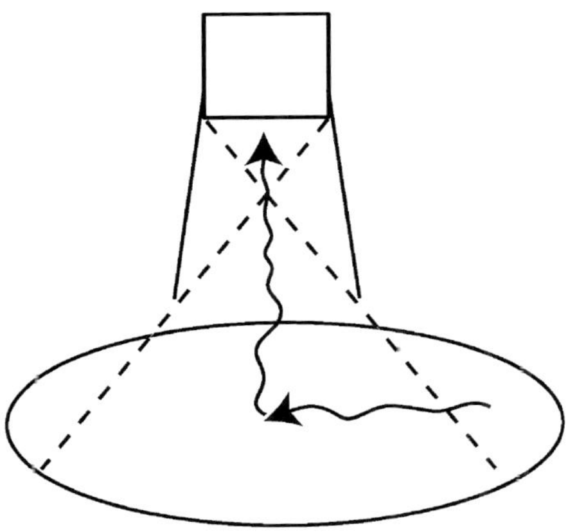

Fig. II.2.6. Demonstration of how a photon, originating outside the field of view of the collimator, can be detected because of the change in its direction imposed during Compton interaction

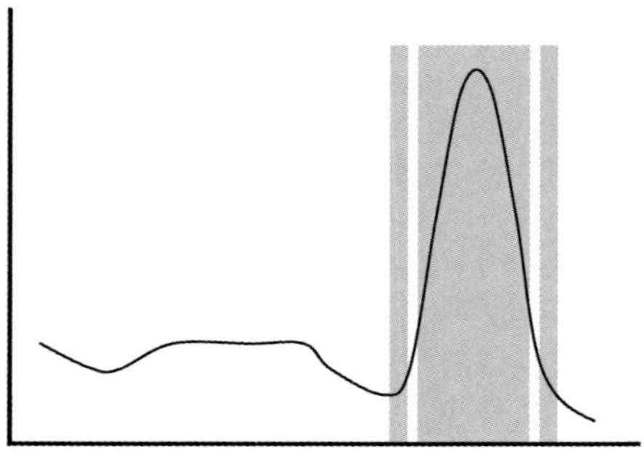

Fig. II.2.7. Pulse height analyser windows used in scatter rejection, a few keV in width and set just below and above the photopeak

peak. Both methods are better than narrowing the photopeak window because primary radiation is not lost.

Attenuation

Attenuation of photons in body tissue is a serious problem, in that a uniform distribution can produce apparently different counting rates at different locations in the body, because photons pass through different thicknesses and possibly different types of tissue before exiting the body (see Fig. II.2.8). It is possible to reduce the effect of attenuation by using radionuclides that emit high energy photons, because the photon loss reduces sharply with increasing energy up to ~100 keV and stabilises with further increases. Therefore, radionuclides emitting photons with energies >100 keV are preferred. Because the sensitivity of the NaI(Tl) detectors decreases with increasing photon energy, the optimum range for in-vivo use is 100–300 keV.

Surface Probes

The measurement of the count rate over different areas of the body can be undertaken with a single probe system, held in an adjustable gantry to allow positioning in relation to the body. These have the advantages of cheapness, small size, and simple operation and maintenance. The diameter of the crystal should be chosen to suite the investigations undertaken, varying from 25 to 50 mm. For thyroid uptake studies, 38–50 mm diameters are used (Becker et al. 1996). In organ uptake investigations, in addition to using radionuclides emitting high energy photons, attenuation effects are corrected by also measuring a known activity in a phantom that reflects the average size, shape and depth of the relevant organ in a standard man.

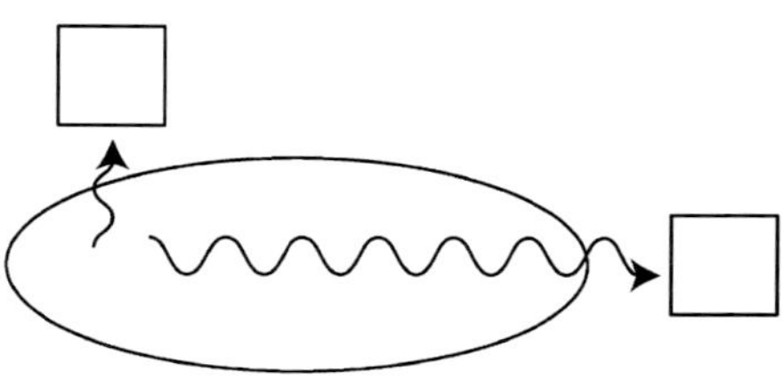

Fig. II.2.8. Attenuation of photons in body tissue; loss of photons by absorption or scattering becomes more likely as the photon path length and the tissue density increase

Collimator

The design of a collimator should provide: a geometric efficiency which is as high as possible, in order to keep the radiation burden to the patient to a minimum; a field of view that is well defined but with sufficient flexibility to accommodate size variations among patients whilst excluding of activity in other parts of the body; and a uniform geometric efficiency across the field of view, i.e. a flat-field. Compromise is necessary, because the geometric sensitivity varies inversely as the square of the separation between the source and the detector but the uniformity improves as this distance increases. For thyroid uptake studies, a flat field collimator providing 10 cm diameter field of view should be used (Becker et al. 1996).

Operation

Acquisitions are undertaken to measure: the total count in a set duration, the time required to accumulate a set total count, or the count rate.

Multiple Probes

A number of probes in a fixed geometric orientation can be used to acquire data from a particular volume of interest, e.g. cerebral perfusion.

Whole Body Counter

These are used to measure total activity in the whole, rather than in parts of, the body, and are designed to achieve a uniform sensitivity to any activity distribution. Various designs exist, a common one of which is the shadow-shield counter, in which four large, ~15 cm diameter, well shielded NaI(Tl) detectors are positioned above and below a patient table, shown in Fig. II.2.9. The field of view is a transverse section of the body, the width of which is determined by collimator slits and activity is measured as the patient is moved, at constant speed, past the detectors (McAlister 1979). Another type involves a shielded room, usually with 10–20 cm thick steel walls surrounded by concrete (McCready 2000).

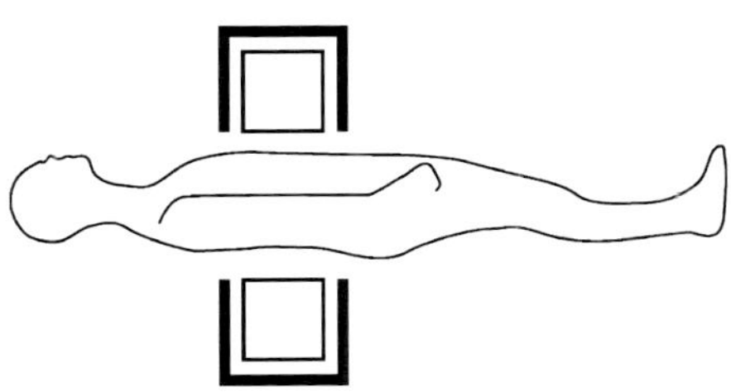

Fig. II.2.9. Shadow-shield whole body counter, showing large shielded detectors above and below the patient, which traverse the complete length of the patient to estimate the total activity in the body

Intra-Operative Probes

Intraoperative probes have been used for more than 50 years (Perkins and Britten 1999); the earliest being Geiger-Mueller (GM) detectors with ^{32}P. In these, the short range of the β^- particles meant that all the detected radiation originated from activity adjacent to the probe and collimation was not required. In the mid 1980s, basic γ detecting intraoperative probes were introduced, which led to the rapid development of radioimmunoguided surgery and sentinel node detection (Hoffman et al. 1999).

These are 10–20 mm diameter probes which are used interactively to identify small activity accumulations. Because of this mode of use, the method of count rate presentation can affect the detectability of small sources (Tiourina et al. 1998; Perkins and Britten 1999; Zanzonico and Heller 2000). Ergonomic characteristics, ease of use (Tiourina et al. 1998), detection efficiency enabling detection of deep-seated nodes or those with low uptake, and collimation are the most important factors in probe design, particularly since the target radionuclide accumulation may be close to other areas of high activity, such as the injection site. To enable directional counting, the probes are shielded at the back and sides, and precise spatial localisation is achieved using a collimator attached to the front of the detector, usually comprising 5–10 mm of tungsten (Tiourina et al. 1998; Perkins and Britten 1999). A good energy resolution is also important to exclude radiation scattered from adjacent areas (Tiourina et al. 1998).

Because of the nature of their use, a backup detector is often recommended. The detectors can be either scintillation crystal, which are ~10 mm thick, or ionisation semiconductors, which are only ~1 mm thick. A problem peculiar to the use of the scintillation crystal detector is the temperature change that can be experienced between ambient room temperature and body temperature. This difference may be many °C and may cause significant changes in detection sensitivity, resulting from PMT gain changes of ~1% per °C which shift the photopeak energy. This can be minimised, by using a remote PMT and transmitting the light pulse from the scintillation crystal by flexible fibreoptic cable, which also allows the detector to be smaller and feel more like a surgical instrument. The price of this is a deterioration in energy resolution, to ~39% for ^{99}Tc^{m}, caused by light loss in transmission. This does not translate into significantly poorer lesion detectability, however. Although semiconductor detectors have much poorer detection sensitivity than scintillation crystal detectors they have much better energy resolution, because they produce more electrons per interacting photon, and hence have better scatter rejection (Zanzonico and Heller 2000).

Imaging probes should be the next development (Hoffman et al. 1999) and, recently, a hand held γ camera for intraoperative use has been introduced (Pitre et al. 2003).

Monitoring

Environmental

For the monitoring of environmental dose rates, instruments should be able to measure over a range 0–10 mSv h^{-1}. Two instruments can be used: the ionisation chamber and the Geiger-Mueller (GM) detector. For low dose rates, a physically smaller GM detector can have the same sensitivity as a larger ionisation chamber.

Geiger-Mueller Detector

These are gas-filled detectors operating in region 4 of the voltage characteristics described in Chapter II.1, and are very suitable for environmental monitoring because they are robust; their output signal being large and requiring only simple electronics. For the detection of β^- particles, there is a small thin aluminised Mylar window either at the end or on one side of the GM tube. For the detection of x or γ photons, this window is generally closed. To be used for the measurement of dose or dose rate, they must have energy compensation to correct for their increased detection sensitivity <200 keV, which is implemented by installing an absorptive filter around the probe. Because the interaction of photons in the gas volume itself is minimal, the detection of these radiations is primarily through the photoelectrons and Compton electrons generated in the inner walls of the GM tube. The incident radiation causes a discharge such that the current produced is independent of both the energy of the radiation and voltage. Once the discharge is established, however, it has to be stopped by chemical quenching. This creates a resolving time of $50–300 \times 10^{-6}$ s and limits the maximum usable count rate to $\sim 10^3$ counts s^{-1}. At high radiation intensities, pulses are detected at such a rate that they cannot be separated and the dose rate can be erroneously indicated to be small.

Contamination

Contamination monitors are hand held units, using either crystal scintillation detectors or Geiger-Mueller detectors.

Personal

Personal monitoring is undertaken using passive devices such film or thermoluminescent dosimeter (TLD), or active monitors.

Film

This is held in a plastic holder with a number of filters, and is suitable for monitoring x, γ, and high energy β^- radiation, over the range 0.1 mSv to several Sv. After processing, the amount of blackening in relation to the pattern of filters, allows the intensity, type and energy of the radiation to be determined. Because the stored information fades with time, the film cannot be used for more than 3 months before being developed.

Thermoluminescent Dosimeter (TLD)

Again, these are held in small plastic holders with a number of filters, and can be used for up to 3 months but it is more sensitive than film and can monitor down to 0.01 mSv. They are also small enough to be used for extremity monitoring. The absorbed radiation is discharged, by heating, for measurement.

Active Detectors

These are often pocket sized GM instruments and provide an immediate assessment of radiation exposure by numerical display or audible warning. They are useful to supplement the passive monitors.

References

Becker D, Charkes ND, Dworking H, Hurley J, McDougall R, Price D, Royal H, Sarkar S (1996) Procedure guideline for thyroid uptake measurement 1.0. J Nucl Med 37:1266–1268

Fleming J, Williams N, Skrypniuk J, Thorley P, Rose M, Driver I (2002) Report 85 radioactive sample counting – principles and practice. Institute of Physics and Engineering in Medicine, York

Hoffman EJ, Tornai MP, Janecek M, Patt BE, Iwanczyk JS (1999) Intraoperative probes and imaging probes. Eur J Nucl Med 26:913–935

McAlister JM (1979) Radionuclide techniques in medicine. University Press, Cambridge

McCready VR (2000) Milestones in nuclear medicine. Eur J Nucl Med 27 [Suppl]: S49–S79

Perkins AC, Britten AJ (1999) Specification and performance of intra-operative gamma probes for sentinel node detection. Nucl Med Commun 20:309–315

Pitre S, Menard L, Ricard M, Solan M, Garbay J-R, Charon Y (2003) A hand-held imaging probe for radio-guided surgery: physical performance and preliminary clinical experience. Eur J Nucl Med 30:339–343

Tiourina T, Arends B, Huysmans D, Rutten H, Lemaire B, Muller S (1998) Evaluation of surgical gamma probes for radio-guided sentinel node localisation. Eur J Nucl Med 25:1224–1231

Todd JH, Short MD, Bessent RG, Hawkins LA, Leach KG, Ottewell D, Morgan DW (1980) HPA topic group report 33 an introduction to automatic radioactive sample counters. Hospital Physicists' Association, London

Tyler DK, Baker M, Woods MJ (2002) NPL secondary standard radionuclide calibrator. Syringe calibration factors for radionuclides used in nuclear medicine. Appl Radiat Isot 56:343–347

Zanzonico P, Heller S (2000) The intraoperative gamma probe: basic principles and choices available. Semin Nucl Med XXX:33–48

CHAPTER II.3

Single Photon Planar Imaging

Abbreviations

ADC	analogue to digital converter
CFOV	central field of view
COV	coefficient of variation
DU	differential uniformity
FFOV	full field of view
FWHM	full width at half maximum
FWTM	full width at tenth maximum
GFOV	geometric field of view
IU	integral uniformity
LE	low energy
LEAP	low energy all purpose
LEGP	low energy general purpose
LEHR	low energy high resolution
LEHS	low energy high sensitivity
LEUHR	low energy ultra high resolution
LSF	line spread function
ME	medium energy
PHA	pulse height analyser
PMT	photomultiplier tube
PSF	point spread function
SPET	single photon emission tomography
UFOV	useful field of view

Imaging of function is a very sensitive indicator of pathology, with changes in function possibly becoming apparent many months before consequential changes in morphology. Such an image can be formed by moving a detector across the body and mapping the variations in emitted photon intensity, using a rectilinear scanner. However, these types of investigation are very time consuming because of the inefficient use of the emitted radiation; whilst information is being acquired at one part of the area of interest, information is being ignored at other parts. The formation of an image over a large area is a significant improvement for diagnosis, in terms of efficiency and data quality, and also because more information can be extracted by subsequent analysis. The image area needs to be sufficiently large to include a whole organ such as the liver or lungs, or a complete area of interest, such as the abdomen (see Fig. II.3.1). The technique necessitates detecting the single photon emissions from all parts of the area of interest simultaneously and with equal efficiency. It is accomplished using the γ camera.

The γ camera has evolved from a relatively simple analogue system into a complicated unit incorporating various digital correction techniques to improve the quality of the data (O'Connor 1996); and whose proper use requires an understanding of its operating characteristics and limitations. Maximal performance demands evaluation of a number of performance indices and application of several corrections; and maximal image quality demands implementation of optimal acquisition parameters for particular investigations. In justification of the radiation exposure to the patient, image quality must be optimised such that the maximum diagnostic information is obtained. Even though a large area is imaged, the process remains inefficient because the detector does not surround the patient. Since photons in the energy range of interest,

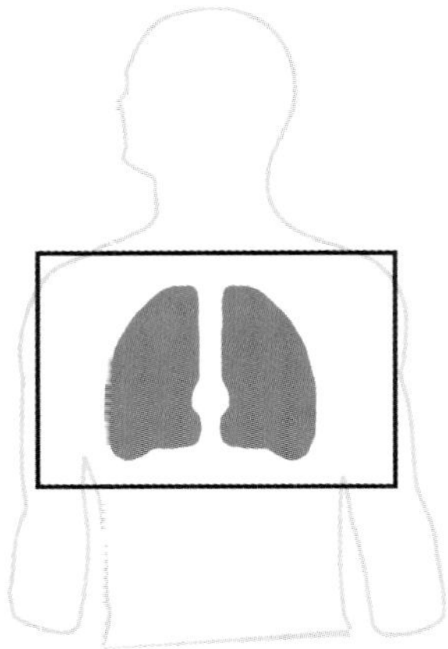

Fig. II.3.1. Planar imaging over an extended area

between 50 and 500 keV, cannot be focused; the sensitivity is further degraded because absorptive collimation must be used to exclude photons travelling in inappropriate directions (Goris 1998). In large part, this also determines spatial resolution. A further degradation in spatial resolution is imposed when photons, scattered by Compton interaction in the patient, are included in the image. The collimator cannot exclude these, and this has to be undertaken by energy discrimination.

Gamma Camera

The central component of the gamma camera is a large thin scintillation crystal which detects single photon events (see Fig. II.3.2). The distribution of these photons, on the crystal, is determined by an absorptive lead collimator which is positioned at the front face of the crystal. Performance measurements undertaken on the detector, without a collimator attached, are termed intrinsic. Those taken with the collimator attached are termed extrinsic or system. The rear face of the crystal is optically coupled to an array of photomultiplier tubes, PMTs, which provide information to position and energy circuits to produce an electronic representation of the activity distribution. These components are mainly contained within the camera head, but with the use of digital electronics, a substantial proportion of the electronic components can be remotely

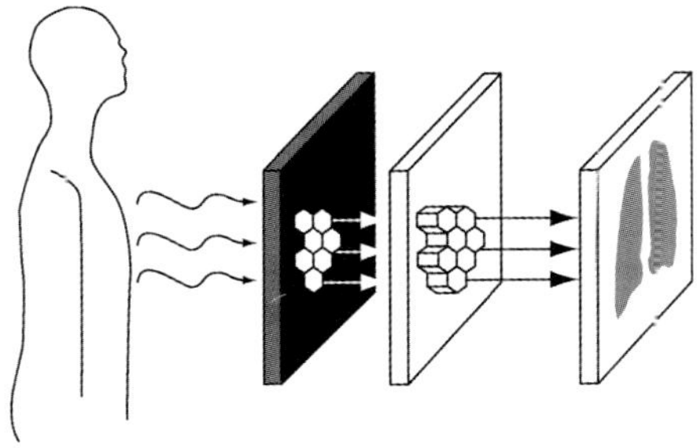

Fig. II.3.2. Components of a gamma camera showing photons passing through a collimator to interact with a crystal, attached to the rear face of which are photomultiplier tubes (PMTs). The information from the array of PMTs is used to localise the position of the photon interactions in the crystal and so form an image of the activity distribution

positioned. This allows a more compact head design with better patient positioning capabilities and less thermal problems. The head is a light-tight, lead-lined protective housing, and is attached to the camera gantry which controls its movements. The thickness of the shielding varies with the maximum energy of the γ photons that may be imaged by the camera, and is usually 25–50 mm. Cameras which are only likely to be used with $^{99}Tc^{m}$, such as mobiles, are lightly shielded, whereas those for use with dual photon emitting radionuclides are more heavily shielded.

Collimator

The collimator determines photon distribution on the crystal by absorptive collimation. It comprises a flat lead plate, containing an array of small closely packed holes, or a lead cone with a single hole at the apex. Of the γ photons incident on the collimator, only those within the geometric field of view of the holes are transmitted to the crystal. The performance characteristics of various collimators are discussed in more detail in Appendix V.3.

Parallel Hole

The most commonly used collimator is the parallel hole type. In this, the hole axes are parallel to each other and perpendicular to the crystal face (see Fig. II.3.3). The image size is the same as the object size and no spatial distortions are introduced. A slant hole collimator is a variant, in which the parallel holes are not perpendicular to the crystal face but are angled, usually by ~25° to the perpendicular. The latter has characteristics similar to the parallel hole collimator but can be positioned closer to the area of interest in some studies.

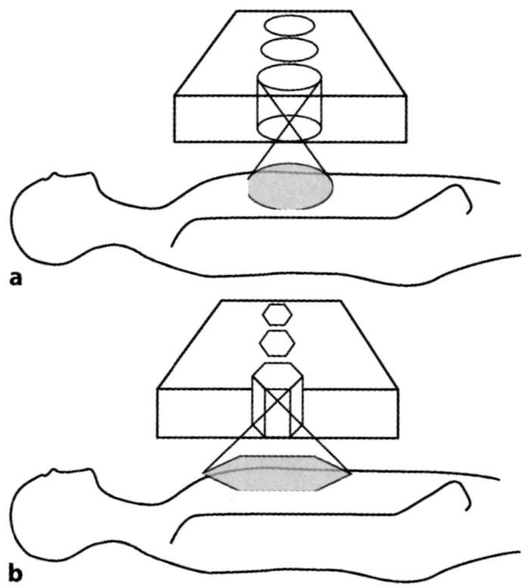

Fig. II.3.3. The field of view of holes in a parallel hole collimator. In reality there are many thousand holes per collimator (see Appendix V.3) and the field of view of each is very small. (**a**) round and (**b**) hexagonal shapes

Energy Range

All collimators are classified according to the γ photons energies over which they can operate, determined by the thickness of the septa between the holes. The energy range categories vary slightly among manufacturers but are approximately: low (LE) 0–200 keV, medium (ME) 200–400 keV, high (HE) 400–600 keV. When too low an energy collimator is used, septal penetration will occur. This generally imposes a uniform background on the image but may produce a star effect when the activity accumulation is focal (Dilsizian et al. 2001). This is particularly apparent when the holes are round because, in this case, the septal thickness varies because of packing considerations. This can be appreciated in Fig. II.3.4. A quantitative estimate of septal penetration can be obtained from the planar spatial resolution full width at tenth maximum, FWTM, a large or unmeasurable value indicating significant septal penetration (Jarritt and Acton 1996). The collimator is very much the weak point in the performance chain of the γ camera because of the geometric field of view of each hole and because of its absorption of most of the incident photons. Both the spatial resolution and the sensitivity of the gamma camera are degraded and, hence, these are the two principal parameters used to describe collimator performance.

Spatial Resolution (Geometric)

This is quantified as the FWHM, which is equal to the minimum separation required between two line sources if they are to be just resolved – the Rayleigh criterion. It is, however, a relative crude expression of resolution and is sometimes supplemented by the FWTM (IPSM 1992). It is a consequence of hole geometry; improving as the hole diameter reduces and as the ratio of hole diameter to length reduces, narrow long holes giving the best spatial resolution (see Fig. II.3.5). Low energy collimators are pro-

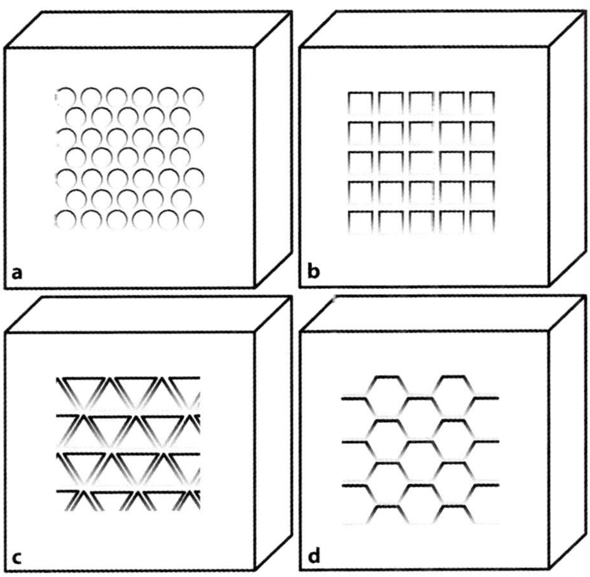

Fig. II.3.4. Variations in packing and septal thicknesses for various hole shapes in parallel hole collimators. (**a**) round, (**b**) square, (**c**) triangular and (**d**) hexagonal

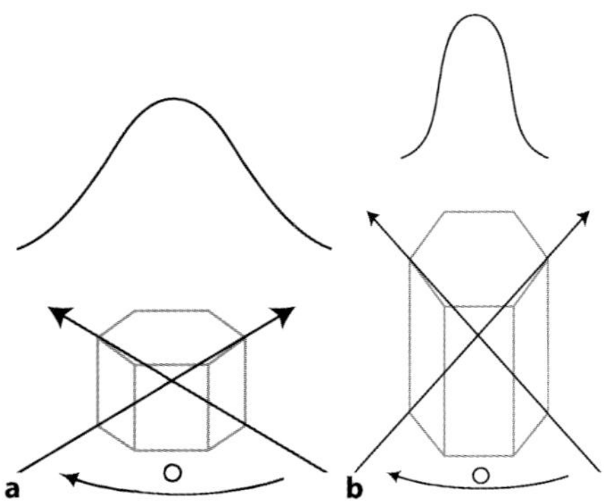

Fig. II.3.5. Geometric spatial resolution variation with collimator hole shape. (**a**) short fat holes are characterised by a wide point spread function and large FWHM. (**b**) long narrow holes are characterised by a narrower point spread function and a smaller FWHM

vided in four spatial resolution categories: high sensitivity (HS), general or all purpose (GP or AP), high resolution (HR) and ultra-high resolution (UHR). The latter two can generally be operated at slightly higher energies than the general purpose, because the extra length of the collimator holes reduces septal penetration by high energy γ photons. The higher the energy and the lower the spatial resolution rating of the collimator, the worse the spatial resolution at the collimator face.

Spatial resolution degrades approximately linearly as the object to collimator separation increases (Moore et al. 1992; Rosenthal et al. 1995), the best being obtained next to the collimator face (see Fig. II.3.6). A separation of only 4–5 cm results in a degradation in spatial resolution by as much as a factor of two. This can be appreciated from the graph shown in Fig. II.3.7. What can also be appreciated from this graph, is that the higher the spatial resolution rating of the collimator, the better the maintenance of spatial resolution with separation, i.e. smaller FWHM (Moore et al. 1992).

Sensitivity

The geometric sensitivity of a collimator is the fraction of incident γ photons which are transmitted to the crystal, when there is only air and no attenuating material between the source and the collimator. It is only ~0.1% and is related solely to hole geometry and the area not covered by lead, with short wide holes and thin septa giv-

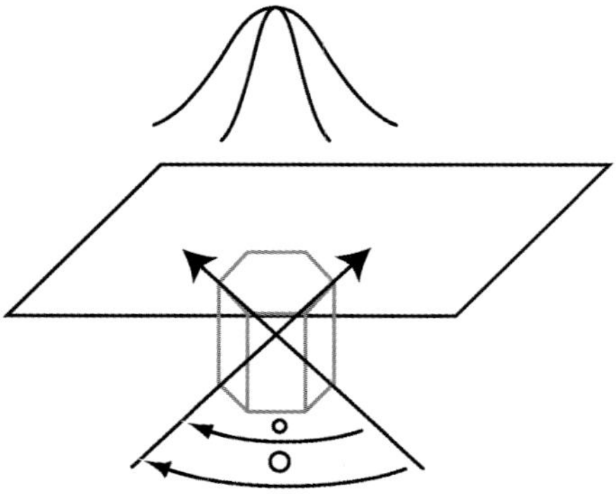

Fig. II.3.6. Variation of collimator geometric spatial resolution with separation. As the source moves away from the face of the collimator, it becomes blurred, its point spread function becomes wider and its FWHM larger

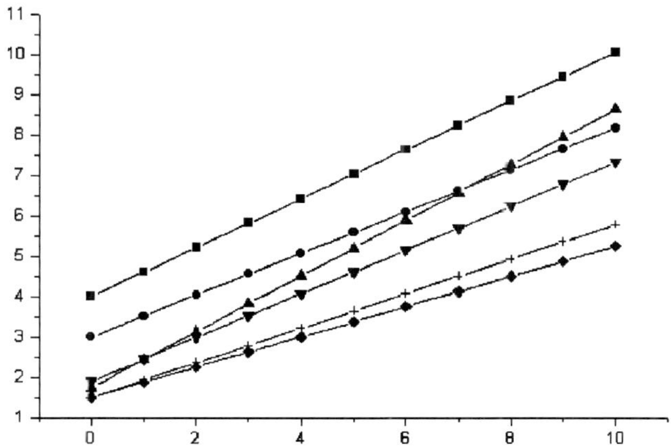

Fig. II.3.7. Linear deterioration of collimator geometric spatial resolution with separation. The data has been calculated, as the FWHM in millimetres, using equation V.3.3 given in Appendix V.3 and values for hole diameter and length, typical of each category of collimator. The ordinate represents FWHM in millimetres and the abscissa represents separation in centimetres. ■ HEGP, ● MEGP, ▲ LEHS, ▼ LEGP, + LEHR, ◆ LEUHR

ing the best sensitivity. Low energy collimators usually have a high sensitivity variety, which operates within the same energy range as the general purpose. Typical approximate relative sensitivities of the various categories of collimator are listed in Table II.3.1. Sensitivity should not depend on the source to collimator separation, if there is minimal septal penetration and if the measurements are made in air and over the whole collimator face. For $^{99}Tc^{m}$, the variation is <5% up to 25 cm but for ^{123}I, the reduction in sensitivity maybe up to 36% because of septal penetration by its high energy photons, particularly with low energy collimators (Fleming and Alaamer 1996). The requirements for better sensitivity and spatial resolution are exclusive, an improvement in resolution being accompanied by a reduction in sensitivity.

Table II.3.1. Typical approximate relative sensitivities of the various categories of collimator

Energy	Type	Sensitivity
Low	High sensitivity	1.6
	General purpose	1.0
	High resolution	0.6
	Ultra-high resolution	0.4
Medium	General purpose	0.6
High	General purpose	0.7
Annihilation	General purpose	0.4

Scatter

As well as absorbing photons, the collimator also scatters photons. These may then be detected by the camera. However with a symmetrical 20% PHA window around 140 keV and a LEGP collimator, the proportion of the detected scattered photons originating in the collimator is only ~1.9%. The problem does become more severe when low or medium energies are being imaged, and the radionuclides also emit higher energy photons which can be scattered and detected in the lower energy PHA windows. For example, a 5% ^{124}I contamination of ^{123}I would cause the scatter contribution to rise to ~6% (Moore et al. 1992).

Non-Parallel Hole

Non-parallel hole collimators are those in which the holes converge to a point or a line in front of the collimator face or diverge from a point or a line behind the collimator face. With these, image magnification or minification is achieved but spatial distortions are introduced.

Pinhole

The pinhole collimator is the only type that is not constructed as a flat lead plate. Instead it is a lead or tungsten cone, usually of length 20–25 cm, with a single hole at the apex (see Fig. II.3.8). It is used to produce a magnified inverted image of a small object; image magnification depending on the object position relative to the physical length of the collimator (see Fig. II.3.9) but, again, spatial distortions are introduced into the image. The increasing magnification and detection sensitivity, and improving system spatial resolution with decreasing separation (see Fig. II.3.10), suggests that the object should be as close as possible to the collimator. However, detection sensitivity degrades towards the periphery of the field of view and this deterioration is particularly acute at small separations (see Fig. II.3.11). This indicates the benefit of accepting reduced magnification and detection sensitivity, and degraded system spatial resolution in favour of obtaining a flatter detection sensitivity response (Connolly et al. 1999;

Fig. II.3.8. Pinhole collimator and its field of view

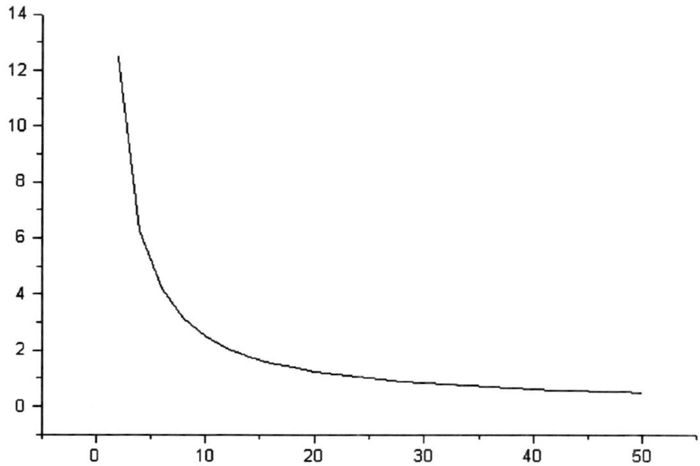

Fig. II.3.9. Variation in magnification with separation for a pinhole collimator. The data has been calculated using equation V.3.10 given in Appendix V.3, an aperture diameter of 4 mm and a collimator length of 25 cm. The ordinate represents magnification and the abscissa represents separation in centimetres. The magnification is very high at small distances and becomes <1 beyond a distance equal to the collimator length

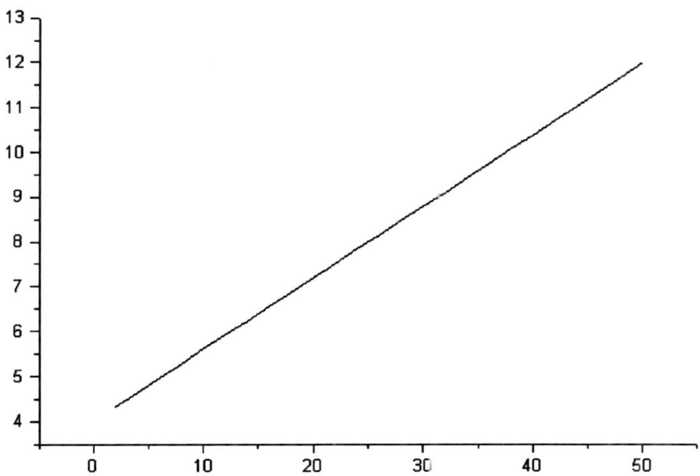

Fig. II.3.10. Linear deterioration of collimator geometric spatial resolution with separation for a pinhole collimator. The data has been calculated, as the FWHM in millimetres, using equation V.3.16 given in Appendix V.3, an aperture diameter of 4 mm and a collimator length of 25 cm. The ordinate represents FWHM in millimetres and the abscissa represents separation in centimetres

Metzler et al. 2002). A hole diameter of 4–6 mm is considered optimum but both the spatial resolution and the sensitivity can be altered by changing the size of the aperture using various inserts, a smaller aperture giving increased spatial resolution and

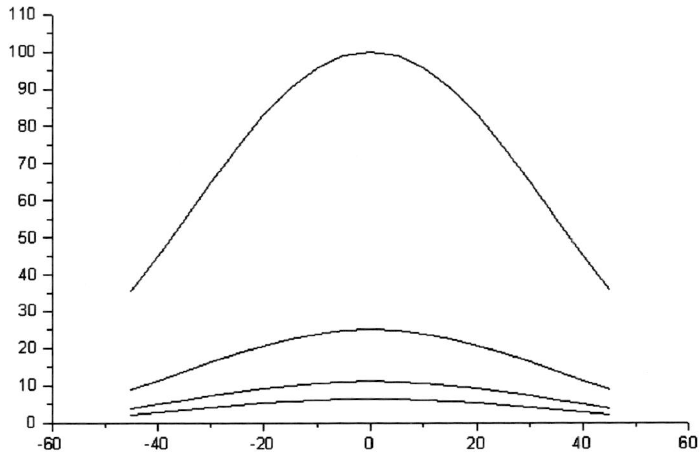

Fig. II.3.11. Variation in detection sensitivity across the field of view for a pinhole collimator. The data has been calculated using equation V.3.17 given in Appendix V.3, an aperture diameter of 4 mm and a collimator length of 25 cm. The four curves represent separations of 5, 10, 15, and 20 cm; relative sensitivity being normalised to 100% on-axis at a separation of 5 cm and decreasing as the separation increases. The ordinate represents relative sensitivity and the abscissa represents position across the field of view in terms of angle (degrees) made with the pinhole

decreased sensitivity and a larger aperture giving decreased spatial resolution and increased sensitivity.

Crystal

Although room temperature semiconductor detectors such CdTe and CdZnTe, either in the form of single detectors or as segmented monolithic detectors, have been investigated for replacing the NaI(Tl) scintillator (Darambara and Todd-Pokropek 2002), the latter still maintains its monopoly in current commercial systems. The advantages of the former would be an improved energy resolution and a smaller dead space at the periphery of the camera which would result in a better imaging geometry.

The size of the crystal constrains the full or geometric field of view of the camera and determines its detection sensitivity. Initially round with a small diameter, crystals are now typically rectangular with a field of view of ~590 × 390 mm which requires a crystal size of ~600 × 450 mm. The thickness determines the detection sensitivity, and must be sufficient to ensure a reasonable probability of photoelectric absorption at the photon energies in use. Because of the dominance of $^{99}Tc^{m}$, the thickness has been optimised for 140 keV, being usually 9.5 mm and giving an optimal energy range of 100–200 keV. Thinner crystals, such as 6–8 mm, are used only when low energy radionuclides are exclusively used, for instance in mobile cameras and in cameras dedicated for nuclear cardiology. The sensitivity of a 9.5 mm thick crystal is nearly 100% for γ energies up to 100 keV and is ~84% at 140 keV. At higher energies this thickness shows a marked reduction in sensitivity, however, achieving only 13% at 500 keV (see Table II.5.6 in Chapter II.5). Therefore, thicker crystals, such as 12.5 mm, are offered for greater sensitivity when higher energy radionuclides are used, but cause a slight degradation in both energy and spatial resolution, and spatial linearity at 140 keV.

The crystal material itself affects the rate at which the camera can detect incident γ photons, without producing a distorted image. This is determined by the length of each scintillation pulse. Because the processing circuits can only deal with one interaction at a time, each interaction can only be detected if there is no overlap from consecutive scintillations. The length of each scintillation pulse in the NaI(Tl) crystal is $\sim 8 \times 10^{-5}$ s.

The light conversion efficiency of the crystal determines the spatial resolution that can be achieved because this is affected by the statistical nature of the scintillation pulse. Each pulse comprises a certain number of light photons. The fewer the photons, the greater the uncertainty is introduced in defining where the point of interaction occurred. Low energy γ photons produce fewer light photons and therefore create a poorer image because of degraded spatial resolution. A crystal with a conversion efficiency higher than the 13% of NaI(Tl) (see 'Crystal' in Chapter II.1), would improve spatial resolution.

The mechanical properties of the crystal affect the conditions under which the camera can be used. It is expensive to produce a large crystal of NaI(Tl), and the final product is hygroscopic, fragile and requires protection from both mechanical and thermal stresses. The crystal has a high coefficient of thermal expansion. The recommended maximum rate of temperature change is ~1°C per hour which corresponds to a room temperature change of ~3–5°C per hour when a collimator is installed (IPSM 1992). Should the crystal seal be damaged, moisture will ingress and damage the operating characteristics of the crystal. The crystal will become yellow, which will result in reduced light transmission to the photomultiplier tubes, PMTs. This increases the statistical uncertainty of detection, thus degrading spatial and energy resolution, and uniformity.

The rear face of the crystal must be sealed by a thin transparent covering to avoid moisture damaging the hygroscopic NaI(Tl), and to allow the transmission of light from the crystal to the multiple PMTs. The number and arrangement of these PMTs are where the camera differs dramatically from probe systems.

Light Guide

Between the crystal and the PMTs is a light guide. This is an optical plate, several centimetres thick, which must be optically coupled to the rear face of the crystal and to the front faces of the photomultiplier tubes. For good spatial linearity and uniformity, the light from each interaction must be distributed among a number of PMTs. This was traditionally achieved using thick light guides but these reduced the total light collection for each interaction, and again led to increased statistical uncertainty of detection. To obtain the best spatial resolution, therefore, the light guides must be as thin as possible. These can now be used, without causing spatial distortion and non-uniformity (IPSM 1992), by employing digital energy and linearity correction circuits and by shaping the light guide between PMTs.

The optical coupling is done by direct bonding or via silicone optical fluid or grease, which must be free of air bubbles. Any deterioration in this coupling will reduce transmission because of total internal reflection, and will again lead to increased statistical uncertainty of detection, thus degrading spatial and energy resolution, and uniformity.

Photomultiplier Tubes

If the only interaction was photoelectric absorption, light emission corresponding to the total energy of the photon would originate from the point of interaction and spread throughout the crystal in an isotropic manner (Scrimger and Baker 1967). For a 140 KeV interaction, this is ~4000 light photons, see 'Light production' in Chapter II.1. The scintillation light photons, from a γ photon interaction, therefore reach a number of PMTs surrounding the point of interaction, and the amplitude of each output pulse is determined by the solid angle made with the point of interaction. Thus, the proportion of the light reaching the photocathode of each PMT varies inversely with the distance from the interaction. A photoelectron is produced for every ~10 light photons which reach the photocathode of a PMT. In total, therefore, ~100 photoelectrons are produced from a $^{99}Tc^{m}$ interaction, although this may be improved using PMTs with prismatic windows. The photoelectric pulse, in each PMT, is then amplified by $\sim 10^{7}$ to give the final output pulse. Even so, a pre-amplifier, attached to each tube, is required before this signal can be transmitted over any appreciable distance.

The PMTs are square, round or hexagonal and vary in size from 5–7.5 cm. Between 37 and 94 are currently used in a detector head; arranged in a close packed, usually hexagonal, array to ensure a minimum separation. Better spatial linearity and intrinsic spatial resolution are achieved with a larger number of smaller PMTs. The former because detection sensitivity varies, depending where on the PMT window the light photon enters. When this is moved from the edge of a PMT towards its centre, the light collection efficiency increases more rapidly than the distance moved. Thus, a light photon entering near the centre of a PMT appears even nearer the centre. This causes the image of a line moving across a PMT to be bowed towards its centre; i.e. pincushion distortion over a PMT and barrel distortion between PMTs.

The gains of all the PMTs and pre-amplifiers, in an array, must be matched so that uniform output pulse heights are obtained for the same light input, and the energies of incident γ photons can be correctly identified. Incorrect gain will be seen as reduced detection sensitivity in the area of the PMT, because some output pulses will be too small to be accepted by the pulse height analyser and some valid γ photon interactions will therefore not be recorded. In the extreme case of PMT failure, there will be a complete absence of counts. Rapid changes in PMT gain are caused by fluctuations in ambient temperature and in the high voltage supply. Slow changes occur because of different aging rates in each tube, which are particularly apparent in the first 6 months of use. Periodically, the deterioration in uniformity must be corrected by acquiring an energy correction matrix, see 'Correction techniques – Energy' later in this chapter.

Processing Logic

The output pulses, from all the PMTs, are processed to provide three values using 'Anger' type logic (Anger 1964). The first, z, represents the total amount of light collected from the scintillation, and is related to the energy deposited in the crystal by the incident γ photon. The other two are x and y. These represent the spatial location of the scintillation, i.e. γ photon interaction, on the crystal.

z Value, Energy

The total light collected, by the many PMTs surrounding the point of interaction, represents the energy of the incident γ photon and to determine this the outputs of all the PMTs are summed. For a number of incident γ photons, this produces a composite photopeak which should have an energy resolution, FWHM, <10% at 140 keV. This is affected by the small gain differences between individual PMTs. Those with gains lower than the average contribute to the low side of the composite photopeak and those with gains higher than the average contribute to the high side. Further processing of the summed PMT output pulse is required to ensure that its amplitude is linearity related to the incident γ photon energy and not dependent on the spatial location of the interaction in the crystal.

x and y Values, Position

The PMTs closest to a scintillation event will collect the greatest amount of light and will therefore produce output pulses of the greatest amplitude. By summing the output pulses of the PMTs in vertical and horizontal halves of the field of view, thus creating four quadrants x^+, x^-, y^+, y^-, it is possible to determine the exact location of the scintillation event. This is the Anger type logic (Anger 1964) and is derived from the relative amplitudes, of the x^+ and x^- summed pulses and the y^+ and y^- summed pulses, being proportional to the distance of the scintillation event from the centre line of the crystal. However, the number of photoelectrons, produced in each PMT, will vary randomly about an average value, which will result in the x and y signals varying randomly about the true value. Thus, the image of a point will be blurred into a disc, determining the intrinsic spatial resolution of the camera.

The calculated positional values must be corrected for the energy of the incident γ photon. For instance, low energy γ photons which generate smaller amounts of light per scintillation event would produce smaller summed output pulses. Without correction, these would translate to smaller positional signals and thus smaller images than a higher energy photon. A linearity correction also has to be imposed on the calculated positional information. As mentioned above, because of the variation in sensitivity of light collection with geometry in the PMT, the output pulses are not perfectly linear with distance. This manifests as straight lines being bowed between tubes, which is one of the main causes of spatial non-uniformity.

Pulse Height Analyser

During an acquisition, the amplitude of the z pulse from each γ photon interaction is tested, by the PHA, to determine whether the energy of the incident photon is within the range of values allowed for the radionuclide being used. If the z amplitude is accepted, then the positional, x and y, information, for that interaction, is recorded; thereby excluding photons which have experienced Compton scatter in the imaged object as well as those from other radionuclides. More than one PHA acceptance window is sometimes used to produce a single image. This may be, for instance, in ^{67}Ga imaging in which images are acquired for 2 or 3 photon energies and are then amalgamated (see Fig. II.3.12), or in producing a subtracted image from two acquired at different photon energies.

Computers

Two computer functions are required: acquisition and processing. There is a tendency for these to be carried out by separate units connected by a high speed link. The acquisition unit tends to be relatively simple and is used to set up the patient details and acquisition parameters, control the gantry and imaging bed movement, and accumulate the image data. When digitised, the image is represented by a matrix of square picture elements, pixels, in which the activity distribution is translated to the number of counts in each pixel. The processing unit is more powerful than the acquisition computer and undertakes complicated data processing. If the three (x, y and z) signals from the camera are analogue, they are digitised by analogue to digital converters (ADCs), before being passed to the acquisition computer. There is a move for this digitisation to be accomplished at the PMT level, some cameras having one ADC for a group of PMTs and some having one ADC for each PMT. This improves data quality and allows flexibility in head design, software being used to implement the processing logic, including energy and linearity correction, and correction for the effect of different γ energies.

Whole Body Mechanism

In this, the camera or bed move, at a user specified constant speed, relative to each other to provide an image of the whole body as a single entity instead of as several separate 'spot' views. It requires the imaging bed or gantry to be motorised and controlled by the acquisition computer (see Fig. II.3.13). The x- and y- positional coordinates are adapted to represent the moving position of the camera. If the heads are not rectangular, activity accumulations at the middle of the scan width will be in the field of view longer than those at the periphery and the weighting given to each location must be adjusted depending on its position across the scan width. These are exposure time corrections. To restrict the distance that the camera or bed has to move, the field of view,

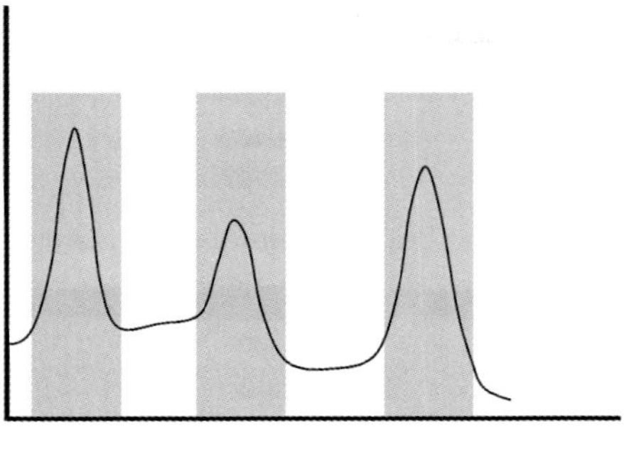

Fig. II.3.12. Three pulse height analyser windows, used to form one image, when the radionuclide emits photons of three energies

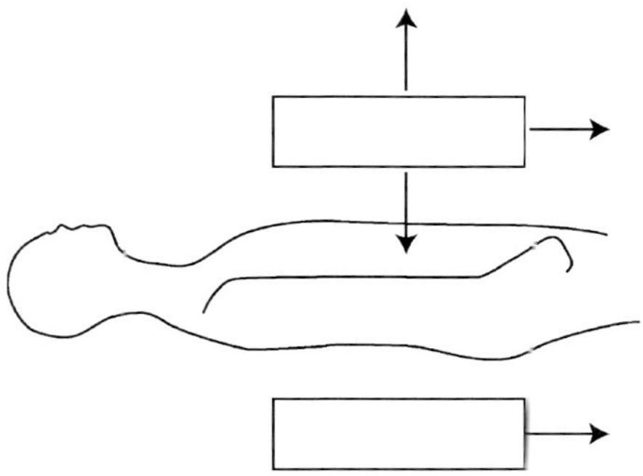

Fig. II.3.13. Whole body acquisition. Both detectors translate along the full length of the patient. The vertical position of the lower detector is determined by the bed, that of the upper detector is adjusted to the shape of the patient in order to minimise separation

along the direction of motion, may be expanded and collapsed at the start and end of the scan. The camera may describe a fixed or a variable vertical distance from the imaging bed. This will translate into a variable or a fixed vertical distance from the anterior surface of the patient. In the latter, the spatial resolution will alter with position along the patient. If the camera does not cover the width of the patient, more than one scan has to be acquired. In these cases, if the fields of view overlap, artefactually increased activity accumulations may be created (Blokland et al. 1997).

Head Configurations

The most obvious variation among γ cameras is the number and the configuration of the heads: single head, dual head fixed at 180°, dual head fixed at 90°, dual head with variable orientation, triple head fixed at 120°, and triple head with variable orientation. The single heads are round, square or rectangular whereas the multiple heads tend to be rectangular, and most rectangular head designs would have the long detector axis across the patient.

Single Head

This is the basic configuration for general purpose single photon planar and tomographic (SPET) imaging (see Fig. II.3.14). These cameras are very versatile for planar imaging but suffer poor sensitivity for SPET.

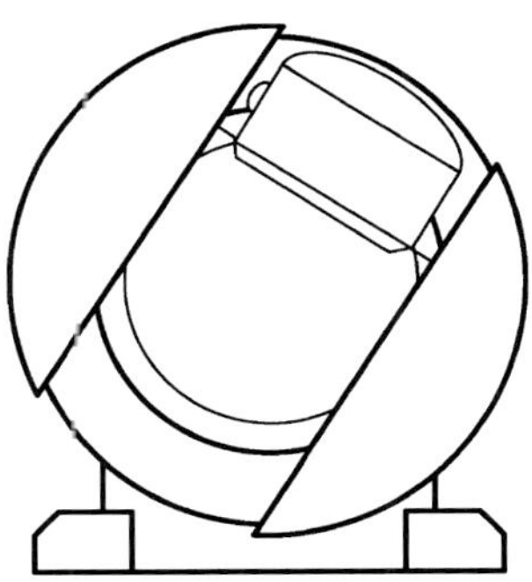

Fig. II.3.14. Single head configuration

Dual Head Fixed at 180°

The dual heads are opposed at a fixed angle of 180°, and are designed primarily for whole-body imaging and general purpose, non-cardiac, SPET (see Fig. II.3.15). The sensitivity is double that of a single head system but this configuration has limited flexibility in general purpose planar imaging. It can be used for dual photon imaging, however.

Dual Head Fixed at 90°

The dual heads are orientated at a fixed angle of 90° (see Fig. II.3.16). They are designed primarily for cardiac SPET but brain SPET can also be performed, and they are not suitable for planar imaging. Sensitivity is again double that of a single head system.

Dual Head with Variable Orientation

The dual heads can be set at any relative orientation and this is the most versatile of designs (see Fig. II.3.17). It facilitates the full range of planar and SPET acquisitions and, in the 90° orientation, cardiac SPET. Sensitivity is again double that of a single head system. It can also be used for dual photon imaging.

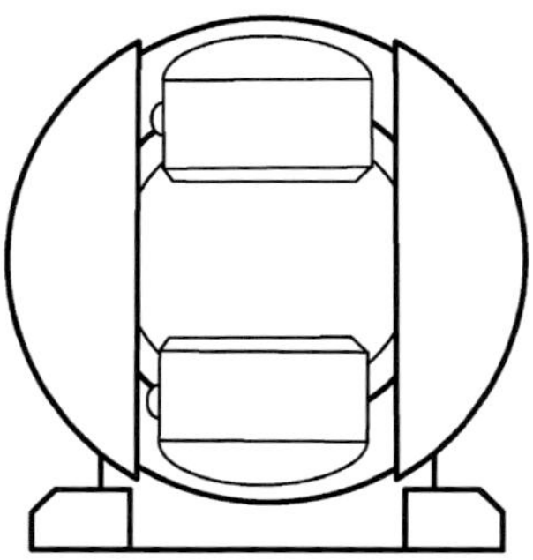

Fig. II.3.15. Dual Head fixed at 180° configuration

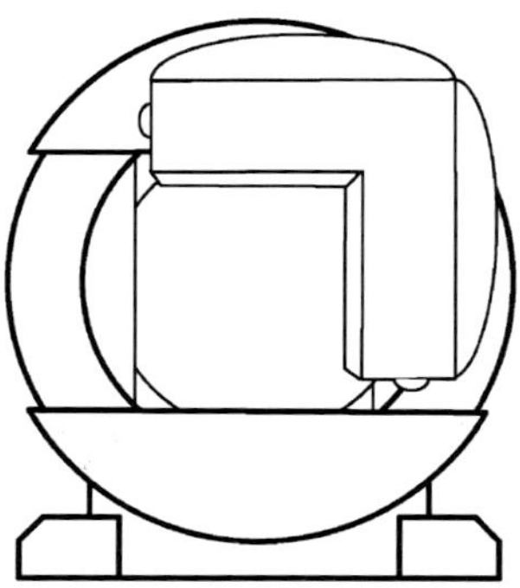

Fig. II.3.16. Dual Head fixed at 90° configuration

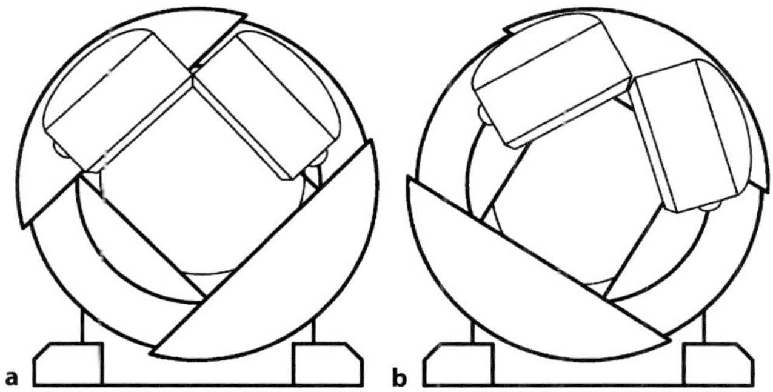

Fig. II.3.17. Dual Head with variable orientation configuration. (**a**) set at 90°, (**b**) set at <90°

Triple Head Fixed at 120°

The sensitivity of this configuration (see Fig. II.3.18) is higher than that of the dual head systems. Although particularly suited to brain, cardiac and general SPET. it can also be used for dual photon imaging.

Triple Head with Variable Orientation

This is the most recently introduced configuration, which seems particularly suitable for SPET and dual photon imaging.

Performance Characteristics

The performance of a γ camera is characterised by six parameters: energy resolution, spatial linearity, uniformity, spatial resolution, sensitivity, and count rate. Most of these are interrelated, however, e.g. a good proportion of non-uniformity being caused by spatial distortion. The performance of the camera varies across the field of view, becoming better towards the centre, and it is therefore divided into three regions. The largest is the geometric or full field of view (GFOV or FFOV). Within this is the useful field of view (UFOV) which is 95% of the GFOV and the central field of view (CFOV)

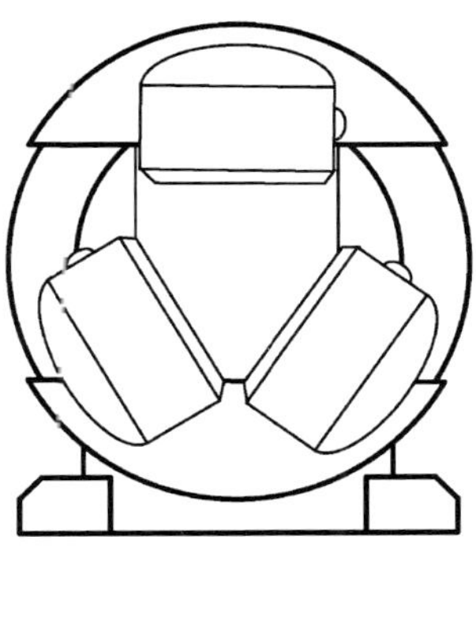

Fig. II.3.18. Triple Head fixed at 120° configuration

which is 75% of the GFOV (see Fig. II.3.19). The variation in performance values among these regions is reducing, however, with digitisation before the x, y and z values are established, and especially with the provision of one ADC for each PMT.

Energy Resolution

This is described in Chapter II.1 and should be less than 10% at 140 keV. The smaller the value of energy resolution, the better the system can be at excluding scattered photons, while maintaining detection sensitivity for unscattered photons. This parameter determines the width of the pulse height analyser window that can be implemented and consequently detection sensitivity and contrast.

Spatial Linearity

The accuracy, of the calculated x and y positions of photon interactions in the crystal, is affected by both random, or statistical, and systematic errors; resulting in interactions being recorded at the wrong position and imposing spatial distortions on the image. The systematic errors comprise major spatial miscalculations and, more subtle, spatial misplacements resulting from detection sensitivity variations over the PMTs. Two effects are involved: pincushion distortion moves interactions detected near the centre of a PMT, further towards the centre; and barrel distortion moves interactions detected between PMTs, further away from the centre of the nearest PMT (IPSM 1985). The result is that straight lines, on the object, appear as curved lines on the image. Non-linearities are also caused by non-uniformities in the light guide. Spatial non-linearity is quantified as the maximum spatial displacement over the field of view, a small value corresponding to high linearity. Also spatial positioning may be dependent on the energy of the incident photons. For a radionuclide, such as ^{67}Ga which is imaged using three PHA windows, this can cause loss of intrinsic spatial resolution. Multiple window spatial registration is a measure of this degradation (NEMA 1994).

Uniformity

The detection sensitivity across the field of view should be constant, and non-uniformity is an indication of any variation in this. It comprises two components: intrinsic

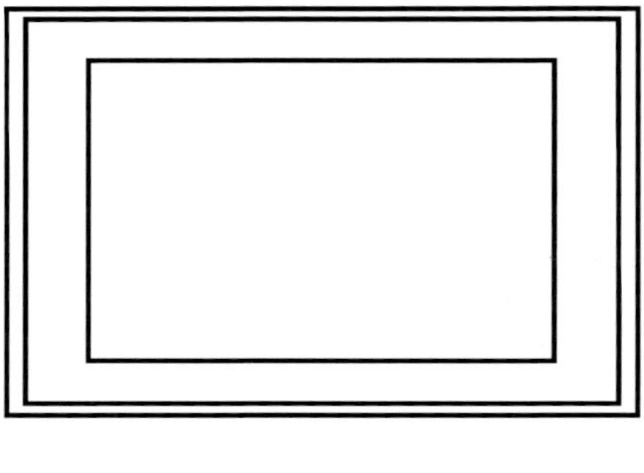

Fig. II.3.19. The divisions of the field of view showing the outermost geometric or full field of view (GFOV or FFOV) and moving inwards to the useful field of view (UFOV) and the central field of view (CFOV)

and collimator which together make up the total or system non-uniformity, and will vary therefore, depending on which collimator is loaded. It can be appreciated using a uniform activity distribution.

Quantification

Non-uniformity is quantified as integral, which is the maximum variation in count density over the whole field of view or as differential, which is the maximum rate of change of count density over a limited distance designed to approximate the size of a PMT (NEMA 1994). Small values correspond to high uniformity, and the values of differential non-uniformity are always less than those of integral non-uniformity.

Integral non-uniformity (%) is defined as (IAEA 1991; NEMA 1994):

$$IU = 100 * (C_{max} - C_{min})/(C_{max} + C_{min}) \tag{II.3.1}$$

Although it is sometimes also given by:

$$IU\,(+) = 100 * (C_{max} - C_m)/C_m \tag{II.3.2}$$

$$IU\,(-) = 100 * (C_{max} - C_m)/C_m \tag{II.3.3}$$

where the following are pixel counts in the field of view being considered:

C_{max} maximum
C_{min} minimum
C_m mean

Differential non-uniformity (%) is defined as:

$$DU = 100 * (C_{hi} - C_{low})/(C_{hi} + C_{low}) \tag{II.3.4}$$

Although it is sometimes also given by:

$$DU = 100 * (C_{hi} - C_{low})/(C_{hi}) \tag{II.3.5}$$

This calculation is performed over the field of view, by finding the 5 contiguous pixels, in any row or column, which has the highest count difference; C_{hi} is the highest and C_{low} the lowest pixel count in these 5 pixels.

The coefficient of variation, of all the pixel counts in the field of view, indicates of the overall variation in the uniformity, and is given by:

$$COV = 100 * \sigma/C_m \tag{II.3.6}$$

where σ is the standard deviation of the pixel counts.

Intrinsic

Currently, values of intrinsic non-uniformity for $^{99}Tc^m$ should be less than those shown in Table II.3.2. Both spatial distortion over the field of view and, to a lesser extent, variations in the positions of the individual PMT photopeaks contribute to regional non-uniformities (Graham et al. 1995). Pincushion distortion produces areas of apparently increased activity over the centre of the PMTs and barrel distortion produces areas of apparently decreased activity between the PMTs (IPSM 1985). Subtle regional uniformity variations are caused when the chosen PHA window does not overlie the photo-

Table II.3.2. Intrinsic non-uniformity values for $^{99}Tc^{m}$, in percent

Integral		Differential	
CFOV	UFOV	CFOV	UFOV
2.5	3.0	1.5	2.0

peaks of all the PMTs, a result of gain changes in the individual PMTs. The detection sensitivity of those which are mismatched, will be reduced and this will worsen as the mismatch increases, with the corresponding areas on the image demonstrating erroneously low count densities. As the chosen PHA window width is narrowed, this effect becomes more pronounced.

Another pattern of non-uniformity is a bright ring around the edge of the image, the edge effect, which is caused by two mechanisms. The first is a higher efficiency of light collection towards the periphery of the crystal resulting from internal reflection. The second is the lack of opposing PMTs beyond the edge of the crystal which causes interaction positions to be displaced towards the centre of the field of view. The peripheral, ~1–2 cm, of the image is masked to eliminate this effect (IPSM 1985; O'Connor 1996).

Collimator

Degradation in uniformity due to collimator effects is caused by non-uniform construction across the collimator and varies with the particular collimator installed on the camera. The value is measured as system uniformity and includes the intrinsic component. Currently, values of system non-uniformity for ^{57}Co should be less than those shown in Table II.3.3.

Whole Body

The uniformity of a whole body image depends, additionally, on the drive mechanism and system calibration. Irregular motion causes artefacts that are apparent on visual inspection but are difficult to quantitate, partly because the counting statistics required for a quantitative assessment would be difficult to achieve with clinically relevant scan speeds (NEMA 1994).

Table II.3.3. Values of system non-uniformity for ^{57}Co, in percent

Integral		Differential	
CFOV	UFOV	CFOV	UFOV
4.0	5.5	2.5	3.0

Spatial Resolution

This describes the error in establishing the precise location of the incident photon interaction, and the ability to resolve two separate points or lines as separate entities (IPSM 1992). Spatial resolution can be quantified as the FWHM and FWTM of the activity profile across a point or a line source (see Fig. II.3.20), called the point spread function (PSF) and line spread function (LSF). The FWHM is determined primarily by the diameter and length of the collimator holes, and the FWTM indicates the degree of septal penetration (Jarritt and Acton 1996). These figures do not provide an indication of the smallest object which can be seen, however, because one smaller than the spatial resolution of the camera may be detected, if its contrast is high enough (IPSM 1992). It is therefore possible for two detectors to have equal spatial resolutions and for one to perform better than the other. The intrinsic spatial resolution (R_i) of the detector and the geometric spatial resolution (R_c) of the collimator combine to give the system spatial resolution (R_s), where $R_i \ll R_c$:

$$R_s^2 = R_i^2 + R_c^2 \qquad \text{(II.3.7)}$$

Intrinsic

Currently, values of intrinsic spatial resolution, in terms of FWHM, for $^{99}Tc^m$ should be near to 3.5 mm for 9.5 mm thick crystals. This is established primarily by the statistical fluctuation in the number of light photons detected by each PMT and hence the light conversion efficiency of the NaI(Tl) crystal, the standard deviation being $\sqrt{n}$ where n is the mean. Intrinsic spatial resolution degrades as the incident γ photon energy decreases, because lower energy γ photons produce fewer light photons, resulting in larger relative statistical fluctuations; intrinsic spatial resolution being ~ proportional to $1/\sqrt{E_\gamma}$, where E_γ is the incident γ photon energy. Thus, using a higher energy γ photon will improve the intrinsic spatial resolution and a lower energy photon will degrade it. This sets an ultimate limit, of ~2 mm FWHM, for $^{99}Tc^m$ (NEMA 1994).

Increasing the crystal thickness will also cause a slight deterioration in intrinsic spatial resolution, by increasing the possibility of multiple Compton scattering before photoelectric absorption (see Fig. II.3.21). Following such an event, if the scattered

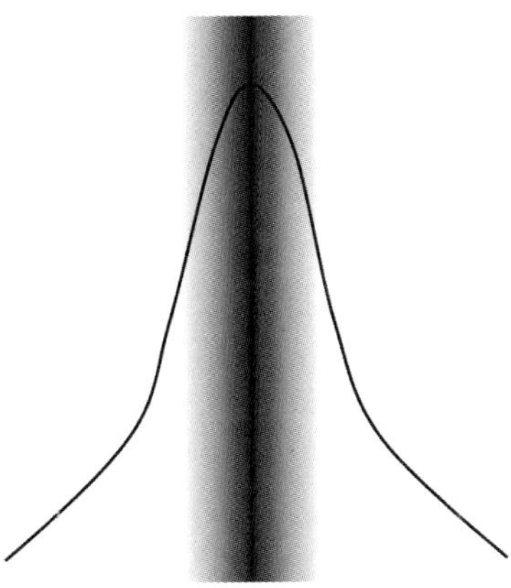

Fig. II.3.20. The spatial resolution of a γ camera can be appreciated by imaging a line source whose diameter is much smaller than the spatial resolution of the camera. The image appears as a blurred line. The count profile across this allows quantification as FWHM or FWTM

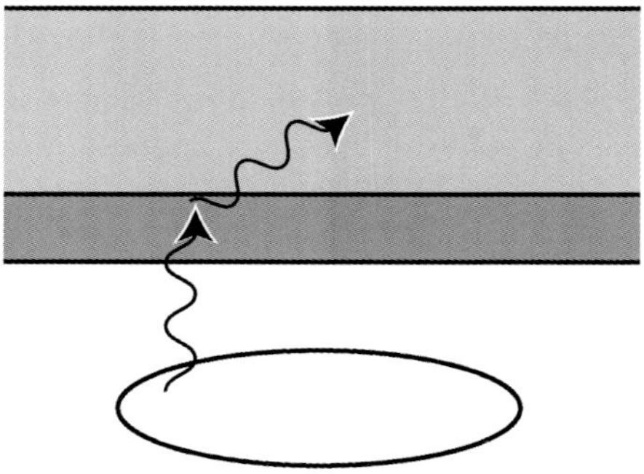

Fig. II.3.21. Multiple interactions in thick crystals. The photon experiencing Compton scattering would escape a thin crystal but stands the chance of being absorbed in a thick crystal

photon remains within the crystal, it may be absorbed elsewhere within the crystal by the photoelectric effect. In this case, the light will be emitted from multiple points and the interaction will be localised between them, displaced from the original point of interaction (IPSM 1992). This degradation in intrinsic spatial resolution in thicker crystals, is worse with low energy photons, which are absorbed within the first few millimetres, because there is a greater spreading of the scintillation light before it reaches the PMTs, which introduces more inaccuracy into the localisation.

Improvements in intrinsic spatial resolution have been achieved, partly as a result of improvements in camera uniformity, occasioned by effective non-uniformity correction techniques. This has overcome the restriction that satisfactory uniformity could only be achieved at the expense of degraded intrinsic resolution, using relatively thick light guides and large diameter PMTs. With the current correction techniques, thinner light guides and larger numbers of smaller PMTs can be used to estimate the interaction positions more accurately and achieve better intrinsic spatial resolution (NEMA 1994).

System

The combination of intrinsic spatial resolution and collimator geometric spatial resolution results in values of system spatial resolution for $^{99}Tc^{m}$ shown in Fig II.3.22. As well as the intrinsic spatial resolution being degraded by the collimator, system spatial resolution also deteriorates as collimator to source separation increases. Objects are imaged much better, the closer they are to the collimator face. At a separation of as little as 5 cm, which is the typical imaging depth, system resolution is dominated by collimator resolution.

Whole Body

As in the case of uniformity, the system spatial resolution of a whole body image is additionally dependent on the performance and alignment of the drive mechanism, and most problems associated with this will affect the spatial resolution. The resolution in the direction of motion, is dependent on the drive mechanism and system calibration, and across the direction of motion it is affected mainly by the alignment of the camera and the table (NEMA 1994).

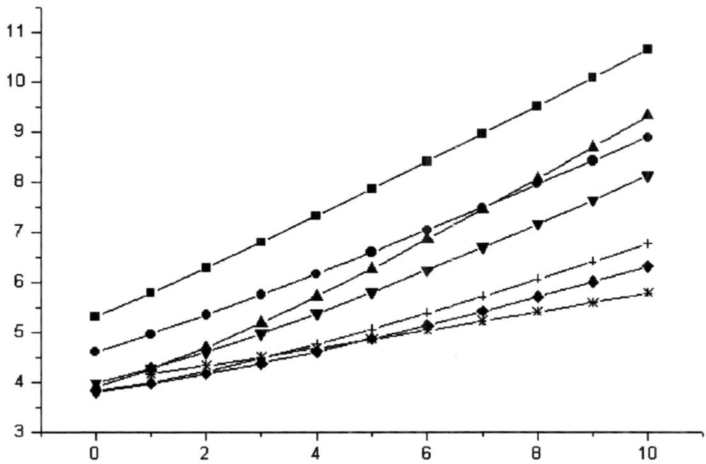

Fig. II.3.22. System spatial resolution with separation for $^{99}Tc^{m}$, calculated, as the FWHM in millimetres, by combining the values already presented for geometric collimator spatial resolution and assuming an intrinsic spatial resolution of 3.5 mm. The ordinate represents FWHM in millimetres and the abscissa represents separation in centimetres. The system values are higher than the individual geometric and intrinsic values and deteriorate with separation, but non-linearly. ■ HEGP, ● MEGP, ▲ LEHS, ▼ LEGP, + LEHR, ◆ LEUHR, × pinhole

Sensitivity

This is the probability of detecting an incident photon, and is quantified as the count rate obtained from a standard flat source at a distance from the camera. System sensitivity is a combination of intrinsic (S_i) and collimator (S_c) sensitivity, with $S_i \gg S_c$. Intrinsic sensitivity increases with increasing crystal thickness and PHA window width, and decreases with increasing photon energy; a 9.5 mm thick crystal being nearly 100% efficient for photon energies <100 keV (Jarritt and Acton 1996). System sensitivity depends on the collimator used. Current values, for LEGP collimators, should be near to 130 counts s^{-1} MBq^{-1}.

Count Rate

Count-rate performance, with increasing activity, has two components: linearity of observed count rate and accuracy of spatial positioning. The activities used clinically are unlikely to cause a significant degradation in image quality but quantitative analysis of the images could be compromised by count rate non-linearity (IPSM 1985).

Count Rate Non-Linearity

With increasing count rates, there is an increasing probability of more than one photon interaction occurring almost simultaneously, in which case the multiple interactions will be interpreted as being a single interaction. The scintillation light pulses will be summed, and the resultant energy signal (z) may be rejected as being out with the PHA window (see Fig. II.3.23). Thus, count loss, and hence non-linearity between observed count rate and activity, will occur. The probability of such pulse pileup is determined by the count rate over the whole energy spectrum, and not just within the photopeak, because all interactions must be processed before being tested by the PHA. A major determinant, of the count rate capability of the detector, is the dead-time. This is a period, initiated by the start of an event processing, during which other events cannot be processed. It is determined by both the resolving time, τ, and the time taken for x, y and z processing; τ depending on the duration of the scintillation pulses and the proportion of this allowed to be used for light collection. The apparent dead-time depends on the window fraction, which is the count rate in the PHA window as a proportion of the count rate in the total spectrum; the apparent dead-time increasing as the window fraction decreases. Scatter increases dead-time because it decreases the window fraction, the smallest effect being when the radionuclide is uniformly distributed and scatter is minimised, the largest when a point activity source is in a scattering medium. At clinical count rates, typical dead-times are $5–10 \times 10^{-6}$ s and losses are not serious. At higher count rates, many cameras behave as paralysable systems, however (IPSM 1992). In these, the occurrence of an interaction during the dead-time period will not just result in the event being undetected, but will initiate a new dead-time, in effect, extending the dead-time. At very high count rates, this results in a reduction of observed count rate with increasing activity (see Fig. II.3.24).

Spatial Misplacement

As well as count rate non-linearity, pulse pileup can cause image artefacts between sources of very high activity. If simultaneous interactions originate from these, the positioning logic will only localise a single event at a position which satisfies the summed

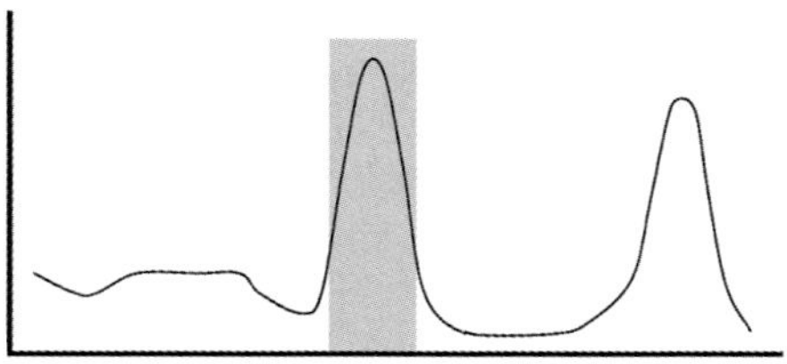

Fig. II.3.23. Rejection of counts at high count rate. Two photon interactions occurring almost simultaneously will appear as a single interaction of higher energy. This will be out with the pulse height analyser window, shown shaded, and will be rejected

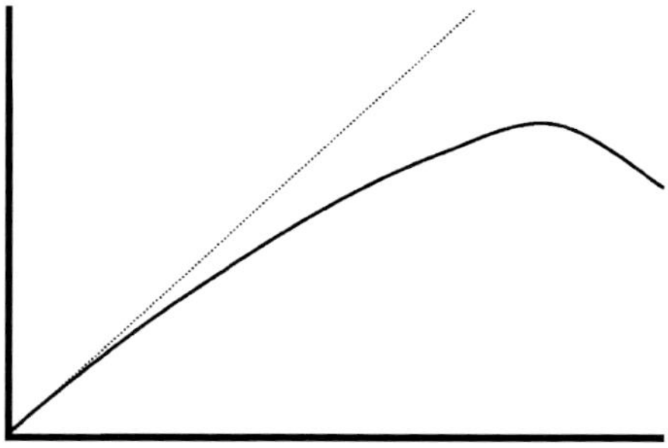

Fig. II.3.24. Count rate non-linearity. The ordinate represents photon interactions that are recorded, the abscissa represents photon interactions that actually occur. As the count rate increases, more photon interactions are not detected

scintillation light pattern. If the interactions were partial energy absorptions by Compton scattering, the summed energy signal may be accepted by the PHA and an erroneous event recorded between the sources. The effect is a loss of image contrast and spatial resolution, which can be avoided by shielding any very high activity areas from the detector.

High Count Rate Mode

Early digitisation improves count rate capability. In addition, pulse pile-up can be reduced by detecting and separating two overlapping scintillation pulses by examining pulse tails. This reduces the integration duration, however, and may cause a deterioration in spatial resolution by reducing the number of light photons recorded per scintillation.

Correction Techniques

A number of the performance characteristics can be improved by correction techniques.

Energy

Energy correction is applied, intrinsically, to correct the variations in photopeak positions that occur, over the field of view, because of gain changes in the PMTs. The detector is exposed to a uniform activity distribution, and a matrix of slight photopeak energy shifts is determined; such that, at all the matrix points, equal count rates are obtained in two narrow windows set on either side of the ideal photopeak. This ensures that the photopeaks, at all points across the detector, are symmetrically positioned between the two windows and will coincide with the chosen PHA window. An energy correction matrix must be acquired for each radionuclide in use, because the errors in photopeak position may not remain constant as the incident photon energy varies.

Spatial Linearity

Spatial linearity is calculated by imaging straight lines, aligned along the x and y axis of the detector. A matrix, of correction factors, is created using the deviation between the true and the indicated positions. This is applied intrinsically, to correct the spatial mispositioning of every detected x and y position.

Uniformity

The energy and linearity corrections reduce intrinsic non-uniformity to ~2–3% and produce an image which is essentially free of spatial non-linearities. A previously dominant scheme of count skimming; adding or subtracting pixel counts over the field of view, has been relegated to a third stage in the correction process; because it does not attempt to correct the causes of non-uniformity. It requires the acquisition of a uniformity correction matrix for each radionuclide used.

References

Anger HO (1964) Scintillation camera with multichannel collimators. J Nucl Med 5:515–531
Blokland JAK, Camps JAJ, Pauwels EKJ (1997) Aspects of performance assessment of whole body imaging systems. Eur J Nucl Med 24:1273–1283
Connolly LP, Treves ST, Davis RT, Zimmerman RE (1999) Pediatric applications of pinhole magnification imaging. J Nucl Med 40:1896–1901
Darambara DG, Todd-Pokropek A (2002) Solid state detectors in nuclear medicine. Q J Nucl Med 46:3–7
Dilsizian V, Bacharach SL, Khin MM, Smith MF (2001) Fluorine-18-deoxyglucose SPECT and coincidence imaging for myocardial viability: clinical and technologic issues. J Nucl Cardiol 8:75–88
Fleming JS, Alaamer AS (1996) Influence of collimator characteristics on quantification in SPECT. J Nucl Med 37:1832–1836
Goris ML (1998) Predictions for nuclear medicine in the next decade. Radiology 208:3–5
Graham LS, Fahey FH, Madsen MT, van Aswegen A, Yester MV (1995) Quantitation of SPECT performance: report of task group 4, nuclear medicine committee. Med Phys 22:401–409
IAEA (1991) Tecdoc 602 quality control of nuclear medicine instruments. International Atomic Energy Agency, Vienna
IPSM (1985) Report 44 an introduction to emission computed tomography. Institute of Physical Sciences in Medicine, London
IPSM (1992) Report 66 quality control of gamma cameras and associated computer systems. Institute of Physical Sciences in Medicine, London
Jarritt PH, Acton PD (1996) PET imaging using gamma camera systems: a review. Nucl Med Commun 17:758–766
Metzler SD, Bowsher JE, Greer KL, Jaszczak RJ (2002) Analytic determination of the pinhole collimator's point-spread function and RMS resolution with penetration. IEEE Trans Med Imaging 21:878–887
Moore SC, Kouris K, Cullum I (1992) Collimator design for single photon emission tomography. Eur J Nucl Med 19:138–150
NEMA (1994) Standards publication NU1 performance measurements of scintillation cameras. National Electrical Manufacturers' Association, Washington
O'Connor MK (1996) Instrument- and computer- related problems and artifacts in nuclear medicine. Semin Nucl Med XXVI:256–277
Rosenthal MS, Cullom J, Hawkins W, Moore SC, Tsui BMW, Yester M (1995) Quantitative SPECT imaging: a review and recommendations by the focus committee of the Society of Nuclear Medicine Computer and Instrumentation Council. J Nucl Med 36:1489–1513
Scrimger JW, Baker RG (1967) Investigation of light distribution from scintillations in a gamma camera crystal. Phys Med Biol 12:101–103

Single Photon Tomographic Imaging

Abbreviations

AM arithmetic mean
COR centre of rotation
CT computed tomography

EM	entropy maximisation
GM	geometric mean
FBP	filtered back-projection
FWHM	full width at half maximum
IFBP	iterative filtered back-projection
LE	low energy
ML-EM	maximum likelihood expectation maximisation
OS-EM	ordered subset estimation maximisation
PHA	pulse height analyser
RMS	root mean squared
ROI	region of interest
SPET	single photon emission tomography

Planar images suffer a reduction in image contrast because of contributing information from activity in over- and underlying parts of the body. Tomographic images, produced by Single Photon Emission Tomography (SPET), overcome this by spatially separating the object activity distributions (Tsui 1996), see Fig. II.4.1. This spatial separation also facilitates improved spatial localisation of activity distributions. The technique also has the potential to quantify the regional distribution of activity and thus better indicate organ function.

The operation of SPET systems differs in several respects from the planar systems described previously. One of the main differences is that, because the final image is created using a considerable amount of processing and bears little obvious relationship to the acquired images, visual quality assurance is extremely difficult. To consistently achieve optimal quality in the final images, therefore, it is essential to appreciate the principles involved in order to implement the appropriate acquisition protocols and to identify any artefacts in the data (Groch and Erwin 2000). There are several techniques for undertaking the reconstructions which have to be performed to obtain the tomographic images; each of which involves various operating parameters, presents different advantages and disadvantages, and imposes restrictions on the accuracy of the reconstruction, which may be alleviated by adding a number of correc-

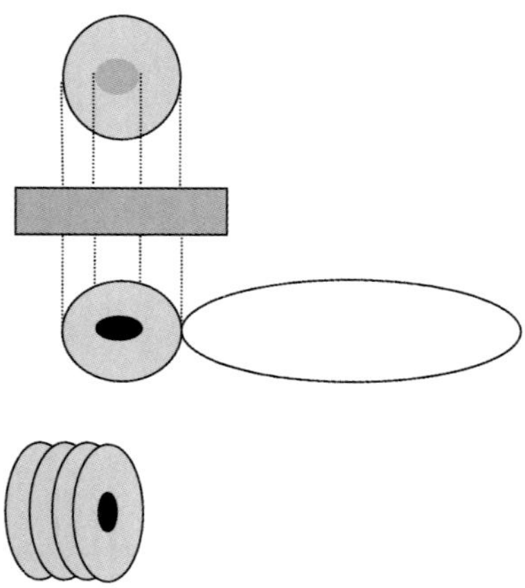

Fig. II.4.1. Comparison of planar SPET images. Image contrast is compromised in planar images, represented at the top of the figure, because of contributing information from activity in over- and underlying parts of the body. Image contrast is improved in SPET images because this information is spatially separated, represented at the bottom of the image

tions. Because visual quality assurance is extremely difficult, a number of factors concerning the quality of the instrumentation, peculiar to this modality, have to be considered to ensure that the systems are working properly. Finally when good quality reconstructed data has been obtained, it may be necessary to reorient it, in order to improve the diagnostic sensitivity.

Gamma Camera Systems

The principle of SPET imaging, is that transaxial slices are reconstructed from a series of planar images, called projection images, acquired sequentially by a γ camera rotating around the patient (see Fig. II.4.2). The images acquired are affected by all the planar limitations explained in Chapter II.3. The motion of the camera around the patient can present a hazard and touch sensitive sensors are required for the protection of the patient. Data is collected over a rotation of 360° for a complete acquisition or over 180° for a restricted acquisition, with rotation in either clockwise or anticlockwise direction.

SPET capable cameras are usually also capable of acquiring planar studies but may be restricted in their movement, such as not allowing caudal tilt. There is pressure for systems to comprise multiple, rather than single detectors, to increase acquisition sensitivity, since administration activity is constrained on radiological protection grounds. In such systems, the detectors acquire data independently and simultaneously, with their data usually being combined to simulate one detector performing a greater rotation. For instance, in a dual head system, a 360° data acquisition may be accomplished with each detector describing only 180°.

The projection images are either acquired as the camera rotates continuously or during regular stops in the rotation at constant angles (see Fig. II.4.3). The latter is a step and shoot technique and no data is collected during movement from one position to the next, which can take several seconds. Continuous acquisition involves a more complicated data management but increases acquisition sensitivity by avoiding the unproductive time between stops. It can also avoid judder in the detector, caused by the repeated stopping and starting. A hybrid combination of the two techniques is sometimes used (Cao et al. 1996). It is more appropriate for continuous acquisition to

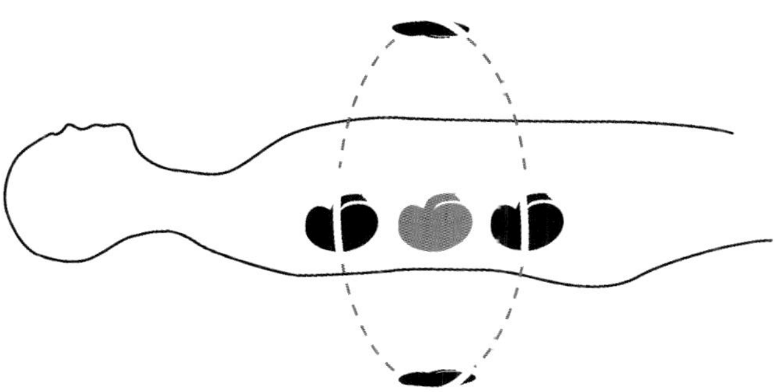

Fig. II.4.2. Sequential planar projection acquisitions at different orientations around the patient, which form the basis of SPET imaging

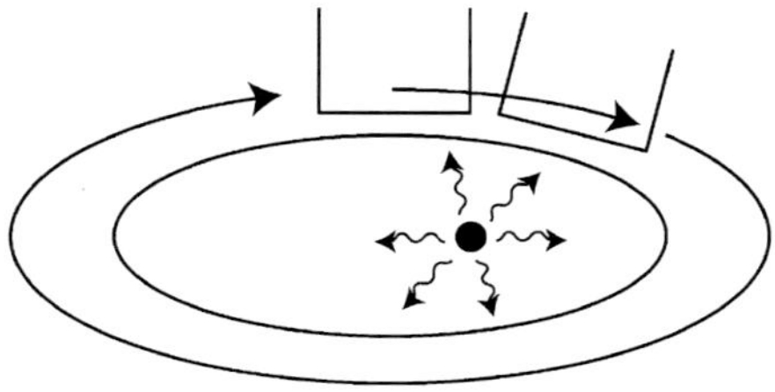

Fig. II.4.3. The γ camera moves around the patient to acquire the planar projection data necessary for SPET reconstruction

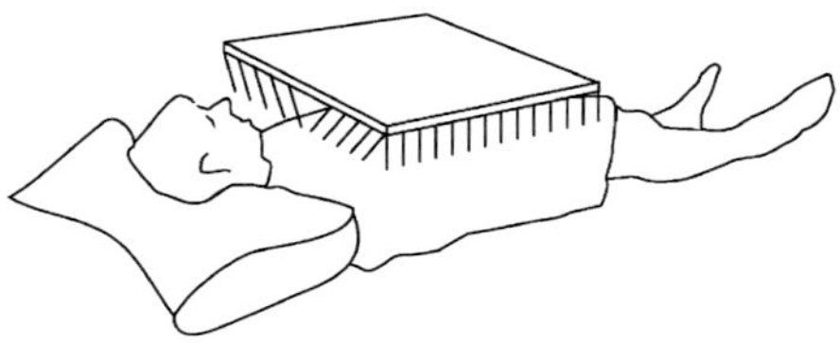

Fig. II.4.4. Fan beam collimator showing holes converging across the patient but remaining parallel along the patient

replace the step and shoot technique, when a higher number of projections is involved, e.g. 120 projections over a rotation arc of 360° defining 3° increments between projections. In this case, continuous acquisition will impose little spatial degradation on the acquired data because the angle between projections is so small, but 120 × ~3 s, i.e. 6 minutes, can be removed from the acquisition duration or used for acquiring more data. However, a 60 projection 360° rotation would involve 6° increments between projections and some spatial degradation would be expected, but the time saved would be 60 × ~6 s, i.e. 6 minutes. The total movement time is reduced when multiple heads are used, for instance from 6 minutes to 3 minutes for a dual head system, because of the fewer projections required for each detector.

The collimators usually used are the parallel hole type, because they produce spatially undistorted images. However, improvement of spatial resolution with depth can be achieved using the fan beam type (Mahowald et al. 1999), which is often used in brain and cardiac SPET imaging. In this, the holes converge across the patient but remain parallel along the patient, thus focussing to a line (Pareto et al. 2001), see Fig. II.4.4.

Descriptors of Image Quality

The quality of tomographic images is described by a number of parameters additional to those used in planar imaging.

Noise

When a tomographic image is reconstructed, the behaviour of the image changes significantly (Rosenthal et al. 1995), e.g. it no longer obeys Poisson statistics, as it does in the planar image. As well as depending on the count density and its distribution, the noise now also depends on the reconstruction process and any corrections used, such as: attenuation, scatter and resolution recovery. It also depends on the smoothing filter. The noise in each reconstructed pixel is no longer independent of that in its neighbours. The reconstruction process causes the noise to be positively correlated at short distances and negatively correlated at longer distances, which means that either more or less noise will be present than predicted by Poisson statistics and a standard deviation calculated for the pixels inside a uniform ROI will depend on the size and shape of the ROI (Rosenthal et al. 1995).

The level of the noise can, however, be estimated from the standard deviation of the counts in an image, reconstructed from a uniform activity concentration (IAEA 1991). The level of noise is indicated by the signal to noise ratio and, in general, the greater the smoothing effected by the back-projection smoothing filter (see 'filtered back-projection' in 'reconstruction techniques'), the better the apparent signal to noise ratio will be. However, this is at the cost of degraded spatial resolution. For a reconstructed image of a uniform activity distribution, the root mean squared (RMS) uncertainty is:

$$\text{RMS uncertainty} = f \times (\text{counts per pixel})^{3/4} / (\text{total counts})^{1/2} \qquad \text{(II.4.1)}$$

where
f = 120 for iterative reconstruction
f >120 for filtered back projection, depending on the filter shape

When the activity distribution is not uniform but concentrated in a particular area, the RMS uncertainty is reduced:

$$\begin{aligned}\text{RMS uncertainty} = f \times (\text{total counts})^{1/4} / \\ (\text{average counts per pixel in area of interest})^{3/4}\end{aligned} \qquad \text{(II.4.2)}$$

If the activity is concentrated in one pixel, then this becomes the usual $1/\sqrt{N}$ multiplied by a factor 1.2 to represent errors in the reconstruction. These equations can be used to predict how many events must be collected for a given uncertainty. To achieve a 10% uncertainty, assuming that the activity is distributed throughout 400 pixels, 10^6 counts must be collected (Tung and Gullberg 1994; Budinger 1996, 1998).

Contrast

Tomographic contrast quantitates the appreciation of a difference in activity concentration in the reconstructed image. It is dependent on the size of the area of interest, and particularly on the relation of this to the tomographic spatial resolution.

Uniformity

Tomographic uniformity assesses the uniformity of an image, reconstructed from a homogeneous activity concentration.

Spatial Resolution

Within slice tomographic spatial resolution describes the sharpness of the reconstructed image, and is defined as the FWHM of a reconstructed point within the slice (see Fig. II.4.5). It is dependent on planar system spatial resolution and so will alter with collimator type. It is always worse than the planar value, but in an well adjusted and operated system should not degrade by more than 2 mm or 10% (IAEA 1991). It will also be affected by acquisition errors, such as vibration, and by the acquisition parameters. The latter includes the radius of rotation, if a parallel hole collimator is used, because of the deterioration in planar system spatial resolution with separation.

Thus, tomographic spatial resolution will vary over the reconstructed image. This is particularly apparent with a 180° acquisition rotation, when the worst spatial resolution will be at the most distant part of the reconstructed image. In a 360° acquisition rotation, there should be little difference in spatial resolution between the horizontal and vertical directions at the axis of rotation for a circular acquisition rotation. For non-circular rotations, particularly in the presence of scatter, there may be differences. At the periphery of the field of view, the radial spatial resolution should be similar to that at the axis of rotation but the tangential spatial resolution should be better. Such variations can cause difficulty in quantitation.

Further influences on spatial resolution are processing parameters such as the back-projection smoothing filter, spatial resolution being progressively degraded with increasing smoothing (IPSM 1985), which is described in Chapter III.3. Between slice tomographic spatial resolution describes the sharpness of the reconstructed image along the patient, over a number of reconstructed slices. It is defined as the FWHM of a reconstructed point along the axis of rotation. To calculate this, the reconstruction must be undertaken using a slice thickness of one pixel, because otherwise larger thicknesses will become the dominant factor.

Sensitivity

Tomographic slice sensitivity is the number of counts, per unit activity, in a reconstructed slice obtained from a cylinder containing a uniform activity concentration. Similarly, tomographic volume sensitivity is the number of counts, per unit activity, in a number of reconstructed slices. These are dependent on a number of factors, including: detector configuration, collimator type, radionuclide type, energy window setting, source configuration (NEMA 1994).

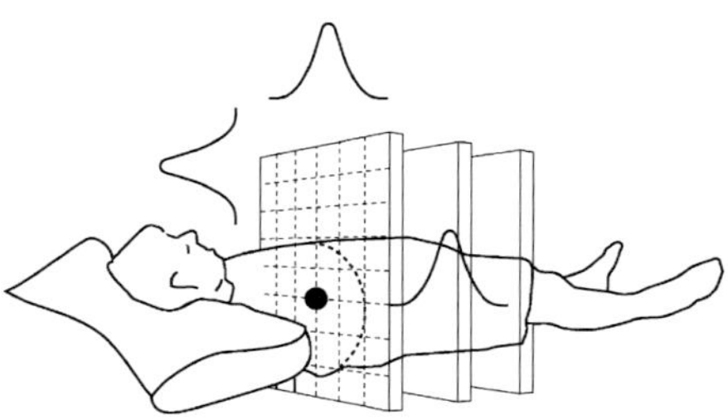

Fig. II.4.5. Tomographic spatial resolution is described by the FWHM of count profiles through a reconstructed point source. These are shown within a transaxial slice in the horizontal and vertical directions, and between transaxial slices

Linearity of Tomographic Activity Estimation

The linearity, of the system, describes the relationship between the counts in the reconstructed image and the object activity. It does not guarantee accuracy or precision, but is important for comparing regions of different activity concentrations, in the reconstructed images. Both scatter and attenuation are significant causes of non-linearities.

Quantitative Accuracy of Tomography

This documents the similarity between an activity estimated, from the reconstructed image, and the object activity. It is affected by several factors, including: scatter, attenuation, reconstruction filter, slice thickness, point spread function and partial volume effects. The precision is the reproducibility with which such an activity is repeatedly estimated.

Reconstruction Techniques

Transaxial images are created from the projection images by reconstruction, making the assumption that only activity along a line, perpendicular to the detector, contributes to the counts detected in the corresponding projection image. The transaxial slices do not necessarily have the same matrix size as the projection images. Two popular techniques are: back-projection and the iterative method; the latter requiring more computation but facilitating easier implementation of reconstruction corrections (Bruyant 2002).

Back-Projection

The traditional method is back-projection. At the start of the reconstruction process, all the transaxial slices are devoid of information. For each projection image, the line of pixels, across the patient and corresponding to the transaxial slice to be reconstructed, is identified (see Fig. II.4.6). For each pixel in this line, the counts are back-projected along the projection ray, perpendicular to the collimator face (see Fig. II.4.7). These counts are placed in all those pixels, of the transaxial matrix, which are intersected by the projection ray. Because this is oriented at an angle to the matrix, the intersection may not correspond to a whole pixel, and counts may have to be assigned by interpolation to adjacent pixels.

This process is repeated for all projection images. Because there is no information about the distribution of radioactivity along the projection ray, all intersected pixels are assumed to contain equal counts; consequently, most are allocated erroneous values which form a non-uniform background activity on the reconstructed image. As the projection images are back-projected, the counts superimpose at various locations and an approximation to the source distribution is generated. Some degradation in spatial resolution is imposed because of the separation of the reconstructed pixels from the detector. If projection images are acquired over a 360° rotation, each projec-

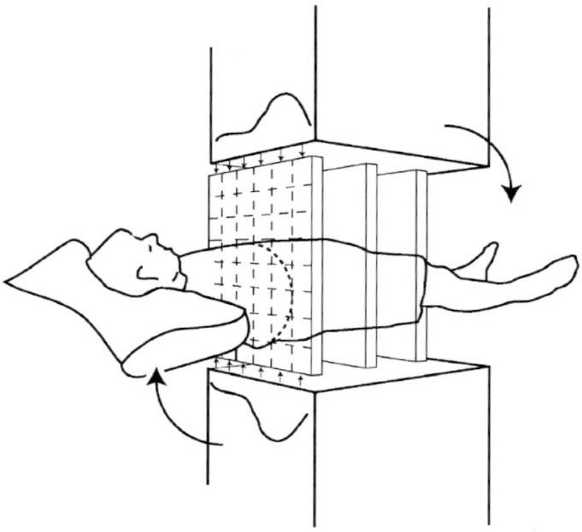

Fig. II.4.6. Each projection image contains information concerning a number of adjacent transaxial slices, which is back-projected to from the transaxial slices

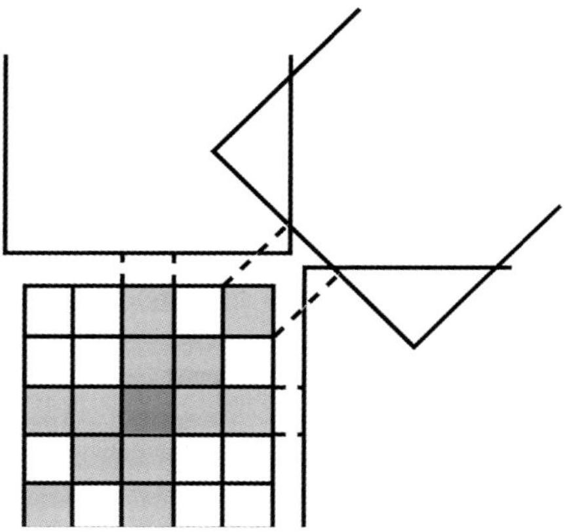

Fig. II.4.7. Reconstruction by back-projection involves the counts in each pixel in the projection images being projected onto the transaxial image along the projection ray, perpendicular to the collimator face

tion ray will receive contributions from opposite directions (see Fig. II.4.8), and the spatial resolution will be degraded over the whole image. The data can be combined using two methods, in which the resultant pixel counts are determined by taking either the geometric (GM) or the arithmetic (AM) means. The AM is usually preferred because, although the GM generally gives slightly more uniform resolution with position and has some advantages when imaging single localised sources, it leads to 'streak' artefacts when there are multiple sources (IPSM 1985). If projection images are ac-

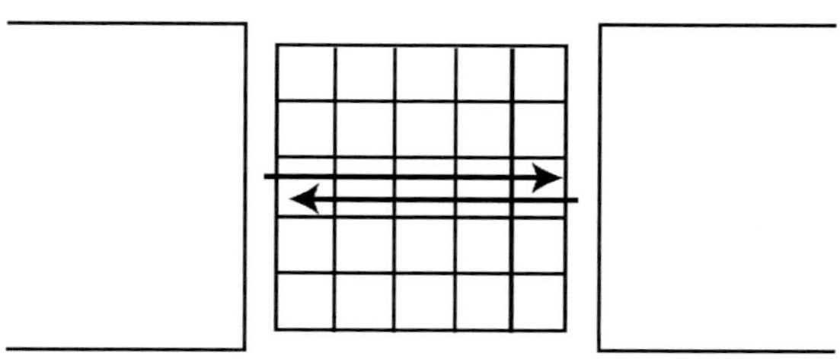

Fig. II.4.8. Construction of a transaxial slice with projection rays whose data was acquired from opposite orientations

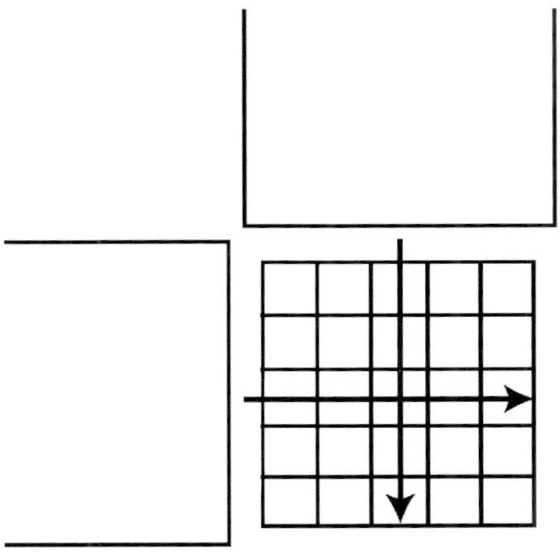

Fig. II.4.9. Construction of a transaxial slice with unopposed projection rays

quired over a 180° rotation, there is no ability to add information to the distal end of the projection ray (see Fig. II.4.9) This means that spatial resolution, further from the collimator face than the centre of rotation, is worse than when data over 360° is available.

Filtered Back-Projection

A modification of the back-projection technique is to filter the projection images before they are back-projected; to remove the erroneous background activity, introduced as a consequence of the assumption of uniform counts along each projection ray. This appears as a star pattern, around each focus of activity accumulation, with as many spikes as there are projection images (see Fig. II.4.10). When there are many projection views, the individual spikes become less apparent. They can be removed by back-pro-

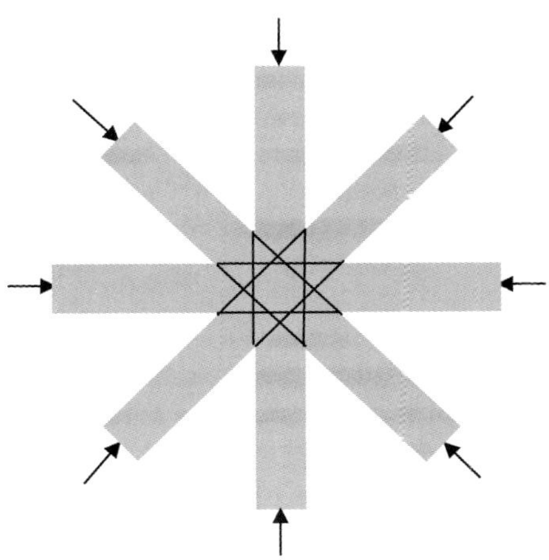

Fig. II.4.10. Star pattern, formed around each focus of activity accumulation, during the back-projection process as a consequence of the assumption of uniform counts along each projection ray

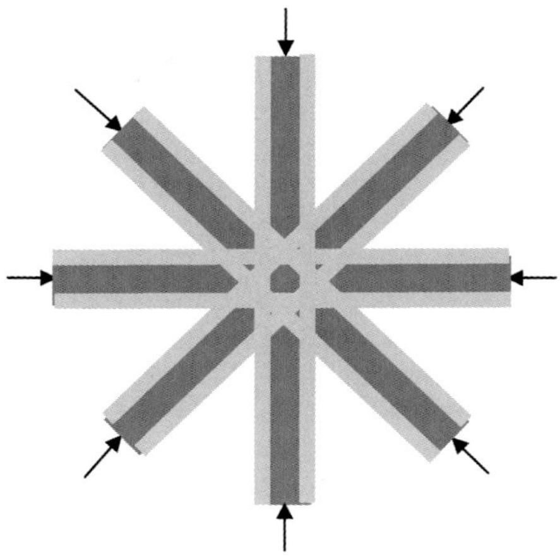

Fig. II.4.11. Reduction of star pattern artefact by applying a back-projection filter, achieved by back projecting negative values along lines adjacent to the projection rays

jecting negative values along lines adjacent to the projection rays, i.e. by filtering the projection data (see Fig. II.4.11). This filter must always be used in back-projection. Its characteristics can be derived by considering a point of activity, reconstructed from an infinite number of projection images. In this, the counts are maximum at the point and decrease, away from the point, as the inverse of the distance (1/r). In frequency space, this can be represented as: amplitude = 1/f, where f is frequency. The filter which will remove this effect is: amplitude = f. When this is used to filter the reconstructed data, all frequencies in the image will have an amplitude of 1. This is the description of an ideal point source. The filter, described as amplitude = f, is a ramp filter shown in Fig. II.4.12.

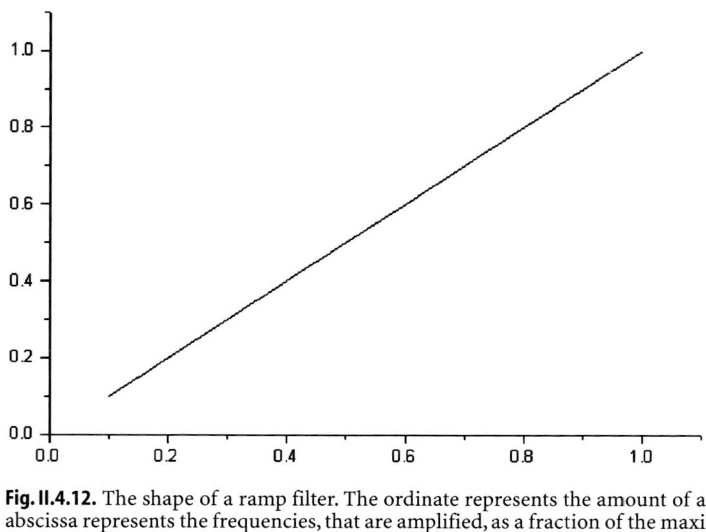

Fig. II.4.12. The shape of a ramp filter. The ordinate represents the amount of amplification and the abscissa represents the frequencies, that are amplified, as a fraction of the maximum frequency. The effect is that higher frequencies undergo higher amplification

Such a filter will amplify the high frequencies in the reconstructed image, producing a result in which the noise is much worse than usually seen in planar images. This is a particular problem when the projection images do not have a large number of counts and random noise is high. A second, smoothing filter, must be applied to control this noise, by setting an upper limit to the frequencies that will allowed through the reconstruction. A number of smoothing filters are available such as: Butterworth, Hamming, Hann, Parzen, Shepp-Logan, the most common being the Butterworth and Hamming. The restorative filters Metz, and Wiener are also used (Hansen 2002). These are described in Chapter II.7. Filtered back-projection only works properly with complete and reasonably consistent data (IPSM 1985).

Iterative Techniques

These involve solutions of linear equations by successive approximations. They demand more computation than filtered back-projection, but offer advantages when the data are incomplete or less consistent (IPSM 1985), in that noise is less pronounced in background regions and streak artefacts are not produced. The improvement in image quality is particularly apparent in low count density reconstructions (Vandenberghe et al. 2001).

An initial activity distribution, in the transaxial slice to be reconstructed, is assumed (Bruyant 2002). This is either completely uniform or is a distribution obtained from a filtered back-projection reconstruction. In the former case, it is often a non-zero uniform image that the same total projection counts as the measured projection data (Tsui et al. 1998). In the latter case, only a ramp filter need be used; a smoothing filter not being required because high frequency noise will be attenuated as part of the iterative reconstruction. The only initial guess that seems to work well with the ML-EM algorithm (see 'statistical reconstruction' below), however, is a flat image (Rosenthal et al. 1995).

The initial activity distribution, in the transaxial slice, is then forward projected as estimated projection images at each acquisition angle (see Fig. II.4.13). The estimated projection images are compared with the actual projection images, and the differenc-

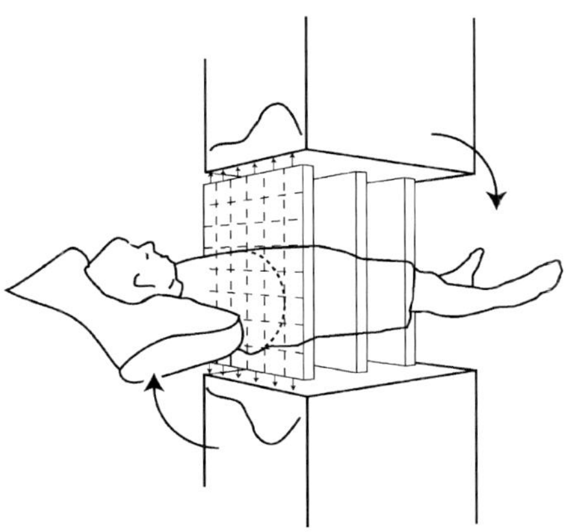

Fig. II.4.13. Iterative reconstruction. The activity distribution, in the estimated transaxial slice, is forward projected to be compared with the projection image at each acquisition angle

es used to suggest adjustments to be made to the transaxial slice to better approximate the imaged activity distribution. The projection, comparison and adjustment sequence is repeated a number of times as the iterations of the calculation. This iterative process is repeated until a required accuracy is achieved.

Another advantage is that, unlike in filtered back-projection, corrections can be incorporated directly into the reconstruction calculations, for instance for non-uniform attenuation, scatter correction, depth-dependent resolution recovery of the effects of collimator blurring (Pareto et al. 2001; Bouwens et al. 2001; Vandenberghe et al. 2001). This is because the forward projection needs an estimate of the distribution of, for example, attenuation throughout the transaxial slice so that an accurate estimate of the projection images can be calculated, and thus correction for non-uniform attenuation is easy to incorporate at the forward projection step (IPSM 1985).

Two categories of methods are available for undertaking the iterative calculations: iterative filtered back-projection and statistical reconstruction (Galt et al. 1999).

Iterative Filtered Back-Projection

These algorithms very rapidly approach a final image and further iterations cause no significant change (Galt et al. 1999). An example is the Chang iterative filtered back-projection (IFBP) algorithm (Tsui et al. 1998).

Statistical Reconstruction

These produce a reconstructed image with lower noise than the iterative filtered back-projection algorithms (Tsui et al. 1998).

Entropy Maximisation (EM)

The most common method used to compare the estimated projection images with actual projection images, and to make the adjustment based on the comparison, is the maximum likelihood method, often referred to as Entropy (or Expectation) Maximisation (EM) or Maximum Likelihood Expectation Maximisation (ML-EM). This converges to a transaxial image that has a maximum likelihood of having projections most similar to the actual projections (Patton and Turkington 1999). It converges slowly, requiring more iterations than the IFBP algorithms, but provides more control over image noise. There is no definite rule on how many iterations are required and this has to be determined by experience (Galt et al. 1999). It may require 50–200 iterations. This slow convergence rate is a major disadvantage and the OS-EM algorithm (see below) has been proposed to accelerate the reconstruction process (Bruyant 2002). This provides similar image quality and noise characteristics but much faster convergence rates compared with the ML-EM algorithm, e.g. up to 15–30 times faster, in as few as 2–5 iterations (Tsui et al. 1998; Zaidi and Hasegawa 2003).

Ordered Subset Estimation Maximisation (OS-EM)

Ordered Set (or Subsets) Entropy (or Expectation) Maximisation (OS-EM) algorithm is similar to the ML-EM algorithm except that the projection data are divided into (n) subsets, each containing 2 or more projections arranged in a specific order. For example, opposite projection images may be associated (Tsui et al. 1998; Patton 2000) or, if there are 64 projections, they may be organised into 16 subsets, each containing 4 images: from subset 1 with projections 1, 17, 33, 49 to subset 16 with projections 16, 32, 48, 64. Convergence is facilitated if each subset contains projections equally distributed around the patient (Bruyant 2002). The ML-EM algorithm is then performed on each subset, with the solution from the first subset being used as the start for processing of the second subset, and repeated until all subsets are processed. For data divided into n subsets, this algorithm is ~n times faster than the standard ML-EM algorithm (Tsui et al. 1998; Patton 2000; Bruyant 2002; Zaidi and Hasegawa 2003).

Limits on Reconstruction Accuracy

In comparison with planar imaging, only image contrast is improved by tomographic reconstruction and all other parameters of image quality are degraded. However, the improvement in image contrast improves diagnostic accuracy to an extent that more than compensates for the deterioration in the other parameters. The aim is to increase the contrast as much as possible whilst limiting the deterioration in the other parameters. The quality of the reconstructed images depends on a combination of the planar performance of the camera, its motion and the reconstruction algorithm. Differences in head tilt, gantry flexing, collimator characteristics, reconstruction algorithm and smoothing filter shape in filtered back-projection will also affect the contrast. Increases in PHA window width and a loss of energy resolution will produce a loss in image quality (Graham et al. 1995). Tomographic contrast deteriorates as the size of the object activity distribution decreases; depending also, to some extent, on the shape of the activity distribution. The effect is to reduce the apparent difference in activity concentration compared to the surrounding area. The software used to undertake the reconstructions has a significant effect of the accuracy of the reconstruction, varying among suppliers and different revisions from the same supplier (Jarritt et al. 2002).

Reconstruction Corrections

A number of corrections can be applied to the tomographic reconstructions to improve the images (Bouwens et al. 2001).

Decay

To avoid significant reconstruction artefacts, the radioactive distribution must remain constant throughout the acquisition. This will only be the case if the rotation duration <10% of the effective half-life of the radiopharmaceutical (Jarritt and Acton 1996). In routine investigations, therefore, it is not usually necessary to correct the projection images for radioactive decay, unless quantification of activity or comparison with a normal database is required. This is because, although apparent sensitivity will be lower towards the end of the acquisition rotation, it will have little discernable effect on the relative activity distributions throughout the reconstructed image. For instance, a typical acquisition duration of 30 minutes will only demonstrate a reduction of 6% by the end of the acquisition, if a biologically stable $^{99}Tc^{m}$ pharmaceutical is used. For longer half-life radionuclides the effect is negligible.

Attenuation

Attenuation correction compensates for counts lost due to attenuation, which would otherwise result in the reconstructed pixel count depending, not only on radioactive concentration, but also on the location of the activity distribution in the reconstructed image. This is because, photons pass through different thicknesses and densities of tissues and therefore are absorbed or scattered by different amounts before reaching the detector as it moves around the body (see Fig. II.4.14). It affects more the deeper parts of the object, which would otherwise appear to have been detected with decreased sensitivity compared to the periphery and, without correction, a hot rim would appear on the reconstructed image. Also without correction, the image would not be quantitatively accurate and volume information would be less accurate, because the edges of organs of interest are often best defined in terms of counts per pixel contour levels (IPSM 1985). Restoration requires a knowledge of the distribution of the attenuating tissue. Two techniques are available: analytical and transmission.

Analytical Methods

These assume a uniform attenuation distribution, and require a definition of the linear attenuation coefficient (μ_l) for the tissue being imaged and the body contour, which can be determined either manually or with automatic edge-detection methods (Zaidi and Hasegawa 2003). Correction has to be undertaken carefully, because small errors in the attenuation map can cause considerable inaccuracy in the reconstructed image (Cao and Holder 1998). Of the different methods of attenuation correction, the two most common are:

Pre-Reconstruction – Sorensen's Method

This is the simplest method and is valid only if acquisition is over a rotation arc of 360°. Each pixel in the projection images is multiplied by a factor calculated on the lin-

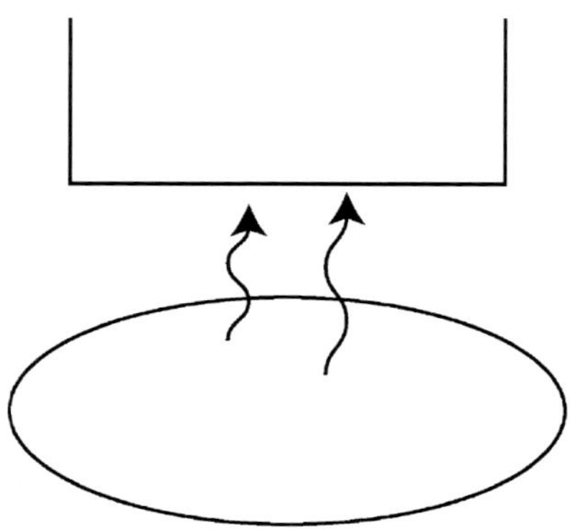

Fig. II.4.14. Attenuation of photons in body tissue; as also occurs with probe detectors (see Fig. II.2.8). Loss of photons by absorption or scattering becomes more likely as the photon path length and the tissue density increase

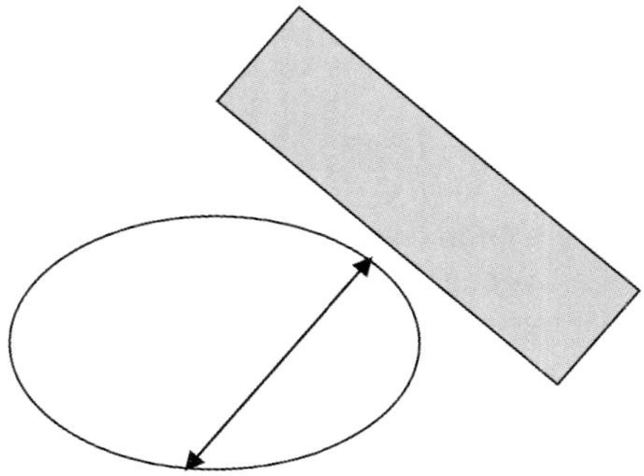

Fig. II.4.15. Projection ray length. The length of the projection ray, in tissue, is manipulated by the attenuation coefficient μ_l to calculate the attenuation factor that must be applied to each pixel in the projection images

ear attenuation coefficient μ_l and the length of its projection ray (L) in the body, see Fig. II.4.15 (Oppenheim and Appledorn 1988):

$$\mu_l L/(1 - e^{-\mu_l L}) \tag{II.4.3}$$

This factor is the fraction of photons which will escape the body and reach the detector and assumes a uniform distribution of activity along the projection rays, because this is initially unknown. The factor is different for the two cases when the arithmetic and geometric means of opposite projections are taken in the reconstruction. The correction gives acceptable results when applied to distributions of uniform activity concentrations containing small defects, similar to its initial assumption of uniform activity distribution along the projection rays. It gives poor results, however, for small sources distributed in the object, a situation dissimilar to that of its initial assumption (IPSM 1985).

Post-Reconstruction – Chang's Method

The expected detection efficiency for each pixel in the reconstructed image is calculated. Each pixel in the reconstructed image is then corrected according to its calculated detection efficiency. This is known as the first-order Chang correction. It produces a good correction for a single point source in an attenuating object but over-corrects for a distributed source, producing an artefactual central bulge in the image. A second-order correction can be used to overcome this (Oppenheim and Appledorn 1988).

Transmission Methods

The analytical methods do not produce good results when the attenuating medium is very inhomogeneous. This is particularly apparent in the thorax, because of the reduced attenuation of the lungs, and therefore significant in myocardial perfusion studies. To overcome this, an actual attenuation map of the object is created, by undertaking a transmission acquisition as well as the emission acquisition. There should be no spatial misregistration between the two, and they should be acquired using the same geometry and without patient movement between them (Patton 2000). When this is not the case, anatomic locations attenuation map do not correspond with equivalent positions in the emission data, and will cause artefactal hot or cold areas to be created in the reconstructed emission images (Zaidi and Hasegawa 2003).

A map of attenuation coefficients is created by reconstructing the transmission projections using filtered back-projection (see Fig. II.4.16). These are then used directly in iterative reconstruction of the emission images. These methods produce the best results but require additional set-up and scan time, and radiation exposure of the patient.

A number of techniques for transmission attenuation compensation are available. Most involve a radionuclide source, including: a scanning line or point source, multiple line source arrays, and a fixed line or point source. In these, the transmission photons are recorded by the emission detectors. The attenuation map must be as accurate as possible to avoid imposing errors during the correction, and there are significant differences in the quality of correction, among commercial systems (O'Connor et al. 2002). However, noise in the emission images remains more important than noise in the transmission images (Tung and Gullberg 1994). A significant problem can arise if the transmission energy is similar to, or slightly less than, the emission energy because of scatter of emission photons into the transmission energy window. This is considered to be responsible for a decrease in measured attenuation coefficients in the vicinity of an emission source, in large but not in small phantoms (Almquist et al. 2001). When idle, the transmission sources are shielded and, in the case of the scanning sources, parked. An alternative, to radionuclide sources, is an x-ray transmission source which requires its own detectors (DePuey and Garcia 2001).

Scanning Line Source or Point Source

Scanning line source systems extend across and translate along the patient and use the 97 and 103 keV photons from ^{153}Gd (see Fig. II.4.17). They are the most common (Zaidi and Hasegawa 2003) because of their implementation on dual detector systems in the 90° configuration, with parallel hole collimators, in which two sources are used with a total activity of ~4 GBq (Almeida at al. 1998). They are used with a spatially scanning PHA window, which follows the source movement, and which allows the estimation of the bidirectional crosstalk between the transmission and emission energy windows (DePuey and Garcia 2001). This produces high quality, narrow beam attenuation factors and offers a very flexible approach in that the radionuclides used can be of higher, lower or the same energy; with the crosstalk between the emission and

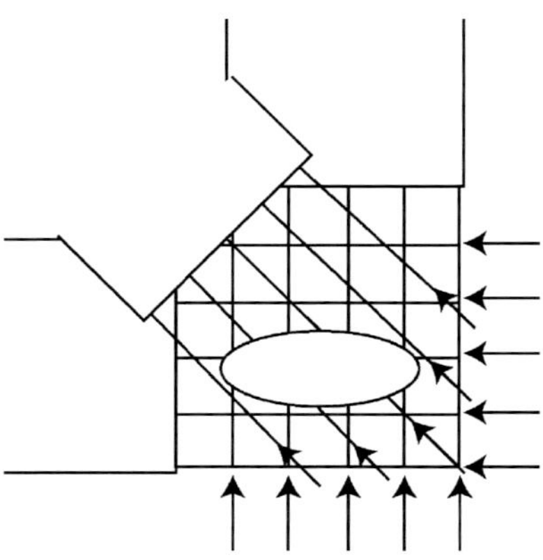

Fig. II.4.16. Principles of transmission attenuation correction. The transmission of photons through the body are detected as planar projection images. A map of attenuation coefficients is then created, in transaxial slices, by reconstructing the transmission projections using filtered back-projection

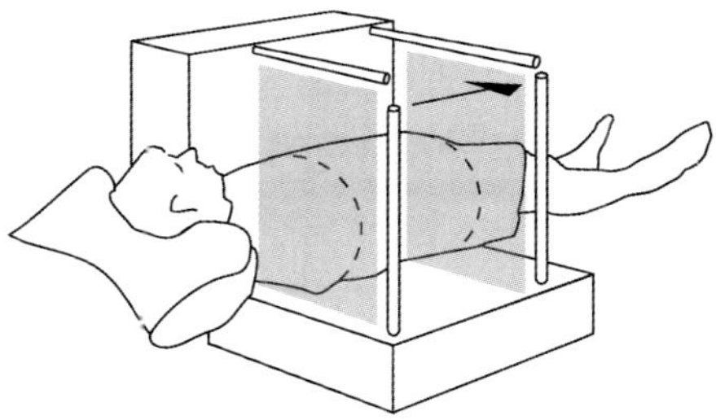

Fig. II.4.17. Scanning line source for transmission attenuation correction. Transmission photons are directed through a transaxial slice of the body. The sources translate along the patient, to cover the field of view, during the time that the detectors are stopped and emission data is being acquired

transmission windows being low at <5%. Because of the scanning PHA window, there is a ~10% reduction in emission count rate, however (Bailey 1998).

The scanning collimated point source system uses the 356 keV photons from ^{133}Ba, and relies on penetration of the collimator septa to obtain a uniform intensity, the high energy photons making it suitable for use with LE collimators (Hendel et al. 2002). Only scatter into the emission window need be estimated and corrected.

Flood Source or Multiple Line Sources

These irradiate the complete crystal during the transmission acquisition on 90° dual detector systems. The multiple line geometry (Celler et al. 1998) uses ^{153}Gd sources opposite the detectors (see Fig. II.4.18), again with parallel hole collimators, and gives a higher emission crosstalk count rate in the transmission window to transmission ratio compared to the scanning line source geometry. The crosstalk is measured using a split energy window about the ^{153}Gd photopeak (DePuey and Garcia 2001).

Fixed Line Source with Fan Beam Collimator

This technique uses the 59 keV photons from an ^{241}Am source on a triple head (Matsunari et al. 1998) or a dual opposed head system (see Fig. II.4.19). It, again, irradiates the complete crystal during the transmission acquisition and can encounter a problem with scattered emission photons. The geometry may be symmetric on triple head systems or asymmetric on dual and triple head systems (Gilland et al. 1998). The magnification of the fan beam, resulting in a reduced field of view, gives a high probability that the transmission images will be truncated for most applications apart from the head (Bailey 1998), which can be reduced using the asymmetric geometry. Truncation

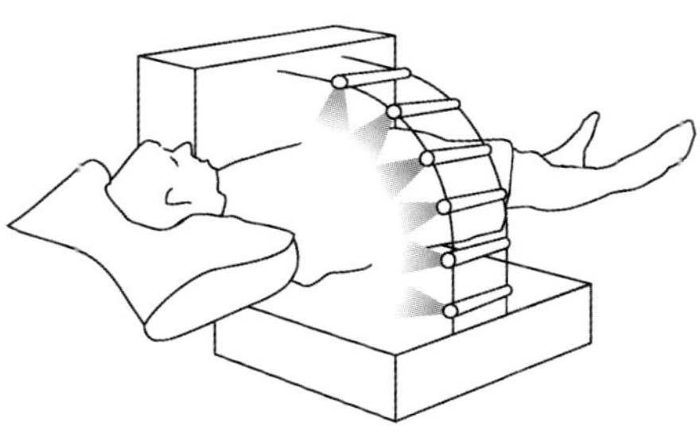

Fig. II.4.18. Multiple line sources for transmission attenuation correction. Transmission photons are directed through the body over the whole field of view simultaneously

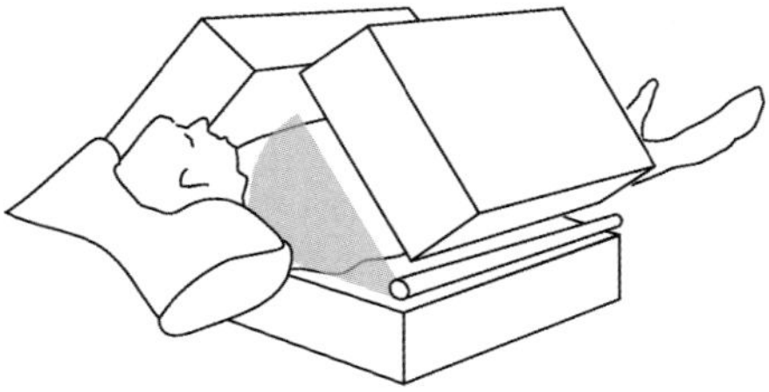

Fig. II.4.19. Fixed line source for transmission attenuation correction. Transmission photons are directed through the body over the whole field of view simultaneously but towards one fan beam collimator

means that there will be missing data at the periphery of the transmission image. In such a case, the attenuation map will be reconstructed from incomplete data and will not be accurate (Bailey and Meikle 2000) and artefacts can be generated in the reconstructed emission image unless corrected for (Galt et al. 1999). However, studies suggest that this only becomes a problem, clinically, for severe truncation of >40% (Gregoriou et al. 1998).

CT

Recently, a SPET system with incorporated CT scanner (Kashiwagi et al. 2002), mounted on a common gantry, has become available (see Fig. II.4.20). A transaxial anatomical transmission map is acquired as the system rotates around the patient similarly to a 3rd generation CT (Bocher et al. 2000). The fan beam is wide enough to include the full width of the patient and to produce slices 1 cm thick. Being precisely registered (Takahashi et al. 2002), it provides attenuation maps that can be used for anatomic correlation as well as attenuation correction (Bocher et al. 2000). Their quality is much less than diagnostic CT images, but is adequate and involves a radiation dose a factor of 10–20 less (Patton 2000); the noise and spatial resolution, of even low quality CT images, still being much better than can be obtained with a radioactive source (Turkington 2000).

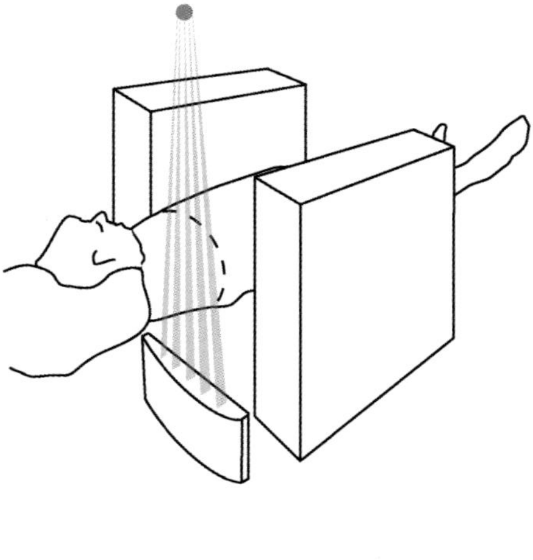

Fig. II.4.20. On-board or in-line CT mounted on a common gantry for transmission attenuation correction. A transaxial anatomical transmission map is acquired as the system rotates around the patient

Scatter

Scatter correction improves image contrast and is essential if absolute quantification is required. A number of correction techniques are available, based on the removal of a fixed proportion of the photons acquired in a pixel; because the limited energy resolution makes it impossible to identify scattered from non-scattered photons.

The earliest and most widely used method involves acquiring scatter images in additional narrow PHA energy windows, either just below e.g. 92–125 keV for $^{99}Tc^{m}$ (dual window technique), or on either side of the main photopeak (triple window technique, see Fig. II.2.7). The latter being used if photons, with energies higher than the target photopeak, are present. The assumption being that scatter recorded in these windows will be similar to the scatter present in the photopeak (Galt et al. 1999). The scatter images are multiplied by a normalising factor, determined experimentally or by computer simulation, and the normalised scatter images are then subtracted from the main image (IPSM 1985). This subtraction will result in an increase in image noise and smoothing is sometimes used before image subtraction. The normalising factor is not constant (Koral et al. 1990), however, and changes as a function of angle around the patient. These techniques are therefore, generally, considered unsuitable for quantitation and, for this, a modified dual-window technique incorporating a normalising factor which varies as a function of angle around the patient, has been used (Rosenthal et al. 1995).

An alternative is to determine the spatial distribution of the contribution of scattered relative to unscattered photons, and remove its effect on the reconstruction by deconvolution using restorative filters (Koral et al. 1990; Tsui et al. 1998). Scatter compensation can also be incorporated into the reconstruction process itself (Galt et al. 1999). This can be accomplished by reducing the values of the attenuation coefficients used for attenuation correction or by incorporating the scatter estimate into an iterative reconstruction algorithm (Tsui et al. 1998). If scatter correction is undertaken, subsequent attenuation correction using analytical methods should involve the narrow beam value of μ_l.

Spatial Resolution

Spatial resolution recovery attempts to overcome the effects of loss of planar system spatial resolution with separation, and can produce contrast and signal-to-noise improvement in the reconstructed image. This is accomplished by enhancing selected spatial frequencies using a deconvolution filter. This process is inappropriate with noisy projection images but, given a high enough count density, it is possible to calculate a filter which will enhance the lower frequencies and suppress the higher frequencies and thus provide some sharpening. Such a technique has the attraction that it can be applied without operator intervention and can be included in iterative reconstruction (Tsui et al. 1998; Galt et al. 1999).

Effect of Imaging Parameters on Image Quality

Whereas sub-optimal image quality, because of a poorly functioning or operated camera should be apparent in planar imaging, this is not the case in SPET imaging. It is possible to reconstruct poor images or images containing artefacts without this being appreciated (IAEA 1991). It is therefore vital to identify the imaging parameters that affect the reconstruction and to optimise them to achieve the highest quality image reconstruction.

Multiple Detector Systems

In multiple detector systems, the performance characteristics of each individual detector must be matched, including: the alignment along the axis of rotation to better than 2 mm, detection sensitivity and pixel size (IPEM 2003). Misalignment can create artefacts in the reconstructed images (DePuey 1994) and any such misalignment is usually minimised by corrections acquired on-site.

Pixel Size

The pixel size must be known accurately so that the appropriate matrix and acquisition zoom can be chosen, and also so that analytical attenuation correction can be implemented correctly; the latter involving calculation of the projection ray length.

Uniformity

The reconstruction process amplifies planar non-uniformities considerably (IPSM 1985) because any artefactual count variations, from intrinsic or collimator origin, appear in the same location on each of the acquired projection images (IAEA 1991), whereas true count distributions change position on the projection images (see Fig. II.4.21).

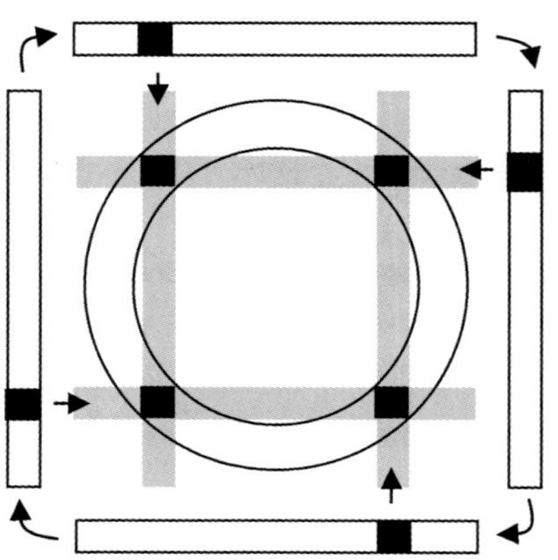

Fig. II.4.21. Principles of non-uniformity amplification in SPET reconstruction. A count artefact caused by a detector problem will appear in the same position on all projection images. When these are back-projected, artefactual rings will be created on the reconstructed images

The planar non-uniformities cause apparent rings (Leong et al. 2001) which are always concentric with the centre of rotation, but this may not correspond to the centre of the image if part of the image has been zoomed (O'Connor 1996). For circular acquisition rotations, the rings appear as concentric circles, with restricted 180° rotation arcs causing half circle artefacts. For non-circular rotations, the artefacts have a more complicated appearance. Thick rings fairly widely spaced are usually caused by camera non-uniformities whereas thin rings, narrowly spaced, can be generated digitally in the camera–computer interface or in the uniformity correction circuitry (IAEA 1991). In clinical studies, these artefacts may appear as incomplete rings which may give the appearance of abnormal activity distributions. The perception of these artefacts depends on the size and location of the non-uniformity in relation to the centre of rotation (O'Connor and Vermeersch 1991); the amplification being inversely related to the distance from the axis of rotation, and increasing towards the centre of rotation (Oppenheim and Appledorn 1988; IPSM 1985) because of the decreasing circumference of the rings. At the centre of rotation, a defect need only be small to create a very noticeable artefact, a 1% non-uniformity in the projection image will be amplified to ~20% non-uniformity in the reconstructed image (IAEA 1991), which may be seen as a discrete cold or hot spot.

The relationship between ring artefact magnitude and planar image non-uniformity is very dependent on the smoothing filter used (O'Connor and Vermeersch 1991) and the uniformity requirements are similar for both FBP and OS-EM reconstructions. Although the ring magnitude in OS-EM reconstructed images is ~1/3 that of FBP images, there is a corresponding reduction in image noise with OS-EM and the ratio of ring magnitude to image noise is relatively similar for both techniques (Leong et al. 2001).

Consequently, planar uniformity requirements are much more stringent for tomographic than planar imaging. Differential uniformity values are smaller than integral but it is the more important parameter for SPET considerations. This is because it describes the maximum non-uniformity found in any small region and is an indicator of focal abnormalities which cause ring artefacts. Also, system uniformity is the important parameter because a small collimator defect can cause reconstruction uniformity artefacts, even if the intrinsic uniformity is good. With multiple detectors, uniformity must be similarly good for each detector. The level of differential uniformity necessary in order ensure that ring artefacts do not affect interpretation depends on the particular investigation, and will be when any ring artefacts are less than the reconstructed image noise. This will vary with the type of study and the pragmatic approach is to ensure that ring artefacts will not be visible, in even the highest count clinical studies undertaken (O'Connor 1996).

Whereas in planar imaging, non-uniformity up to 5% is acceptable, the level suggested for projection images in SPET is 1% (Rogers et al. 1982; Rosenthal et al. 1995). Many current cameras, with intrinsic energy, linearity and flood corrections, can approach the limit (O'Connor 1996). Many systems cannot, however. For these count skimming, using a high count density planar system flood image, has to be employed. This procedure should not be used to compensate for cameras operating at less than peak performance or with serious collimator damage. Acquisition must be undertaken using the radionuclide, PHA window, collimator and scatter conditions expected during the SPET study. Great care must be taken in this procedure, because clinical studies contain large variations in count density, both within a given study and for different types of study. With very low count studies, there is a possibility that application of the uniformity correction map to the planar data can cause increased noise and ring artefacts in the reconstructed images (O'Connor and Vermeersch 1991). The most difficult task is to produce a sufficiently well mixed flood source to achieve better than 1% non-uniformity, which necessitates at least 10^4 counts/pixel, and translates to 30×10^6 counts in a 64×64 acquisition matrix and 120×10^6 counts in a 128×128 matrix.

Variations of Uniformity and Sensitivity with Rotation

It is possible that systems can experience variation in sensitivity and uniformity over the rotation arc. One of the causes of this is a gain shift of photomultiplier tubes because of orientation changes with respect to an external magnetic field. Inclusion of magnetic shielding has excluded this and current SPET systems should not experience significant effects if strong magnetic fields are not present in the vicinity, particularly since the smaller tubes used in current generation systems are easier to shield than the older larger ones. Other causes are change of thermal gradients in the detector as a function of angle (Graham et al. 1995) and poor optical coupling of a PMT to the light guide or crystal (O'Connor 1996).

Gantry Alignment

As the camera rotates it must view the same region in space. The only way to do this is if the crystal is parallel to this axis. Failure will result in blurring of the reconstructed images (see Fig. II.4.22). Alignment of the gantry, and hence the detector crystal, is paramount. For instance, a 1° gantry misalignment with the detector head at a 20 cm radius of rotation will cause a 3.5 mm displacement of image data along the axis of rotation (O'Connor 1996). The patient does not have to be aligned with the camera's axis of rotation although this will aid interpretation of the images by providing consistent display.

The gantry must be aligned perpendicular to the camera's axis of rotation. If not the crystal cannot be aligned parallel to this axis. The easiest way to set up a system is to

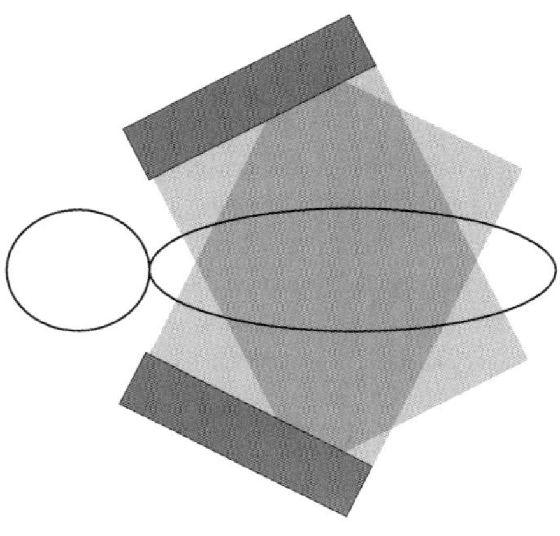

Fig. II.4.22. Principles of gantry misalignment. If the detectors are not parallel to the axis of rotation, they will image different regions during rotation causing blurring of the reconstructed images

make the axis of rotation perpendicular to gravity. The gantry should then be parallel to the force and the crystal perpendicular. A mechanism to help in achieving this is the use of self levelling floor material. A spirit level or plumb line can then be used to make sure the system is properly adjusted. There is a tendency for the spirit level to be placed on the back of the camera head or on the front of the collimator. This is only good enough if they are themselves in alignment with the face of the crystal.

Collimator Hole Angulation

The collimator holes must be precisely perpendicular to the axis of rotation throughout rotation; an angulation error of 0.6° will cause a 1% degradation in reconstructed spatial resolution (Malmin et al. 1990), therefore an angulation error of only 1° is allowed (Eckholt and Bergmann 2000), otherwise, spatial distortion and a loss of contrast in the reconstructed image, will result (see Fig. II.4.23). The effect of non-orthogonality across the patient is a varying centre of rotation, COR, offset at different radii of rotation, and the effect along the patient is equivalent to camera head tilt with a single projection line viewing more than one transaxial plane. Non-orthogonality may be caused by hole misalignment in the collimator and by incorrect mechanical movement of the detector head (Busemann-Sokole 1987).

Rotation Smoothness

In a step and shoot acquisition, the camera must be stationary when data is being acquired and must not be juddering during this period. In a continuous acquisition, the rotation must be smooth and must not vary in speed. Problems can be caused by cable or slip ring changes, and can result in significant variation in resolution, sensitivity, and uniformity as a function of rotation angle, which can degrade reconstruction resolution and uniformity (Hines et al. 1999a,b, 2000).

Centre of Rotation

The projection image matrix must be centred on the mechanical COR, otherwise there will be a loss of spatial resolution and uniformity with each reconstructed point being blurred with a circular motion (see Fig. II.4.24). This is because an activity distribution

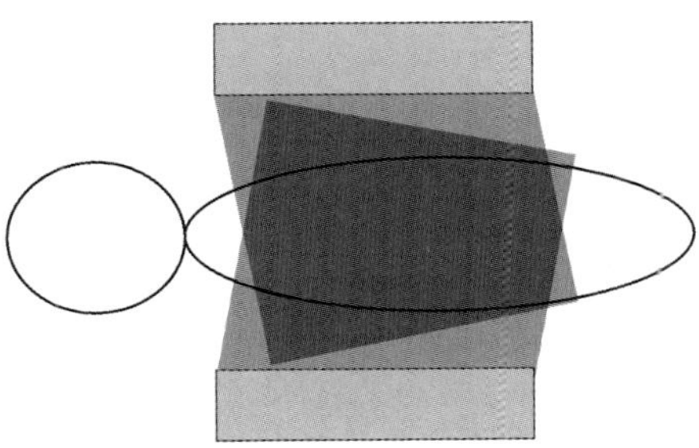

Fig. II.4.23. Principles of collimator hole misalignment. If the detectors are parallel to the axis of rotation but the collimator holes are not perpendicular to the detector face, they will image different regions during rotation causing blurring of the reconstructed images

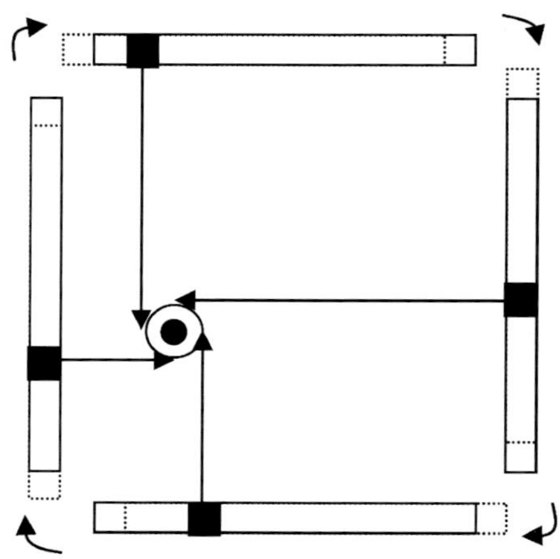

Fig. II.4.24. Principles of centre of rotation offset correction. If the projection image is not centred on the mechanical centre of rotation, a detected activity distribution will be offset from its true position, in the same direction, on each projection image. When these are then back-projected, a small circle will be created, the diameter of which will depend on the offset error

will be projected onto different parts of each projection image. In extreme cases, in a circular rotation a single point of activity will no longer be reconstructed as a small point but it will form a circle. In a 180° rotation arc, points are blurred in a semicircular fashion. With non-circular orbits, the blurring will occur in a more complex manner.

The centre of the projection image, itself, need not be at the mechanical COR, but the difference, called the COR offset, must be accurately known to the reconstruction software so that the matrix can be adjusted. Errors of as little as 0.5 pixel in a 128×128 matrix can degrade image quality perceptibly (DePuey 1994; O'Connor 1996). This offset must be stable with time, otherwise erroneous corrections will be implemented. If the COR offset is large, the tomographic field of view will be reduced because of the applied matrix shift.

Drifting electronics can cause a change in offsets, resulting in a shift of the image position relative to the electronic axes, which alters the COR of the system. The offset may vary with the angle of rotation but this usually only occurs if a mechanical fault, such as a collimator shift or gantry torsion, is present. Very few systems implement correction for such second order effects. If any of the motions, required to maintain the camera heads pointing at the COR, leave adjustment, the system COR will change. In the case of non-circular rotations, the same problem occurs if the detector head is not pointing towards the point, typically off-axis, where the computer software expects it to be (Hines et al. 1999a,b, 2000).

References

Almeida P, Bendriem B, de Dreuille O, Peltier A, Perrot C, Brulon V (1998) Dosimetry of transmission measurements in nuclear medicine: a study using anthropomorphic phantoms and thermoluminescent dosimeters. Eur J Nucl Med 25:1435–1441

Almquist H, Norrgren K, Palmer J, Jonson B, Wollmer P (2001) Performance of simultaneous emission-transmission systems for attenuation-corrected SPECT: a method for validation applied to two camera systems. Nucl Med Commun 22:759–766

Bailey DL (1998) Transmission scanning in emission tomography. Eur J Nucl Med 25:774–787

Bailey DL, Meikle SR (2000) Does attenuation correction work? J Nucl Med 41:960–961

Bocher M, Balan A, Krausz Y, Shrem Y, Lonn A, Wilk M, Chisin R (2000) Gamma camera-mounted anatomical X-ray tomography: technology, system characteristics and first images. Eur J Nucl Med 27:619–627

Bouwens L, van de Walle R, Nuyts J, Koole M, D'Asseler Y, Vandenberghe S, Lemahieu I, Dierckx RA (2001) Image-correction techniques in SPECT. Comput Med Imaging Graph 25:117–126 Br Lib 24.6.3

Bruyant PP (2002) Analytic and iterative reconstruction algorithms in SPECT. J Nucl Med 43: 1343–1358

Budinger TF (1996) Single photon emission computed tomography. In: Sandler MP, Patton JA, Coleman RE, Gottschalk A, Wackers FJ Th, Hoffer PB (eds) Diagnostic nuclear medicine, 3rd edn. Williams and Wilkins, Baltimore, pp 121–138

Budinger TF (1998) PET Instrumentation: what are the limits? Semin Nucl Med XXVIII:247–267

Busemann-Sokole E (1987) Measurement of collimator hole angulation and camera head tilt for slant and parallel hole collimators used in SPECT. J Nucl Med 28:1592–1598

Cao Z, Holder LE (1998) Effects of the attenuation map used in the Chang algorithm on quantitative SPECT results. J Nucl Med Technol 26:178–185

Cao Z, Maunoury C, Chen CC, Holder LE (1996) Comparison of continuous step-and-shoot versus step-and-shoot acquisition SPECT. J Nucl Med 37:2037–2040

Celler A, Sitek A, Stoub E, Hawman P, Harrop R, Lyster D (1998) Multiple line source array for SPECT transmission scans: simulation, phantom and patient studies. J Nucl Med 39:2183–2189

DePuey EG (1994) How to detect and avoid myocardial perfusion SPECT artifacts. J Nucl Med 35: 699–702

DePuey EG, Garcia EV (2001) Updated imaging guidelines for nuclear cardiology procedures, part 1. American Society of Nuclear Cardiology. J Nucl Cardiol 8:G1–G58

Eckholt M, Bergmann H (2000) Angulation errors in parallel-hole and fanbeam collimators: computer controlled quality control and acceptance testing procedure. J Nucl Med 41:548–555

Galt JR, Cullom SJ, Garcia EV (1999) Attenuation and scatter compensation in myocardial perfusion SPECT. Semin Nucl Med XXIX:204–220

Gilland DR, Jaszczak RJ, Greer KL, Coleman RE (1998) Transmission imaging for nonuniform attenuation correction using a three-headed SPECT camera. J Nucl Med 39:1105–1110

Graham LS, Fahey FH, Madsen MT, van Aswegen A, Yester MV (1995) Quantitation of SPECT performance: report of task group 4, nuclear medicine committee. Med Phys 22:401–409

Gregoriou GK, Tsui BM, Gullberg GT (1998) Effect of truncated projections on defect detection in attenuation-compensated fanbeam cardiac SPECT. J Nucl Med 39:166–175

Groch MW, Erwin WD (2000) SPECT in the year 2000: basic principles. J Nucl Med Technol 28: 233–244

Hansen CL (2002) Digital image processing for clinicians, part II: filtering. J Nucl Cardiol 9:429–437

Hendel RC, Corbett JR, Cullom SJ, DePuey EG, Garcia EV, Bateman TM (2002) The value and practice of attenuation correction for myocardial perfusion SPECT imaging: a joint position statement from the American Society of Nuclear Cardiology and the Society of Nuclear Medicine. J Nucl Cardiol 9:135–143 and J Nucl Med 43:273–280

Hines H, Kayayan R, Colsher J, Hashimoto D, Schubert R, Fernando J, Simcic V, Vernon P, Sinclair RL (1999a) Recommendations for implementing SPECT instrumentation quality control. Nuclear Medicine Section – National Electrical Manufacturers' Association (NEMA). Eur J Nucl Med 26: 527–532

Hines H, Kayayan R, Colsher J, Hashimoto D, Schubert R, Fernando J, Simcic V, Vernon P, Sinclair RL (1999b) National Electrical Manufacturers' Association recommendations for implementing SPECT instrumentation quality control. J Nucl Med Technol 27:67–72

Hines H, Kayayan R, Colsher J, Hashimoto D, Schubert R, Fernando J, Simcic V, Vernon P, Sinclair RL (2000) National Electrical Manufacturers' Association recommendations for implementing SPECT instrumentation quality control. J Nucl Med 41:383–389

IAEA (1991) Tecdoc 602 quality control of nuclear medicine instruments. International Atomic Energy Agency, Vienna

IPEM (2003) Report 86 quality control of gamma camera systems. Institute of Physics and Engineering in Medicine, London

IPSM (1985) Report 44 an introduction to emission computed tomography. Institute of Physical Sciences in Medicine, London

Jarritt PH, Acton PD (1996) PET imaging using gamma camera systems: a review. Nucl Med Commun 17:758–766

Jarritt PH, Whalley DR, Skrypniuk JV, Houston AS, Fleming JS, Cosgriff PS (2002) UK audit of single photon emission computed tomography reconstruction software using software generated phantoms. Nucl Med Commun 23:483–491

Kashiwagi T, Yutani K, Fukuchi M, Naruse H, Iwasaki T, Yokozuka K, Inoue S, Kondo S (2002) Correction of nonuniform attenuation and image fusion in SPECT imaging by means of separate X-ray CT. Ann Nucl Med 16:255–261

Koral KF, Swailem FM, Buchbinder S, Clinthorne NH, Rogers WL, Tsui BMW (1990) SPECT dual-energy-window Compton correction: scatter multiplier required for quantification. J Nucl Med 31: 90–98

Leong LK, Kruger RL, O'Connor MK (2001) A comparison of the uniformity requirements for SPECT image reconstruction using FBP and OSEM techniques. J Nucl Med Technol 29:79–83

Mahowald JL, Robins PD, O'Connor MK (1999) The evaluation and calibration of fan-beam collimators. Eur J Nucl Med 26:314–319

Malmin RE, Stanley PC, Guth WR (1990) Collimator angulation error and its effect on SPECT. J Nucl Med 31:655–659

Matsunari I, Boning G, Ziegler SI, Kosa I, Nekolla SG, Ficaro EP, Schwaiger M (1998) Effects of misalignment between transmission and emission scans on attenuation-corrected cardiac SPECT. J Nucl Med 39:411–416

NEMA (1994) Standards publication NU1 performance measurements of scintillation cameras. National Electrical Manufacturers' Association, Washington

O'Connor MK (1996) Instrument- and computer-related problems and artifacts in nuclear medicine. Semin Nucl Med XXVI:256–277

O'Connor MK, Vermeersch C (1991) Critical examination of the uniformity requirements for single-photon emission computed tomography. Med Phys 18:190–197

O'Connor MK, Kemp B, Anstett F, Christian P, Ficaro EP, Frey E, Jacobs M, Kritzman JN, Pooley RA, Wilk M (2002) A multicenter evaluation of commercial attenuation compensation techniques in cardiac SPECT using phantom models. J Nucl Cardiol 9:361–376

Oppenheim BE, Appledorn CR (1988) Single photon emission computed tomography. In: Gelfand MJ, Thomas SR (eds) Effective use of computers in nuclear medicine. McGraw-Hill, New York, pp 31–74

Pareto D, Pavia J, Falcon C, Juvells I, Cot A, Ros D (2001) Characterisation of fan-beam collimators. Eur J Nucl Med 28:144–149

Patton JA (2000) Instrumentation for coincidence imaging with multihead scintillation cameras. Semin Nucl Med XXX:239–254

Patton JA, Turkington TG (1999) Coincidence imaging with a dual-head scintillation camera. J Nucl Med 40:432–441

Rogers WL, Clinthorne NH, Harkness BA, Koral KF, Keyes JW (1982) Field-flood requirements for emission computed tomography with an Anger camera. J Nucl Med 23:162–168

Rosenthal MS, Cullom J, Hawkins W, Moore SC, Tsui BMW, Yester M (1995) Quantitative SPECT imaging: a review and recommendations by the focus committee of the Society of Nuclear Medicine Computer and Instrumentation Council. J Nucl Med 36:1489–1513

Takahashi Y, Murase K, Higashino H, Mochizuki T, Motomura N (2002) Attenuation correction of myocardial SPECT images with X-ray CT: effects of registration errors between X-ray CT and SPECT. Ann Nucl Med 16:431–435

Tsui BM (1996) The AAPM/RSNA physics tutorial for residents. Physics of SPECT. Radiographics 16:173–183

Tsui BM, Frey EC, LaCroix KJ, Lalush DS, McCartney WH, King MA, Gullberg GT (1998) Quantitative myocardial perfusion SPECT. J Nucl Cardiol 5:507–522

Tung CH, Gullberg GT (1994) A simulation of emission and transmission noise propagation in cardiac SPECT imaging with non-uniform attenuation correction. Med Phys 21:1565–1576

Turkington TG (2000) Attenuation correction in hybrid positron emission tomography. Semin Nucl Med XXX:255–267

Vandenberghe S, D'Asseler Y, van de Walle R, Kauppinen T, Koole M, Bouwens L, van Laere K, Lemahieu I, Dierckx RA (2001) Iterative reconstruction algorithms in nuclear medicine. Comput Med Imaging Graph 25:105–111

Zaidi H, Hasegawa B (2003) Determination of the attenuation map in emission tomography. J Nucl Med 44:291–315

Dual Photon Imaging

Abbreviations

2D	two dimensions
3D	three dimensions
ACF	attenuation correction factor
ADC	analogue to digital converter
BGO	bismuth germanate oxide
C-OS-EM	coincidence-list-ordered sets expectation-maximisation
CT	computed tomography
DPET	dual photon emission tomography
FBP	filtered back-projection
FDG	fluorodeoxyglucose
FORE	Fourier re-binning
FOV	field of view
FWHM	full width at half maximum
GSO	gadolinium oxyorthosilicate
LEHR	low energy high resolution
LSO	lutetium oxyorthosilicate
MSRB	multi-slice re-binning
NEC	noise equivalent count
OS-EM	ordered subset estimation maximisation
PHA	pulse height analyser
PMT	photomultiplier tube
SPET	single photon emission tomography
SSRB	single-slice re-binning
YSO	$Y_2SiO_5(Ce)$

Dual photon emission tomography (DPET) overcomes one of main limitations of SPET, the effect of absorptive collimation; and enhances the accuracy of the activity representation in the tomographic image by allowing accurate correction for attenuation. Its main operational difference to SPET is that it relies on two photons, from the same annihilation event, being detected in coincidence. Removal of the absorptive collimation results in better detection sensitivity and spatial resolution which is independent of depth in the patient. Although spatial resolution, in SPET, can approach comparability with DPET using focussed, e.g. fanbeam, collimators; the detection sensitivity of SPET, even with multi-detector systems, cannot approach that of dedicated 3D DPET. A major barrier, to clinical implementation of DPET, is the high cost of dedicated systems (Dilsizian et al. 2001). This can be overcome using a coincidence γ camera, which can operate as both a single photon and a coincidence system (Budinger 1998). These may be the most innovative devices introduced in nuclear medicine during the last few years but are complements to, rather than competitors of, dedicated systems (Kunze et al. 2000). To achieve the required functionality, the single photon γ camera requires a number of modifications.

Coincidence Imaging

The position of a β^+ annihilation can be localised to a projection line between two opposed detectors because the two γ photons, which are created in the annihilation, appear at the same time and travel in opposite directions (see Fig. II.5.1). Ensuring that the detected photons originate from the same disintegration, requires them to be identified as coincident. Technically, this can be established to within 5×10^{-10} s, but the timing window, used, is wider at ~1.2×10^{-8} s for bismuth germanate oxide (BGO) based dedicated systems (see 'Dedicated systems – detectors'). This is because the annihilation will not usually be equidistant from the two detectors, and the discrepancy in the distances will result in the photons being detected at different times. In the torso, this is of the order of 2×10^{-9} s (Burger and Berthold 2000). The wider time window is also required to accommodate the processing time of the detectors, mainly due to the light emission characteristics of the crystals. The window can be reduced to 0.8×10^{-8} s for gadolinium oxyorthosilicate (GSO) and NaI(Tl), and to 0.6×10^{-8} s for lutetium oxyorthosilicate (LSO) (Tarantola et al. 2003).

Detected Events

The timing window is opened when the first event is detected and then kept open for the set time. This results in detected events being either single events, when only one event is detected in the time window, or coincidence events.

Single Events

Single events occur when one photon is lost due to detection geometry, scatter, attenuation or low detector efficiency (see Fig. II.5.2). Depending on the detector, the coincidence count rate may be only 0.1–2% of the singles rate in patients (Budinger 1998; Blockland et al. 2002). Therefore, to enable sufficient coincidences to be acquired in a reasonable time, detection systems must have a high count-rate capability. The deter-

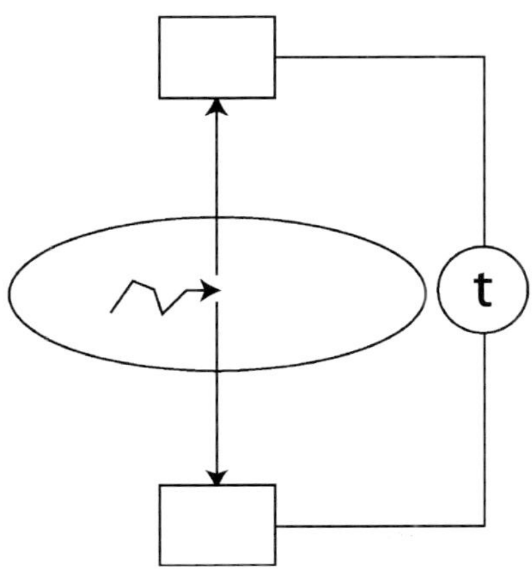

Fig. II.5.1. Principles of dual photon detection. The position of a β^+ annihilation can be localised to a projection line between two opposed detectors, which are linked by coincidence circuitry

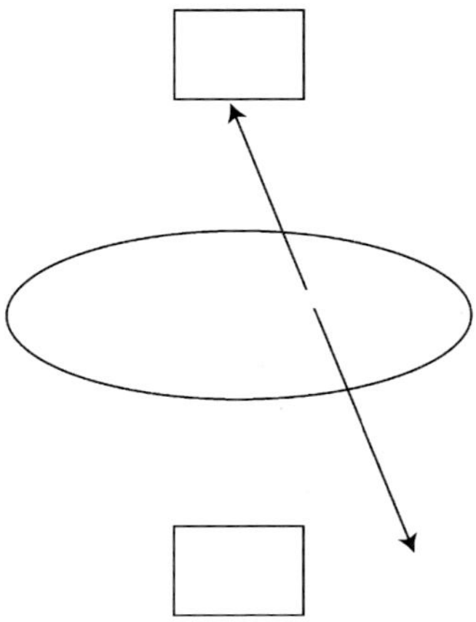

Fig. II.5.2. Singles events occur when only one, of the two photons created in a β^+ annihilation, is detected

minants of the singles rate, for a point source in air, can be appreciated from Jarritt and Acton (1996):

$$R_{si} = 2\,A\,\varepsilon\,d\Omega \tag{II.5.1}$$

where
A is the source activity
ε is the detector efficiency
dΩ is the solid angle subtended by the detectors

Coincidence Events

These are classified as true, scatter and artefactual coincidences. In scatter coincidences, one or both of the detected photons experiences Compton scattering before being detected. Artefactual coincidences are: random or accidental, in which the two detected photons originate from different disintegrations; and multiple, in which more than two photons are detected in coincidence, but this is unlikely. All but true coincidences involve errors in determining the disintegration position, which degrades the spatial resolution of the reconstructed image.

True Coincidences

True coincidences occur when two photons are detected in coincidence, originate from the same annihilation event and reach the detectors without deflection. The determinants of the true coincidence rate, for a point source in air, can be appreciated from:

$$R_{tr} = A\,\varepsilon^2\,d\Omega \tag{II.5.2}$$

The solid angle, dΩ, is present only as the first power because, in true coincidences, if one of the emitted photons travels in a direction suitable for detection, its opposing photon must travel in a direction to coincide with the opposing detector. Because of the solid angle involved, the true coincidence rate decreases in an inverse-square relationship with detector separation. The detector efficiency is present as ε^2 because both photons must be detected to register a coincidence. A reasonable coincidence detection rate demands high-efficiency detectors, and the ε^2 term reveals the penalty for the low-efficiency devices, such as wire chambers and NaI(Tl) crystals (Jarritt and Acton 1996).

Scatter Coincidences

These result in displacement of the projection line, and in most instances involve only one of the photons being scattered (see Fig. II.5.3). Unlike single photon imaging, the origin can appear to be outside the radioactive distribution (Patton and Turkington 1999).

The consequence is that a significant low spatial frequency background is added to the reconstructed images, which degrades spatial resolution and image contrast (Lewellen 1998) by locally raising the apparent activity and imposing a resolution blur, particularly between high and low activity accumulations (Budinger 1998). The description of the degradation is complicated, depending on the detector geometry, source distribution, and the type and distribution of scattering material (Jarritt and Acton 1996).

Collimation

Although absorptive collimation is not mandatory, it can be introduced to reduce these scatter coincidences. To do this, absorptive septa are installed in the transaxial plane of the detectors. This restricts photon direction axially, but not transaxially, in what is termed two dimensional (2D) acquisition mode (Fig. II.5.4). System design considerations are that the scatter fraction is lower for a larger detector–patient separation and longer septa. 2D acquisition mode significantly reduces detection sensitivity and for some, particularly low count rate, studies therefore it is not appropriate and fully three dimensional (3D) acquisitions must be undertaken (Budinger 1998), see Fig. II.5.5. In both acquisition types, large end shields, are installed to restrict detection of radiation originating from outside the field of view (see Fig. II.5.6).

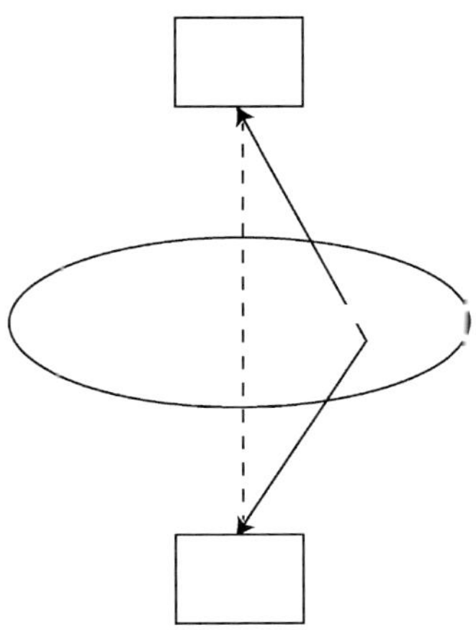

Fig. II.5.3. Scatter, usually of one photon, can cause a mispositioning of the annihilation event when the projection line is created between the two detectors

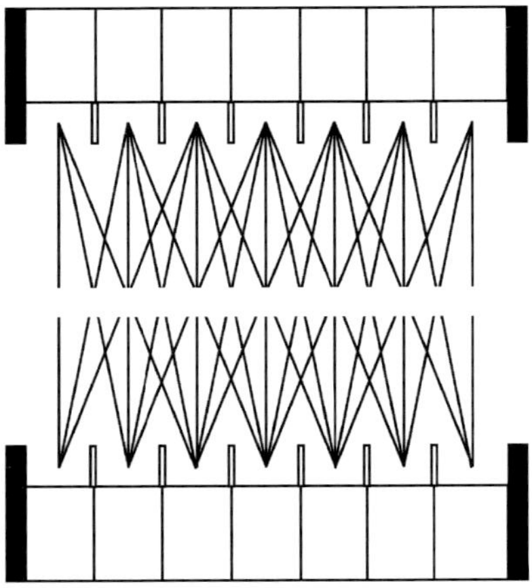

Fig. II.5.4. 2D collimation by absorptive septa located between the detectors. Detection of annihilation photons, in transaxial planes away from where the annihilation occurred, is restricted; in the example shown, to transaxial planes on either side of the annihilation plane

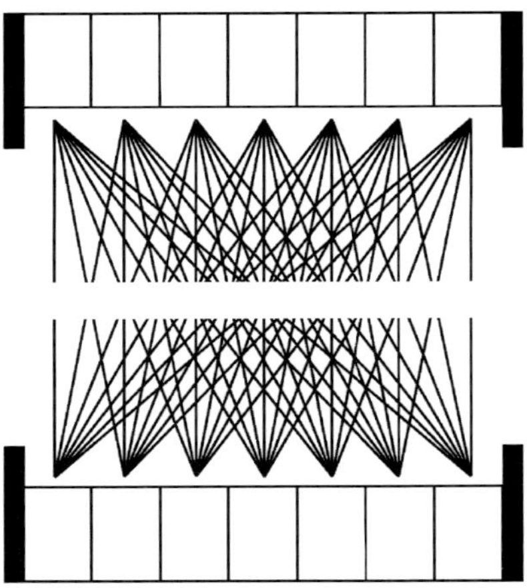

Fig. II.5.5. 3D collimation by removing inter-plane septa. Detection of annihilation photons, in transaxial planes away from where the annihilation occurred, is unrestricted and many more projection lines are possible

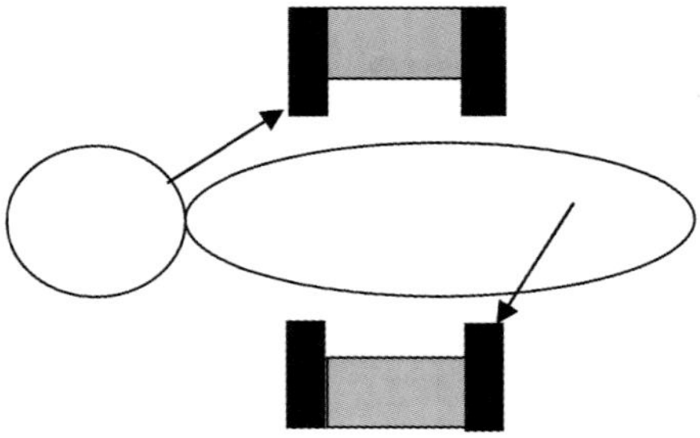

Fig. II.5.6. Large end shields, installed to restrict detection of radiation originating from outside the field of view

Energy Discrimination

Some scatter can be rejected by raising the lower PHA window setting. The energy loss for most forward scattered annihilation photons is small, however, even for significant deflections (see equation I.3.27 in Chapter I.3). Therefore, photons scattered through small angles cannot be discriminated against. Increasing the low energy threshold also reduces the detection sensitivity by excluding true coincidences (Budinger 1998).

Correction

Background subtraction can improve image contrast. Although there is no accurate solution for scatter correction, a number of correction techniques have been developed, including: dual-energy window, convolution-subtraction, and Monte Carlo methods (Grootoonk et al. 1996; Zaidi 2000). Increasingly, sophisticated scatter correction procedures are under investigation, particularly those based on accurate scatter models. Scatter correction works well for 2D acquisitions but less well for 3D acquisitions when the activity is not well concentrated within the field of view (Burger and Berthold 2000). Currently, the preferred compensation strategy is the incorporation of modelling the degradations, caused by scatter, into an interactive reconstruction method. Scatter correction software is fundamental on dedicated systems but is only being introduced on coincidence γ camera systems (Zaidi 2001).

Random Coincidences

These occur when two photons are recorded as one coincident event, but originate from two singles events that have been detected simultaneously (see Fig. II.5.7). The result is the creation of an artefactual projection line with consequent degradation in contrast caused by the imposition of a background activity. This is uniform in dedicated systems and non-uniform in coincidence γ cameras because, in γ cameras, it is impractical to use long end shields and contributions from activity outside the field of view are significant (IPEM 2003). Random coincidences are a consequence of the width of the coincidence timing window (τ). The rate is proportional to τ and to the

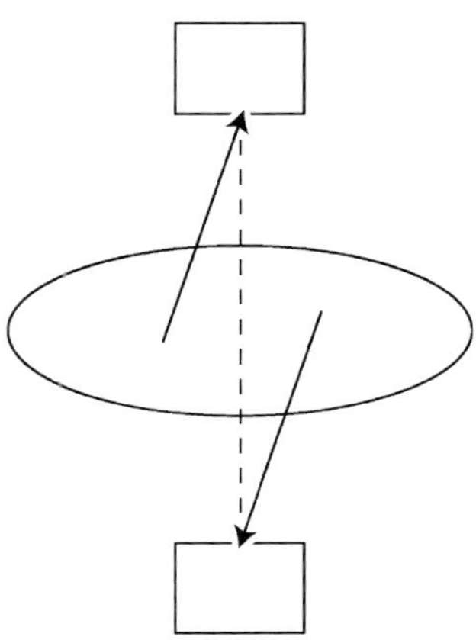

Fig. II.5.7. Random coincidences occur when two photons are recorded as one coincident event, but originate from two singles events which have been detected simultaneously. This creates an artefactual projection line between the two detectors

product of the singles rates at each of the two detectors (c_1 and c_2), i.e. ~activity^2 in the field of view between the two detectors.

$$R_r = 2\tau c_1 c_2 \tag{II.5.3}$$

The randoms fraction of true coincidences increases as the singles rate increases. The determinants of the relationship between the random coincidences rate (R_r) and the true coincidences rate (R_{tr}) can be appreciated from (Jarritt and Acton 1996):

$$R_r/R_{tr} = 8 A \tau d\Omega \tag{II.5.4}$$

which demonstrates that the random coincidences rate increases linearly relative to the true coincidences rate for increasing activity. This is because the true coincidences rate increases only in proportion to the activity whereas the random coincidences rate increases ~activity^2. The problem, of randoms, is therefore particularly severe at high count rates. The equation also illustrates the importance of keeping the timing window as narrow as possible. However, the slower the decay time of the crystal, the wider the window has to be and the more likely a random event will be registered. The crystal usually used in dedicated systems, BGO, has a slower decay time than that used in coincidence γ cameras, NaI(Tl), but both are much slower than some of the newer materials, such as LSO and GSO (see 'Detectors' in 'Dedicated systems').

Correction

There is no way to distinguish between true and random coincidences. The contribution of random coincidences can be reduced by reducing the width of the coincidence timing window, but an increasing number of true coincidences will also be rejected.

Correction can be undertaken using either the delayed window or the singles correction method. The delayed window technique is the most usual, and is based on the randoms rate being constant and independent of when the singles events are recorded on each detector. Two coincidence windows are used (see Fig. II.5.8). The first, usual, window measures the combined true and random coincidences. These are referred to as prompt coincidences. The second window is delayed by several 10^{-7} s from the first, not to detect a coincidence but another singles event. As only random coincidences are detected, simultaneously, in both windows; apart from statistical fluctuations, the ran-

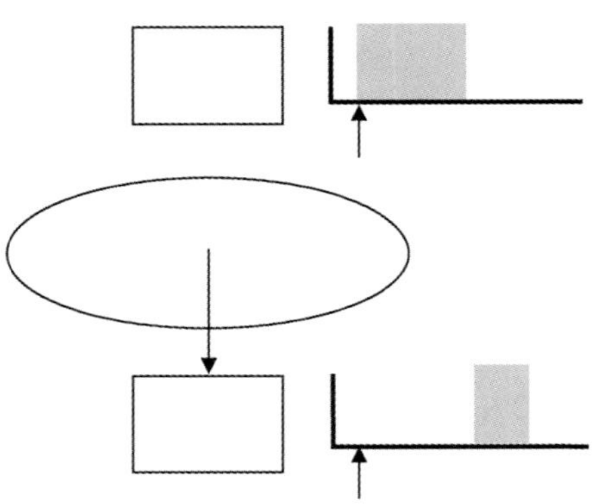

Fig. II.5.8. The delayed window technique for correction of random coincidences. When a photon is detected, a coincidence window is opened, as usual, in the other detectors. This is followed by a second window in the first detector which will detect further singles events

dom rate can be established and subtracted from the acquired data (Budinger 1998). Statistical fluctuations do, however, increase noise in the image. An alternative method involves measuring the singles rate for each detector, c_1 and c_2, and estimating the rate of random coincidences using equation II.5.3 for R_r. Because the singles rates are high, this generates little noise but does require an accurate value for the coincidence timing window, τ, which can vary slightly during acquisition (Fleming et al. 2000).

Sensitivity

Sensitivity is the number of true coincidences detected from a unit of activity. It is higher in DPET than in SPET because, without the necessity of intra-slice absorptive collimation; photons travelling at any angle to the detector, within a transaxial slice, are accepted. Sensitivity is further increased in 3D acquisition, when even the inter-slice collimation is removed. Two main factors determine sensitivity: the detection geometry and the absorption efficiency. The geometric factor is determined by the solid angle, of the detectors, around the imaging volume. The best geometry is a sphere surrounding the entire patient, the most practical is a cylinder surrounding part of the patient (see 'Sensitivity' in 'Dedicated systems'). The efficiency of the detector depends mainly on the density and thickness of the scintillation crystal.

Spatial Resolution

Besides the effects of the instrumentation (intrinsic): two factors concerned with the annihilation event place a limitation on the minimum achievable detection spatial resolution. These are the range of the β^+ particle before annihilation (annihilation effects) and the non-colinearity or angle of the annihilation photons (coincidence effects).

$$\text{total detection spatial resolution} = \sqrt{(\text{intrinsic}^2 + \text{range}^2 + \text{angle}^2)} \qquad \text{(II.5.5)}$$

These set the ultimate limit on detection spatial resolution (Jarritt and Acton 1996; Budinger 1998). The spatial resolution, achieved in the reconstructed image, depends additionally on the reconstruction process

Annihilation Effects

The projection line identifies the β^+ annihilation position rather than the position of its emission. The difference is the range of the β^+, which introduces an uncertainty as to the exact position of the annihilation event and this uncertainty increases with the emission energy of the β^+ particle (see Fig. II.5.9). The maximum range varies from 2.4 mm for ^{18}F up to 20 mm for ^{82}Rb. However, because the particle describes a tortuous path, the actual range is much smaller and characterised by an exponential function with FWHM values 0.2–2.6 mm (Burger and Berthold 2000). Hence, only a small fraction of β^+ particles achieve their maximum range. The actual ranges are listed for common radionuclides in Table II.5.1.

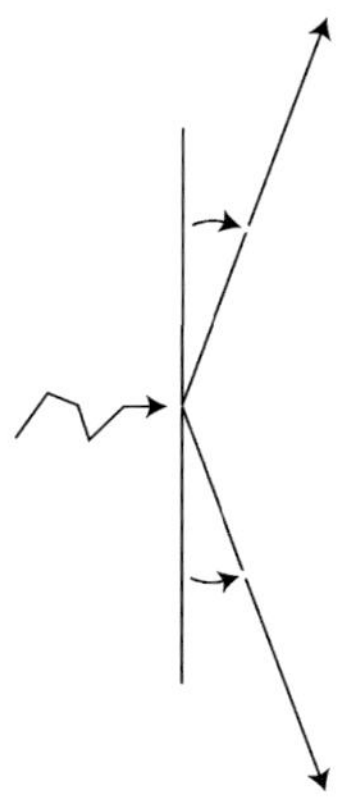

Fig. II.5.9. Uncertainty in localising the annihilation event due to the range of the β^+ particle before annihilation and the angular diversion of the annihilation photons after annihilation

Table II.5.1. Maximum energy and actual range of β^+ particles emitted from usual radionuclides (adapted with permission from Cherry and Phelps 1996)

Radionuclide	β^+ energy E_{max} (MeV)	Range in water FWHM (mm)
^{18}F	0.64	0.22
^{11}C	0.96	0.28
^{13}N	1.19	0.45
^{15}O	1.72	1.04
^{68}Ga	1.90	1.35
^{62}Cu	2.94	2.29
^{82}Rb	3.35	2.60

Non-Colinearity Effects

The directions of the annihilation photons are assumed to be exactly opposite. However, at annihilation the β^+ can retain a small momentum and, because of its conservation, this results in a small angular variation between the opposing photons, which has a Gaussian angular spread with a FWHM of 0.5° (see Fig. II.5.9). The error in spatial resolution, caused by this, increases with the separation of the detectors (d) according to $0.0022 \times d$ (Tarantola et al. 2003), i.e. ~0.9 mm for a 40 cm separation and ~2.2 mm for a 100 cm separation. Thus for the low energy β^+ emitters, this effect is the dominant factor in determining the spatial resolution limit (Jarritt and Acton 1996; Budinger 1998).

Time of Flight Correction

Spatial resolution could be improved by identifying the position of the annihilation event, along the projection line, using the relative detection times of the photons. Such a time of flight system, with perfect time resolution, would identify the exact location of the event and a reconstruction algorithm would not be required. The presently achievable temporal resolution, of 5×10^{-10} s, only permits localisation to within 9 cm,

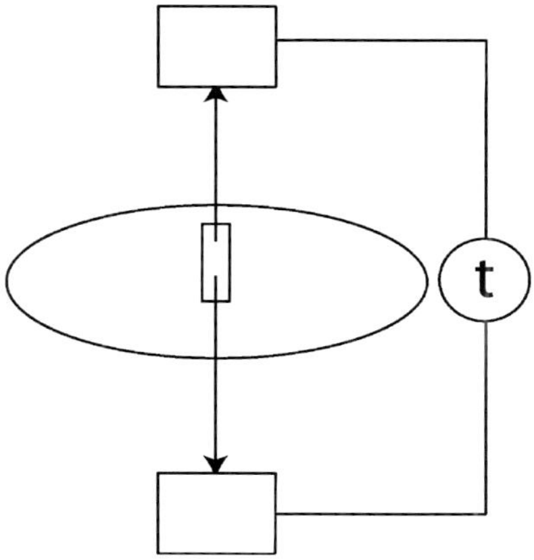

Fig. II.5.10. In a perfect system the position of the annihilation event, along the projection line, could be determined precisely by measuring the exact time difference between the annihilation photon interactions. In reality, the error is large

however (see Fig. II.5.10). A faster scintillator could be used to improve this but these have much lower detection efficiencies, which reduce the detection sensitivity and negate the performance gain. The development of new fast scintillators which maintain their detection efficiency, offer the possibility of time of flight in the future (Lewellen 1998; Yamaya et al. 2000).

Noise Equivalent Count Rate (NEC)

The majority of systems demonstrate similar spatial resolution, and the best discriminator of their relative performance is the coincidence count rate. Count rate itself is not an adequate descriptor, however, because of the contributions from random and scattered coincidences; it is only the true coincidences that are of concern (Fleming et al. 2000). The parameter used is the NEC, or noise equivalent sensitivity, which is proportional to the signal-to-noise ratio in the reconstructed images, and the variation in NEC, with activity, allows a good comparison of the performance of different systems. However for this, the peak true coincidences rate and the activity, at which the peak true coincidences and peak NEC rates occur, must also be considered. A very sensitive system may saturate at a relatively low activity, but this may be higher than that of a system with a lower sensitivity at a higher activity (Daube-Witherspoon et al. 2002). The NEC is thus the count rate after corrections are made for random and scattered coincidences. Its value is dependent on the size and shape of the radioactivity distribution.

$$NEC = R_{tr}^2/(R_{tr} + R_s + R_r) \tag{II.5.6}$$

where

R_{tr} is the true coincidences rate

R_r is the randoms coincidences rate

R_s is the scatter coincidences rate, from within the imaged activity distribution

As the imaged activity increases, the NEC will eventually decrease. This is because R_r increases with activity^2 but R_{tr} only increases with activity (Budinger 1998).

Attenuation

The effects of attenuation are more severe in coincidence imaging than in SPET, because both annihilation photons must be detected for a coincidence to be recorded (Wahl 1999). The probability of this is much less than for the single photon situation, even considering the lower attenuation coefficient of the 511 keV photons. Only ~1 in 5 photon pairs is not attenuated in brain imaging and only ~1 in 30–40 when imaging the torso (Cherry and Phelps 1996). The attenuation correction factors involved are therefore much larger than in SPET, see Table II.5.2. The effect of attenuation is, however, much easier to deal with in DPET than in SPET because it depends only on the total thickness of tissue between the two detectors. It does not depend on the position of the annihilation relative to each detector, because both photons must be acquired in order to register a coincidence. Loss of either, due to absorption or scattering, means that such a coincidence will not occur. Therefore, the probability of detecting an annihilation depends only on the probability of both photons reaching the detectors and this depends only on the total tissue thickness between the detectors (see Fig. II.5.11). Attenuation effects become apparent because, although the position along a projection line is not important, the combination of all the projection lines from all acquisition angles exhibits greater attenuation when the origin is at depth than when it is shallow (Delbeke and Sandler 2000). The effect of attenuation is also much easier to

Table II.5.2. Attenuation correction factors for SPET and DPET in various tissues

	Head	Thorax	Body
SPET ($^{99}Tc^{m}$)	2.5	7	5–10
DPET	10	30	60–100

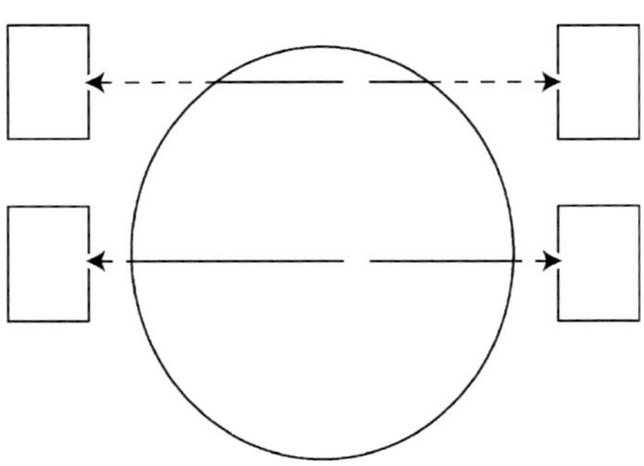

Fig. II.5.11. Attenuation depends only on the total thickness of tissue between the two detectors. It does not depend on the position of the annihilation relative to each detector

Table II.5.3. Narrow beam linear attenuation coefficients (μ_l) cm^{-1} for $^{99}Tc^{m}$ and at 511 keV in various tissues (King et al. 1995; Dilsizian et al. 2001)

	Lungs	Soft tissue	Bone
$^{99}Tc^{m}$	0.051	0.153	0.286
511 keV	0.031	0.095	0.151

deal with in DPET than in SPET because there is less variation in the linear attenuation coefficient, μ_l, across the body, for the 511 keV energy of the annihilation photons, than there is for the lower energy photons used in SPET, see Table II.5.3.

Correction

Attenuation correction is the most important data correction applied, if quantitation is required (Fleming et al. 2000). Additionally, non-uniformity and spatial distortions are introduced in the reconstructed images if attenuation is not corrected (Turkington 2000); areas of high activity extending along the lines of lower attenuation. The emission data can be corrected before reconstruction because all radiation measured along a projection line is attenuated by the same factor, regardless of the position of the annihilation along it. This is not the case, in SPET, except when assumptions are made about the position of the radioactivity along the projection ray, because the attenuation depends on the depth. Because of this, both filtered back projection and iterative reconstruction can be used to reconstruct the corrected images; the attenuation factor, along each projection line, being calculated using analytical techniques or determined directly using transmission measurements.

Analytical

This requires an assumption of uniform attenuation throughout the imaged volume, and so cannot be used in the thorax, or elsewhere where bone or gas is present. However, it is frequently used in brain studies because of the relatively low additional attenuation of 511 keV photons in the thin cortical bone of the skull, but only when accurate quantitation is not necessary. An outer contour of the imaged object must be defined, either from the raw projection data or from uncorrected reconstructed images, which is often an ellipse. Even though there is no noise associated with the calculation, it does amplify the measurement errors in deep lying low activity regions, thus altering the noise distribution in the reconstructed image (Turkington 2000).

Transmission

The attenuation of the body must be measured when: the attenuation is not uniform, the outer contour cannot be determined, or quantitative accuracy is required. Transmission data is acquired, as shown in Fig. II.4.16 of Chapter II.4, and an attenuation map for each transaxial slice, of the emission data, is reconstructed (Fleming et al. 2000). Attenuation factors, used to correct the emission data, are calculated for each projection line by summing the attenuation values over all the pixels along that projection line.

Methods

The attenuation, of annihilation photons along each projection line, is measured directly, using an external source. The calculation, of the attenuation factors for each projection line, is more complicated in 3D mode than in 2D, because of the difficulty in defining the projection line path lengths. A simplification is to acquire transmission data in 2D mode and use the reconstructed attenuation map to correct the 3D emission data. Errors of 15–20% can be caused, in the attenuation map, by scatter coincidences (Budinger 1998), and these impose significant noise on the attenuation corrected emission data.

Segmentation

This is also known as hybrid attenuation correction (Bartenstein et al. 2002). The noise in the reconstructed attenuation map can be constrained by considering which tissue is represented by the measured attenuation coefficients (Xu et al. 1990). These can take on only a few possible values, for: lung, soft tissue and bone; and segmenting the attenuation map using these categories can reduce noise, and can also allow reduced transmission acquisition durations (Burger and Berthold 2000) whilst maintaining the quantitative accuracy of longer transmission acquisition durations (Bettinardi et al. 1999). Care must be exercised if contrast agent has been used because this can give the impression of bone.

CT

This appears to be the best technique for undertaking attenuation correction. The attenuation maps are generated using short, several second, duration CT acquisitions (Patton 2000; Budinger 1998); the noise and spatial resolution of even low quality CT images being much better than can be obtained with a radioactive source (Akhurst and Chisin 2000; Burger and Berthold 2002). This data can also be used for scatter correction.

Transformation

If the transmission data is acquired using a photon energy other than 511 keV, e.g. 40–140 keV of CT data, the attenuation coefficients in the reconstructed map must be adjusted before correcting the emission data. As attenuation depends on both the photon energy and the absorbing material, an accurate theoretical relationship cannot be calculated (Burger and Berthold 2002). However, because most of the body is water, most pixels in the attenuation map can be corrected as (Turkington 2000):

$$a_c = a_t * \mu_c / \mu_t \tag{II.5.7}$$

where
a are attenuation coefficients
calculated at the energy of the transmission photons, a_t
corrected to photons of energy 511 keV, a_c
μ are linear attenuation coefficients
of the transmission photons in water, μ_t
of 511 keV photons in water, μ_c

Pixels representing bone do not transform as simply, because of the higher component of photoelectric absorption involved in the attenuation.

Decay

Correction for the decay of the radionuclide is usually implemented during image reconstruction.

Dedicated Systems

The very high total count rates, which are required to record acceptable true coincidences rates, are achieved by having a very large number of individual detectors, each operating at lower counting rates; and by surrounding the patient completely with rings of detectors, see Fig. II.5.12. The latter improves the true coincidences to singles ratio, by decreasing the probability that the partner of a detected photon will be lost. The ratio is further improved by shielding the field of view from activity in other parts of the body, using a large diameter ring of up to 100 cm and substantial end shields (Lewellen et al. 1999). Dedicated systems mainly use crystal scintillation detectors. Multiwire proportional chambers can achieve high spatial resolution but suffer from the low absorption efficiency of gases, which may be only 10–20% (0.1–0.2) of that of a crystal detector. When the combined detection efficiency of both opposed detectors is considered, this translates to an overall detection efficiency of 0.1^2–0.2^2 (1–4%), which is 80–20 times less than that achieved using crystal detectors (Budinger 1998).

Detectors

The crystal used, almost exclusively, for the last 10 years has been bismuth germanate oxide (BGO). This is not the ideal scintillator in that its light output is low, 0.15 times that of NaI(Tl), and its decay time is long, 1.3 times that of NaI(Tl). These result in a high random coincidences fraction at high count rates and an inability to effectively separate scatter coincidences from true coincidences. It is used because it is very dense, ~2 times that of NaI(Tl), and is, therefore, more efficient at stopping high energy photons; 90% of incident photons interacting in 3 cm thick crystals and giving a total detector efficiency of 0.9^2 (81%). To equal this, a NaI(Tl) detector would have to have a thickness of 8.4 cm. It is also non-hygroscopic, which simplifies construction, allows close packing of detector blocks and extends detector life (Bar Shalom et al. 2000). NaI(Tl) crystals, on the other hand, would be very difficult to construct to guarantee avoiding the damaging effect of water absorption.

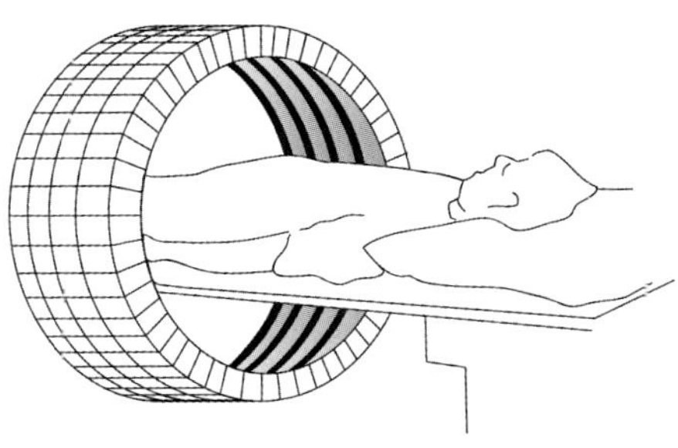

Fig. II.5.12. Dedicated PET showing adjacent rings of individual detector blocks surrounding the patient. Septa are shown as black extensions on the inside of the detector ring

Other crystals are under consideration (Bendriem et al. 1999). Barium fluoride (BaF_2) has the fastest decay time, ~80 times that of BGO, but a very low density, only 0.7 times that of BGO, and therefore low efficiency. The ideal scintillator would have the speed of BaF_2, the density of BGO and the light output of NaI(Tl). Lutetium oxyorthosilicate (LSO) is near to this ideal (Budinger 1998); it has a slightly higher density than BGO, emits up to 5 times as much light and decays nearly 8 times as rapidly (Bendriem et al. 1999). The higher light output results in better energy resolution and reduced scatter fraction (Bar Shalom et al. 2000), improved NEC and reduced scanning time, which could be as little as 10% of the present duration. Gadolinium oxyorthosilicate (GSO) is another crystal that also approaches this ideal (Muehllrhner et al. 2002). Both LSO and GSO have recently been introduced into commercial systems.

Compton interaction within the crystal is a problem particularly in certain crystals, such as BaF_2. In these, if the lower PHA threshold is set high, to reject scatter coincidences, then a loss of true coincidences results which lowers the sensitivity of the system (Budinger 1998; Fahey 2001).

With BGO and LSO, 2–8 adjacent rings are used, each ring being constructed of up to ~70 detector blocks. The number of block rings determines the axial length imaged, which is up to 25 cm (Daube-Witherspoon et al. 2002), and which can be stacked to provide data for whole-body imaging. Each detector block comprises a crystal cube, with a 3–5 cm square or rectangular face and a thickness ~3 cm, coupled to four square PMTs, see Fig. II.5.13. The face is partially sliced into an array of 8 × 8 or 6 × 6 crystal elements of dimensions ~0.4–0.6 cm, resulting in 16–64 adjacent crystal rings, each comprising up to ~600 crystal elements (Fahey 2001). The depth of the slices determines the distribution of the scintillation light among the PMTs; the light produced in each crystal producing a unique combination of signals in the PMTs, which allows an interaction to be associated to a particular crystal element, and the energy to be determined using 'Anger' type logic described in 'camera head' in Chapter II.3. This crystal segmentation is implemented in order to improve the intrinsic spatial resolution of the system, explained later in "parameters of image quality", and this type of construction is used to reduce cost and provide improved reliability compared to individual crystal construction. The arrangement of coupling one large PMT to 16 crystal ele-

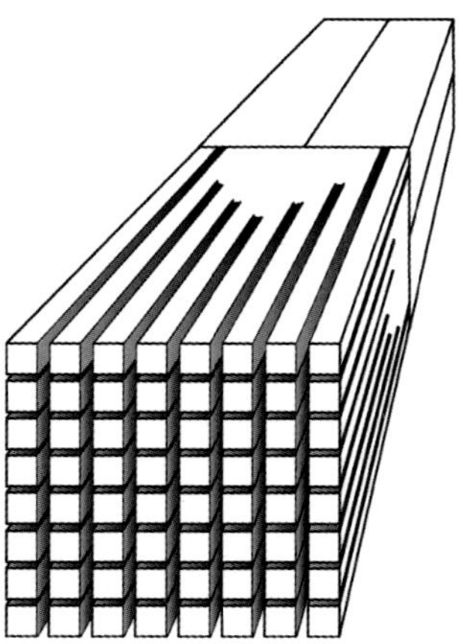

Fig. II.5.13. Detector block, partially sliced into an array of up to 8 × 8 crystal elements, coupled to four square PMTs

ments produces faster and less variable timing than using many small tubes, and reduces signal processing by having 2 rather than 64 outputs.

Different geometries are used for NaI(Tl) and GSO. For NaI(Tl), 6 curved crystals are used, each 2.5 cm thick and coupled to 43 PMTs. For GSO, 28 modules are packed hexagonally, each consisting of ~600 crystals attached to a continuous light guide and coupled to 15 PMTs (Tarantola et al. 2003).

Cost Limiting Detectors

Dedicated systems are expensive because of the large number of detectors needed (Burger and Berthold 2000). To reduce the cost, alternative designs, other than using the complete ring of detectors, have been implemented (Fahey 2001). A partial ring is created by two opposed and slightly offset BGO detector banks which rotate during acquisition.

Field of View

The FOV is defined electronically as the width of a fan, which limits the number of crystals that can be coincident with an opposing crystal, see Fig. II.5.14. This limits the effect of astigmatism, which is discussed later in "spatial resolution".

Operating Mode

2D mode acquisition is implemented by introducing absorptive, e.g. tungsten, septa, ~1 mm thick and ~60–120 mm long (Adam et al. 1997; Ramos et al. 2001), between each detector block ring. These are shown in Fig. II.5.12 as black extensions on the inside of the detector ring. This mode provides a low, ~10–20% for BGO, fraction of scattered events (Burger and Berthold 2000; Daube-Witherspoon et al. 2002) and a typical clinical true coincidences rate of ~4×10^4 counts s^{-1} (Lewellen et al. 1999; Paul et al. 2001). The septa are retracted for 3D mode acquisition, which improves detection sen-

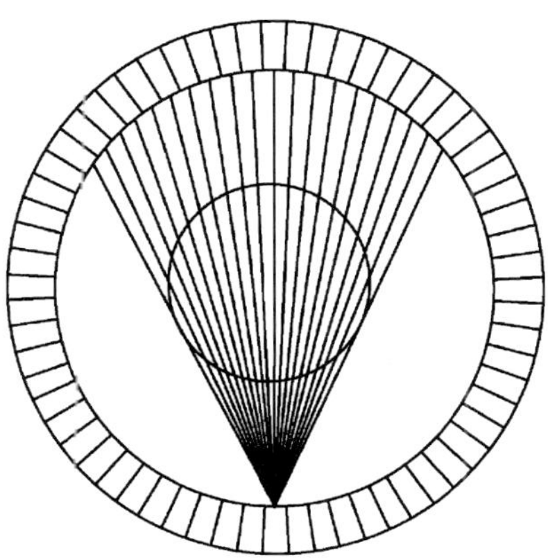

Fig. II.5.14. Electronic definition of the field of view as the width of a fan

sitivity and allows shorter acquisition durations, by increasing the number of projection lines. The typical clinical true coincidences rate is ~3×10^5 counts s^{-1} (Lewellen et al. 1999). This mode is particularly appropriate for low activity accumulation studies when signal-to-noise (NEC) gains of ~4 (Budinger 1998) can be achieved. There are a number of data manipulation problems associated with the 3D technique, however: the acquisition data sets become very large; attenuation correction and system calibration are more complicated; and the fraction of scatter coincidences increases, to ~30–40% for BGO, which makes scatter correction mandatory (Burger and Berthold 2000; Daube-Witherspoon et al. 2002; Tarantola et al. 2003). However, at least one manufacturer presently produces only 3D systems. In both modes, substantial end shields are used to reduce the effect of activity outside the field of view.

Parameters of Image Quality

Sensitivity

As well as the crystal absorption efficiency, ε, detection sensitivity also depends on the source–detector geometry. In 2D imaging, this is determined by the solid angle subtended by each block ring, see Fig. II.5.15, because coincidences are detected between any two crystal elements in this ring (Budinger 1998):

$$S = \varepsilon^2 * \text{block axial dimension}/2\,r \tag{II.5.8}$$

Using BGO, a ring diameter of 100 cm and a detector block axial dimension of 5 cm, this is 0.81 × 5/100=0.04. Thus, in 2D acquisition, even with efficient crystals, the detection sensitivity is very poor because a large proportion of annihilation events are not detected. Thus the spatial resolution of the reconstruction is noise limited, being more dependent on the number of counts acquired rather than on the intrinsic spatial resolution of the detector. In 3D acquisition, detection sensitivity rises by a factor of ~4.5–5.5 (Lewellen et al. 1999), determined by the change in the source–detector solid angle.

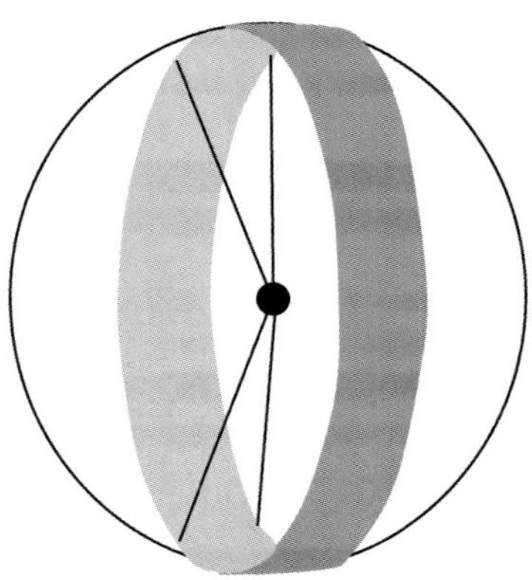

Fig. II.5.15. Detection geometry of point source and ring of detectors. In 2D imaging, geometrical detection efficiency depends on the solid angle subtended by each block ring

Variation Along the Axial Field of View

Sensitivity is highest at the centre of the axial field of view, where more crystal elements are available for detection, and decreases towards the edges; this being more severe in 3D than 2D mode, see Fig. II.5.16. To obtain a uniform response, correction factors are determined by acquiring calibration data using a phantom containing a uniform distribution of activity (Burger and Berthold 2000).

Spatial Resolution

The major determinant of the overall spatial resolution, of detection, is the intrinsic resolution of the system, especially for the lower energy β^+ emitters. This depends on the size of the crystal elements used because, at the centre of the field of view, the line spread function of two detectors operating in coincidence is a triangle with a FWHM of half the detector dimension, see Fig. II.5.17. Thus for a 6 mm detector element, the intrinsic spatial resolution will be 3 mm. This worsens towards the periphery of the FOV. There is also a 'block effect' which is an uncertainty in positioning due to the detector block being sliced into crystal elements and 'Anger' logic used to localise the interaction (Tarantola et al. 2003).

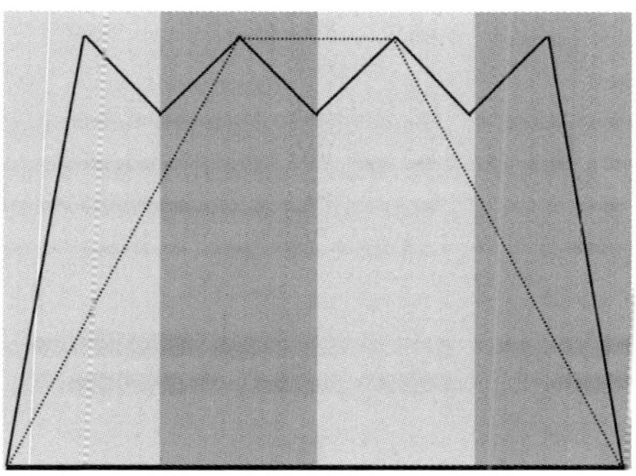

Fig. II.5.16. Variation in sensitivity along the axial field of view for a 4 block ring system. The extent of each block ring is marked as a different shade of grey. The solid line represents the sensitivity variation in 2D acquisition mode and the broken line represents the 3D mode

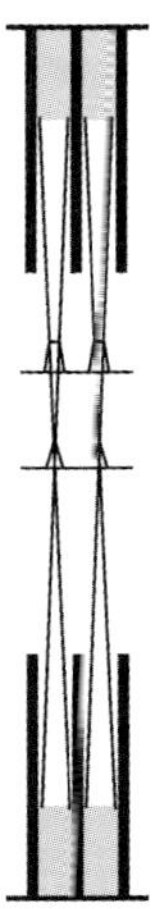

Fig. II.5.17. The intrinsic spatial resolution of the system depends on the size of the detectors, the line spread function being a triangle with a FWHM of half the detector dimension (reprinted from Budinger 1998 p. 259, with permission from Elsevier)

For such very high resolution systems, the contributions from annihilation and non-colinearity effects will also be significant. Taking the annihilation effects from Table II.5.1, the non-colinearity effects for a 100 cm diameter ring and the intrinsic spatial resolution for a 6 mm detector element, the total spatial resolution can be calculated to be the values shown in Table II.5.4.

A further deterioration in spatial resolution is caused, however, by the low light output of the scintillation crystal. For example with BGO, each annihilation photon interaction produces only ~150 photoelectrons at the PMT photocathodes. This results in significant fluctuation in position estimation because of the statistical variation in distribution among the 4 PMTs of the detector block. This causes deterioration in the spatial resolution of each detector of ~2 mm, which results in a further deterioration of the reconstructed spatial resolution, see Table II.5.5. Most systems reconstruct clinical images with reconstructed spatial resolutions that are 20–100% worse than the intrinsic resolution (Mullani 2000). It is the reconstruction process that imposes the further degradation on the spatial resolution; which is further degraded if acquisitions are undertaken in 3D mode rather than in 2D, determined not only by the geometry but also by the methods of scatter correction and the reconstruction strategy (Budinger 1998).

Table II.5.4. Spatial resolution values for usual radionuclides, including consideration of annihilation and non-colinearity effects, and detector element size

Radionuclide	Total resolution FWHM (mm)
^{18}F	3.70
^{11}C	3.73
^{13}N	3.75
^{15}O	3.86
^{68}Ga	3.96
^{62}Cu	4.37
^{82}Rb	4.54

Table II.5.5. Spatial resolution for usual radionuclides, including consideration of statistical fluctuation in event position estimation

Radionuclide	Total resolution FWHM (mm)
^{18}F	4.23
^{11}C	4.23
^{13}N	4.25
^{15}O	4.35
^{68}Ga	4.43
^{62}Cu	4.80
^{82}Rb	4.95

Variation Across the Transaxial Field of View

Towards the edge of the transaxial field of view, there is an increasing probability that a photon, incident on a detector crystal, will experience an acute angle of incidence. This is not the case at the centre of the field of view, where all photons will intersect the crystal perpendicular to its transaxial edge. A similar effect occurs axially, but this also depends on the separation of the detecting crystal rings. As the crystal thickness and the angle of incidence increase, there is an increasing probability that the interaction will occur in a detector adjacent to the one which first intersected the photon, see Fig. II.5.18. This results in an error in determining the photon entry point and a mispositioning of the projection line, in the radial and axial directions, causing a degradation in spatial resolution. Thus, radial and axial spatial resolution degrade with increasing distance from the centre of the field of view. Because the adjacent crystal is always nearer to the centre of the transaxial field of view, the spread of the resolution function will be towards the centre of the field of view. It is the opposite axially. For example, from a FWHM of ~4.5–6.3 mm at the centre of the transaxial field of view: transaxial spatial resolution deteriorates at a radial distance of 20 cm, to a radial resolution ~8.3–8.9 mm and a tangential resolution ~4.7–5.8 mm; and 3D axial spatial resolution to ~7.8–8.1 mm. The axial slice width measured in the 2D mode varies between ~4.0–4.5 mm FWHM at the centre of the transaxial field of view to ~5.0–6.0 mm at 10 cm offset (Adam et al. 1997; Brix et al. 1997; Tarantola et al. 2003).

Temporal Resolution

Nearly all systems allow data collection at 1–5 s intervals (Bergmann 1998).

Count Rate Capability

The count rate capability is ultimately determined by the light decay time of the crystals and the integration time, which is the time taken for ~90% of the scintillation light to be collected by the PMTs. The integration time is the major disadvantage of the sliced crystal block system, in that many detecting elements are coupled to only a few

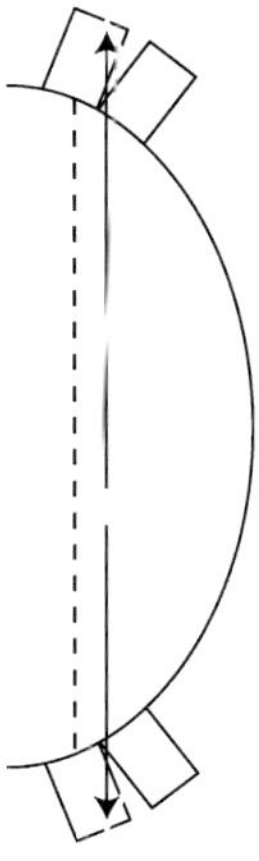

Fig. II.5.18. At acute interaction angles, the photons may be detected in crystals adjacent to the ones which first intersected the photons

PMTs. Because each crystal block has only one 'Anger' type processing circuit, events within the integration time of the unit cannot be separated even if they occur in different crystal elements. This causes both loss of detection and mispositioning of events (Budinger 1998). The integration time is usually 2–3 times the decay time of the crystal, which for BGO is $\sim 10^{-6}$ s, and the detected count rate (C) is:

$$C = C_0 e^{-\sigma C_0} \qquad \text{(II.5.9)}$$

where
C_0 is the incident count rate on the detector
σ is the dead-time, determined by the integration time

Thus, for each detector block, dead-time losses of ~10% are introduced at a singles rate of $\sim 3 \times 10^4$ counts s^{-1}. From this sort of calculation, predictions regarding the total singles count rate and hence total true coincidences rate can be made. In BGO systems, typical values for peak true coincidences rate are: for 2D acquisition, 1.89×10^5 counts s^{-1} at an activity concentration of 5.8×10^4 Bq cm^{-3}; and for 3D acquisition, 1.09×10^5 counts s^{-1} at 1.97×10^4 Bq cm^{-3}. Typical values for peak NEC rate are: for 2D acquisition, $>1.25 \times 10^5$ counts s^{-1} and for 3D acquisition, 1.92×10^4 counts s^{-1} at 7.15×10^3 Bq cm^{-3} (Daube-Witherspoon et al. 2002).

Correction

In most systems, the dead-time is measured and used to correct the acquisition data for detection loss. Correction techniques cannot, however, re-position those events which were mispositioned.

Acquisition

Acquisition modes are two dimensional with inter-slice absorptive septa in place and three dimensional with the septa removed. Whole body acquisition is an extension of these modes which extends the imaged volume along the axis of the patient by undertaking separate acquisitions in sequential axial positions. During an acquisition, coincidences are registered by many detector pair combinations, see Fig. II.5.19. These rep-

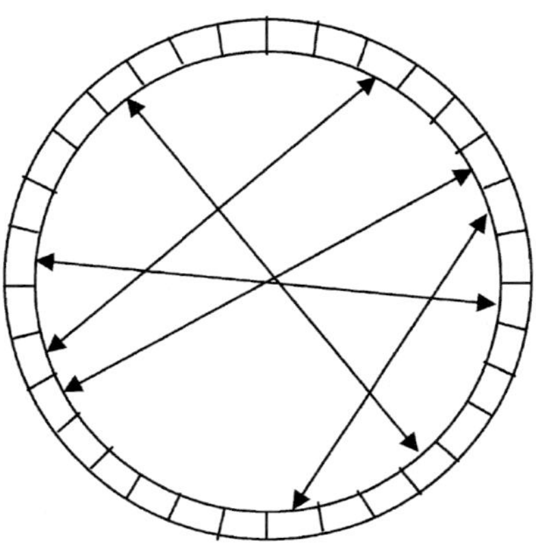

Fig. II.5.19. Coincidences registered between detector pairs

resent annihilations occurring in the many possible projection lines between various detector pairs. The counts registered in individual detector pair combinations are ordered into a sinogram matrix; within each row of which, the sequential counts stored represent the coincidences in detector pairs which form parallel projection lines across the imaged volume at a particular angle to the ring of detectors, see Fig. II.5.20. The stacked rows represent different angles around the detector ring.

2D Slice

A sinogram is created for each transaxial image plane, i.e. each crystal ring or plane in which the annihilation occurred.

Sensitivity is improved by accepting coincidence interactions occurring in other than the annihilation plane. These interactions must be in planes equal distances either side of a crystal ring. For example, if coincidence interactions are recorded in crystal rings 8 and 10, the event is added to the coincidences recorded for ring 9. The maximum such separation is limited because eventually, with increasing separation, the septa absorb the photons originating a number of crystal rings away. Even with such techniques, however, sensitivity is still low, which is apparent in Fig. II.5.4.

Axial spatial sampling is improved by also recording coincidence interactions in adjacent crystal planes and in planes equal distances either side of the space between two crystal rings. For example, if coincidence interactions are recorded in crystal rings 7 and 8, the event is classified as a coincidence occurring in ring 7.5. Thus a system with n crystal rings will generate 2n−1 sinograms and transaxial image planes. That is, a 16 crystal element system, would form 31 transaxial slices. Again the maximum separation of detection planes is limited because, with increasing separation, there is increasing axial blurring of the image which is exaggerated towards the periphery of the

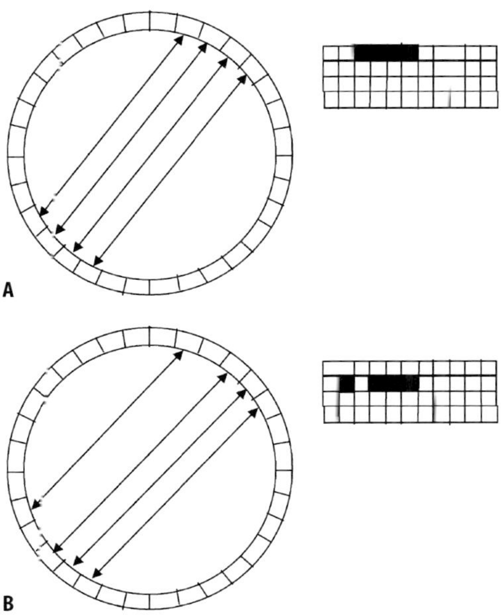

Fig. II.5.20. A and **B**, examples of sinograms. In each row, the sequential counts stored represent the coincidences in detector pairs which form parallel projection lines across the imaged volume at a particular angle to the ring of detectors. The stacked rows represent different angles around the detector ring

field of view, see 'Variation across the transaxial field of view' in 'Spatial resolution' above. This limits the maximum separation of rings contributing to an image to ~9, e.g. rings 3 and 12.

In interleaved mode, a bed moving technique can be used to improve the axial spatial sampling, similar to altering the pixel size in SPET. The imaging bed is moved a distance of a 1/4 of a crystal element width between two acquisitions. This interleaves two datasets, such that the separation between adjacent transaxial images is only 1/4 of a crystal element width instead of 1/2 of the width, discussed above.

2D Whole-Body

The axial image length can be increased by moving the imaging bed, between sequential acquisitions, a distance slightly less than the axial length of the system to create a small overlap between image volumes, see Fig. II.5.21. A combination of this and the interleaved mode, can be used to image with high axial spatial resolution over an extended axial distance of the patient. Such an acquisition can result in several hundred transaxial images.

3D Volume

This mode, shown in Fig. II.5.5, allows a finer axial sampling interval, not defined by the spacing between the crystal planes. This has the effect of minimising partial volume effects and reducing the variation in detection sensitivity along the axial field of view. 3D acquisition data can be re-ordered into sinogram matrices and then treated in the same way as 2D acquisition data, but this causes a deterioration in axial spatial resolution because of the uncertainty in positioning coincidences recorded in widely separated detectors. This is discussed later in "coincidence γ cameras", on which it is implemented. In dedicated systems, a 3D processing technique is used to prevent artefacts appearing in the images (Jarritt and Acton 1996). This technique considers the precise direction of each projection line and produces, for n detector rings, n^2 sinograms rather than the 2n–1 produced in 2D processing. There are two general approaches, the exact and the approximate method (Defrise et al. 1994). The former has the better signal to noise ratio, but the latter can be accomplished faster.

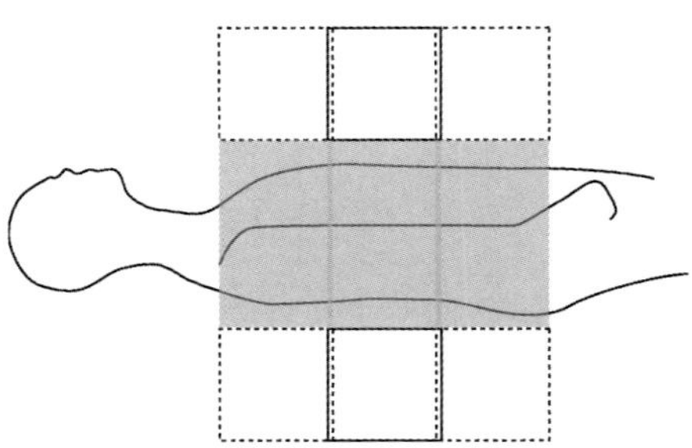

Fig. II.5.21. Whole-body acquisition is undertaken by moving the imaging bed, between sequential acquisitions, a distance slightly less than the axial length of the system to create a small overlap between image volumes

Scatter Rejection

Scatter rejection, by energy discrimination, is difficult using BGO, because of its poor energy resolution, of only 20–30%, which is a consequence of its inefficiency in light production. This means that the low energy threshold of the PHA window has to be set very low to maintain detection efficiency at reasonable levels. Typically, the PHA window is set at 350–650 keV (Adam et al. 1997). The difficulty of scatter rejection is compounded because the energy of scattered annihilation photons is not much less than 511 keV; the lower limit of 350 keV would correspond to a 57° deflection (see 'Scatter coincidences – energy discrimination' above). NaI(Tl) or GSO based systems have much better scatter rejection properties because of their superior energy resolution, i.e. 10–15%, which allows the PHA window to be set at 435–665 keV (Tarantola et al. 2003).

Attenuation Correction

Because the accuracy of attenuation correction is one of the major advantages of DPET, transmission attenuation correction is standard on dedicated systems; the coincidences recorded along a particular projection line indicating the amount of attenuation along that line (Turkington 2000). The transmission data will contain statistical noise depending on the number of counts acquired, and this noise will propagate into the corrected reconstructed image. It is the transmission statistics which limit the corrected reconstructed image quality (Xu et al. 1990). Therefore, the transmission acquisition duration must be sufficient to reduce the statistical noise to an acceptable level. This additional imaging time can be demanding on patient cooperation, especially if whole body acquisitions are required. The disadvantage of additional noise must therefore be balanced against the length of the total acquisition duration.

A number of geometries are in use. One is a ring source for each block ring, usually containing the β^+ emitting radionuclides ^{68}Ga/^{68}Ge in secular equilibrium. This was used in early systems. Coincidences are detected between a detector adjacent to the annihilation event and one following transmission through the patient (Zaidi and Hasegawa 2003). This design was later reduced to the present most common technique of 1–3 thin continuously orbiting ^{68}Ga/^{68}Ge axial rods (Bailey 1998; Zaidi and Hasegawa 2003), see Fig. II.5.22.

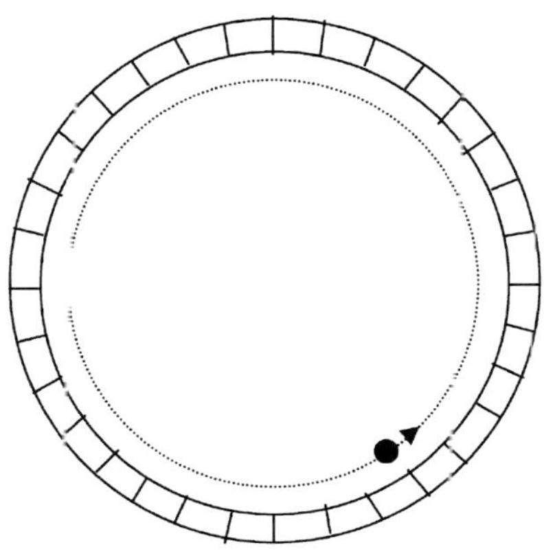

Fig. II.5.22. Transmission attenuation correction using an axial rod orbiting just inside the detector ring

The detector block adjacent to the source experiences a high count rate which can cause a high dead-time. It is avoided by electronic collimation which only accepts projection lines that pass through the known position of the rods (Budinger 1998; Lewellen et al. 1999; Zaidi and Hasegawa 2003). This reduces the number of scattered and random coincidences recorded, as well as the number of true coincidences from any radionuclide administered to the patient. The transmission acquisition can be carried out before, at the same time as or after the emission acquisition (Turkington 2000), the latter two reducing the imaging time by allowing the radiopharmaceutical accumulation phase to occur away from the machine.

Because of the potential dead-time problems and the disadvantage that, with a half-life of 288 days, $^{68}Ga/^{68}Ge$ sources must be replaced approximately every year, some manufacturers have recently started using singles sources such as ^{137}Cs (Lewellen et al. 1999; Beyer et al. 2000, 2002; Zaidi and Hasegawa 2003). In this, a rod rotates just inside the field of view (Turkington 2000). Higher acquisition rates can be tolerated but the data then contains significant amounts of scatter (Lewellen et al. 1999). In addition, the higher energy photon improves object penetration (Zaidi and Hasegawa 2003).

CT

Transmission data provided by CT (Beyer et al. 2000, 2002) is the best option. If the unit is not incorporated with the DPET system, both transmission and emission images must be accurately co-registered (Jarritt and Acton 1996). When incorporated with the DPET system, the CT unit is at the front of the gantry, before the DPET detectors (Townsend and Beyer 2002). This is the opposite configuration to that used in SPET. The CT images must be smoothed to achieve the spatial resolution of DPET, often using a Gaussian filter with FWHM of 8 mm. The attenuation coefficients must also be transformed to represent the attenuation experienced by the annihilation photons (Kinahan et al. 1998).

Reconstruction

The sinograms, created from the 2D or 3D acquisitions, are corrected for attenuation using the attenuation correction factors (ACF), calculated for different parts of the image. These are the $\log_e$ ratios of the blank/transmission values. The corrected sinograms are then reconstructed to form transaxial images, using filtered back-projection (FBP) or iterative reconstruction. It has been shown that iterative methods produce better results than FBP (Hsu 2002).

Gamma Camera Systems

The main obstacle to the implementation of dedicated systems is the cost of the equipment. The clinical information, achieved using such systems particularly with ^{18}FDG, has encouraged the use of γ cameras for imaging annihilation radiation. This can be undertaken in two ways. Very high energy collimators can be installed and the camera operated in single photon detection mode or modifications can be made to a multiple head system and the camera operated in dual photon detection mode. An advantage is that the axial field of view is much larger on γ camera systems, ~40 cm, than on dedicated systems, up to 25 cm (Budinger 1998; Daube-Witherspoon et al. 2002; Gemmell et al. 2002), which is valuable for whole body acquisitions.

Collimated Gamma Cameras

This is an inexpensive option but the spatial resolution and sensitivity improvements, promised by DPET imaging, are not realised because ultra high-energy collimators are used and photons are detected as 'singles'. The technique was initially applied in oncology but its lesion detecting capability was disappointing. It is still used successfully in cardiology, for assessing myocardial viability (Kuge and Katch 2001), where the spatial resolution demands are not as critical (Jarritt and Acton 1996; Dilsizian et al. 2001). The collimator septa must be thick to avoid penetration (Dilsizian et al. 2001) and, to achieve a spatial resolution FWHM of 10 mm, lead septa would cover ~50% of the face (Blockland et al. 2002). For approximately comparable spatial resolution, ultra-high energy collimators yield only ~1/4 the detection sensitivity of low energy collimators (Budinger 1998). Thus, a FWHM of only 15–20 mm is expected (Bar Shalom et al. 2000), but even so the detection sensitivity is low, a typical clinical count rate being only 10^3 counts s^{-1}. The low level background caused by septal penetration will be likely to reduce clinical efficacy and the weight involved can cause gantry problems (Dilsizian et al. 2001). When compared with dedicated systems, even operating in 2D mode, the sensitivity reduction factor is ~40 (Lewellen et al. 1999). It is also not possible to produce collimators with different spatial resolution characteristics to address the various detection requirements. To detect a slight reduction in activity over a large area, low noise images and a high sensitivity, poor resolution collimator are required. To detect a small area of increased activity, a poor sensitivity, high resolution collimator is required (Rousset et al. 1998). These limitations have transferred enthusiasm from collimated to coincidence systems (Jarritt and Acton 1996).

Coincidence Gamma Cameras

The more expensive but better option is to use the camera without conventional collimation and operated in coincidence mode, offering the potential to detect lesions of only a few millimetres in size (Delbeke and Sandler 2000). The principles will be reviewed in a dedicated report (IPEM 2004). The cost, additional to a SPET system, remains low compared to a dedicated system, and the system can be used in either single or dual photon mode. Because of this, these systems have stimulated an increase in clinical β^+ imaging. The technique requires a dual detector system with the heads operated at a relative orientation of 180°, or a triple head system (Kuikka et al. 2002). The dual head systems use a coincidence window of 1.2×10^{-8} s and the triple head system use a narrower 1.0×10^{-8} s, to reduce the increase in random coincidences occasioned by the extra detector (Tarantola et al. 2003).

Modifications

For coincidence mode imaging to be successful using γ cameras, a number of construction alterations are necessary. The priority is to improve the count rate capability and this has been achieved for investigations using ^{18}F. It is unlikely that this will be repeated for other radionuclides, because these all have much shorter half-lives and demand much higher count rate capability to acquire the counts necessary in clinical investigations (Lewellen et al. 1999). In order to provide coincidence detection imaging of acceptable quality, the trend for γ cameras to be optimised for low energy (<200 keV) imaging has had to be reversed.

Detector Shielding

The detector head shielding has had to be thickened significantly, in order to shield the detector from radiation outside the field of view. The magnitude of the problem can be appreciated when the required amount of lead is considered, the half-value thickness is 0.3 mm for ^{99}Tc^{m} photons and 4.1 mm for annihilation photons, and the tenth value thicknesses are 0.9 mm and 13.5 mm, respectively. This means that the gantry must be more substantial to maintain controlled tomographic movement whilst carrying this extra weight.

Crystal Thickness

NaI(Tl) crystals are used exclusively (Lewellen et al. 1999) and one of the two important factors, which affect the performance of a coincidence γ camera, is the thickness of these (Fleming et al. 2000), see Table II.5.6. The usual value in a single photon system is 9.5 mm because this is the optimal for ^{99}Tc^{m} photons, for which it has an absorption efficiency of 84%. However, the absorption efficiency is only 13% for annihilation photons, which is comparable with wire chambers, but can be improved by increasing the thickness (Jarritt and Acton 1996). The improvement in total detection efficiency is dramatic, because this is proportional to the product of the absorption efficiencies of each detector, see 'True coincidences' above (Jarritt and Acton 1996; Bar Shalom et al. 2000; Fleming et al. 2000). Thus, although the absorption efficiency of a 19.1 mm crystal shows an improvement of 85% over a 9.5 mm crystal, the total detection efficiency shows an improvement of 241%. Thicknesses of ~25 mm are under consideration. However, even at these, absorption efficiencies at 511 keV are very low in comparison with the 0.9 achieved by dedicated systems. Again, the difference is especially dramatic when the total efficiency is considered, being 81% (0.9^2) in dedicated systems and

Table II.5.6. Absorption efficiencies for different thicknesses of NaI(Tl) crystals

Thickness (mm)	511 keV	^{99}Tc^{m}
9.5	0.13	0.84
12.7	0.17	0.91
15.9	0.21	0.95
19.1	0.24	0.98

only 5.8% (0.24^2) with the 19.1 mm crystal, i.e. a difference of nearly 14 times (Delbeke and Sandler 2000).

Disadvantages of Increasing Crystal Thickness

Intrinsic spatial resolution deterioration imposes a limit on the maximum thickness of crystal which can be used, see 'Spatial resolution – intrinsic' in Chapter II.3. This is a particular problem with low photon energies such as from ^{201}Tl, but is less severe at higher energies; for instance with ^{99}Tc^{m}, the FWHM is 3.8 mm in a 9.5 mm crystal and 4.3 mm in a 15.9 mm crystal. The effect of this difference is reduced by the collimator which dominates the system spatial resolution, particularly at depth (Fleming et al. 2000), to such an extent that at 10 cm in air, the system spatial resolution differs by only 0.2 mm for the above two crystals when a LEHR collimator is used (Patton and Turkington 1999). The spatial resolution degradation is not clinically significant, therefore (Patton 2000; Bird et al. 2001), and clinically similar whole body bone scans have been obtained with a dual-head coincidence gamma camera and a conventional dual-head gamma camera (Inoue et al. 2001).

A number of modifications to the crystal have been proposed, one of which is that, in the thicker crystals, the face nearer the PMTs is scored orthogonally, see Fig. II.5.23. These cuts restrict the light dispersion, which improves spatial resolution, at the lower energies, when the camera is used in single photon mode (Patton 2000).

Alternative Crystal Designs

Improved absorption efficiencies, for both low and high photon energies, can be achieved using lutetium oxyorthosilicate (LSO) and NaI(Tl) (Bendriem 1999; Patton 2000) or LSO and Y_2SiO_5(Ce) (YSO) crystals sandwiched together in a "phoswich" type detector (Fahey 2001). In the latter, 1.5–2 cm of YSO is placed in front of LSO in a block detector design. The density and light output of LSO gives a good efficiency for coincidence detection but because it has a high natural radioactivity, it cannot be used for detection of single photons (Blockland et al. 2002). These are detected by the low density but very high light output YSO. The scintillation decay times in the two crystals are different, and events can be localised to each by pulse shape discrimination. The refractive index of both crystals are similar, which allows efficient transfer of the light generated by the low energy photons to the PMTs. Also the use of many small block detectors, gives significant gains in count rate capability (Budinger 1998; Bendriem et al. 1999) over a single large area detector, because of parallel signal processing.

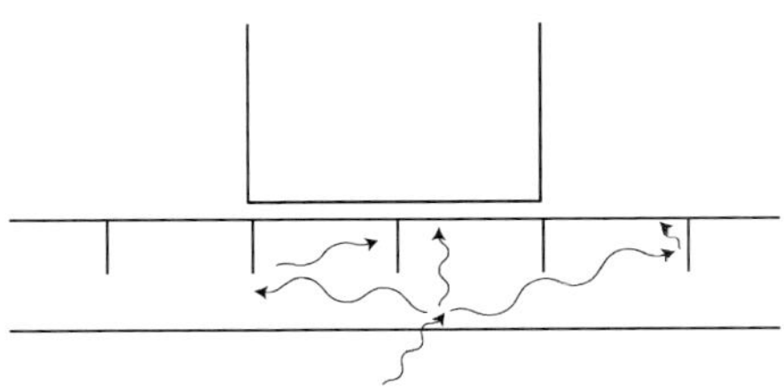

Fig. II.5.23. Scoring of crystal, at the face adjacent to the PMT, to restrict light dispersion

Pulse Height Analyser

The energy range of the PHA has been expanded, from ~400 keV, to include the 511 keV of the annihilation photons. Because of the dependence of camera uniformity on incident photon energy, this has involved extending the linearity and uniformity correction maps (Patton 2000). The energy windows used in coincidence imaging are wider, at ~30% (Kunze et al. 2000), than those used in SPET, to increase detection sensitivity. However, this reduces the true coincidences fraction by increasing the scatter and randoms fractions (Fleming et al. 2000) from ~20% to ~25% for 2D and from ~30% to ~38% for 3D acquisition (Kunze et al. 2000).

Coincidences could only be recorded when the detected energy, of both photons is in the photopeak of the annihilation radiation, each photon being totally absorbed by photoelectric interaction in the crystal (see Fig. II.5.24). To compensate for the poor absorption efficiency of NaI(Tl) however, these systems optionally register additional coincidences in which one of the coincident photons undergoes sub-total absorption by Compton interaction in the crystal (Kunze et al. 2000). The valid coincidence combination is a total absorption in one detector and a sub-total absorption in the opposing detector, which increases absorption sensitivity by a factor of ~2. The additional interactions are detected, in the Compton region of the PHA spectrum, by setting a second energy window around 320 keV. This energy level decreases the contribution from scattering in the patient, which is further reduced by installing filters over the crystal face to limit the number of scattered photons reaching the crystals. As well as increasing the scatter fraction, from ~25% to ~33%, this technique significantly increases the randoms fraction; together accounting for >50% of total coincidences. It also slightly degrades the spatial resolution (Lewellen et al. 1999; Patton and Turkington 1999). A further increase in sensitivity is achieved by recording coincidences in which both interactions are detected in the Compton window, but this exaggerates the loss of contrast. The need to use this technique diminishes with thicker crystals, because of the increased probability of total absorption (Fleming et al. 2000).

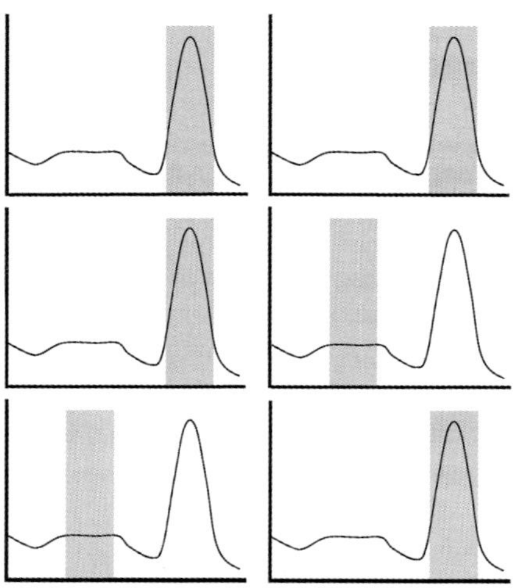

Fig. II.5.24. Various coincidence interaction types: top, photopeak–photopeak; middle, photopeak–Compton; bottom, Compton–photopeak

Count Rate Capability

One of the two important factors, which affects the performance of a coincidence γ camera, is the count rate capability; because only ~1% of the interactions detected at each head result in a coincidence, the rest being singles events which are not included in the image data (Fleming et al. 2000). The latter are particularly prevalent, because it is impractical to use long end shields and contributions from activity outside the field of view are significant (IPEM 2003). The ratio of true to single coincidences can be increased by installing slat collimation transaxially, which need only be coarse, thus reducing the acceptance of single and scattered events by limiting the axial acceptance angle (Kunze et al. 2000). Because of the prevalence of singles, to achieve a clinically acceptable coincident counting rate, a very high total counting rate must be accommodated. The typical clinical coincidence rate, without loss of image quality, is only ~7–10 $\times 10^3$ counts s^{-1} in 2D acquisition mode (Lewellen et al. 1999; Patton and Turkington 1999) and ~20×10^3 counts s^{-1} in 3D acquisition mode (Budinger 1998). At these true coincidences rates, the expected singles rate will be ~$7–20\times 10^5$ counts s^{-1}, and the count rate capability of each head must be equal to this (Patton and Turkington 1999; Fleming et al. 2000). When compared to the typical clinical true coincidences rate in dedicated systems of $3–6\times 10^5$ counts s^{-1} (Budinger 1998; Lewellen et al. 1999), these figures demonstrate how much longer the acquisition time must be to acquire statistically acceptable data. As the singles rate increases, the probability of random coincidences and count losses due to dead-time also increases. The latter to >25% at the peak of the true coincidences rate curve. The best image contrast is achieved in the portion of the true coincidences rate before the peak and the image starts to degrade as the peak is approached. In 3D imaging, the true coincidences rate curve rises more rapidly and peaks earlier than in 2D imaging (Patton and Turkington 1999), see Fig. II.5.25.

To achieve the required increase in the maximum count rate capability, all coincidence γ cameras have a high count rate mode. Two main techniques have been used to obtain this: pulse clipping and local centroiding. Both techniques degrade energy resolution and, to a lesser extent, intrinsic spatial resolution (Fleming et al. 2000).

Pulse Clipping

The integration time, for the collection of scintillation light after photon interaction in the crystal, is reduced, by a factor of ~5 from the usual 10^{-6} s used in single photon de-

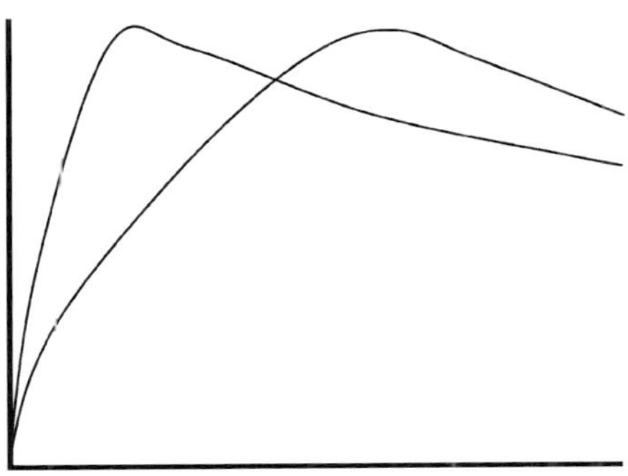

Fig. II.5.25. True coincidences rate curve. The ordinate represents the true coincidences rate and the abscissa represents the singles rate. As the single rate increases, the true coincidences rate increases to a peak and then decreases. The true coincidences rate curve which rises more rapidly and peaks earlier is obtained in 3D imaging, the curve which rises less rapidly and peaks later is obtained in 2D imaging

tection (Blockland et al. 2002). This reduces the dead-time and is feasible because the amount of light produced by the total absorption of an annihilation photon is ~4 times as much as that produced by a $^{99}Tc^{m}$ photon. The energy resolution is not as good as it could be, because not all of the scintillation light is collected, which introduces some uncertainty into the measurement (Patton and Turkington 1999).

Local Centroiding

The identification of the interaction position is accomplished using only a subset of the PMTs (Budinger 1998), restricted to the closest one and the 6 adjacent (Patton 2000), see Fig. II.5.26, which allows detection of an interaction in one part of the crystal to proceed independently of events in other parts of the crystal (Patton and Turkington 1999), making it possible to register several events simultaneously and significantly increase count rate capability. Digital cameras, in which each PMT has an ADC, should enable an even higher count rate capability whilst maintaining image quality and minimising count losses, by permitting more effective segmentation of the crystal.

Performance Parameters

Sensitivity

Tomographic sensitivity is a very difficult performance characteristic to compare among systems, being dependent on both phantom size and detector geometry (Jarritt and Acton 1996), such as number of heads, head configuration, spacing between gantry stops, radius of rotation and coincident head acceptance angle (Stodilka and Glick 2001). The lack of absorptive collimation means that the detection sensitivity is high, at least 10^4 higher than collimated mode. However since both photons must be detected, the effective sensitivity gain is only about 10^2. Furthermore, since both photons must exit the body and this combined attenuation is greater than single photon attenuation at 140 keV, the overall sensitivity gain is <10. The sensitivity for detection of true coincidences ~650 counts s^{-1} kBq^{-1} ml^{-1} in 2D acquisition and ~3000 in 3D acquisition (Kunze et al. 2000). This means that the activity in the field of view must be much less than used in collimated mode to avoid dead-time losses (Lewellen et al. 1999).

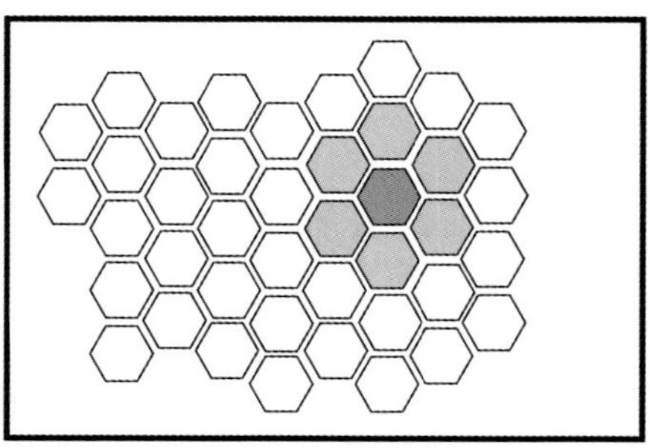

Fig. II.5.26. Local centroiding. Count rate capability is increased by restricting the PMTs, which are used to collect the scintillation light, to the one closest to the interaction and the 6 adjacent

Variation Across the Transaxial Field of View

There is a variation in sensitivity across the transaxial field of view; the maximum sensitivity occurring at the centre and decreasing towards the periphery (Fig. II.5.27). This is caused by the solid angle geometrical effect, whereby there is a reduction in solid angle within which the annihilation photons can reach both the detectors, see Fig. II.5.28. This necessitates correction of the sinogram data, using a sensitivity profile, during image reconstruction (Patton and Turkington 1999). These correction factors are large, tending to infinity at the edge of the field of view, and cause noise amplification (Lewellen et al. 1999). This sensitivity map can be measured or calculated, a calculated map being preferable because the radius and the head orientation often change between different acquisitions (Vandenberghe et al. 2002).

Spatial Resolution

Removal of absorptive collimation means that the spatial resolution of the system is dominated by the intrinsic spatial resolution of the detector, and this does not deteriorate with depth in the patient. In air, the axial FWHM is 6–6.5 mm for 2D acquisition with an axial acceptance angle of 2° and ~10 mm with an angle of 16°. The transaxial FWHM is 4.5–7 mm at the centre of the field of view (Delbeke and Sandler 2000; Bar Shalom et al. 2000; Fleming et al. 2000; Kunze et al. 2000), which is very close to that of dedicated systems. Unlike dedicated systems, however, there is less deterioration in the spatial resolution towards the periphery of the transaxial field of view, approximating 8 mm. This is because of the extended flat nature of the crystals. The photons, from pe-

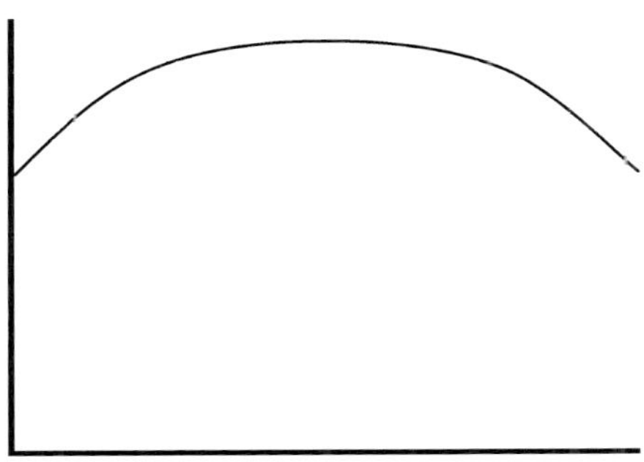

Fig. II.5.27. Sensitivity variation across the transaxial field of view, showing decrease towards the periphery. The ordinate represents detection sensitivity and the abscissa represents the position across the field of view

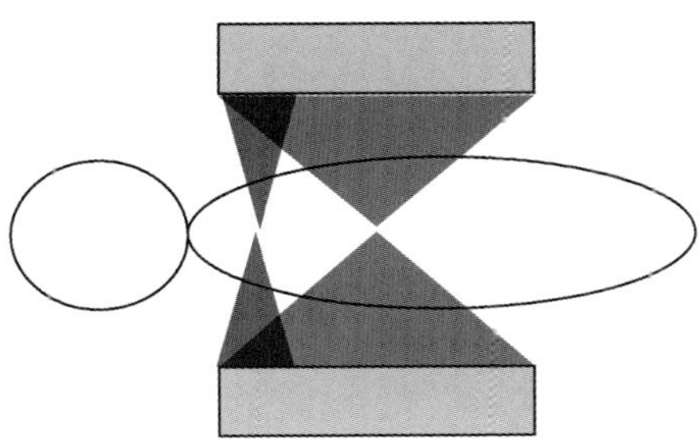

Fig. II.5.28. Solid angle effect. The variation in sensitivity across the transaxial field of view, shown in Fig. II.5.27, is caused by the solid angle geometrical effect. Sources at the periphery have a reduced number of possible projection paths compared to those at the centre of the field of view

ripheral annihilations, remain incident on the crystals nearly perpendicularly and there will be little uncertainty in determining their entry point. The loss in spatial resolution being a result of the reduced detection sensitivity (Kunze et al. 2000).

An effect not experienced in dedicated systems is again caused by the extended flat nature of the crystals. Because a circle of detectors is not formed around the patient orientated towards the centre of the field of view, a photon originating at the centre of the field of view will be incident at an acute angle if it strikes the crystal at its edge. This will cause a deterioration in spatial resolution because of the uncertainty in determining the entry point (see Fig. II.5.29). However, because it is necessary to increase the sensitivity as much as possible, it is necessary to accept all photon angles of incidence and the consequential degradation in spatial resolution, which may be as high as 40% (Patton and Turkington 1999).

Acquisition

Data acquisition times are extended also because of the need to rotate the detectors around the object, which limits the radionuclide which can be used to the longer lived ^{18}F. Data is acquired in a step and shoot fashion. Cameras with slip-ring technology use multiple rotations, e.g. 3 min per rotation to minimise the effect of radionuclide decay whereas cameras without this technology perform a single slow rotation (Patton and Turkington 1999; Delbeke and Sandler 2000).

Acquisition Mode

The systems are operated either in the 3D mode without any absorptive collimation or with absorptive inter-plane collimation across the axis of rotation, similar to the 2D mode of dedicated systems.

3D Mode

The filters installed over the crystal faces in 3D acquisition mode, to reduce the number of scattered photons reaching the detectors, comprise a sheet of lead with a backing of lower atomic number materials. The lead absorbs the lower energy scattered photons originating from the field of view and the backing absorbs x photons pro-

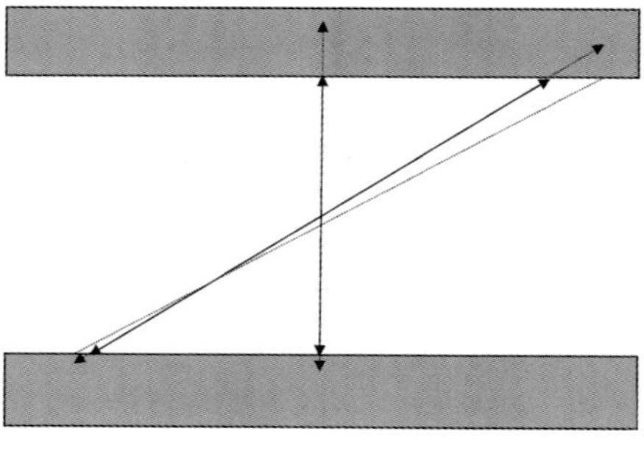

Fig. II.5.29. Acute incidence causing a deterioration in spatial resolution because of the uncertainty in determining the entry point of the photon into the crystal

duced by interactions in the lead. The sheet is surrounded by a thick lead frame, which reduces the number of scatter coincidences originating from outside the field of view (Fleming et al. 2000). The latter is particularly necessary in body imaging, but less so in brain imaging. With this collimation, the scatter fraction is similar to that experienced in 3D dedicated systems, ~26% using photopeak–photopeak interactions only and ~38–45% using both photopeak–photopeak and photopeak–Compton interactions (Lewellen et al. 1999; Patton and Turkington 1999; Kunze et al. 2000).

2D Mode

In this mode; collimators, with parallel lead septa arranged perpendicularly to the axis of rotation, define transaxial slices (Fleming et al. 2000). The coarseness varies but examples are: 57 × 5 or 40 × 4 mm septa, spaced at 10 or 12 mm intervals. By restricting the detection solid angle, the septa reduce the fraction of events that are due to randoms and scatter, and significantly reduce the singles rate from activity outside of the field of view (Delbeke and Sandler 2000). This restriction in the axial acceptance angle leads to improved NEC rates for torso imaging (IPEM 2003). The scatter fraction is reduced to ~14% using photopeak–photopeak interactions only and to ~25–29% using both photopeak–photopeak and photopeak–Compton interactions. However, these improvements are again achieved at the expense of a reduction in true coincidences sensitivity (Lewellen et al. 1999; Patton and Turkington 1999; Fleming et al. 2000; Kunze et al. 2000).

Not all out-of-plane coincidences are removed, but the maximum separation between the planes and the origin of the annihilation event, is often restricted, as in dedicated systems (Patton and Turkington 1999). There is no intra-slice collimation and transaxial coincidences may detected at any angle. In this way, the arrangement is similar to that used in 2D acquisitions in dedicated systems.

Attenuation Correction

The first systems did not provide transmission techniques for attenuation correction, but these have recently been introduced and measure the distribution of attenuation coefficients in a manner similar to used on dedicated systems. Both radioactive source and CT based systems are available, the CT systems also providing anatomic detail to aid interpretation (Turkington 2000).

Requirements

The transmission acquisition duration should be short compared with the emission acquisition duration, which is already long, but the transmission data should have high enough counts that no substantial statistical noise is introduced into the reconstruction. In dedicated systems it is difficult to fulfil both, but with coincidence γ cameras the emission count rate is much lower and the quality of transmission data can therefore be lower. Transmission acquisitions should be able to be acquired after administration, to maximise departmental efficiency and patient cooperation, whilst avoiding patient motion between transmission and emission acquisitions. These acquisitions must not, therefore, be influenced by the emission radiation.

^{137}Cs Source

The usual transmission source used is ^{137}Cs (Fleming et al. 2000), which emits 662 keV photons and which, together with the good energy resolution of NaI(Tl), allows clean transmission data to be acquired in the presence of emission radiation (Gemmell et al. 2002). The maps of attenuation coefficients, reconstructed from this data, have to be transformed to those applicable at the annihilation energy, however. ^{137}Cs also has a half-life of 30 years which means that the system does not have to accommodate a range of source activities, unlike ^{68}Ge used on dedicated systems and ^{153}Gd used on SPET which have half-lives of less than a year. However, high activity sources have to be installed because the detector efficiencies are even lower, at 662 keV, than they are at annihilation energies. This requires heavier shielding which puts more strain on the gantry.

Implementation

The sources can be arranged as, one or more, fixed point sources attached to one camera. They can be collimated by a local shield or axially collimated, into multiple fan beams, by the septa on the opposing camera (Laymon 2000; Turkington 2000). They are positioned as close to the edge of the camera as possible to maximise the field of view. Alternatively, a single collimated ^{137}Cs source can be translated along the edge of each camera, exposing the opposing camera to a fan beam of radiation. These arrangements can achieve a count rate which is sufficient to allow the transmission acquisition duration to be much shorter than the emission acquisition duration. However, the acquisition geometries require the transmission data to be either re-binned into parallel projections and reconstructed as usual or reconstructed using an iterative algorithm that includes a consideration of the acquisition geometry.

Other Sources

^{133}Ba and ^{153}Gd are also used; ^{133}Ba having the advantage of long half-life and reduced replacement rates (Zaidi and Hasegawa 2003). β^{+} emitting sources are not used because the coincidence count rate is severely limited and the technique is not practical. The detection sensitivity is lower for coincidence detection than the single photon detection of ^{137}Cs, and the activities of the sources must also be restricted because of their closeness to one camera and the probability of compromising the count rate capability of the system (Turkington 2000; Dilsizian et al. 2001).

CT

As in dedicated systems, a spiral CT can be installed on the system (Patton et al. 2000), in this case between the camera heads and the main gantry. Post administration transmission acquisition is possible because the count rate from the CT is much higher than the emission count rate. The attenuation map needs to be transformed to be applicable at the annihilation energy, and because of this, a relatively high tube voltage (140 kVp) must be used. At lower voltages photoelectric absorption is significant, which makes this transformation difficult. Even low CT currents yield adequate attenuation maps (Kamel et al. 2002). Again, as well as providing an attenuation map to correct for attenuation, the CT image can be used to help interpret the DPET image (Turkington 2000), which improved lesion localisation in one-third of the patients studied (Delbeke et al. 2001).

Other Corrections

Other factors, including scatter and random coincidences, produce less severe problems than attenuation and, although corrections are essential for accurate quantitation, they primarily increase the background level, which can be manipulated by display adjustment. It is likely, however, that they will be implemented (Turkington 2000). Both scatter (Kunze et al. 2000) and random coincidences can constitute a significant fraction of total coincidences, and because randoms are concentrated along the axis of rotation in accordance with the sensitivity profile, they impose a non-uniform background. To maximise image quality, it is important that this is subtracted (Lewellen et al. 1999).

Reconstruction

2D reconstruction is appropriate on data acquired when slit absorptive collimation is implemented on acquisition, in that it is designed for independent transaxial slice data. With 3D mode acquisition data, a 3D reconstruction is more appropriate and results in improved image quality, by optimising the inherent spatial resolution in the 3D data set. However, a popular technique is to re-bin the 3D data into 2D transaxial slice data and then reconstruct using 2D filtered back-projection or iterative algorithms. The re-binning is accomplished by allocating each projection line, which is detected by coincidence interactions in different transaxial planes over the axial field of view, into appropriate transaxial slice(s).

The re-binning introduces an approximation into the data which is dependent on the method used. There are three re-binning techniques: single-slice SSRB, multi-slice MSRB and Fourier FORE. The simplest and currently the most widely used, is SSRB, in which the annihilation event is positioned in the transaxial slice mid-way between the slices in which the interactions actually occurred (Fig. II.5.30). Axial and, to a lesser extent, tangential spatial resolution deteriorate as the activity moves away from the centre of the field of view, reducing the detectability of low contrast lesions at the periphery of the transaxial field of view. This can be improved by restricting the axial angles

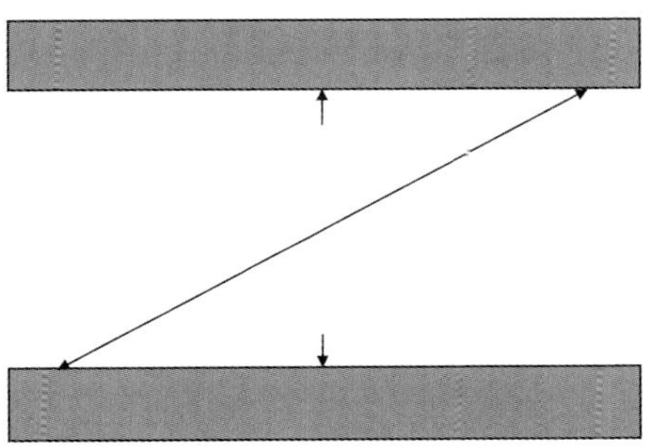

Fig. II.5.30. SS re-binning, the annihilation event is positioned in the transaxial slice mid-way between the slices in which the interactions actually occurred

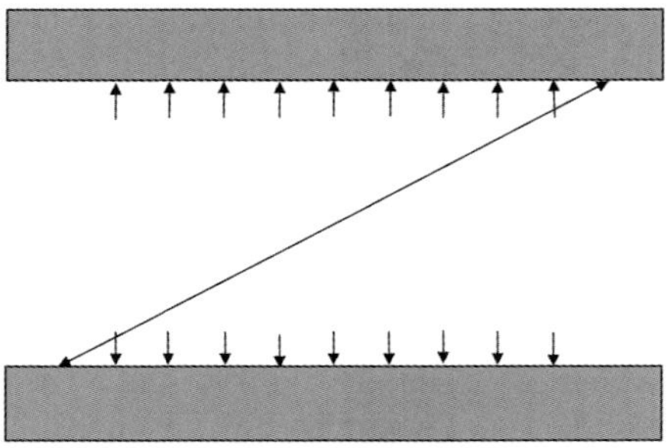

Fig. II.5.31. MS re-binning, an event is allocated to every transaxial plane between the detecting slices

of the projection line allowed to be re-binned. MSRB allocates an event to every transaxial plane between the detecting slices, which degrades axial spatial resolution (Fig. II.5.31). This can be corrected using recovery filters, but these introduce noise into the reconstructed image.

FORE achieves a quality of reconstructed image similar to produced by 3D reconstruction techniques (Lewellen et al. 1999; Patton and Turkington 1999; Fleming et al. 2000), and is especially effective when combined with attenuation correction and OS-EM techniques (Lartizien et al. 2003). The reconstruction process is then essentially the same as that for SPET (Delbeke and Sandler 2000). Typically, on the axis of rotation, the FWHM axial resolution is 6–10 mm and the transverse resolution is 6–6.5 mm (Kunze et al. 2000).

Iterative

The emission data can be reconstructed using an iterative reconstruction algorithm based on the ordered subsets estimation maximisation (OS-EM) algorithm, which results in improved spatial resolution and reduced statistical noise compared with filtered back-projection (Lewellen et al. 1999; Delbeke and Sandler 2000; Fleming et al. 2000; Patton 2000). Also 3D reconstruction algorithms can be incorporated directly, and any corrections can be implemented during the reconstruction process (Patton and Turkington 1999). A modification of OS-EM, the coincidence-list-ordered sets expectation-maximisation (C-OS-EM) (Levkovitz et al. 2001; Delbeke and Sandler 2000) can be used if the data is acquired in list-mode. It is suitable for two rotating planar detectors (Levkovitz et al. 2001), and is implemented on systems using slip ring technology (Patton 2000). Each rotation of the detector heads defines a subset of data and therefore the subsets are ordered in time instead of space as with the conventional OS-EM algorithm (Patton 2000; Delbeke and Sandler 2000). It has been suggested that C-OS-EM should replace FBP, as it becomes available (Delbeke et al. 2001).

References

Adam L-E, Zaers J, Ostertag H, Trojan H, Bellemann ME, Brix G (1997) Performance evaluation of the whole-body PET scanner ECAT EXACT HR following the IEC standard. IEEE Trans Nucl Sci 44:1172–1179

Adam LE, Karp JS, Daube-Witherspoon ME, Smith RJ (2001) Performance of a whole-body PET scanner using curve-plate NaI(Tl) detectors. J Nucl Med 42:1821–1830

Akhurst T, Chisin R (2000) Hybrid PET/CT machines: optimized PET machines for the new millennium? J Nucl Med 41:961–962

Bailey DL (1998) Transmission scanning in emission tomography. Eur J Nucl Med 25:774–787

Bailey DL, Young H, Bloomfield PM, Meikle SR, Glass D, Myers MJ, Spinks TJ, Watson CC, Luk P, Peters AM, Jones T (1997) ECAT ART – a continuously rotating PET camera: performance characteristics, initial clinical studies, and installation considerations in a nuclear medicine department. Eur J Nucl Med 24:6–15

Bar Shalom R, Valdivia AY, Blaufox MD (2000) PET imaging in oncology. Semin Nucl Med XXX: 150–185

Bartenstein P, Asenbaum S, Catafau A, Halldin C, Pilowski L, Pupi A, Tatsch (2002) European Association of Nuclear Medicine procedure guidelines for brain imaging using [^{18}F]FDG. Eur J Nucl Med 29:B43–B48

Bendriem B, Casey M, Eriksson L, Schmand M, Eriksson M, Frey J, Nutt R (1999) From PET to PET/-SPECT. Rev Acomen 5:156–159

Bergmann SR (1998) Cardiac positron emission tomography. Semin Nucl Med XXVIII:320–340

Bettinardi V, Pagani E, Gilardi MC, Landoni C, Riddell C, Rizzo G, Castiglioni I, Belluzzo D, Lucignahi G, Schubert S, Fazio F (1999) An automatic classification technique for attenuation correction in positron emission tomography. Eur J Nucl Med 26:447–458

Beyer T, Townsend DW, Brun T, Kinahan PE, Charron M, Roddy R, Jerin J, Young J, Byars L, Nutt R (2000) A combined PET/CT scanner for clinical oncology. J Nucl Med 41:1369–1379

Beyer T, Townsend DW, Blodgett TM (2002) Dual-modality PET/CT tomography for clinical oncology. Q J Nucl Med 46:24–34

Bird NJ, Old SE, Barber RW (2001) Gamma camera positron emission tomography. Br J Radiol 74: 303–306

Blokland JA, Trindev P, Stokkel MP, Pauwels EK (2002) Positron emission tomography: a technical introduction for clinicians. Eur J Radiol 44:70–75

Brix G, Zaers J, Adam LE, Bellemann ME, Ostertag H, Trojan H, Haberkorn U, Doll J, Oberdorfer F, Lorenz WJ (1997) Performance evaluation of a whole-body PET scanner using the NEMA protocol. J Nucl Med 38:1614–1623

Budinger TF (1998) PET Instrumentation: what are the limits? Semin Nucl Med XXVIII:247–267

Burger C, Berthold T (2000) Physical principles and practical aspects of clinical PET imaging. In: von Schulthness GK, Buck A, Engel-Bicik I, Steinert HC (eds) Clinical positron emission tomography. Correlation with morphological cross-sectional imaging. Lippincott Williams and Wilkins, Philadelphia

Cherry SR, Phelps ME (1996) Positron emission tomography: methods and instrumentation. In: Sandler MP, Patton JA, Coleman RE, Gottschalk A, Wackers FJT, Hoffer PB (eds) Diagnostic nuclear medicine, 3rd edn. Williams and Wilkins, Baltimore, pp 139–159

Daube-Witherspoon ME, Karp JS, Casey ME, DiFilippo FP, Hines H, Muehllehner G, Simcic V, Stearns CW, Adam LE, Kohlmyer S, Sossi V (2002) PET performance measurements using the NEMA NU 2-2001 standard. J Nucl Med 43:1398–1409

Defrise M, Geissbuhler A, Townsend DW (1994) A performance study of 3D reconstruction algorithms for positron emission tomography. Phys Med Biol 39:305–320

Delbeke D, Sandler MP (2000) The role of hybrid cameras in oncology. Semin Nucl Med XXX: 268–280

Delbeke D, Martin WH, Patton JA, Sandler MP (2001) Value of iterative reconstruction, attenuation correction, and image fusion in the interpretation of FDG PET images with an integrated dual-head coincidence camera and X-ray-based attenuation maps. Radiology 218:163–171

Dilsizian V, Bacharach SL, Khin MM, Smith MF (2001) Fluorine-18-deoxyglucose SPECT and coincidence imaging for myocardial viability: clinical and technologic issues. J Nucl Cardiol 8:75–88

Fahey FH (2001) Positron emission tomography instrumentation. Radiol Clin North Am 39: 919–929

Fleming JS, Goatman KA, Julyan PJ, Boivin CM, Wilson MJ, Barber RW, Bird NJ, Fryer TD (2000) A comparison of performance of three gamma camera systems for positron emission tomography. Nucl Med Commun 21:1095–1102

Gemmell HG, McKiddie F, Davidson J, Chilcott F, Welch A, Egred M (2002) The use of a dual-headed gamma camera to image myocardial perfusion using $^{13}NH_3$ – comparison with full-ring PET. Eur J Nucl Med 29:S176

Grootoonk S, Spinks TJ, Sashin D, Spyrou NM, Jones T (1996) Correction for scatter in 3D brain PET using a dual energy window method. Phys Med Biol 41:2757–2774

Hsu CH (2002) A study of lesion contrast recovery for iterative PET image reconstructions versus filtered backprojection using an anthropomorphic thoracic phantom. Comput Med Imaging Graph 26:119–127

Inoue T, Oriuchi N, Koyama K, Ichikawa A, Tomiyoshi K, Sato N, Matsubara K, Suzuki H, Aoki J, Endo K (2001) Usefulness of dual-head coincidence gamma camera with thick NaI crystals for nuclear oncology: comparison with dedicated PET camera and conventional gamma camera with thin NaI crystals. Ann Nucl Med 15:141–148

IPEM (2003) Report 86 quality control of gamma camera systems. Institute of Physics and Engineering in Medicine, London

IPEM (2004) The basics of gamma camera PET. Institute of Physics and Engineering in Medicine, London

Jarritt PH, Acton PD (1996) PET imaging using gamma camera systems: a review. Nucl Med Commun 17:758–766

Kamel E, Hany TF, Burger C, Treyer V, Lonn AHR, von Schulthness GK, Buck A (2002) CT vs ^{68}Ge attenuation correction in a combined PET/CT system: evaluation of the effect of lowering the CT tube current. Eur J Nucl Med 29:346–350

Kinahan PE, Townsend DW, Beyer T, Sashin D (1998) Attenuation correction for a combined 3D PET/CT scanner. Med Phys 25:2046–2053

King MA, Tsui BMW, Pan T-S (1995) Attenuation compensation for cardiac single-photon emission computed tomographic imaging, part 1. Impact of attenuation and methods of estimating attenuation maps. J Nucl Cardiol 2:513–524

Kuge NTY, Katch C (2001) Current status and future of metabolic cardiac imaging (editorial). Nucl Med Commun 22:847–850

Kuikka JT, Sohlberg A, Husso-Saastamoinen M (2002) PET imaging using a triple-head gamma camera. Clin Physiol Funct Imaging 22:328–331

Kunze W-D, Baehre M, Richter E (2000) PET with a dual-head coincidence camera: spatial resolution, scatter fraction and sensitivity. J Nucl Med 41:1067–1074

Lartizien C, Kinaham PE, Swensson R, Comtat C, Lin M, Villemagne V, Trebossen R (2003) Evaluating image reconstruction methods for tumor detection in 3-dimensional whole-body PET oncology imaging. J Nucl Med 44:276–290

Levkovitz R, Falikman D, Zibulevsky M, Ben-Tal A, Nemirovski A (2001) The design and implementation of COSEM, an iterative algorithm for fully 3-D listmode data. IEEE Trans Med Imaging 20:633–642

Lewellen TK (1998) Time-of-flight PET. Semin Nucl Med XXVIII:268–275

Lewellen TK, Miyaoka RS, Swan WL (1999) PET imaging using dual-headed gamma cameras: an update. Nucl Med Commun 20:5–12

Muehllrhner G, Karp JS, Surti S (2002) Design considerations for PET scanners. Q J Nucl Med 46: 16–23

Mullani NA (2000) Comparing diagnostic accuracy of γ camera coincidence systems and PET for detection of lung lesions. J Nucl Med 41:959–960

Patton JA (2000) Instrumentation for coincidence imaging with multihead scintillation cameras. Semin Nucl Med XXX:239–254

Patton JA, Turkington TG (1999) Coincidence imaging with a dual-head scintillation camera. J Nucl Med 40:432–441

Patton JA, Delbeke D, Sandler MP (2000) Image fusion using an integrated, dual-head coincidence camera with X-ray tube-based attenuation maps. J Nucl Med 41:1364–1368

Paul AK, Tatsumi M, Fujino K, Hashikawa K, Nishimura T (2001) Feasibility of a short acquisition protocol for whole-body positron emission tomography with fluorine-18 fluorodeoxyglucose. Eur J Nucl Med 28:1697–1701

Ramos CD, Erdi YE, Gonen M, Riedel E, Yeung HWD, Macapiniac HA, Chisin R, Larson SM (2001) FDG-PET standardized uptake values in normal anatomical structures using iterative reconstruction segmented attenuation correction and filtered back-projection. Eur J Nucl Med 28: 155–164

Rousset OG, Ma Y, Evans AC (1998) Correction for partial volume effects in PET: principle and validation. J Nucl Med 39:904–911

Stodilka RZ, Glick SJ (2001) Evaluation of geometric sensitivity for hybrid PET. J Nucl Med 42: 1116–1120

Tarantola G, Zito F, Gerundini P (2003) PET instrumentation and reconstruction algorithms in whole-body applications. J Nucl Med 44:756–769

Townsend DW, Beyer T (2002) A combined PET/CT scanner: the path to true image fusion. Br J Radiol 75:S24–S30

Turkington TG (2000) Attenuation correction in hybrid positron emission tomography. Semin Nucl Med XXX:255–267

Vandenberghe S, d'Asseler Y, Kolthammer J, van de Walle R, Lemahieu I, Dierckx RA (2002) Phys Med Biol 47:289–303

Wahl RL (1999) To AC or not to AC: that is the question (editorial). J Nucl Med 40:2025–2028

Xu EZ, Mullani NA, Gould KL, Anderson WL (1991) A segmented attenuation correction for PET. J Nucl Med 1990 32:161–165

Yamaya T, Obi T, Yamaguchi M, Ohyama N (2000) High-resolution image reconstruction method for time-of-flight positron emission tomography. Phys Med Biol 45:3125–3134

Zaidi H (2000) Comparative evaluation of scatter correction techniques in 3D positron emission tomography. Eur J Nucl Med 27:1813–1826

Zaidi H (2001) Scatter modelling and correction strategies in fully 3-D PET. Nucl Med Commun 22:1181–1184

Zaidi H, Hasegawa B (2003) Determination of the attenuation map in emission tomography. J Nucl Med 44:291–315

Image Presentation

Appropriate display and hardcopy are as important to proper interpretation as any other part of the imaging chain, and any degradation in their quality will prevent optimisation of the imaging process. Photographic film was the traditional hardcopy medium; originally being used to accumulate counts during an acquisition to create an analogue image and then subsequently being used to record the results of a digital acquisition for interpretation. This is, effectively, becoming obsolescent. With digital acquisition becoming almost universal, hard copy is increasingly being produced by networked digital printers of various kinds and much interpretation is being undertaken directly from the display. This requires the digital data to be translated into colour or monochrome representation and this must be chosen carefully, particular translations suiting certain data and others degrading the representation.

Display

Analogue image acquisition directly to a film, produced an image which was fixed and could not be manipulated. Digital acquisition now enables an almost limitless amount of post acquisition processing. It does, however, require the numerical data to be transformed into a recognisable image. This is done by a colour or grey scale translation that maps a display intensity or colour level to the counts in each pixel. The simplest form of image processing involves changing the appearance of the image. The main adjustments to an image on a monitor are the threshold and saturation levels, which are used interactively to adjust the image contrast. These levels are indicated as number of counts or as the percentage of the count range. Pixels containing counts below the threshold level are not displayed and those containing counts above the saturation level are set to the highest colour or monochrome shade. Initially, the full range of the scale is implemented by associating the saturation level to the highest pixel count and the threshold level to 0 counts. Between these levels, the pixel counts are associated with colour or monochrome shades according to tables of translations, which are discussed in Chapter III.5. Many such translation tables are used, the common ones being the linear and non-linear grey and pseudo colour scales, and isocount mapping which attaches a particular intensity or colour to pixels within count bands. An alternative to simple translation from numerical data to display scale is histogram equalisation, in which the same number of pixels are allocated each colour or grey increment. Some of these tables have become particularly associated to certain investigations. For instance, with the translation table typically used for myocardial perfusion images, subtle variations within the myocardium are exaggerated whilst those within the background, around the heart, are suppressed.

Often, several image groups may need to be reviewed. In this case, the image representation has to be normalised in some way so that they can be displayed consistently. There are two approaches. In the first, the images are adjusted to the brightest pixel in all the images. However, in this method one artefactual pixel may then control the whole series. In the second, the maximum range is used in each image. In this method, the relationship among the images may not be appreciated.

Hardcopy

The proportion of clinical images transferred to hardcopy is reducing and, at the same time, the type of media used is changing. This comprises: monochrome or colour, and transmission or reflective. Copying the display directly to hardcopy will not usually result in the same image appearance, because the visual characteristics of the hardcopy medium are different to those of the display. A further translation has, therefore, to be imposed on the data. This is chosen experimentally for optimal user preference. Since the tendency is for a hardcopy device to service multiple acquisition and processing units, via a network, it may be necessary to implement a number of different translation tables, each appropriate to a particular acquisition or processing station and user.

When film is used for hardcopy, laser engines are replacing cathode ray tubes as the method of exposure. In these, a thin, $\sim 8.5 \times 10^{-5}$ m, laser beam scans the film, and its intensity is modulated to determine the optical density. The photographic film used is single sided, which means that the active emulsion is coated on only one side of a plastic base. On the opposite side of this is an antihalation layer, which prevents light from being reflected back from the base to the emulsion, resulting in sharper dots. Very high resolution images are produced as a result of the narrow beam. However, the line spacing is constant which means that multiple small images, displayed on one film, suffer some loss of spatial resolution. Being digitally controlled, these systems can maintain the film density at very a constant level, by exposing a sensitometry film and measuring the densities using a digital densitometer, with this being fed back to the laser engine to control its output power.

The digital data undergoes geometric manipulations when the pixel data is converted into dots for creating the hardcopy, these include: sharp, bilinear and cubic transformations. The sharp, or replication, transformation converts the pixel count to a dot density; this can produce images in which the pixels are very apparent. The bilinear, cubic or cubic spline transformations interpolate the dot density between adjacent pixels, which results in the pixels being less apparent.

Darker areas on the film represent a higher radioactivity concentration, and lighter areas a lower concentration. This darkening is quantified, as optical density, by:

$$\log_{10} [\text{intensity of light incident on film/intensity of light transmitted}]$$

Thus, an optical density of 2 represents a 1% transmission of incident light, and will appear almost black, whereas a density of 0 represents a 100% transmission and will appear clear, even an unexposed film will demonstrate a density of ~0.12. In routine imaging, densities up to ~0.3 appear equally light, transmitting ~50% of incident light, and densities >2 appear equally dark. The aim is to maintain the characteristic curve for optical densities between ~0.2 and 2.0 consistent over time (O'Connor et al. 1991). A number of alternatives to film, which can use transmissive or reflective media not sensitive to light, now exist. The latter suffer a limitation of maximum density, usually of ~1.3–1.7, however.

Processing

There are two processing techniques, used according to the film construction. The traditional chemical, or wet, processing involves developing, fixing, washing and drying. Over-developing, by increasing the duration or temperature, will darken a film and under-developing will have the opposite effect.

Dry laser imagers uses a laser beam to remove carbon pigment from the media according to the length of the exposure and creates each pixel independently, rather than scanning across the media. Processing uses heat. One type uses a silver-based photothermographic film which is sensitive to light. This is exposed by an infrared laser diode and developed using a temperature of ~120°C. Unlike wet processing, all the chemicals remain in the film during storage, which renders them susceptible to on-going heat and light exposure especially immediately after printing. This type of film should not be exposed for protracted periods on the viewing screen and should not be overlapped on the screen since image transfer can occur. Also the printer itself should not be under bright light or near sources of heat because of the sensitivity of the films immediately after printing. Another uses thermal dye diffusion which means that these films are not light sensitive and not as susceptible to the above problems (Lu et al. 1999). Hot wax or dye sublimation techniques are used in colour imagers.

There is no significant difference in diagnostic accuracy when interpretation is using wet or dry processing (Krupinski 1996).

References

Krupinksi EA (1996) Clinical assessment of dry-laser-processed film versus traditional wet-processed film with computed tomography, magnetic resonance imaging, and ultrasound. Acta Radiol 3:855–858

Lu ZF, Nickoloff EL, Terilli T (1999) Monthly monitoring program on DryView laser imager: one year experience on five Imation units. Med Phys 26:1817–1821

O'Connor MK, Brown ML, Hung JC, Hayostek RJ (1991) The art of bone scintigraphy – technical aspects. J Nucl Med 32:2332–2341

Computers and Communications

Peter J. Riley

Abbreviations

ADC	analogue to digital converter
AGP	accelerated graphics port
ALU	arithmetic logic unit
API	application programming interface
ATA	advanced technology attachment
BIOS	basic input output system
CD	compact disc
CD-R	writeable cd-rom
CD-RW	re-writeable cd-rom
CPU	central processing unit
DDE	dynamic data exchange
DICOM	digital imaging and communications in medicine
DVD	digital versatile/video disc
EPR	electronic patient record
EPROM	erasable programmable ROM
FDD	floppy disk drive
FDDI	fibre distributed data interface
FLOP	floating-point operation
FPU	floating point unit

FTP	file transfer protocol
HDD	hard disk drive
HIS	hospital information systems
HL7	health language
i/f	input/output interface
IP	internet protocol
IPX	internet packet exchange
LAN	local area network
LED	light emitting diode
LLC	logical link control
MAC	media access control
MSW	Microsoft Windows
NAS	network attached storage
NIC	network interfacing card
OD	optical disc
OLE	object linking and embedding
OS	operating system
OSI	open systems interconnection
PACS	picture archiving and communications systems
PC	personal computer
PCI	peripheral component interconnect/interface
RAID	redundant array of independent/inexpensive discs
RAM	random access memory
RIP	routing information protocol
RIS	radiology information systems
ROI	region of interest
ROM	read only memory
SAP	service access point
SCSI	small computer system interface
SMTP	simple mail transfer protocol
SQL	structured query language
TCP	transfer control protocol
UDP	user datagram protocol
UPS	uninterruptible power supply
USB	universal serial ports
UTP	unshielded twisted pair
WAN	wide area network
WWW	world wide web

Computers

Nuclear medicine was one of the earliest imaging modalities to use computers for acquisition, processing, storage, and display of digital images. In the most general sense, computers are designed to:

1. Receive information from various sources, the INPUTS to the computer, e.g. characters typed from the keyboard, movements and button presses from the mouse, streaming multimedia data from CDs and the internet, etc.

2. PROCESS the received data, e.g. interpret a keypress as an alphanumeric character or as a control code, interpret a mouse click to activate a procedure associated with a symbolic control on the monitor screen, convert the multimedia stream into separate audio and video signals, etc.
3. OUTPUT the processed data, e.g. add a character typed at the keyboard to a simulated page in a wordprocessor or to open a Menu in the application, move a mouse pointer around a monitor screen to point at and activate various controls, play audiovisual media on the monitor and speakers, etc.

Computer systems employed in nuclear medicine have specific tasks which can also be broken down into Data Acquisition, and Image and Data Processing.

1. Data Acquisition: INPUT instrument data from the γ camera's Analogue to Digital Converters (ADCs), PROCESS the ADC data by scaling the patient voxel intensity into a pixel intensity and formatting it into a matrix array, and OUTPUT the matrix to the monitor to display it as a medical image.
2. Image and Data Processing: INPUT regions of interest (ROI) in the medical image with a lightpen, PROCESS the image data within the ROI e.g. sum the activity in the region, and OUTPUT the region statistics to the monitor e.g. the cumulative sum, the mean and the standard deviation of the region activity.

Previous chapters have examined specific details of the data acquisition systems, and the following chapter will detail the processing algorithms most commonly employed in nuclear medicine. Here, we examine other characteristics of the computer system which can impact a nuclear medicine service. It is convenient to consider the computer as comprising two major subsystems, the hardware and the software.

Hardware

Components

The physical components of the computer are referred to as the hardware and consist of various electronic components. As the analysis above suggests, computer hardware falls into three main categories:

1. Input Devices: keyboard, mouse, lightpen.
2. Processor Components: the central processing unit (CPU), random access memory (RAM), read only memory (ROM), basic input output system (BIOS).
3. Output Devices: monitor, printer, film recorder.

To which must be added the following:

1. Input/Output Interfaces, to connect the input and output devices to the processor: PS/2 ports for the keyboard and mouse, serial ports (RS 232,432) for mouse and fax/modem, universal serial ports (USB), parallel/printer port, small computer system interface (SCSI), firewire (IEEE1394), video/graphics card, Ethernet.
2. Storage Devices, to store software and data:
   Hard disk drive (HDD), floppy disk drive (FDD), writeable and re-writeable cd-rom drives (CD-R, CD-RW) and digital versatile/video disc drives (DVD), other types of optical discs (OD), magnetic tape, Network Attached Storage (NAS).

The input, output and storage devices are usually modular units with cables to attach to the processor module via specific plugs and sockets. The processor itself is a large electronic board with various slots to accept the input/output interface (i/f) boards/cards, and sockets to accept the cable connections of the modular components.

The interfaces listed are a small sample of those available; represent those most commonly found at the time of writing, but standards change rapidly. There has been a convergence of plug-in board standards across multiple platforms, i.e. on motherboards designed for various type of CPU such as Intel, Sun, SGI, Motorola, Apple. The slot-types currently found on many motherboards include:

1. PCI: Peripheral Component Interconnect/Interface, nearly every type of i/f card is available in the PCI format. Input and output devices are often referred to as Peripherals.
2. AGP: Accelerated Graphics Port, followed by a multiplier e.g. X4 to designate its speed relative to the original AGP slot, is the video/graphics card interface.
3. ISA/EISA: are older (legacy) slots rapidly disappearing from the scene.
4. ATA66/ATA133: Advanced Technology Attachment, the HDD, FDD and CD/DVD drives interface (replacing IDE/EIDE).
5. RAID5: for implementing a Redundant Array of Independent/Inexpensive Discs system at level 5, which provides failsafe data protection on the HDDs.

These components are found in two main configurations in nuclear medicine equipment. They may be embedded in a rack system within a console, where they are only accessible to the vendor engineers. Or they are enclosed in the more familiar Personal Computer (PC) case as a stand-alone system and may be sited away from the acquisition hardware, and may be referred to as a workstation. This latter configuration is becoming more common as smaller dedicated CPU boards take over the role of equipment controllers to provide distributed intelligence in the system.

Console systems used to employ industrial CPUs for which there was a limited software library available for in-house developments on the system. The user was then dependent upon the vendor to provide all their software requirements. Similar CPUs may be employed as mini-boards in a distributed network. The reasoning here may be one of patient safety; the vendor did not wish to encourage software tools which may act directly upon equipment performance, e.g. it might be possible to accidentally override a collision sensor.

Workstation systems usually employ more common "consumer" CPUs for which there are numerous software vendors and which are more open for software development. The workstation may issue commands to the equipment CPUs but it is the local intelligence which controls the actions of the equipment, thereby avoiding accidental user interventions.

Mixed console-workstation systems may be found. There has been much international pressure to standardise image, patient and network data formats and protocols between vendors. It is now essential that equipment conform to these standards (e.g. DICOM, HL7, Ethernet *q.v.*) to enable data to be readily and safely exchanged between systems and institutions. We shall address safety later, but it here refers to secure communications and error checking. Thus, even subsystems may communicate using standardised protocols, as between a Console and a Workstation.

Specialist Devices in Nuclear Medicine Computers

Arithmetic Logic Unit (ALU), Array Processor, Floating Point Unit (FPU), Math Co-processor are names given to special chipsets/boards for implementing fast numeric calculations. They are less common now in conventional nuclear medicine equipment as generic processing speeds of around 1 GFLOP s^{-1} are common, i.e. computers can perform one Billion FLoating-point OPerations/sec, as many CPUs have incorporated math co-processors on-chip. Other medical imaging systems, which require fast multislice volume reconstructions, may still employ these devices.

Hardware Operation

The Booting Process

When the computer is powered on it undertakes the bootstrap procedure, also referred to as "booting up" or booting. This arises from the necessity for the hardware to access the software instructions and to be able to interact with the user; a problem is that it requires instructions as to the process to acquire instructions (an apparent impossibility, like a person trying to lift themselves off the ground by pulling on their boot-straps). The BIOS is a hardwired set of instructions on the motherboard which gets the computer started and directs the loading of the main Operating System (OS) from the appropriate storage medium, usually specific magnetic tracks on the master primary HDD. These hardwired BIOS instructions are stored in ROM, i.e. they cannot be overwritten (or at least not easily overwritten, Erasable Programmable ROMs, EP-ROM, are now commonly used for BIOSes enabling system software upgrades).

Associated with the BIOS is another type of non-volatile memory, often referred to as the "CMOS", actually a CMOS-ROM, in which hardware settings may be customised according to user requirements. This also contains information on security settings to prevent access to the system without the appropriate passwords. The date and time are also updated here between successive boots. The CMOS draws a small amount of power from an onboard battery, which may require replacement after years of operation. This CMOS-ROM is often targeted by viruses, in which case it must be discharged, by removing the battery, in order for the virus code to be erased – together with the customisation data! Hence, it is wise to backup the CMOS to retrieve the previous settings. The CMOS is accessed at boot time, typically with <ctrl><alt><esc>, or the necessary key-presses should be shown on the monitor during the boot process.

The software instructions are loaded from the storage media into RAM where the CPU can more quickly access them. The term RANDOM ACCESS is used in various contexts to refer to a system where data can be accessed directly, rather like going to a specific pigeon-hole, e.g. data on a CD can be accessed almost immediately by moving the reading laser straight to the required track. This is in contrast to SERIAL ACCESS in which data becomes available sequentially, e.g. on a magnetic tape one must play all the tape in front of the data required before it is accessible, which can be a slow process.

Access times from storage media are typically several milliseconds, compared to access times from RAM which takes nanoseconds only. The CPU is typically running at several gigahertz so it is executing several instructions per nanosecond, so even RAM access can cause delays in the execution cycle whilst instructions are fetched. Typically, the CPU will have a much faster cache RAM in which RAM instructions can be prefetched in bulk in order to minimise memory access delays.

Memory Organisation

Computers are systems of very fast switches which can shuttle data in much the same way as a railway switching yard reconfigures freight cars. A typical computer processor chip may contain hundreds of millions of switches in a package of only a few square centimetres. The state of these switches can be either ON or OFF and this is the fundamental unit of measurement in computing, referred to as a BIT. Switches are most commonly organised and stored in memory in groups of 8 BITS referred to as a BYTE. Since each bit in a byte can be either on or off then the 8 bits in a byte can have $2\times2\times2\times2\times2\times2\times2\times2=2^8=256$ possible configurations. The configurations are shown using 0 to represent off and 1 to represent on. So a memory byte with all switches off would be 00000000, with all on 11111111, and everything in between. These patterns look like numbers and indeed they are numbers in the BINARY number system:

Just as 359 means 3 hundreds plus 5 tens plus 9 ones, see Table II.7.1, numbers can be represented in the Binary System (see Table II.7.2). At the most fundamental level these 256 configurations (0 to 255 binary) could represent 256 different types of INSTRUCTIONS which could be stored in memory. The processor could have 256 built-in procedures and it will execute one specific instruction when it reads that instruction as an input from memory. Instructions are stored in a memory area, where each unit of memory is one byte. There are special storage places in the processor referred to as REGISTERS.

The binary number system has become the convention in specifying computer related measurements. Since 2^{10} is 1024 there is an approximate equality with our usual decimal numbers of thousands, millions, billions which are referred to as kilo, mega, gaga representing 1024, 1024^2, 1024^3 and so on. Other conventions are not always followed, e.g. uppercase B to stand for Byte(s) and lowercase b to stand for bit(s). Memory is now usually specified in megabytes (MB), whilst communication speeds are usually specified in kilobits per sec (kb/s) and Megabits/sec (Mb/s). HDD and DVD storage capacities are usually in gigabytes (GB), whilst portable storage media like floppies, CDs and flash memory are in MB.

Digital Images

Just as the BIT is the basic unit of memory, so the PIXEL is the basic unit of measurement in a digital image appearing on the computer monitor. If each pixel is allocated just one bit of memory then the image can consist of only black or white pixels (on or off). The simplest medical images have one byte of memory allocated to each pixel and so could have only 256 grey levels. In nuclear medicine, images may be made up of an

Table II.7.1. Number representation in the decimal system

100	10	1
3	5	9

Table II.7.2. Number representation in the binary system

128	64	32	16	8	4	2	1	Total	From
0	0	0	0	0	0	0	0	0	
0	0	0	0	0	0	0	1	1	
0	0	0	0	0	0	1	0	2	
0	0	0	0	0	0	1	1	3	2+1
0	0	0	0	1	0	1	0	10	8+2
0	1	1	0	0	1	0	0	100	64+32+4
1	1	1	1	1	1	1	1	255	etc

array of 128 pixels across the monitor × 128 pixels down the monitor, and each pixel a 2 byte "depth" i.e. 16 bits or 4096 grey levels; this would be represented as 128×128×16. So each image requires this much memory in the video RAM and on the storage media. Whilst an nuclear medicine image may require only 128×128×16 bits = 32kB, a digital chest x-ray may require 2000×2500×16=10 MB (without compression[1]). So, image size and number of images per examination will determine many of the computer and monitor characteristics required in a specification.

Hardware Recommendations

Some general comments on the selection and use of computers in a nuclear medicine environment:

Input Devices

1. Choose an optical mouse, which has no moving parts and is easily cleaned.
2. If computers are employed in wet labs, e.g. isotope preparation or in-vitro counting, then it is wise to use sealed keyboards.
3. Lightpen lenses should be cleaned routinely to ensure accurate detection.

PC Case

4. When specifying equipment it is wise to leave room to expand the system:
   i. Leave a free slot on the motherboard e.g. PCI.
   ii. Leave sufficient room in the rack for an extra storage device e.g. another HDD.
5. Ensure that the power supply has spare capacity to be able to supply expansion board upgrades e.g. 400 W.

[1] In this example the actual image depth requirement is only 12 bits for 4096 grey levels. But as memory is organised in units of 8-bit bytes there would be wastage of 4-bits with conventional storage of 1 pixel in two bytes. Additional software could make use of those 4-bits to store a part of another pixel, thereby reducing the overall image storage by 25%. This constitutes a form of lossless compression, lossless because all the original pixel data is recoverable. There are more efficient forms of lossless compression using various algorithms which may compress memory storage requirements by factors of around 2.5. Higher lossy compressions are possible, but their use in medical imaging is still under assessment.

6. Ensure that the interior of the PC case is well cooled with correctly positioned fans.
7. An Uninterruptible Power Supply (UPS) can prove a useful investment. In the event of power failure they provide a few minutes grace to save data and perform a smooth system shutdown.
8. A power-line filter to remove high voltage transients and interference can help to keep systems stable. A manual reset in the event of power failure is advisable to prevent rapid off–on fluctuations.
9. Ensure that the chassis and the power cable are properly earthed at the wall outlet.

Output Devices

10. Monitor screens should be cleaned routinely, especially if used with a lightpen.
11. Ambient light levels should be adjustable when viewing medical images and the monitor positioned to avoid light reflections.
12. All cables attached to port sockets should be securely screwed into place, but beware of over-tightening as these small screws strip their threads very easily. Thumb-tightening screws on cables are preferable.
13. All peripheral cables should be neatly coiled, labelled and tied to prevent a mare's nest. When multiple computers are used in the same area the labels on cables can save time during fault-finding or reconfiguring.

Communications Devices

14. Network cards, fax/modem cards, network switches/hubs: it is advisable to choose cards or modules which have separate light emitting diodes (LED) to indicate Link status and Activity status, which can aid in fault diagnosis.

Computer Hardware Safety

The hazards to personnel working with computers are easily managed. The following are recommended procedures to minimise potential hazards:

15. Choice of keyboard is sometimes limited by the function key assignments utilised by the system software. However, if a choice is available then ergonomically designed devices are preferable to prevent repetitive stress injuries to personnel e.g. keyboards with wrist-guards.
16. Personnel using the computers may also be handling patients and radionuclides. It is possible that small levels of contamination can accrue to input devices such as keyboard, mouse and lightpen. These devices should be monitored and cleaned routinely, the isopropyl alcohol swabs commonly used for disinfecting are effective.
17. Most nuclear medicine areas have prohibitions against eating and drinking, and the practice should be extended to the workstation area to prevent spills onto input devices.
18. It is preferable to operate the computer in a room without carpet (most of the nuclear medicine environment is without carpet). It is very easy to generate an electrostatic charge when working on carpet, which can damage the computer.

19. If moving boards in the main chassis (PC case):
    i. Switch off the power at both the computer power supply and at the wall outlet – if there is no switch at the wall then remove the power cable, though it is preferable to retain an earth connection to the chassis.
    ii. It is preferable to work on a non-carpeted floor. If the area is carpeted then it is advisable to have a non-static mat or an earthed wrist link to prevent static discharge whilst handling electronic boards or chipsets.
    iii. If soak-testing the system after upgrade or repair then, after visually confirming that there is no arcing or burning, close the case as the fans are designed to cool the components within a confined environment.

Computer Hardware Security

20. Unauthorised access to a computer is most readily obtained by rebooting the system and interrupting the normal boot sequence to bypass security software. This can be prevented by changing the CMOS settings such that the boot order is to HDD only and by-passes the floppy and other drives.
21. Access to the CMOS must also be protected by toggling on the CMOS password facility.
22. It is not normally necessary, but for full security the PC case itself would have to be lockable to prevent access to the motherboard and the storage drives/cables. The computer modules may also be secured to the desk or benchtop.

Software

Operating Systems

The computer user is seldom made aware of the differences in hardware platforms that are available for commercial nuclear medicine systems. The user interacts with input and output devices which are essentially the same products connected to all these computers. So the underlying details of the processor module become irrelevant, as there has been a convergence in interfaces and in underlying performance characteristics. The one area where the user encounters differences is on the software side, and most notably with the Operating System (OS).

As previously indicated, there are still console systems running proprietary operating systems for which there is little support beyond the supply vendor. In such systems it is essential that the vendor provides standardised data format and communication protocols to enable data exchange with Workstations; see "Communications, network protocols".

Workstations are primarily based upon Microsoft Windows or Unix OSes, and possibly others for Apple and other platforms.

Microsoft Windows (MSW) comes in various versions, the unsecure versions are 3, 95, 98 and ME, and the more secure versions are NT, 2000 and XP at the time of writing. It is highly recommended that a secure OS version is specified in order to comply with the Data Protection Acts that have been implemented in various countries. MSW is most commonly found on Intel-Pentium-compatible platforms, but is also available for other CPUs such as the Alpha.

Unix OSes are numerous across multiple hardware platforms. This should mean that the user is able to jump from one platform to another with little difficulty, and this is largely true. However, there are multiple command line shells and windows available with subtle differences that can cause problems. Unix OSes have good security capabilities.

Other platforms have a variety of OSes which may have good support, but there appears to be a convergence towards Unix variants on these also.

The underlying OS largely determines what other application programs can be run on the nuclear medicine workstation to extend your processing power. This is a question of availability, the more common an OS the more third party software that will be available for it. This may also affect pricing, the greater the user base in general the less costly an application should be.

Since Unix has been around for several decades there tends to be a large installed base of software available to run on this OS. However, porting software between platforms usually requires a detailed knowledge of the command line interface and of compiling and linking source files and libraries. Such porting assumes that source and libraries have been made available. This is true for many application programs developed in academia, but is not the case for commercial products, in which case the user is again dependent upon the supply vendor.

Another perceived advantage of Unix was the built-in scripting language, which enables rapid program development. A typical application would be to grab an image file and reformat it from some proprietary format into a data file which could be used in other generic software e.g. reformat to DICOM to read into a dicom medical image processing package. However, with the advent of standards in data formats and network protocols this user intervention has become largely unnecessary.

Application Programs

Nuclear Medicine Vendor Applications

Most applications now employ Graphical User Interfaces (GUI) for ease of use, with window, icons and mouse pointer (WIMP). The nuclear medicine vendor must supply the software necessary for the following minimum tasks:

1. Data Acquisition from the nuclear medicine instrumentation, which may include acquisition times, camera and patient table movements, and energy window settings.
2. Correction Routines to apply to the data for attenuation and sensitivity factors.
3. Reconstruction routines for 3D data visualisation of imaged patient organs.
4. Image Processing software to define ROIs in an image and to determine the change in activity in those regions across multiple images acquired in sequential time intervals. Image enhancement routines.
5. Artificial Intelligence routines may be incorporated to detect and correct for patient movement.
6. Data Processing software to determine areas under dynamic activity curves, to apply various filters to smooth data, to fit model equations to the experimental data.

Additionally, the vendor may provide customised software for:
- Examination reporting
- Patient booking and billing
- Radiopharmaceutical ordering and inventory
- Radioisotope disposal recording
- Quality assurance statistics
- General statistics on patients and examinations

Office Applications

More often than not however, these latter tasks are left to the user to implement, and this is the benefit of having an open OS. A generic word processor can be used for the examination report. Most word processor applications employ template documents, so it is easy to design standardised forms for reporting purposes. The remaining functions may be incorporated into spreadsheet or database applications. Most OSes have so-called office software suites which include these applications and in which documents and functions may be interlinked with Dynamic Data Exchange (DDE) and Object Linking and Embedding (OLE). Accessing the Application Programming Interface (API) can automate these tasks. But these tasks require a computerate user to establish the customised suite for nuclear medicine.

Graphic design and presentation applications are often bundled into the office suite and can serve a useful role in education.

Scientific Applications

Apart from the routine applications in nuclear medicine discussed above, an open workstation environment can also be used for research and development:
1. Mathematical analysis and modelling of data may be undertaken with numerical analysis or symbolic analysis software e.g. Mathematica, MATLAB, MathCad, Maple. Macsyma and Reduce are frequently found in academic research and development.
2. Computer aided diagnosis. The applications above are often augmented with specialist packages for implementing Image Processing, Neural Networks, Fuzzy Logic and Genetic Algorithms. These techniques may be combined to produce artificial intelligence aids in diagnosis.
3. Statistics packages may be employed not only in QA matters, but also in clinical trial design and analysis of patient data. SPSS is a common standard in academia, but there are numerous packages available.

Software Recommendations

Some general comments on the selection and use of software in an nuclear medicine environment:

Operating System

1. Use a secure OS which requires password access. There are usually multiple levels offering different degrees of access to users. There should be one system administrator who is responsible for the maintenance of the OS and the hardware, the installation of software, the determination of privileges to different level users, and the establishment of communication protocols. A logbook must be used to record all such settings in the event of the system administrator becoming unavailable. All changes must be meticulously recorded. All errors must be recorded, their origin identified and their cure detailed.
2. A password protocol must be established. Individual users must have their own password so errors can be traced and corrected. Users can be grouped into workgroups with the same access privileges. In extreme cases it may be necessary to implement key logging if errors persist without an identifiable cause. Passwords should be updated regularly. Passwords must not be written down in the working environment.
3. Administrators, and programmers granted access, must agree on the programming languages to be installed. Unix system shells and scripting languages in other OSes should also be agreed upon. For scientific programming consideration must be given to:
   i. Floating-point arithmetic implementation. Many script interpreters have only integer arithmetic available.
   ii. The availability of an "evaluate" command can be crucial. This enables text (i.e. string) input of a mathematical function to be parsed for immediate evaluation. It is surprising how few languages have implemented this feature.
4. System backup must be undertaken on a regular basis. A system crash can cause long delays if the whole system has to be rebuilt. Drive mirroring can be implemented by installing a second HDD with the appropriate software. In a large system it may be wise to implement a RAID system including a spare replacement drive.
5. Patient data and reports should be archived daily to backup media or Network Attached Storage (NAS).

Application Programs

6. Most programs should be DDE and OLE compliant.
7. Word processors should implement document templates, have easy form design capabilities and be able to embed nuclear medicine images.
8. Database applications should be Structured Query Language (SQL) compliant and have good data import and export capabilities, including delimited plain text.
9. Spreadsheet applications should have a good range of scientific functions implemented and the ability to define new functions. It should have good data import and export capabilities, including delimited plain text.
10. Presentation applications should have a good range of scientific and medical symbols in the symbol library and be able to create correctly formatted mathematical equations.
11. Image processing applications should have Digital Imaging and Communications in Medicine (DICOM) import and export capability, sizeable image stacks, grey level window controls, movie display, ROI statistics, a convolution kernel library and customisation ability, and image arithmetic. More advanced processing sta-

tions should have multiple thresholding with regional growth and erosion, multiplanar image reconstruction, 3D surface and volume rendering.

Communications

One of the principal areas of development in computing over the years has been the advancement in standardising communications between computer systems. This refers not just to improvements in networking technology, but also the development of communication protocols to allow multi-platform systems, with various Operating Systems, to be able to negotiate and exchange data intelligently and reliably. The Structured Query Language (SQL) for database negotiations is probably the most familiar example. In the medical and diagnostic imaging arena there are the Digital Imaging and Communications in Medicine (DICOM) and Health Language 7 (HL7) protocols. These are now widely deployed to facilitate patient data record and image exchanges across Picture Archiving and Communications Systems (PACS), Radiology Information Systems (RIS) and Hospital Information Systems (HIS). When purchasing new equipment it is imperative to consider the depth of communications required. The PACS, RIS, HIS networks are rapidly becoming a standard requirement in healthcare. Even if there is only a stand-alone requirement for the immediate future, one should consider the utility of having DICOM and HL7 compliant systems to avoid obsolescence.

Network Hardware

Local Area Network (LAN) designs are commonly defined by the cable type and connectivity topology (see Table II.7.3):

1. Cables used were originally thick coaxial type RG08 with BNC connectors and 50 ohm terminators. Thin coax type RG58 were then deployed but over smaller distances due to signal degradation. Unshielded Twisted Pair (UTP) and UTP category 5 cables with RJ-45 plugs were then used over smaller distances. Optical Fibre or Fibre Distributed Data Interface (FDDI) cables permit faster speed over longer distance at greater expense. There is also a move to wireless LAN technologies employing radio links.
2. The topologies most commonly used are:
   a. the Bus type, which is a linear backbone linking all the network nodes (workstations etc) and transmissions are propagated and viewed by all nodes.

Table II.7.3. LAN configurations

	10 Base 5	10 Base 2	10 Base T	100 Base T	Gigabit
Topology	Bus	Bus	Star	Star	Bus/pt-pt
Media type	Thick coax	Thin coax	UTP	UTP cat5	Optic fibre
Transfer rate (Mb s^{-1})	10	10	10	100	1000
Distance (m)	500	185	100	100	2100

b. The Star type, which employs a central hub or switch from which connections radiate out to the nodes. Each hub/switch has an uplink port to other networks and multiple ports for connection to the workstations.
c. The Point-to-Point type, which is a linear connection from one node to the next.

These configurations are most typically found in Ethernet networks, the IEEE 802.3x standards. The 10 BaseT is commonly referred to as Ethernet, and the 100 BaseT as Fast Ethernet. There are other configurations and topologies, such as the IBM Token Ring, the IEEE 802.5 standard. The WiFi networks are becoming more accessible and are governed by the IEEE 802.11x standards.

Each station in the network has a Network Interfacing Card (NIC) which connects to the network by the appropriate cable and connector. The most common NIC is a 10/100 Ethernet card with RJ-45 connector for use with UTP cat5 cable on a star topology. This type of LAN is typically found in an office area, and may be connected to other remote LANs by a more robust network configuration to form a Wide Area Network (WAN). The 10 BaseT system is rapidly being superseded, but is typically found connected to a 10/100 BaseT hub/switch with a 100 BaseT uplink to the WAN.

Network Protocols

The International Standards Organisation has defined the Open Systems Interconnection (OSI) layered model of network communications. The hardware discussed in the previous section is defined in Layer 1 of the OSI model. The various communication protocols are defined in the other layers:

- Layer 7: Application interface, defines common services such as File Transfer Protocol (FTP), Simple Mail Transfer Protocol (SMTP) and World Wide Web (WWW).
- Layer 6: Presentation of data with encoding and decoding formats, e.g. JPEG, MPEG, ASCII.
- Layer 5: Session establishment for data coordination between nodes, e.g. Novell's Service Access Points (SAPs) and NetBUEI.
- Layer 4: Transport protocols, e.g. User Datagram Protocol (UDP), Transfer Control Protocol (TCP).
- Layer 3: Network routing protocols and addressing, e.g. Internet Protocol (IP), Internet Packet Exchange (IPX), Routing Information Protocol (RIP).
- Layer 2: Data Link, subdivided into Media Access Control (MAC) and Logical Link Control (LLC).
- Layer 1: Physical layer, as previously discussed, the network hardware specification.

These layers may be configured in many possible combinations appropriate to the networking task. The most common designation on Ethernet networks is the TCP/IP protocol. It is the most popular as it is the protocol used by the World Wide Web internet.

A system administrator with considerable experience would be required to customise these settings. For the stand-alone user the workstation's communications will be set up according to the vendor's data exchange requirements to control the acquisition equipment and to transfer data to the workstation.

Network Recommendations

Some general comments on the selection and use of networks in a nuclear medicine environment:

1. Ensure that the image transfer protocol is DICOM compliant.
2. If connecting to a RIS or HIS ensure that the Electronic Patient Record (EPR) is HL7 compliant.
3. Choose a cable technology appropriate to the environment. Beware of unshielded cables in an electrically noisy or RF noisy locale.
4. Fast Ethernet with a central hub and radial nodes is the de facto standard for a small LAN at the time of writing.
5. Network security should not be difficult to ensure in a LAN with a secure OS. But if the LAN has internet connectivity then a secure software firewall must be installed.
6. Various peripherals e.g. printers and Network Attached Storage (NAS) devices, have their own built-in NICs and can be accessed over the network. This ability to share resources can reduce inefficiencies in unnecessary duplication and empower more people with access to sophisticated devices.
7. A large user base can also benefit from server based applications. Here, the server runs the applications locally whilst the user's workstation becomes a client terminal to view and interact with that application remotely.
8. The OS and its network settings should be backed-up regularly. It is advisable that all network configurations are recorded and all changes to those settings are detailed.

PART III

Clinical Procedures

Non-Imaging Studies

Abbreviations

BSA body surface area
DTPA diethylene triamine penta acetic acid
EDTA ethyl diamine tetra acetic acid
ERPF effective renal plasma flow
GFR glomerular filtration rate
HSA human serum albumin

MAG3 mercaptoacetylglycine
PHA pulse height analyser
RBC red blood cell
TPC total plasma clearance
TER tubular extraction ratio

These studies comprise in vitro and external probe in vivo investigations. Unusually in nuclear medicine, the in vitro studies can estimate absolute values of physiological parameters. The probe studies can assess the in vivo distribution of radionuclides, that would be difficult to image. Both require only small administration activities, except in the case of intraoperative probes. Additionally, a type of external probe investigation which requires even less activity, is one using whole body counters. This can be used instead of, for instance faecal counting, to estimate the amount of tracer retained in the whole body. Non-imaging studies are the simplest investigations to undertake. This simplicity, however, should not obscure the need for accuracy at every part of the process. Since the result is a number rather than an image, the potential for post-investigation quality assessment is very limited.

In Vitro Studies

In these studies, the physiological parameter to be estimated is determined by combination of radiopharmaceutical and route of administration. The parameter, itself, is calculated by the entry of various measurements into a mathematical model, which is specific for a particular investigation. The performance of the studies requires sampling of body fluid, at various times following administration, the activities of which are then measured on a well type scintillation detector or, less frequently, on a β^- counter. Because several samples are counted in each investigation, if a number of investigations are undertaken per day, an automatic sample changer is useful.

The measured activities must be correlated with the administered activity, by calibration with a standard activity diluted to effectively simulate the distribution in the patient; all activities being measured accurately, including consideration of any residue activities during the radiopharmaceutical manipulations. For measurement, aliquots of both the body fluid and the diluted standard solution should be contained in standard plastic counting tubes and should be constrained to a standard volume, to ensure that the geometrical efficiency of the detection system is constant. The optimal volume, for this, should be determined during acceptance testing of the detection system, as a compromise between being large enough to contain sufficient activity for accurate measurement and small enough that geometric efficiency is not significantly reduced. If it is not possible to obtain enough body fluid for a particular sample, this should be made up to the standard volume with water. The sample volume figure entered into the mathematical model must be the undiluted figure, however. If more than one radionuclide is used in the investigation, a dual channel technique is used to measure the sample, with correction for the scatter of the higher energy photons into the lower energy window. This can be done simply because, usually, cross-talk of the lower energy photons into the higher energy window can be ignored.

$$C_{LC} = C_L - C_H S_L/S_H \qquad \text{(III.1.1)}$$

where
for the sample,
C_{LC} is the corrected count rate in the lower energy window
C_L is the uncorrected count rate in the lower energy window
C_H is the count rate in the higher energy window

for the standard containing the radionuclide with the higher energy photons,
S_H is the count rate in the higher energy window
S_L is the count rate in the lower energy window

All activity measurements must be corrected for background count rate of the detection system and may need to be corrected for the physical decay of the radionuclide. The body fluid samples contain very low activities and, even allowing for the high sensitivity of the detection systems, require a long measurement duration to achieve acceptable statistical errors. However, since the counting is undertaken away from the patient, this duration can be extended. The termination conditions, of counts or time, are therefore determined solely by the required statistical accuracy of the measurements and not by patient cooperation. Thus, if termination is set on counts, the condition can be established from a consideration of the counting statistics, a 1% error requiring 10,000 counts and a 5% error requiring 400 counts. In this case, various measurement durations will be experienced for the various samples. If termination is set on a fixed time, this should be calculated for the lowest count rate likely to be experienced in a particular set of samples. In both cases, the figures input into the mathematical model are usually those presented, by the detection system, in counts per second or counts per minute, which effectively makes the calculation independent of the type of termination condition. This should incorporate a correction for system dead-time, which can be significant. All of the investigations require accurate values of:

1. administration activity and time
2. standard activity
3. sample time, volume and activity

Thus, other than activity measurement, volume and time estimations are also sources of error; all of which propagate through the calculation. The sequence of counting is usually: background, standard, samples, standard, background; and each measurement is usually undertaken twice, the second immediately after the first. This is to give some confidence that the measurements have not been affected by equipment variation or temporary proximity of extraneous radioactive materials. If the two measurements vary by less than three standard deviations, their mean can be input to the mathematical model. If their values vary by more than this, the sample should be recounted. The usual routine investigations include: effective renal plasma flow, glomerular filtration rate, evaluation of vitamin B_{12} absorption, red cell volume and plasma volume, red blood cell survival, and breath analysis.

The value of the physiological parameter is often normalised to a standard value of body surface area of 1.73 m^2 using equation I.5.1 in Chapter I.5.

Effective Renal Plasma Flow (ERPF)

This is usually measured using ^{123}I ortho-iodohippurate (^{123}I–OIH). Recently ^{99}Tc^{m} mercaptoacetylglycine (MAG3) has been used which gives slightly different values, usually quoted as MAG3 clearance, tubular function or tubular extraction ratio for MAG3, TER(MAG3). Following intravenous administration, the radiopharmaceutical describes a biexponential clearance curve, the first exponential of which is completed in ~30 min. Measurement is either by multiple or by single blood samples (Peters and Myers 1998), taken from the arm not used for administration, to avoid contamination of the blood samples. The plasma is counted.

Multiple Samples

Samples are taken every 5 minutes for 30 minutes and then every 20 minutes for approximately another 60 minutes.

$$\begin{aligned}\text{ERPF} &= \text{injected activity/area under the clearance curve} \\ &= \text{injected activity}/(A_1/\alpha_1 + A_2/\alpha_2)\end{aligned} \qquad \text{(III.1.2)}$$

where A_1 and A_2 are the intercepts of the two exponential curves with the activity axis at the administration time and α_1 and α_2 are the rate constants of the exponential curves.

Single Samples

In order to simplify the technique, various single sample models have been proposed. These create an imaginary volume (V) throughout which the administered activity is distributed at the time of sampling:

$$V = \text{injected activity/activity concentration of the sample} \qquad \text{(III.1.3)}$$

This is then related to ERPF, determined by multiple sample, using a regression equation. The following are the techniques which are recommended by the radionuclides in nephrourology committee on renal clearance (Blaufox et al. 1996).

^{123}I Ortho-iodohippurate

The recommended technique (Tauxe et al. 1982) optimally requires a sample at 44 minutes but a sample taken between 39–49 minutes will yield acceptable results. For a sample taken at 44 minutes:

$$\text{ERPF} = 1126.2\,(1 - e^{-0.008(\text{ID}/\text{Cn} - 7.8)})\ \text{ml min}^{-1}/1.73\ \text{m}^2 \qquad \text{(III.1.4)}$$

where

ID is the administered activity (counts per second)

C is the activity concentration in the plasma at the time of sampling (counts per second per litre)

C_n is the activity concentration normalised to a body surface area of 1.73 m^2 (counts per second per litre per 1.73 m^2) = C × BSA/1.73

For a sample taken between 39–49 minutes:

$$\text{ERPF} = F_{max}\,(1 - e^{-\alpha\,(\text{ID}/\text{Cn} - \text{Vlag})})\quad \text{ml min}^{-1}/1.73\ \text{m}^2 \qquad \text{(III.1.5)}$$

where

F_{max} $= 2501.3 - 108.1\,t + 2.656\,t^2 - 0.0206t^3$

α $= 0.0236 - 0.00035\,t$

V_{lag} $= 3.897 + 0.3\,t - 0.0048\,t^2$

t is the sampling time post-injection (minutes)

$^{99}Tc^{m}$ MAG3

Although the optimal sampling time is stated to be ~60 min for MAG3 with an increased time when the ERPF is low, the recommended technique (Bubeck 1993) suggests a sampling time of 44 minutes, with a sample time between 39–49 minutes yielding acceptable results.

For a sample time of 44 min:

$$\text{TER (MAG3)} = -318.6 + 145.9\,\ln(\text{ID}/C_n)\quad \text{ml min}^{-1}/1.73\ \text{m}^2 \qquad \text{(III.1.6)}$$

For a varying sample time:

$$\text{TER(MAG3)} = \alpha + \beta\,\ln(\text{ID}/C_n)\quad \text{ml min}^{-1}/1.73\ \text{m}^2 \qquad \text{(III.1.7)}$$

where $\alpha = -517e^{-0.011t}$ and $\beta = 295e^{-0.016t}$

As an alternative, a paediatric equation developed by the European Paediatric Task Group (Piepsz et al. 1993) is in use, for children above 1 year of age, with blood sampling between 30 and 40 minutes post-administration.

$$\text{MAG3 clearance} = 665.89/[P(t)\ e^{-0.0298512\,(t-35)}] + 1.89 \qquad \text{(III.1.8)}$$

where P(t) is the plasma activity concentration, at the time of sampling in relation to the administered activity (%ID litre^{-1})

This result must then be corrected for BSA.

Glomerular Filtration Rate (GFR)

This is usually measured using ^{51}Cr (EDTA) or $^{99}Tc^{m}$ (DTPA). The latter, although generally thought to be less accurate than ^{51}Cr EDTA, does in fact provide good precision and accuracy. Following intravenous administration, the radiopharmaceutical describes a biexponential clearance between 10 minutes and 4 hours, the first exponential of which is completed in ~2 hours (Peters and Myers 1998). Measurement is either by multiple or by single blood samples, taken from the arm not used for administration, to avoid contamination of the blood samples. The plasma is counted. The following are the techniques which are recommended by the radionuclides in nephrourology committee on renal clearance (Blaufox et al. 1996).

Biexponential Analysis

The most accurate technique requires ~11 blood samples to be taken over a period of 4–5 hours, to properly assess the clearance of the ^{51}Cr-EDTA from the plasma. The area under the plasma curve is usually estimated from a best-fit biexponential (Sapirstein et al. 1955). For routine use, the number of blood samples has been reduced. It has been demonstrated that the number of samples necessary for the definition of the plasma activity–time curve could be reduced to five without significant loss of accuracy (Brochner-Mortensen et al. 1974), or to six taken at 10, 20 and 30 minutes and 2, 3 and 4 hours (Peters and Myers 1998).

$$\begin{aligned} \text{GFR} &= \text{injected activity/area under the clearance curve} \\ &= \text{injected activity}/(A_1/\alpha_1 + A_2/\alpha_2) \end{aligned} \tag{III.1.9}$$

where A_1 and A_2 are the intercepts of the two exponential curves with the activity axis at the administration time and α_1 and α_2 are the rate constants of the exponential curves.

This actually gives total plasma clearance (TPC). TPC can be corrected to renal clearance and then to inulin clearance (Brochner-Mortensen 1978, 1985), incorporating also a correction for sampling from the venous system rather than the arterial (Brochner-Mortensen and Rodbro 1976). More recently, such corrections are not being implemented and raw TPC values are being reported. This avoids the introduction of any institutional variation in correction factors.

Monoexponential Analysis

A more convenient but less accurate method is to consider only the second exponential, taking 3–4 blood samples over a period of 2–5 hours.

$$\text{ie GFR} = \text{injected activity}/A_2/\alpha_2 = \text{injected activity} \times \alpha_2/A_2 \tag{III.1.10}$$

This method underestimates the area under the curve and thus overestimates the GFR. As renal function deteriorates, however, this error reduces because α_2 becomes smaller and A_2/α_2 larger (Peters and Myers 1998). A correction factor can be applied to the monoexponential value GFR_m, to make allowance for ignoring the first exponential component and to obtain the value that would have been obtained using the biexponential method.

Adults

- Chantler et al. (1969):

$$\text{GFR} = 0.93\ \text{GFR}_m \tag{III.1.11}$$

- Brochner-Mortensen (1972):

$$\text{GFR} = 0.990778\ \text{GFR}_m - 0.001218\ \text{GFR}_m^2 \tag{III.1.12}$$

Children

- Chantler and Barratt (1972):

$$\text{GFR} = 0.87\ \text{GFR}_m \tag{III.1.13}$$

- Brochner-Mortensen et al. (1974):

$$\text{GFR} = 1.01\ \text{GFR}_m - 0.0017\ \text{GFR}_m^2 \tag{III.1.14}$$

Two Sample Analysis

Further simplification is achieved by reducing the number of blood samples to two (Russell et al. 1985).

For $^{99}Tc^m$ DTPA:

$$GFR = \{[D \ln(P_1/P_2)/(T_2 - T_1) \exp(T_1 \ln P_2 - T_2 \ln P_1)/(T_2 - T_1)]\}^{0.979} \quad (III.1.15)$$

where

D is the administered activity (counts per minute)

P_1 is the plasma activity concentration at time T_1 (~60 min) (counts per minute per millilitre)

P_2 is the plasma activity concentration at time T_2 (~180 min) (counts per minute per millilitre)

This is also recommended, for use in children, by Piepsz et al. (2001).

Body Surface Area Correction

The final correction applied to the multi-sample techniques, is to correct the calculated GFR to a standard BSA of 1.73 m^2. In the monoexponential technique this correction must be applied before the correction for missing the first exponential. This is imposed in order to account for variations in body size and especially to compare children with adults.

$$GFR_{BSA} = GFR * 1.73/BSA \quad (III.1.16)$$

Single Sample Analysis

Ultimate simplification is achieved by taking only one blood sample. The technique is particularly attractive for use with paediatric patients and was developed using principles originally proposed for effective renal plasma flow estimation (Fisher and Veall 1975). In this, a mathematical model is used to estimate, from the activity concentration of the radiopharmaceutical in the single plasma sample, a GFR value that would be obtained using the multiple sample technique. The models in use are listed in appendix V.4, with the particular multi-sample model from which they are derived, referenced.

Adults

Watson (1992) modification of Christensen and Groth (1986) at t=240 referenced to biexponential analysis

$$709.6265 - \sqrt{[72792 - 0.7604 \ln\{ECV/(Q_0/C(t))\} ECV]} / 0.3802 \quad (III.1.17)$$

where ECV = 8116.6 BSA–28.2

Children

Ham and Piepsz (1991) at t=120 referenced to slope-intercept values with 0.85 correction (Chantler et al. 1969)

$$[2.602\ Q_0/(1000\ C(t))] - 0.273 \qquad \text{(III.1.18)}$$

where

BSA is the body surface area

t is the sample time, in minutes, following administration

Q_0 is the injected activity in counts per minute

C(t) is the plasma sample activity concentration at time t, in counts per minute

This is also recommended by Piepsz et al. (2001).

BSA Correction

In the single sample method of GFR estimation, BSA correction should be implemented by adjusting the plasma activity concentration rather than by altering the final estimate of the GFR, except when the model of Jacobsson (1983) is used. This technique allows single sample models derived for adults to be used in both adults and children, thus affording an age independent method of GFR estimation (Hamilton and Miola 1999).

Comparison with Normal Values

The final figure of GFR obtained is then compared to a normal range (Hamilton et al. 2000), which is usually listed in terms of one and two standard deviations for children (Piepsz et al. 1994; Postlethwaite 1986) and adults (Granerus and Aurell 1981).

Vitamin B_{12} Absorption

This is a test of urinary excretion of cyanocobalamin (vitamin B_{12}), which requires urine collection over 24 hours, rather than blood sampling. The method first described by Schilling is the most widely used, and is the one recommended by the ICSH (1981). Its advantages include good accuracy and ease of performance. In the dual radionuclide version, capsules of ^{58}Co B_{12} and ^{57}Co B_{12} bound to human intrinsic factor are administered orally and a flushing dose of, a relatively large amount of, non-radioactive B_{12} is given intramuscularly within two hours of the capsules. Standard solutions, of both radionuclides, are provided. These are diluted and aliquots prepared for measurement. A complete 24 hour urine collection is undertaken and aliquots prepared for measurement. Because of the mix of ^{58}Co and ^{57}Co in the urine, a dual channel measurement technique is used, with correction for the scatter of the higher energy ^{58}Co photons into the lower energy ^{57}Co window.

The calculations involved are:

- %^{57}Co excretion = 100 × ^{57}Co urine activity concentration × total urine volume/ (^{57}Co standard activity concentration × dilution factor)
- %^{58}Co excretion = 100 × ^{58}Co urine activity concentration × total urine volume/ (^{58}Co standard activity concentration × dilution factor)

$$\text{excretion ratio} = \%^{57}\text{Co excreted}/\%^{58}\text{Co excreted} \qquad \text{(III.1.19)}$$

Red Blood Cell Survival

This is measured by monitoring the removal of radiolabelled RBCs from the circulation over a period of ~50 days. ^{51}Cr is used to label the cells because its elution, from them, is only ~1% per day. Following administration, the radiopharmaceutical describes an almost monoexponential clearance curve which reflects both the removal of RBCs and the elution of the ^{51}Cr from them. Blood samples are taken three times a week until the activity concentration reduces to 50% of its initial value. It is the whole blood that is counted. Extrapolating the monoexponential clearance curve back to the administration time, the half-life of the ^{51}Cr labelled RBCs can be derived as the time when the sample activity concentration reaches 50% of its initial value. Because of the two clearance effects, the measured RBC half-life is shorter than the 60 days that would be expected physiologically.

Breath Analysis

The sample in this investigation is exhaled air, rather than body fluid. Also, the radionuclide used, ^{14}C, is a β^- emitter rather than a γ emitter. The radiopharmaceutical, ^{14}C urea, is administered orally (Balon et al. 1998). Breath samples are taken 20 minutes after administration by exhaling through a trapping liquid for a duration determined chemically. Liquid scintillant is added to the sample vial and the cocktail is measured on a β^- counter. Comparison, of the samples, is with a standard solution. The results are given as the percentage of the administered activity recovered per mmol of CO_2, multiplied by body weight.

$$\text{The amount of } ^{14}\text{C exhaled} = \text{sample activity } y\, w / \text{standard activity} \quad \%\ \text{kg mmol}^{-1} \qquad \text{(III.1.20)}$$

where
y = 100 × proportion of administered activity contained in the standard source/ (volume of collecting solution × concentration of hyamine)
w is the patient weight in kilograms

Gastrointestinal Iron Absorption

Three radionuclides of iron are available, of which the most suitable for diagnostic nuclear medicine is ^{59}Fe.

^{59}Fe

This has a long half-life of 45 days and is, therefore, suitable for protracted studies. It emits high energy γ photons of 1.1 and 1.3 MeV, which are suitable for measurement by in vitro sample or by in vivo whole body counting.

^{55}Fe

This has a very long half-life of 2.7 years. It emits low energy β^- and thus requires measurement by in vitro sample counting in a liquid scintillation system.

^{52}Fe

This has a short half-life of 8.3 h and is, therefore, unsuitable for protracted studies. It emits γ and β^+ radiation which allows it, at low activities, to be imaged by positron emission tomography.

Gastrointestinal iron absorption is assessed, over 6–7 days, following oral administration of ^{59}Fe citrate. This can be estimated by whole body counting as the percentage difference between the activity measured immediately after administration and that remaining in the body at the end of the study. Alternatively the radio-iron not absorbed can be estimated by in vitro measurement of the activity in a complete faeces collection over the duration of the study. This is given as the percentage of administered radio-iron excreted over the duration of the study. In both cases, daily blood sampling over the duration of the study allows assessment of the utilisation of the absorbed radio-iron. The plasma and the RBCs are counted. This is given as the percentage of administered radio-iron appearing in the plasma and RBCs over the duration of the study.

Gastrointestinal Protein Loss

Gastrointestinal protein loss is assessed, over 5 days, following intravenous administration of ^{51}Cr albumin or ^{111}In chloride. It is estimated by the measurement of the activity in a complete faeces collection over the duration of the study, and is given as the percentage of administered radionuclide excreted over the duration of the study. Because ^{111}In chloride is not also excreted in the urine, this is the preferred radiopharmaceutical to ensure that the faecal samples are uncontaminated.

Dilution Analysis

A number of measurements are based on the principle of dilution analysis, a technique attributed to George de Hevesy (McCready 2000), in which a known activity of tracer is introduced into the diluent pool under investigation and the volume of the pool is estimated from the extent of tracer dilution by assessing the activity concentration of the pool. Before a sample is withdrawn from the diluent pool, enough time must be given for the tracer to be thoroughly mixed ie to come to equilibrium. At equilibrium,

$$A = c\,v \qquad \text{(III.1.21)}$$

where

A is the activity of tracer introduced into the diluent pool

c is the sample activity concentration

v is the volume of the diluent pool (unknown)

Incomplete equilibration can result in falsely low values. Equilibration of several tracers is delayed in the presence of oedema (Siegel 1996).

Red Cell Volume and Plasma Volume

Red Cell Volume

For the measurement of red cell volume, the tracer introduced into the diluent pool is radiolabelled red blood cells and the diluent pool itself is the volume of RBCs in the body. $^{99}Tc^{m}$, ^{51}Cr and ^{111}In can be used to label the RBCs. The half-lives of ^{51}Cr and ^{111}In are unnecessarily long for this but $^{99}Tc^{m}$ does not maintain its binding as well. Following intravenous administration, the radiopharmaceutical concentration should remain constant. Blood samples are taken at 10 and 20 minutes. It is the whole blood that is counted, but the sample volume is multiplied by the haematocrit to obtain an activity concentration of the RBCs without the diluting effect of the plasma. If the time for the radiolabelled RBCs to mix is not unduly long, as might happen in congestive heart failure (Siegel 1996), and if significant elution of the radionuclide from the RBCs has not occurred, as might happen with $^{99}Tc^{m}$, the activity concentrations at both times should be identical. If they are not identical, the two values can be used to extrapolate the activity concentration back to administration time, to give the initial sample activity concentration (Peters and Myers 1998).

$$\mathrm{RCV} = A/c \qquad \text{(III.1.22)}$$

Plasma Volume

For the measurement of plasma volume, the tracer introduced into the diluent pool is radiolabelled protein, such as ^{125}I human serum albumin (HSA), and the diluent pool itself is the volume of plasma in the body. There is, however, an extravascular pool of albumin, and so the measured albumin space is somewhat larger than the plasma volume. Following intravenous administration, the radiopharmaceutical describes a monoexponential clearance curve because the radiolabelled protein leaves the circulation slowly to enter the extravascular pool of albumin. Blood samples are taken at 10, 20 and 30 minutes. It is the whole blood that is counted, but the sample volume is multiplied by (1–haematocrit) to obtain an activity concentration of the plasma without the diluting effect of the RBCs. Extrapolating the monoexponential clearance curve back to the administration time, the activity concentration in the plasma, at this time, can be established and a true estimate of the plasma volume achieved.

Combined Study

Red cell volume and plasma volume may be estimated in one investigation by simultaneously administering the two radiopharmaceuticals and using the dual energy counting technique. Correction is made for the scatter of ^{51}Cr activity into the ^{125}I window, ^{125}I activity in the ^{51}Cr window can be ignored.

Apparent Exchangeable Sodium

Apparent exchangeable mass is estimated by dividing the activity of the tracer retained in the body by the specific activity of the tracer in the plasma over a particular study duration. This investigation is undertaken for the calculation of total exchangeable sodium or potassium (Siegel 1996). Exchangeable sodium is measured using either ^{22}Na, which has a half-life of 2.6 years or ^{24}Na, which has a half-life of 15 hours. The investigation is undertaken over 24 hours following oral or intravenous administration of the radiopharmaceutical, a blood sample being taken at 24 hours. The plasma is counted. However, during the investigation time the radionuclide is also excreted. The administered activity must be reduced by this amount before estimating the dilution. The sodium volume is:

$$V = (\text{administered activity} - \text{excreted activity}) / \text{plasma activity concentration at 24 hours} \qquad \text{(III.1.23)}$$

$$\text{exchangeable sodium} = V \times \text{native sodium concentration in the pool}$$

When ^{24}Na is used, the excreted activity is determined from a total urine collection, and a total faecal collection if the patient suffers diarrhoea. ^{22}Na, however, is a γ emitting radionuclide, and when this is used excretion can be measured by whole body counting. This is calculated as:

$$\text{excreted activity} = \text{administered activity } W_{24}/W_0 \qquad \text{(III.1.24)}$$

where
W_{24} is the whole-body measurement at 24 hours
W_0 is the whole-body measurement at 2 hours after oral or immediately after intravenous administration.

Even such a prolonged mixing time will not result in complete equilibration, however, because exchangeable sodium only accounts for ~70% of total body sodium. The qualifier "apparent" is included because, although a one-compartment system is implied, peripheral compartments can exist resulting in a changing apparent exchangeable mass and the necessity for careful choice of sampling time (Siegel 1996).

Calcium Volume

A blood sample is taken at 5 hours after oral administration of ^{45}Ca. The plasma is counted.

$$\text{calcium volume} = A/c \qquad \text{(III.1.25)}$$

Probe Studies

Probe studies are undertaken by taking external measurements at various areas of interest over the body, at various times following administration of a radiopharmaceutical. Unlike tracer studies, the measurements cannot be prolonged because of consideration of patient comfort and cooperation. Thus the termination conditions and therefore the accuracy of the measurement depend on patient time. The error in the counting rate obtained from a probe system can be estimated from equation II.1.14 presented in 'measurement statistics' in Chapter II.1.

$$\text{standard deviation of the net count rate, } \sigma = \sqrt{[(C_{RT} + C_{RB})/t]} \quad \text{(III.1.26)}$$

where
C_{RT} is the count rate over the area of interest (target) in counts per second
C_{RB} is the count rate over a background area (counts per second)
t is the acquisition duration (seconds)

The diagnostic necessity is to determine when the net count rate over the area of interest is statistically different to the count rate over adjacent areas. This requires a difference of at least 2σ, so that there is a 95% probability that any measured difference is not due solely to random count rate fluctuations. The acquisition duration required to achieve this is given by Zanzonico and Heller (2000):

$$t_U = 100\, r\, (r+1)/[U\, C_{RT}\, (r-1)^2] \quad \text{seconds} \quad \text{(III.1.27)}$$

where
$r = C_{RT}/C_{RB}$
U is the uncertainty (i.e. 0.05 for a 5% uncertainty)

This duration reduces as target count rate and target to background count rates increase, and the equation demonstrates the difficulty of detecting low contrast targets. Thus when acquisition duration is limited, for instance using intraoperative probes to investigate a number of sites, detector sensitivity becomes extremely important (Zanzonico and Heller 2000). This can be maximised by degrading the absorptive collimation or by widening the pulse height analyser window which is typically set at 20–30%. As with the tracer studies, the activity in a region of the body can be compared to a standard activity. This may be done using an internal standard, where the measurement is compared to another fixed area of the body, or by using an external standard, where the measurement is compared to one taken using a phantom. For example, an internal standard is usually used in surface counting and an external standard in thyroid uptake measurements. The measurement of the external standard can be prolonged without regard to patient cooperation, but it is not necessary to extend it hugely.

Surface Counting

Surface counting studies involve activity measurements over various organs following administration of the radiopharmaceutical. A probe with a short-bore flat-field collimator in direct contact with the body surface is used. An example, is the RBC sequestration investigation.

Red Blood Cell Sequestration

This is used to determine sites of RBC destruction or sequestration and is assessed, over ~2 weeks, following intravenous administration of ^{51}Cr RBCs. It is undertaken by surface counting over the liver, spleen, heart and thigh. The measurement duration should be the same at each site, and of the order of 300 s. The thigh count is used as background and is subtracted from each of the other three. The measurements begin one day after administration. On the first day at each site, several measurements are

taken until the maximum count is found is found. The position of the probe on the skin is then marked with a waterproof marker, which the patient must reinforce after washing. Measurements are taken every second day thereafter, and continue until the count rate over the heart reduces to below half of its initial value. Changes in the liver and spleen counts with time indicate the role of each organ in any RBC destruction. The results are presented as three ratios, plotted against time, on a linear scale:
1. liver to heart
2. spleen to heart
3. spleen to liver

Thyroid Uptake

In this, the count rate obtained from the thyroid gland is compared with that obtained from a standard. The latter is an accurately known activity of the same radionuclide contained in a Lucite neck phantom. The measurement on the standard should be performed as close in time as possible to those on the patient, in order to minimise any change in system sensitivity. The measurements are made using a probe with a long-bore flat-field collimator. Unlike in the splenic sequestration study, it is essential that the entire thyroid be within the field of view of the probe. Since the field of view of the probe increases with increasing source to detector separation, the probe collimator is not placed in contact with the skin. Instead, it is positioned over the thyroid region, a short distance away from the skin. This also reduces sensitivity variations with separation changes, see 'Activity distribution in a patient' in Chapter II.2. However, it is also essential that the geometry of the patient and standard measurements be identical. Therefore, to enable reproducible positioning, a spacing bar is attached to the collimator, and it is this that is placed in direct contact with the skin or standard. This results in a thyroid-to-crystal distance of 25–30 cm (Becker et al. 1996). Background measurements are necessary and must be subtracted from both the patient and the standard measurements. For the patient, this is obtained over the lower thigh with the spacing bar touching the skin. For the standard, this is obtained in the measurement room without any sources present.

Typical measurement termination conditions are:
- patient 10×10^3 counts or a maximum time of 600 s
- standard 10×10^3 counts
- background 60 s

The measurements are converted to counts per second and the calculation of thyroid uptake is:

$$\text{Uptake (in \%)} = 100 * \text{CR} * \text{AR} \qquad \text{(III.1.28)}$$

where
CR = (thyroid–thigh background)/(standard–room background)
AR = (standard activity)/(administered activity)

Care has to be taken in the use of probe systems because of their susceptibility to dead-time losses at the high count rates associated with counting the capsule standard. If not, differences can be observed between thyroid uptake values measured using a probe and a γ camera; the latter recording counts in direct proportion to the activity of ^{123}I in the phantom, but the probe systems exhibiting a non-linear relationship and yielding uptakes higher than the γ camera (Lee et al. 1995).

Intraoperative Counting

Although most widely accepted in radioimmunoguided surgery RIGS (Hoffman et al. 1999), intraoperative probes are becoming increasingly important in the localisation of sentinel lymph nodes and visually indistinct lesions. Usually photon radiation is detected. However, for reliable localisation, the probes must be placed within ~1 cm of the lesion. This is mainly due to the deterioration in spatial resolution with distance from the detector and, although absorptive collimation can reduce this effect, it is achieved at the expense of reducing sensitivity. The detector is often rotated to improve access to such lesions, care must be exercised however to avoid unrecognised detection of more distant activity accumulations such as residual activity at the injection site. β^- radiation can be detected using plastic scintillation detectors, but these have to be placed immediately adjacent to the lesion to be useful. Also for reliable localisation, the probe must be maintained at each measurement site long enough for sufficient data to be accumulated according to equation III.1.27 given above (Zanzonico and Heller 2000).

Sentinel Lymph Nodes

This enables the sentinel lymph node, of a melanoma or breast tumour, to be identified intraoperatively. Because the radioactivity is almost completely confined to the lymphatic system, the nodes exhibit a very high object contrast (Hoffman et al. 1999; Alazraki et al. 2002). They are also situated superficially and, therefore, poorer energy resolution can be tolerated because scattered photons are not a huge problem. The exception is when adjacent to the injection site and side shielding is important, to reduce the effect of this (Perkins and Britten 1999). The probes require good spatial resolution to enable precise positioning of the small target, and can accept reduced sensitivity and energy resolution to achieve this. The effects of the reduced sensitivity can be mitigated somewhat by increasing the acquisition duration at each site, since these measurements are undertaken before surgical incision; and by restricting the number of sites sampled, using prior approximate localisation. The latter is undertaken using planar γ camera imaging with, possibly, ^{57}Co transmission images for anatomical orientation (Zanzonico and Heller 2000). A 20% symmetric PHA window centred on the photopeak is usually used and a positive accumulation is identified as 3 standard deviations above background (Perkins and Britten 1999).

Visually Indistinct Lesions

As with sentinel lymph nodes, identification of visually indistinct lesions on preoperative planar γ camera images with help intraoperative detection (Perkins and Hardy 1996). Probes for localisation of visually indistinct lesions require good sensitivity to enable a large number of sites to be evaluated using a reasonably short acquisition duration and yet achieving a reasonably low statistical uncertainty. The probes also require good energy resolution, to reduce the effect of scattered radiation from background activity, which may exhibit a variable spatial distribution. Detector characteristics should be chosen to suit the application; for instance, CsI(TI) with its high detection efficiency allows a shorter acquisition time, NaI(TI) is appropriate for deep-lying lesions, and CdTe is well suited for superficial lesions with a high activity concentration in underlying tissue (Benjegard et al. 1999).

References

Alazraki N, Glass EC, Castronovo F, Valdes Olmos RA, Podoloff D (2002) Procedure guideline for lymphoscintigraphy and the use of intraoperative gamma probe for sentinel lymph node localization in Melanoma of intermediate thickness 1.0. J Nucl Med 43:1414–1418

Balon H, Gold CA, Dworkin HJ, McCormick VA, Freitas JE (1998) Procedure guideline for carbon-14-urea breath test. Society of Nuclear Medicine. J Nucl Med 39:2012–2014

Becker D, Charkes ND, Dworking H, Hurley J, McDougall R, Price D, Royal H, Sarkar S (1996) Procedure guideline for thyroid uptake measurement: 1.0. J Nucl Med 37:1266–1268

Benjegard SA, Sauret V, Bernhardt P, Wangberg B, Ahlman H, Forssell-Aronsson E (1999) Evaluation of three gamma detectors for intraoperative detection of tumors using 111In-labeled radiopharmaceuticals. J Nucl Med 40:2094–2101

Blaufox MD, Aurell M, Bubeck B, Fommei E, Piepsz A, Russell C, Taylor A, Thomsen HS, Volterrani D (1996) Report of the radionuclides in nephrourology committee on renal clearance. J Nucl Med 37:1883–1890

Brochner-Mortensen J (1972) A simple method for the determination of glomerular filtration rate. J Clin Lab Invest 30:271–274

Brochner-Mortensen J (1978) Routine methods and their reliability for assessment of glomerular filtration rate in adults. Dan Med Bull 25:181–202

Brochner-Mortensen J (1985) Current status on assessment and measurement of glomerular filtration rate. Clin Physiol 5:1–17

Brochner-Mortensen J, Rodbro P (1976) Comparison between total and renal plasma clearance of [^{51}Cr]ETDA. Scand J Clin Lab Invest 36:247–249

Brochner-Mortensen J, Haahr J, Christoffersen J (1974) A simple method for the accurate assessment of the glomerular filtration rate in children. Scand J Clin Lab Invest 33:139–143

Bubeck B (1993) Renal clearance determination with one blood sample: improved accuracy and universal applicability by a new calculation principle. Semin Nucl Med 23:73–86

Chantler C, Barratt TM (1972) Estimation of glomerular filtration rate from plasma clearance of ^{51}Cr edetic acid. Arch Dis Child 47:613–617

Chantler C, Garnett ES, Parsons V, Veall N (1969) Glomerular filtration rate measurement in man by the single injection method using ^{51}Cr-EDTA. Clin Sci 37:169–180

Christensen AB, Groth S (1986) Determination of ^{99m}Tc-DTPA clearance by a single plasma sample method. Clin Physiol 6:579–588

Fisher M, Veall N (1975) Glomerular filtration rate estimation based on single blood sample. Br Med J 3:542

Granerus G, Aurell M (1981) Reference values for ^{51}Cr-EDTA clearance as a measure of glomerular filtration rate. Scand J Clin Lab Invest 41:611–616

Ham HR, Piepsz A (1991) Estimation of glomerular filtration rate in infants and in children using a single-plasma sample method. J Nucl Med 32:1294–1297

Hamilton D, Miola UJ (1999) Body surface area correction of single sample methods of GFR estimation. Nucl Med Commun 20:273–278

Hamilton D, Riley P, Miola U, Mousa D, Popovich W, Al Khader A (2000) Total plasma clearance of ^{51}Cr-EDTA: variation with age and sex in normal adults. Nucl Med Commun 21:187–192

Hoffman EJ, Tornai MP, Janecek M, Patt BE, Iwanczyk JS (1999) Intraoperative probes and imaging probes. Eur J Nucl Med 26:913–935

ICSH (1981) International Committee for Standardization in Haematology: recommended methods for the measurement of vitamin B_{12} absorption. J Nucl Med 22:1091–1093

Jacobsson L (1983) A method for calculation of renal clearance based on a single plasma sample. Clin Physiol 3:297–305

Lee KH, Siegel ME, Fernandez OA (1995) Discrepancies in thyroid uptake values. Use of commercial thyroid probe systems versus scintillation cameras. Clin Nucl Med 20:199–202

McCready VR (2000) Milestones in nuclear medicine. Eur J Nucl Med 27 [Suppl]:S49–S79

Perkins AC, Britten AJ (1999) Specification and performance of intra-operative gamma probes for sentinel node detection. Nucl Med Commun 20:309–315
Perkins AC, Hardy JG (1996) Intra-operative nuclear medicine in surgical practice. Nucl Med Commun 17:1006–1015
Peters AM, Myers MJ (1998) Physiological measurements with radionuclides in clinical practice. Oxford University Press, Oxford
Piepsz A, Gordon I, Hahn K, Kolinska J, Kotzerke J, Sixt R (1993) Determination of technetium-99m-mercaptoacetylglycine plasma clearance in children by means of a single blood sample: a multicentre study. Eur J Nucl Med 20:244–248
Piepsz A, Pintelon H, Ham HR (1994) Estimation of normal ^{51}Cr EDTA clearance in children. Eur J Nucl Med 21:12–16
Piepsz A, Colarinha P, Gordon I, Hahn K, Olivier P, Sixt R, van Velzen J (2001) Guidelines for glomerular filtration rate determination in children. Eur J Nucl Med 28:BP31–BP36
Postlethwaite RJ (1986) Clinical paediatric nephrology. Wright, Bristol
Russell CD, Bischoff PG, Kontzen FN, Rowell KL, Yester MV, Lloyd LK, Tauxe WN, Dubovsky EV (1985) Measurement of glomerular filtration rate: single injection plasma clearance method without urine collection. J Nucl Med 26:1243–1247
Sapirstein LA, Vidt DG, Mandel MJ, Hanusek G (1955) Volumes of distribution and clearances of intravenously injected creatinine in the dog. Am J Physiol 181:330–336
Siegel BA (1996) Principles of isotope dilution. In: Sandler MP, Patton JA, Coleman RE, Gottschalk A, Wackers FJT, Hoffer PB (eds) Diagnostic nuclear medicine, 3rd edn. Williams and Wilkins, Baltimore, pp 823–826
Tauxe WH, Dubovsky EV, Kidd TJ, Diaz F, Smith LR (1982) New formulas for the calculation of effective renal plasma flow. Eur J Nucl Med 7:51–54
Watson WS (1992) A simple method of estimating glomerular filtration rate. Eur J Nucl Med 19:827
Zanzonico P, Heller S (2000) The intraoperative gamma probe: basic principles and choices available. Semin Nucl Med XXX:33–48

Single Photon Planar Imaging

Abbreviations

EANM European Association of Nuclear Medicine
ECG electrocardiogram
EDV end-diastolic volume
EF ejection fraction
ERNA equilibrium radionuclide angiography
ERPF effective renal plasma flow
FPRNA first pass radionuclide angiography
FWHM full width at half maximum
GFR glomerular filtration rate
LEGP low energy general purpose
LEHR low energy high resolution
LEHS low energy high sensitivity
LEUHR low energy ultra high resolution
MUGA multiple gated acquisition
PER peak emptying rate
PFR peak filling rate
PHA pulse height analyser
PMT photomultiplier tube
ROI region of interest
SV stroke volume
SVC superior vena cava
tPER time to peak emptying rate
tPFR time to peak filling rate

The strength of nuclear medicine lies in its unique ability to image physiological function (Tindale and Barber 2000). Imaging the distribution of a radionuclide provides much more information than can be provided by in vitro or probe investigation. It cannot easily, however, provide the absolute values of the physiological parameters that in vitro investigation can and it is not as sensitive to high energy γ photons as probe investigations. It is also considerably more expensive and difficult to perform. Planar imaging provides a two dimensional representation of a three dimensional object, by integrating the activity along a projection line. The images are thus a composite of overlying contributions from various activity accumulations, which degrades detectability of abnormal activity distributions by reducing object contrast and causing reduced image contrast. The detection of a lesion is proportional to the size and contrast of the lesion and inversely proportional to system spatial resolution and statistical noise in the image (Mullani 2000). One of the great advantages of imaging is to be able to follow variations in the activity distribution over time.

Consistent optimisation of image quality demands an appreciation of both the proper operation of the γ camera and the required acquisition and processing parameters for particular investigations. Careful attention to these procedures is essential to avoid suboptimal investigation and production of artefacts (Forstrom et al. 1996). As system complexity increases, it becomes more important to recognise the various potential artefacts and to appreciate their impact on clinical efficacy (O'Connor 1996).

To extract the maximum amount of diagnostic information from the acquisition data, in various investigations, often involves a number of data processing manipulations, additional to visual inspection of the images. An ideal quantitative parameter should be accurate, reproducible and physiologically representative. It should also be independent of the department in which it is calculated since any such variation is likely to result from differences in technique. This will allow the end user to have confidence in the results and for these to have diagnostic impact. It can only be achieved if the user is aware of the behaviour and limitations of the analysis. To help in this, a number of recommendations regarding both acquisition and processing protocols specific to a number of clinical investigations have been presented by a number of bodies, which are intended to improve harmonisation in the modality. Most analyses involve region of interest construction and calculation of target to background ratios (Tindale and Barber 2000), although many more complicated techniques are readily available such as: convolution, deconvolution, Fourier analysis, factor analysis, ROC analysis, artificial intelligence and artificial neural networks (Chandler and Thomson 1996; de Lima 1996; Lawson 1999). It is anticipated that advanced analysis tools will become more prevalent in routine nuclear medicine practice in the future since continuing rapid improvements in hardware performance have made many, previously computationally unattractive, methods feasible (Todd-Pokropek 2002).

Clinical Use of the Gamma Camera

The gamma camera is a sensitive piece of equipment and has to be treated carefully, to consistently achieve image optimisation. The immediate environment is very important. The temperature, humidity, dust concentration and electrical supply have to be well controlled. The ambient temperature, in particular, must not be allowed to change rapidly. The maximum change per hour will be specified by the manufacturer, and exceeding this could cause fracture of the crystal. The collimator provides a large heat sink which helps to minimise the transfer, to the crystal, of any shift in ambient temperature. A suggested maximum room temperature change is ~3–5°C per hour when the collimator is installed, this will correspond to a crystal temperature change of 1°C per hour (IPSM 1992). Thus a collimator should always be loaded onto a camera, except if a particular quality assurance investigation requires otherwise (IAEA 1991). This also provides mechanical protection against physical damage to the crystal. Another component which is very sensitive to temperature change is the PMT. Gain changes can be caused and this alters the position of the photopeak (O'Connor 1996).

Particular care must be taken with the manipulation of the collimator. The rear face is designed to be immediately adjacent to the crystal surface. If anything is left on the crystal surface and an attempt is made to load a collimator, this will almost certainly result in crystal damage. The collimator itself is very delicate, particularly the low energy category. A scratch or a dent is easy to inflict and will usually cause a focal reduction in sensitivity. Such damage can easily be caused in use and during changing, the latter being reduced by the present generation of automatic and semiautomatic collimator changers, however (O'Connor 1996). When the camera is not in use, the advice given for single head cameras is that the head should be positioned with the face horizontal, this avoids temperature variation across the field of view; and the head should be directed downwards, this helps to prevent separation of the PMT-light guide assembly from the crystal. The camera should also be positioned over the imaging table, to protect the collimator from mechanical damage. For dual and triple head systems, the advice varies among manufacturers.

During use, the radiation environment must be controlled (Hines et al. 1999). This is not straightforward because effects may be transient. Radioactive contamination may interfere with the image when the camera is only in one particular orientation, and a radioactive patient or source in transit may interfere only for a limited time. Such effects may not be immediately apparent on the image, particularly in the presence of abnormal activity distributions. The only solutions are careful working practices, regular contamination surveys and comprehensive risk assessments.

Acquisition Considerations

The aim of the technique is to reveal change in the overall accumulation of a radiopharmaceutical in a whole organ or the variation in relative accumulation throughout an organ or an area of interest. Such accumulation may be higher, commonly called 'hotter', or lower, 'colder or photopenic', than adjacent areas. It may be diffuse or focal, extensive or minute. Such variations demand different acquisition considerations, which significantly alter the quality of the image produced.

Patient Motion

Each image requires a protracted duration to accumulate the required data, during which any artefactual motion of the activity distribution relative to the γ camera will reduce object contrast, and result in reduced image contrast. The first consideration is patient motion. The intended acquisition duration has to be established by considering, as well as the requirements for achieving diagnostic information, the time that the patient can comfortably remain cooperative and avoid body movement. The second consideration is the involuntary motion of individual organs relative to the whole body, for example the heart and lungs. These can move by as much as 2 cm, during a normal cycle and, considering that the system spatial resolution might be $<1/3$ of this, will have a significant effect on the image.

Collimator

The collimator is the only component of the camera that can be changed by the operator. It dramatically affects the images obtained and correct selection is extremely important. Wherever possible, a parallel hole collimator should be used because this provides a 1 : 1 image:object representation without spatial distortion varying across the field of view. Initial selection is in terms of the energy of the γ photons being detected. Because the detection sensitivity and spatial resolution are better with lower energy collimators, the lowest energy collimator possible should be chosen. If too high an energy collimator is chosen, the septa will be visualised. If, however, too low an energy collimator is chosen, the septa will be too thin for the energy of the γ photons present, and a background activity will be imposed on the image, resulting in a loss of image contrast (Forstrom et al. 1996). The energy category must be chosen by considering all the photons involved; many radionuclides emit photons of several energies and some radiopharmaceuticals contain contaminant radionuclides which emit γ photons of a higher energy than the useful radionuclide. In such cases, experimental evaluation of various collimators, using phantoms and cooperative patients, is required. For example, in dual radionuclide studies involving $^{99}Tc^{m}$ and ^{123}I, a medium energy collimator is preferred over a low energy high resolution because of its higher contrast accuracy (de Geeter et al. 1996). Energy categories vary among manufacturers and the choice, for a particular radionuclide, may therefore be different for different systems.

In selecting the appropriate low energy collimator for a particular clinical acquisition, a compromise must be achieved between spatial resolution, which can limit the contrast of the reconstructed image, and detection sensitivity, which determines the noise in the image (Lau et al. 2001). The general purpose type tends to be used for most applications, only being replaced in special circumstances. To detect a small area of increased activity, a poor sensitivity, high resolution collimator is required. To detect a slight reduction in activity over a large area, low noise images and a high sensitivity, poor resolution collimator are required (Rousset et al. 1998). This is true also for some fast dynamic studies.

With parallel hole collimators, spatial resolution deteriorates dramatically with increasing separation between the collimator face and the body surface, see Fig. II.3.22. Purely for this reason, imaging must always be undertaken with the collimator as close to the patient as possible. Failure to minimise the patient to collimator distance is probably one of the most common causes of suboptimal image quality (O'Connor et al. 1991). If there is no septa penetration, detection sensitivity is not affected by such separation and is not a consideration in positioning the camera.

Diverging and converging collimators are unusual on present systems. Their minification and magnification properties are no longer essential and it is preferable to avoid the problems associated with spatial distortion. Slanted parallel hole collimators are sometimes still used, however, to reduce collimator–heart separation in cardiac gated blood pool studies. The only collimator type, other than parallel hole, that is often used is the pinhole; the system spatial resolution of the camera being improved, but at the expense of reducing detection efficiency and introducing some spatial distortion. Small lesions not detected using a parallel hole collimator are often seen using a pinhole, e.g. in thyroid and skeletal joint imaging.

Occasionally, activity is assessed using the camera without a collimator attached. In this situation, spatial distribution is unimportant and the interest is only the total activity in the body. Examples include bile acid retention using ^{75}SeHCAT and gastrointestinal protein loss using ^{111}In chloride (Peters and Myers 1998).

Pixel Size

The pixel size of the acquisition matrix is altered by defining the size of the matrix and the acquisition zoom. It should be determined initially by considering the system spatial resolution, at the part of the body nearest to the collimator face, where it will be the highest. According to the Nyquist theory, the pixel size should not be greater than half the system spatial resolution (Rosenthal et al. 1995), in terms of the FWHM this is pixel size FWHM/3. This pertains to noise free data. For clinical images, it is usual to relax this sampling requirement to reduce the perception of the noise in the image (Sorensen and Phelps 1987) to ~FWHM/2. Pixel sizes larger than this will degrade spatial detail by loosing the higher spatial frequencies, sizes smaller than this will not provide any extra information since the system spatial resolution cannot be improved by digitisation. Parts of the body, further away from the collimator than the surface, will be imaged with a poorer spatial resolution for which the defined pixel size will therefore be adequate. For example, a camera and low energy general purpose collimator with a system spatial resolution at its face of 6 mm, when placed close to the body surface will result in a limiting system spatial resolution of 6 mm. The pixel size, therefore, should not be more than 3 mm. For a 500 mm field of view camera, this can be achieved by setting the matrix to 256 which will give a pixel size of 2 mm. Alternatively, the matrix can be set to 128 and a zoom 1.3 of applied, which will give a pixel size of 3 mm. With distance, the system spatial resolution might deteriorate to 12 mm, in which case each pixel is 1/4 the size of the resolution. In this case, no extra spatial information is provided but, more importantly, none is lost at the surface.

A number of other considerations affect the choice of matrix and acquisition zoom. The first is whether zoom can be applied, since it will reduce the field of view by the same amount, which may be contraindicated if a large area of interest is involved. The next is the expected number of counts. On the basis of Poisson statistics, the statistical error in the image increases as the count in each pixel reduces. Thus in, say, a dynamic study, a much larger pixel is used than would be indicated on the basis of spatial resolution considerations, because the count in each pixel would otherwise be so low. Also, often in these types of study, spatial detail is not as important to the interpretation as reliable counting statistics. Thirdly, sometimes a much smaller pixel size is used than would be warranted on a consideration of spatial resolution. This would be where the counts are high and there would be no concern about statistics. The main reason would be to avoid the appearance of pixels in the image and to give a smoother variation over the image, in say bone scans (Sharp et al. 1982). Post acquisition magnification is also used to improve the appearance of the displayed images. In this case, there is no alteration in of acquisition pixel size, however.

Partial Volume Effect

This occurs when the size of the activity distribution is $<2 \times$ FWHM of the system spatial resolution, ie the volume that can be resolved. A small activity source will appear larger and its activity will be underestimated; e.g. if the spatial resolution is 10 mm, a 2 mm source will appear to have a dimension of 10 mm. This is because objects $<2 \times$ FWHM only occupy part of the resolution volume and contributions from other areas are included in the detected distribution. Large objects will also demonstrate this effect at their borders. The total counts are preserved but they are spread over an image area which is larger than the object activity distribution. Thus the image representation of total activity is accurate but that of concentration distribution is not. The correction factors required to obtain the correct values are known as the recovery coefficients. A more accurate image can be produced, when simple shapes are involved, by making calibration measurements and calculating these recovery coefficients (IPSM 1985; Zito et al. 1996; Hutton and Osiecki 1998; Zaidi and Hasegawa 2003).

Pulse Height Analyser

Lesion detection is compromised by Compton scatter in the object which reduces the object contrast. The pulse height analyser (PHA) is used to reduce this effect by narrowing the photon energy acceptance window, to reject the lower energy scattered photons. Because the energy resolution of NaI(Tl) is only ~10% at 140 keV (Zanzonico and Heller 2000), a significant proportion of detected scattered photons will still be accepted, which degrades image contrast (Gustafsson et al. 2000) by imposing a fairly uniform background activity. It also results in a significant proportion of unscattered photons being rejected.

The PHA window width is, therefore, a compromise; a narrow window improving the image contrast but at the expense of detection sensitivity and imposing an unacceptably long acquisition duration; and a wide window causing loss of image contrast. Typically a 20% PHA window is used. With this ~25–35% of the counts in the image will be from scattered photons but a sufficient proportion of unscattered photons will be detected to achieve a reasonable acquisition duration. With energy resolution improving in newer cameras, the energy window can be reduced whilst still maintaining a reasonable count rate. Usually, a symmetric PHA energy window of 20% is used with $^{99}Tc^{m}$ (Forstrom et al. 1996) but this can be reduced to 15%, with the latest energy resolution cameras, without significant loss in detection sensitivity and with some improvement in image contrast (DePuey and Garcia 2001). As mentioned in Chapter II.1, a rule of thumb is to set a PHA window width to be about twice the energy resolution of the detector. The benefit can be assessed by measuring the intrinsic spatial resolution with various windows. The penalty can be assessed by noting the increased acquisition durations at these windows. Care must be exercised when using a different window widths, however, because the intrinsic camera uniformity may deteriorate as the window is narrowed. The window may also be set asymmetrically high on the photopeak, e.g. 133–161 keV instead of 126–154 keV for $^{99}Tc^{m}$, on the basis that photons detected with energies above the photopeak are less likely to be scattered than those with energies below it. Again, the uniformity may be sensitive to such changes (IAEA 1991).

Multiple PHA windows are used for radionuclides with multiple photopeaks, e.g. ^{67}Ga. This gives a better detection sensitivity with no increase in scatter component. For ^{67}Ga, the usual PHA keV settings are 93±20%, 185±15%, 300±15%; detection increasing by 9–12% using three, rather than two, energy windows (El Fakhri et al. 2002). However, each window effectively produces a different image and all images have to be summed. Care must be taken that the images are accurately aligned, to within 1–3 mm (O'Connor 1996), otherwise loss of spatial resolution and contrast will occur. Multiple PHA windows are also used for investigations involving two radionuclides. In this case, two images are formed but remain as two images. In such acquisitions, care must be taken about the amount of downscatter from the higher energy photon to the lower energy window. An example of this is lung ventilation–perfusion imaging where $^{99}Tc^{m}$ and $^{81}Kr^{m}$ are used (Hastings et al. 1995).

Before imaging is commenced, the position of each PHA window, in use, in relation to its intended photopeak must be checked; a relatively common mistake being to image $^{99}Tc^{m}$ on a ^{57}Co setting (Forstrom et al. 1996). It is preferable to do this using a source rather than using radionuclide in a patient because of the scatter from the tissue, which will deform the photopeak.

Count Rate

It is very unusual, in clinical investigations, to experience a count rate that is anywhere near the level at which the spatial resolution of the camera will start to deteriorate; quantification may be compromised however, e.g. in a first pass cardiac study. Good practice is to operate a camera at a count rate less than would cause a 10–20% input count rate loss; this would be at ~3–5×10^{4} counts s^{-1} (IAEA 1991).

Count Density

Image contrast quantifies the differences in count density over an image. This depends on the activity concentration differences in the object, the object contrast; the higher these are, the easier it is to detect differences in the image.

$$\text{object contrast, } C_o = (A_a - A_n)/A_n \qquad \text{(III.2.1)}$$

where
A_a is the activity concentration in abnormal distribution
A_n is the activity concentration in normal distribution

When there is increased activity accumulation in an abnormal distribution, $A_a > A_n$, high values of object contrast may be obtained, especially when the radiopharmaceutical enjoys a particularly good accumulation. In theory, C_o tends to infinity as A_a accumulates more activity thus depriving A_n, which tends to 0. When there is decreased activity accumulation in an abnormal distribution, $A_a < A_n$, however, there is a limit on the object contrast that may be obtained. This occurs when there is no radiopharmaceutical accumulation in the abnormal area, in which case $C_o = -1$. This discrepancy means that it is often much easier to detect an increased accumulation than a decreased accumulation; even with dimensions less than the system spatial resolution, the higher the object contrast, the smaller the abnormality on the image that can be detected.

Background activity concentration, A_b, reduces object contrast according to:

$$\text{object contrast, } C_o = (A_a - A_n)/(A_n + 2\,A_b) \qquad \text{(III.2.2)}$$

A complicating factor is the information density in the image, defined as the number of counts in the image per centimetre squared of the object. This varies over the image, e.g. in a bone scan, the areas in the image representing bone will have a high information density, whilst those representing soft tissue will generally have a low information density. When information density is small, there is uncertainty as to whether one activity distribution is different from another, and only large activity distributions with high object contrasts can be differentiated from each other. As the information density increases, there is less uncertainty and smaller activity distributions with smaller object contrasts can be differentiated, which is appreciated as finer structural details becoming more apparent in the image.

There is therefore a limit to the maximum count density that is necessary, however, because above a certain density, the system spatial resolution becomes the limiting factor for portraying the finer structural details of the object. If the counts in the background area of the image are n and the increased counts in the target area are (n+Δn); for the target to be reliably detected, Δn must be at least 10% of n, ie $\Delta n = 0.1n$. The statistical uncertainty in each area:

$$\sigma_n = \sqrt{n} \quad \text{and} \quad \sigma_a = \sqrt{(n+\Delta n)} \qquad \text{(III.2.3)}$$

where
σ_n is the sd in the 'normal' area
σ_a is the sd in the 'abnormal' area

and the combined uncertainty:

$$\sigma^2 = \sigma_n{}^2 + \sigma_a{}^2, \text{ when the two activity distributions have equal areas}$$

$$\sigma \sim \sqrt{(2n)}, \text{ if } \Delta n \ll n \qquad \text{(III.2.4)}$$

A difference of 2σ or 0.1n, between the counts in two activity accumulations, will be sufficient for a 95.5% probability that the difference is significant, because (Parker et al. 1984):

$$2\sigma = 2\sqrt{[2n]} = 0.1\,n$$
$$\text{ie } n = 800 \qquad \text{(III.2.5)}$$

The area over which the count density must be calculated is determined by the system spatial resolution. For a system spatial resolution of 8 mm at, say, 5 cm in scatter, assuming a LEHR collimator is installed, the count density must be calculated over an area (a) of 0.8×0.8 cm^2, ie 0.64 cm^2. In this case, the count density required to ensure a 95.5% probability that the difference, between two areas, is significant:

$$n/a = 800 \text{ counts}/0.64 \text{ cm}^2 \text{ or } \sim 1250 \text{ counts cm}^{-2} \qquad \text{(III.2.6)}$$

This is the approximate information density required, in the area of interest, for reliable differentiation of count densities. The following calculations are very approximate but are instructive. For example, in a 500×400 mm posterior thorax bone image, in which most of the counts are in the skeleton and ~40%, or 800 cm^2, of the image represents bone: the image should contain ~10^6 counts. Acquisition of less counts than this may result in abnormal activity distributions, with low object contrasts, being undetected. Counts higher than this may not contribute additionally to detectability because the system spatial resolution may be the limiting factor. If a LEGP collimator, with a system spatial resolution of say 9 mm, is used, instead of the LEHR, the target image count should be ~7.9×10^5. Taking into account the better sensitivity of the LEGP collimator, see Table II.3.1, the acquisition duration for the LEGP would be only ~47% of that of the LEHR. Therefore, optimising the benefit of an increased system spatial resolution has a significant impact on the acquisition duration, which is the main advantage of dual head camera systems for whole body bone imaging.

Multihead Systems

Dual head gamma cameras, particularly with rectangular heads, are particularly suitable for whole body scanning in, e.g. $^{99}Tc^{m}$ bone or ^{67}Ga investigations. For the same acquisition duration, determined on the length of time the patient can cooperate, they allow a doubling of count density, and should not be used to reduce the acquisition period.

Attenuation

Abnormalities in the activity distribution are detected more easily near the body surface than at depth. This is due to attenuation, by tissue containing the radionuclide and by overlying tissues, reducing the number of photons reaching the detector.

Types of Study

Five types of study are possible: static, dynamic, whole body, gated and list mode. The static acquisition is the simplest and, unlike the rest, can be undertaken without a computer. In all of the studies the images take a considerable time to acquire, varying from a few minutes for a static image to, perhaps, an hour for a dynamic study. The patient must remain stationary for the complete acquisition otherwise the image will be blurred. It is usually this cooperation that determines the maximum duration that an image can be acquired for and therefore the maximum count density that can be achieved.

Static

This type of study can only be undertaken on a radiopharmaceutical distribution that does not change, in the area of interest, during the investigation. It comprises a single image, or frame, with an acquisition matrix superimposed on the full field of view, see Fig. III.2.1. To fully document the radiopharmaceutical distribution, a complete investigation consists of a number of such images acquired over different areas of the body and, possibly, from different orientations; providing different perspectives on distribution and some relief from overlying or underlying activity accumulations. The information in each image must be accumulated until the various activity distributions

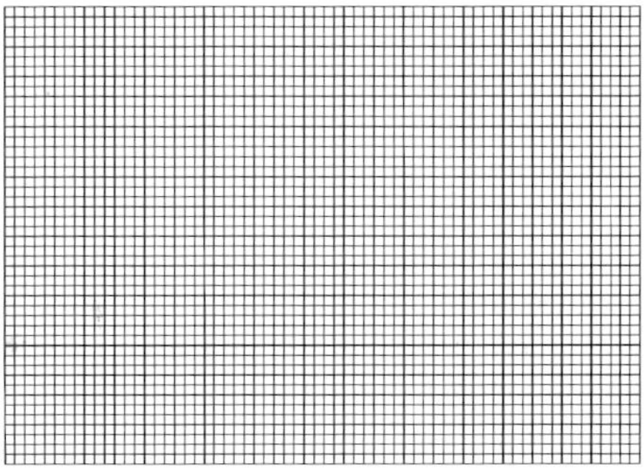

Fig. III.2.1. Full field of view of the γ camera with superimposed acquisition matrix of 64 × 64 pixels

can be differentiated from one another. The controlling factor in achieving this is the count density which determines the terminating conditions of: total counts, total duration or count density over the whole or part of the image.

For organs such as the lung which produce images containing little anatomical structure, acquisitions terminated on total count ensure similar statistical uncertainty in all views. For organs such as the skeleton, however, the proportion of bone to soft tissue or background varies considerably in different views. Thus, if images were taken for the same acquisition duration there would be different count densities in different bone structures and therefore different statistical uncertainty. Usually, therefore, an initial view is taken of, say, the posterior thorax for a total count. This allows comparison among different patients. Thereafter, the views are terminated at the same duration as the first image. This facilitates comparison among the different views. Sometimes, there will never be enough counts in the image and the acquisition termination is set at the maximum duration that it is thought the patient will remain cooperative.

Dynamic

This type of study is undertaken when the radiopharmaceutical distribution alters, in the area of interest, over the duration of the investigation. The investigation comprises a series of static images, see Fig. III.2.2, usually terminating on time duration, and necessitates the camera remaining over one area and in one orientation during the entire study, so that different perspectives on the activity distribution are not possible. There are two main considerations in this type of study. The frame rate or frame duration must be determined on a consideration of the expected activity variation, and from the Nyquist theory, sampling should be at twice the rate of change. Frame durations can, therefore, vary from less than a second to over a minute. Within a study, multiple frame rates are often used, e.g. in a renogram study the perfusion phase might have a rate of 1 frame per second, and the subsequent filtration and excretion phases, frame durations of 10 s and 20 s, respectively. The count density in each image is much lower, than in a static study, and the statistical uncertainty much higher, thus spatial information is limited and the pixel size is usually larger than half the system spatial resolution, with matrices of 64 × 64 or 128 × 128 usually being used. Within a dynamic study, the matrix size does not usually change and it is unusual for collimators other than low energy general purpose or high sensitivity to be used.

Whole Body

This type of study is designed to produce an image of the whole body as one item, most usually being used for $^{99}Tc^{m}$ bone and ^{67}Ga scans. It is the only time that the acquisition matrix is not square, being rectangular in a 4:1 ratio, ie 2048 × 512, 1024 × 256

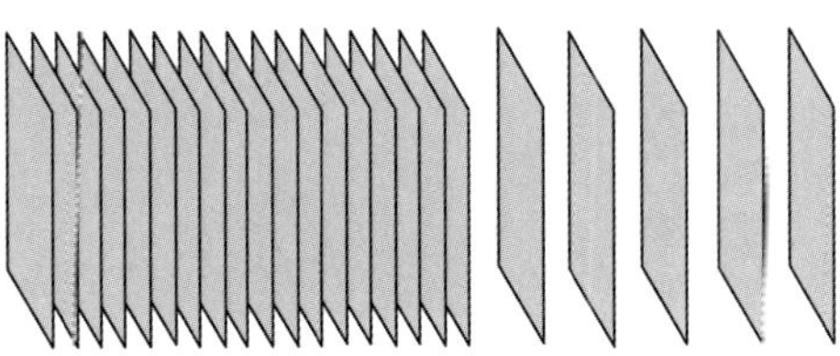

Fig. III.2.2. Dynamic study comprising a series of static images, which can document the variation in activity concentration with time. The figure is drawn to represent an initial phase of short frame durations followed by a phase of long frame durations

or 512 × 128 pixels. The pixels, themselves, are square however. The modern rectangular head cameras lend themselves very well to this type of imaging. The long axis of the camera lies across the patient and covers the transverse body completely, thus accumulating data from the whole body in one pass. If the width of the body were not completely covered, two passes would be required. In such a situation, it is diagnostically very important not to position the join along the middle of the body because this will cause a 'zipper' effect and interfere with imaging of the spine. Small gaps do not cause interpretation errors but large gaps, >0.5 × FWHM spatial resolution, may cause artefactual decrease in high accumulation areas. Small overlaps may create artefactually increased activity accumulations (Blokland et al. 1997).

Acquisition is usually in both the anterior and posterior orientations. The anterior scan is undertaken either by maintaining the camera in a fixed vertical position or varying it with distance along the body. In the former, the camera is positioned as close as possible to the patient but this results in some parts being nearer to the camera than others. In the latter, there is often some form of proximity detection whereby the vertical position of the camera is adjusted as it passes over the body. The posterior aspect is defined by the imaging bed and the camera is positioned as close to this as possible. Usually, a high or ultra high resolution collimator is used for $^{99}Tc^{m}$ radiopharmaceuticals. The variable acquisition parameter, in this study, is the scanning speed, for the usual pass length would be up to 190 cm the typical value would be 10 cm min^{-1}. This would give a total scan duration of 19 minutes which doubles, if the anterior and posterior aspects are imaged separately, using a single head camera.

Gated

This type of study is designed to produce a series of sequential images, synchronised over the cycle of a physiological variable, and is used when the cyclical acquisition duration is short and the activity variation repetitive. The most usual investigation is in the heart, where the R wave of the ECG is used for synchronisation, and in which multiple images, of good statistical quality, are produced between systole and diastole. When first introduced, two images were obtained, one at end-diastole and one at end-systole. This developed to multiple sequential images, the so called MUGA or multiple gated acquisition study.

The first action is the determination of the average duration of the physiological cycle, by sampling a number of sequential cycles. From this, and the required number of individual images in the series, the average individual frame duration is determined. Starting at the beginning of the physiological cycle, data is acquired into sequential time frames until the total acquisition duration is exceeded or the beginning of a new cycle is detected, see Fig. III.2.3. This acquisition process is repeated with subsequent data being added to the appropriate frames, acquired previously, and continues until a large number of cycles have been summed, e.g. up to ~1000 for a MUGA study. The counts in each individual acquisition frame are limited, and exhibit a large statistical uncertainty with poor spatial detail, e.g. in a clinical study, it is unusual to exceed a count rate of 20×10^{3} counts s^{-1} and, therefore, a maximum of only 460 counts would be acquired in a typical 23×10^{-3} s cardiac frame. The final accumulated data demonstrates good statistical certainty, however; the total acquisition duration being set to achieve this, usually either on total acquisition duration or total cycles acquired. A major problem occurs when the duration of the physiological cycle varies, such that the

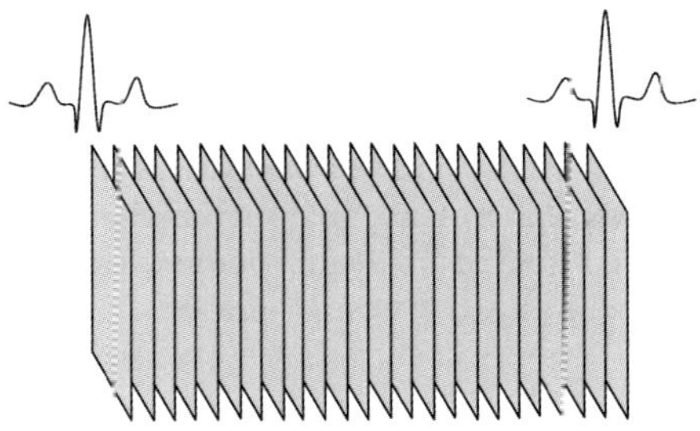

Fig. III.2.3. Gated acquisition. The first QRS complex of the ECG is used to start the acquisition. A series of images are then acquired. The second complex initiates acquisition of a second series of images

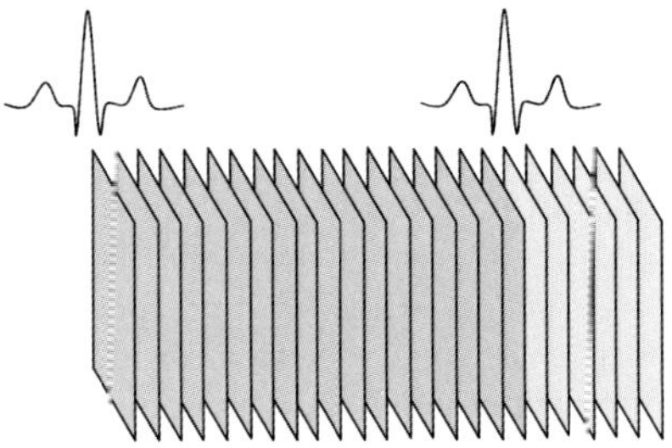

Fig. III.2.4. The physiological cycle repeats before all images are acquired. This causes counts to be lost from the later images

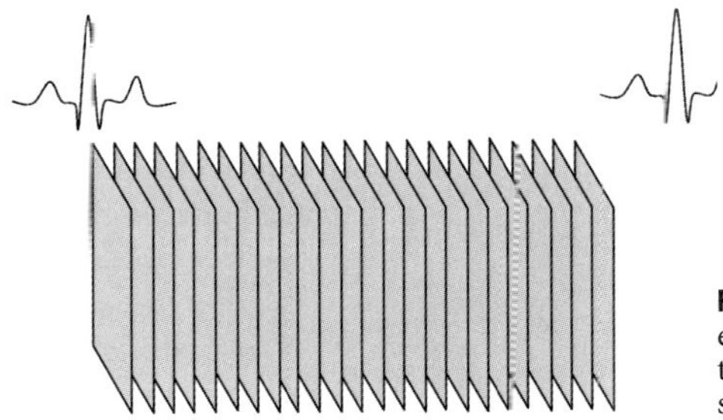

Fig. III.2.5. All images are acquired before the end of the physiological cycle. This means that the later part of the cycle is not represented

start of the next cycle does not correspond with the end of the present acquisition. If the next cycle begins earlier than the end frame the intervals remaining, in the present cycle, will contain lower counts than earlier frames, see Fig. III.2.4. If the next cycle begins later than the last frame, useful data will be ignored (Fig. III.2.5). In both cases, there will be a temporal blurring of the data, in that a frame will contain data that should have been allocated to a previous or a later frame. A low energy general purpose collimator is usually used. The pixel size is set somewhere between those found in a static and a dynamic study. The spatial information in the image being better than a dynamic but not as good as a static.

List Mode

This is an alternative to gated acquisition, which can be used to reduce temporal blurring by avoiding establishing a frame duration at the beginning of the study. This is, therefore, a more flexible form of acquisition, which produces the same end product as the gated study but by reformatting the data after the acquisition, according to the most suitable time interval. The acquired data is stored in temporal sequence with a resolution up to $\sim 10^{-3}$ s, by recording: the physiological signals, and the positional x and y coordinates of each scintillation event and, possibly, the energy (z) of the event. Once reformatted, list mode data is treated in exactly the same way as gated data. The disadvantage of the technique is that the storage requirements are much higher than for matrix mode acquisition. A typical study, consisting of 5×10^6 counts, requires 10 MBytes, 40 times more than the same data acquired as 32, 64×64 images.

Processing Considerations

The raw data can be examined visually or analysis of the data can be undertaken and processed images or derived numerical data used for interpretation. Analysis is a much more complex task than acquisition.

Data Manipulation

The image data can contain a large amount of information which renders image analysis mandatory, particularly in the types of study other than statics. For instance, displaying the images created in a dynamic or gated acquisition can give an impression of how the activity is varying with time but accurate interpretation is difficult. Analysis, therefore, aims to reduce the raw data to a manageable amount, from which useful information can easily be appreciated. One example is a single image which presents condensed data from all the images. Another example is parametric analysis, in which certain parameters are extracted from a series of images, and displayed as a single parametric image or presented as a number.

Static

It is often sufficient to review the raw static images; using a variety of formats, display sizes and orientations. However a number of processing steps may be required.

Smoothing

A simple form of image manipulation is to reduce the effect of counting statistics, which appears as high frequency noise, or 'speckle', in the image. These frequencies are higher than those associated with structure in the image and smoothing is achieved by modifying each pixel value using a consideration of the data in the surrounding pixels, see Chapter III.5. The degree of smoothing is determined by the weights applied to the surrounding pixel data but, in all cases, the spatial resolution is degraded since edges in the image are blurred. Changing the way the surrounding data is weighted can enhance these edges at the cost of increased image noise. More sophisticated filtering techniques can be used to better control the frequencies which are removed from the image.

Matrix Manipulation

The matrix that the data was acquired into may need to be altered. For example, if too small a pixel size was used and problems with statistical noise are apparent, the data can be transformed into a larger pixel size by summation of the counts in the relevant pixels. A more difficult problem is if the pixel size was originally too large. Any loss of spatial resolution that occurred as a result of this cannot be recovered but the cosmetic appearance of the image can be improved by increasing the number of pixels, and thus reducing the pixel size and the coarseness of the pixelation effect. This is done by interpolating the counts in adjacent pixels, to create intervening pixels.

Image Arithmetic

Images may be summed, subtracted, multiplied, and divided with other images or with mathematical constants. The images may also be rotated and shifted linearly. These manipulations form the basis of some sophisticated image comparisons where images are normalised ie adjusted to have comparable count densities, aligned and undergo arithmetic operations to reveal activity distributions of interest by removing unwanted and confusing activity distributions, such as background.

Distribution Based Techniques

A number of processing techniques are based on the activity distribution in the image rather than just on the image itself. For this, some knowledge of the expected physiological activity distribution and anatomy is required. With this knowledge, regions of interest can be created on the image to determine the number of counts in a particular area, usually in relation to those in another part. This provides a simple relative assessment of changes within an image and also with time, when a series of images is available.

Construction of Regions of Interest (ROIs)

There are two aspects to the construction of ROIs. The boundary of the area of interest has to be defined so that the ROI can be drawn along it. This is very difficult because the activity profile at the edge of an organ is not abrupt, but changes slowly. The difficulty comes in deciding whether a particular pixel should be included in an ROI or not. Secondly, only part of a pixel may need to be included, which is a consequence of the partial volume effect. This is usually creates more of a problem when the activity accumulation in the area of interest is low and when there is significant activity adjacent to the ROI rather than when the accumulation is high and it is surrounded by a low background activity.

The boundary of the ROI can be defined manually, which can include drawing or manipulating predefined shapes such as: circles, ellipses and rectangles. Manual ROI construction tends to be very time consuming but is often the most reliable (Houston et al. 1998), if not the most reproducible. The alternatives, semiautomatic and automatic construction, improve reproducibility. The former requires some operator interaction, such as restricting the area of interest. They use edge detection algorithms, which include: calculating count thresholds, isocontours or maximum slope, or identifying where the count gradient is a maximum, ie where the second derivative is zero. The latter produces a tight ROI (Lawson 1999). Edges defined mathematically, like this, are often jagged because of the image noise, but this can be reduced by smoothing. Snake techniques with constraints such as 1st and 2nd derivatives produce smooth boundaries. A more intractable problem is that anatomic variations in activity distribution may cause inappropriate areas to be included in the ROI, which means that most methods are investigation specific. It also means that careful evaluation, and possible interactive modification, of all the boundaries, must be undertaken by the operator, before they are used for further analysis (Rosenthal et al. 1995).

Alteration of Extracted Information

The counts in ROIs can be used to perform manipulations of the data that are clinically relevant.

Background Subtraction

The effect, of overlying and underlying activity distributions, on the number of counts in a target ROI can be removed by drawing another ROI in close proximity. There are many ways to define the shape and position of this background ROI, and these are often investigation specific. The counts in the background ROI are then normalised to the number of pixels in the target ROI and subtracted from it.

$$C = C_R - C_B \times P_R/P_B \tag{III.2.7}$$

where
C is the counts in the target ROI, corrected for background
C_R is the counts in the target ROI
C_B is the counts in the background ROI
P_R is the area of target ROI, in pixels
P_B is the area of background ROI, in pixels

Background correction relies on the assumption that the activity concentration distributions overlying and underlying the target area are similar to that measured by the background ROI. This may not be so, and will introduce an error. More uncertainty is introduced when the activity in the background area is low and subject to large statistical error. For this reason, the size of the background area should not be too small, otherwise during normalisation any statistical or construction errors will be magnified.

Correction for Depth

The counts obtained from ROIs, created on an image obtained using a parallel hole collimator, can only be accurately compared if their activity distributions are at the same depth in the patient. Otherwise, they will suffer differential attenuation which may need to be corrected using an attenuation factor:

$$C_0 = Ce^{+\mu_l d} \tag{III.2.8}$$

where

C is the counts in the ROI

C_0 is the counts, corrected for attenuation

μ_l is the linear attenuation coefficient of the tissue between the activity accumulation and the body surface

d is the distance from the body surface to the middle of the activity accumulation

For $^{99}Tc^m$, the narrow beam value of μ_l for soft tissue is 0.153 cm^{-1} for attenuation by either absorption or Compton scattering. However, some of the scattered photons will be detected and the attenuation will appear to be lower, therefore an effective or broad beam linear attenuation coefficient of 0.11–0.12 cm^{-1} is used. These values are higher in bone and lower in lungs. A reasonable estimation of d, to an accuracy of 0.5 cm (Prigent et al. 1999), can be obtained from a lateral image. An alternative is to acquire images in the anterior and the posterior projections, take the geometric mean of the counts in each ROI and correct for attenuation:

For ROI_1

$$c_1 = \sqrt{(c_{1a}c_{1p})}\ e^{+\mu D/2} \tag{III.2.9}$$

For ROI_2

$$c_2 = \sqrt{(c_{2a}c_{2p})}\ e^{+\mu D/2} \tag{III.2.10}$$

where

c_a is the counts in the ROI on the anterior projection

c_p is the count in the ROI on the posterior projection

D is the total thickness of the body at the ROIs

When D is the same for each ROI, ie at the same part of the body, the exponential factors can be cancelled and the two ROIs can be compared using only the geometric means (see Fig. III.2.6). Otherwise, a lateral image is required to measure D.

Estimation of Activity Within an ROI

This is undertaken so that changes in absolute activity can be followed over time. As well as requiring compensation for attenuation and scatter, absolute quantification of activity in a particular area, also requires correction for activity in overlying and underlying tissues; which has the opposite effect to that of attenuation, increasing the counts in the volume of interest (Macey et al. 1999). The accuracy of the technique de-

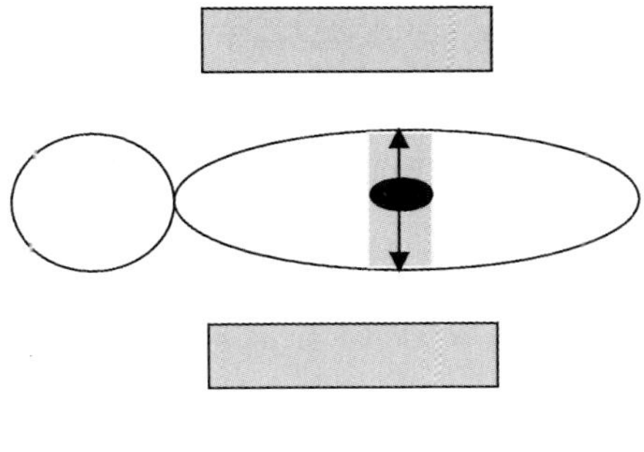

Fig. III.2.6. Depth correction using anterior and posterior projections

pends on several factors, including: background correction method which can introduce an inaccuracy of –26% to +16%, value of effective attenuation coefficient in which a change of 0.01 cm^{-1} can cause a 15% change in calculated activity, and body thickness estimation for which a 2 cm deviation can cause a change of ~10% in calculated activity (Norrgren et al. 2003).

Comparison with Standard Activities

In order to compare a radiopharmaceutical accumulation to a normal value, a standard activity is imaged in a similar geometry using a phantom e.g. thyroid, to simulate the attenuation and scattering characteristics of the body tissue. This takes into account: the activity administered, the camera detection sensitivity and the geometric sensitivity of the detection system. Such phantoms are made of tissue equivalent materials with similar dimensions to a standard patient, into which a source of activity is introduced. If a parallel hole collimator is used and there is no septal penetration, the separation in each case is not critical because sensitivity changes little with separation in air. If, however, a pinhole collimator is used, the spacing is critical and must be exactly the same for the patient and the phantom, because sensitivity varies significantly with separation. By measuring the activities administered to the patient and introduced into the phantom, the percentage of the administered activity accumulating in the area of interest can be calculated as:

$$\%\,UT = 100 * (CM_R/A)/(CM_P/A_P) = 100 * CM_R * A_P/(A * CM_P) \qquad (III.2.11)$$

where

$\%\,UT$ is the % of administered activity accumulating in the area of interest
CM_R is the mean counts in area of interest
CM_P is the mean counts from phantom
A is the activity administered to the patient
A_P is the activity introduced into phantom

The ROI counts have to be corrected for background as discussed above and physical decay, if appropriate. Additionally, corrections for variations in patient size, may be necessary; which is particularly important when investigating paediatric patients.

Dynamic

The images produced in a dynamic study are initially displayed as a cine, in which the images are presented, in sequence, as a continuous loop, with control over the rate of display. The sequence should be inspected for patient movement which would cause problems for analysis. In complicated studies, simultaneous display of multiple cines, each controlled separately, may be required. These studies are difficult to interpret visually and are generally processed to reveal the required information (Burger and Berthold 2000). Initially, this is often similar to statics, such as smoothing which may be extended to temporal filtering by including a consideration of adjacent frames. The advantage of the latter is that spatial resolution is not degraded. In addition, adjacent frames may be summed to, effectively, reduce the frame rate and increase the frame duration. This is often done to help in drawing ROIs by showing the structure in the images better, which may not be obvious at the original frame rate because of poor counting statistics.

The variation in activity over time is best appreciated from activity-time or activity-frame curves, which are constructed by summing the counts under a ROI in each of the frames in the dynamic study. The ROIs are constructed in the same way as in statics but it is essential to cine through all the images to check that there is no significant patient motion and that the area of interest always remains within the constructed ROI. In the case of subtle patient movement, it may be possible to draw a large ROI or use motion correction to reposition the data. For major movement, however, the acquisition usually has to be repeated.

The activity-time or activity-frame curves can be manipulated mathematically similarly to image manipulation, including: smoothing, addition, subtraction, multiplication and division with other curves and with mathematical constants. The ordinate (y-axis) is usually counts or counts per second, however, it may be calibrated as a functionally significant value, such as the % of administered activity (Cosgriff et al. 1992). Usually the plots are on a linear scale but sometimes the ordinate is transformed to logarithmic. Curve arithmetic and line fitting using, for example linear, exponential, gamma-variates, polynomials and splines can be undertaken; and such curves may be used to calculate functionally significant parameters such as: relative function, transit time and washout indices. The series of images may be condensed into one image, which is particularly useful in investigations of motility such as in the gastrointestinal tract. Each image in the series is condensed into one column of pixels, by summing all the pixel counts in each row. These columns are then attached to each other, in sequence, to form one image (see Fig. III.2.7). This image then represents time along the x-axis and distance in the body along the y-axis.

Truly parametric images are produced by attaching some significance to the variation in activity in each pixel, of the images, over time. For instance, assuming that the activity decreases monoexponentially, a decay constant can be calculated and installed in that pixel. Thus the single image becomes a representation of the rates of decay of the various parts of the field of view. Those parts which have extreme values are then able to be differentiated from those that are around the average, because they are immediately visible. Parameters other than decay constants can be used as appropriate, again such that each pixel becomes a visually coded element of a single image.

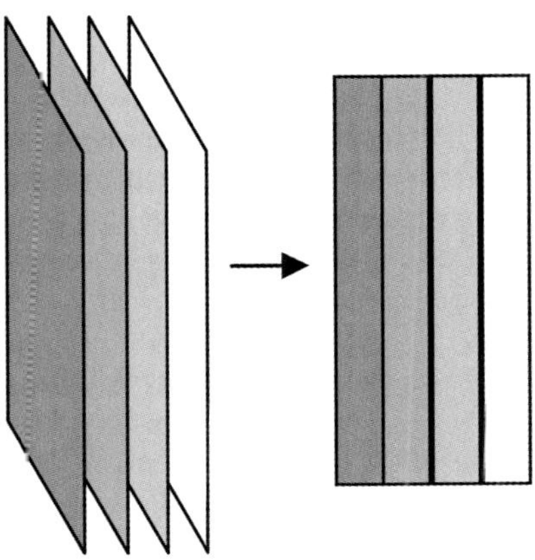

Fig. III.2.7. Condensed image. Each image in a series is condensed into one column of pixels, by summing all the pixel counts in each row. These columns are then attached to each other to form one image, which represents time along the x-axis and distance in the body along the y-axis

Gated

Processing of gated studies is undertaken in the same manner as in dynamic studies, in terms of planar image processing, ROI drawing, activity-time curve construction and formation of parametric images. Because gated studies are used in a very limited number of investigation types, the usual activity distributions expected are well known. This has allowed automatic and semiautomatic ROI drawing to be implemented, which has reduced the amount of interoperator variation imposed at this stage in the processing. The error most likely to be encountered, in such operator independent techniques, is whether to include a particular pixel or group of pixels in the ROI under construction. Most errors occur because the wrong structure is located rather than the wrong pixel, but the ROI can become ragged when the count statistics are poor.

Clinical Protocols

There has been a lack of standardisation in both acquisition and processing protocols, with different computer systems producing distinctly different results for identical input data (Cosgriff 1997). Variations have been found in acquisition protocols, data analysis and presentation of clinical results; and it has been concluded that harmonisation of these would help ensure the overall quality of the modality (Bergmann et al. 1995), and be a simple means of engendering greater confidence among referring clinicians, particularly those that refer patients to more than one centre (Prigent et al. 1999). Recently a number of bodies have published guidelines and the EANM Task Group on Quality Assurance and Standardisation has been established to develop harmonised clinical procedures within Europe; the clinical priorities being considered to be brain, cardiac and renal procedures (Kuyvenhoven et al. 2000a,b). Typical acquisition and processing parameters are listed in appendix V.5, and some clinical investigations, which involve particular requirements or complex acquisition and processing protocols, are discussed below.

Bone

Bone scintigraphy accounts for 25–60% of workload, and optimised parameter values, to ensure maximum sensitivity of the procedures, have been suggested (Kelty et al. 1997). A study can consist of two phases: early and late.

Early Study

This comprises dynamic perfusion and static blood volume phases. In both, spatial resolution considerations are secondary to those of count statistics and, consequently, a LEGP collimator would usually be used. A typical perfusion phase would consist of ~90, 1 s 64 × 64 frames; and blood pool phase would be a ~300 s acquisition on a 128 × 128 or 256 × 256 matrix.

Late Study

Use of the highest resolution collimator available is important and the best images would be obtained with an ultra high-resolution (LEUHR) collimator. However, its poor detection sensitivity and hence long imaging duration requirement generally precludes its routine use. For most planar and whole body imaging, the LEHR collimator appears to offer the best compromise between spatial resolution and imaging duration. Compared to a LEGP, the increase in spatial resolution will frequently allow a more definitive diagnosis to be made. For imaging of the hands, wrists or feet, a LEGP or LEHS collimator provides somewhat degraded spatial resolution but with a 2- to 3-fold increase in detection sensitivity. High resolution images of a small region, such as the wrist or hip, is best obtained using a pinhole collimator. A 256×256 matrix is recommended for static views and 1024×256 or 2048×512 for whole body scans at 10 cm m^{-1}. To reliably detect abnormalities the recommended counts per view are: 800–1500 $\times 10^3$ in the axial skeleton, and $400–800 \times 10^3$ in the extremities. Typically, this translates to 10^6 cts over the thoracic spine with other images being acquired for a similar time per image, and $2.5–3.5 \times 10^6$ cts per view in whole body scans. Scatter reduction techniques are recommended, particularly in obese patients, by implementing an asymmetrically high PHA window (O'Connor et al. 1991; Kelty et al. 1997).

The sensitivity and specificity, of this part of the study, can be increased by digital filtering the images (Starck and Carlsson 1997).

Cardiac

Two types of study are undertaken to assess left ventricular function: equilibrium radionuclide angiography, ERNA, and first pass radionuclide angiography, FPRNA. The latter also being used to assess right ventricular function, for which ERNA would indicate erroneously low ejection fractions because of interference from other cardiac structure activity. A large field of view camera is not a requirement for these studies because a small field of view will encompass enough area, even when lung activity must be assessed during shunt evaluation. For the evaluation of left ventricular function, <2 min imaging duration is required for an ERNA study to achieve the same statistical precision as a FPRNA study (Green et al. 1991).

Comprehensive guidelines have been published by the Quality Assurance Committee of the American Society of Nuclear Cardiology, regarding these investigations. These are DePuey and Garcia (2001) which is an updated version of Garcia (1996), made necessary by improvements in instrumentation and software; Port (1999). Guidelines have also been published by the Society of Nuclear Medicine (Wittry et al. 1997). One of the recommendations that covers both investigations is that the PHA energy window for both studies should be symmetric.

Equilibrium Study

Acquisition

A LEHR collimator should be used unless the data is acquired during exercise, in which case the higher detection sensitivity of a LEGP collimator will be required. In order to maintain comparison consistency, this should then be used if a rest study is subsequently acquired.

Pixel Size

A 64×64 matrix is most often used and the appropriate pixel size achieved by zooming, usually between 1.5 and 3. Assuming the minimum depth of the myocardium to be ~5 cm and the system spatial resolution of a LEHR collimator ~8 mm, at this depth; the pixel size does not need to be <4 mm. It should not, however, be >5 mm.

An appreciation of the various statistical uncertainties, as a consequence of pixel size, can be gained by considering a typical study, in which there might be as many as 5×10^5 counts in each frame of the gated series. The left ventricle is likely to occupy ~10% of the whole field of view, increasing as the zoom factor is increased. This is assumed as systole, the worst case for the heart, both the size and the counts will increase towards diastole. In such a study the background is high, and is likely to be 30–70% of the activity in the heart, on a per pixel basis. Since the different magnifications do not change the count in the heart this will be, 3.75×10^5 and in, the best case, the background 1.25×10^5. Table III.2.1 lists the statistical error per pixel in different imaging conditions, for a 500 mm field of view. It indicates that a 64×64 matrix with zoom of 1.5 has the lowest statistical error of the three combinations considered.

Table III.2.1. Lists the statistical error per pixel in different imaging conditions, for a 500 mm field of view

Matrix	Zoom	Pixel size (mm)	Heart area (%)	Heart area (pixels)	Counts per pixel	Error (%)
128	1	3.9	10	1638	76	11
64	1.5	5.2	15	614	204	7
64	3	2.6	30	1229	102	10

Gating

This is undertaken, in either frame or list mode acquisition, by triggering on the R-wave of the ECG. The more usual is frame mode and the standard technique is to organise the frames forward from the start of the cycle, but any irregularity in the heart rate will cause temporal blurring. This occurs mainly at the end of the cycle, ie diastole, because of cumulative errors towards the end of the cycle and because the duration of the diastolic phase is more variable than that of the systolic phase. Thus, end systolic frames usually remain fairly well synchronised even with moderate variations in the R–R interval, but the end diastolic frames are less so. If the diastolic phase is the more important, it can be better assessed using backwards gating or list mode, the latter offering more flexibility in allocating the image data to appropriate frames. When the variability in the heart rate is severe, data from very short or very long R–R intervals can be excluded, which minimises the temporal blurring but increases the study duration. The range, of R–R intervals, which is acceptable is established by setting a window, e.g. ±10%, around an average value determined by sampling a number of cycles before imaging is started. If the heart rate is varying badly, the loss of data, due to rejection of R–R intervals, may result in poor acquisition count statistics or an unacceptably long acquisition duration. In this situation the acceptance window can be widened, but, if forward framing is used, it will compromise diastolic measurements because the frames at the end of the cardiac cycle will receive variable acquisition contributions, and should be deleted before further analysis.

If the data from such bad beats were not rejected, the calculated ejection fraction would be erroneously low. If the beat was too short, there would be too little data in the frames at end diastole, reducing the difference in counts between end systole and end diastole and reducing ejection fraction. If the beat was too long, the activity in the frames at end diastole would be more end systolic, again reducing the difference in counts between end systole and end diastole and reducing ejection fraction.

Frame Rate

The number of frames required per cardiac cycle is determined by considering the temporal resolution necessary for accurate calculation of the parameter of interest. For ejection fraction only, a low number of intervals can be used, e.g. 24, because the activity changes little for a period at both end-diastole and end-systole. If instantaneous rates of change, such as peak emptying or filling rates, are to be calculated, however, a higher number, e.g. 32, are needed.

Acquisition Duration

The acquired counts are the most important determinants of image quality and parameter accuracy; and, with a LEHR collimator, acquisition should continue until a count density of 2×10^4 counts cm^{-2} is achieved at the centre of the LV, when all frames are added together. This figure should be doubled if a LEGP or LEHS collimator is used. For an exercise acquisition, enduration is the limiting factor and this should be at least 2.5 min.

Processing

Before processing, the images should be displayed in cine mode, to assess the consistency of the gating and the adequacy of the acquired counts. The former should also be evaluated by reviewing the R–R interval histogram. Statistical noise should be reduced by spatially and temporally smoothing the image frames; then, ROIs and activity-time curves are constructed and various parameters calculated from them.

Ventricular ROI

The left ventricular ROI can be defined in many ways, two of which are described. The first involves drawing a rough outline of the ventricle, on an end diastolic frame, just outside its border being careful not to include the right ventricle. The actual edges of the blood pool, defining the inner wall of the ventricle, are then detected by searching inwards. This technique is operator dependent, in the drawing of the loose ROI, but is reproducible by usually managing to avoid including the right ventricle. Another technique is to find the centre of the left ventricle and to then search outwards to detect the actual boundaries of the left ventricle. This technique is operator independent but can include the right ventricle. Both techniques create varying sizes of ROIs that define the edges of ventricular endocardial borders throughout the cardiac cycle. As the heart moves from diastole to systole these should become smaller to define only the ventricle and to avoid including too much background area. Some techniques use just two ROIs, one at diastole and one at systole, which may underestimate ejection fraction. Once constructed, the ROIs must be viewed in relation to the images, to confirm correspondence with the left ventricle and to exclude overlap with left atrial activity, at systole.

Background ROI

Correction for extracardiac background activity is very important in blood pool studies in which, typically, the mean background count is 30–70% of the mean left ventricular end diastolic count. Various techniques have been suggested, the most usual being a 2–3 pixel wide crescent shaped background ROI positioned 5–10 mm, i.e. 1–3 pixels, medial to the lateral or inferolateral wall. Obvious vascular structures must be avoided and the positioning must be consistent. If created automatically, the construction must be visually verified.

Activity-Time Curves

The ROIs are used to create activity-time curves representing the variation in activity, over an average cardiac cycle, from diastole to systole and back to diastole. The background activity-time curve must be inspected and should be flat. Any intrusion from an extraneous vascular structure will give high spots on the curve dictating that the background ROI must be repositioned otherwise the ejection fraction will be erroneously high. This means that the background counts can be averaged over all the frames and the average subtracted from the left ventricular activity-time curve, resulting in a reduction in statistical error. The background subtracted ventricular activity-time curve is then proportional to volume changes over the cardiac cycle, see Fig. III.2.8. The shape of the activity-time curve, over the cardiac cycle, should be inspected to determine whether it appears appropriate. There should be a well defined systolic trough with an appropriate increase in counts during diastole, which should be similar at the beginning and at the end of the curve and with no pronounced drop off in the last few points.

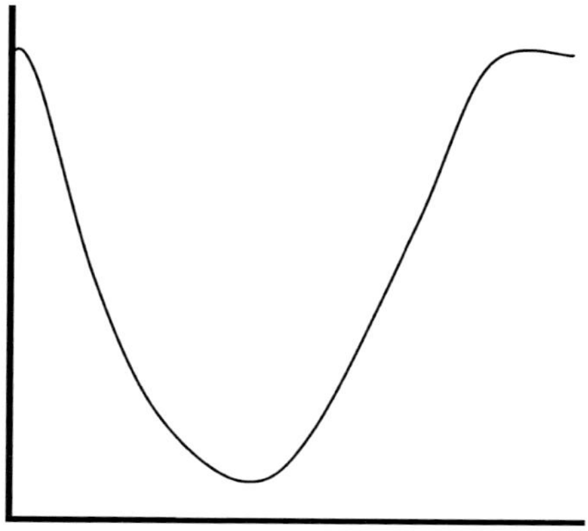

Fig. III.2.8. Activity-time curve of left ventricle. Ordinate represents activity variations in the ROI over the left ventricle and the abscissa represents time during a single cardiac cycle

Ejection Fraction

This is the most important parameter derived from the ventricular activity-time curve and is calculated as:

$$EF = 100 * (ED - ES)/ED \qquad (III.2.12)$$

where
ED is the end diastolic counts
ES is the end systolic counts

This estimation is very reproducible because it is purely count based and involves no geometric assumptions.

Stroke Volume

The stroke volume (SV) image is obtained by subtracting, pixel by pixel, the end-systolic from the end-diastolic frame so that for each pixel (Schiepers and Almasi 1988):

$$SV = EDC - ESC \qquad (III.2.13)$$

where EDC and ESC are end-diastolic and end-systolic counts.

Volumes

Volumes can be estimated using techniques based on counts or geometry. They are not widely used, however, because of the complexities and inaccuracies involved. The count based method requires comparison with activity in a reference blood sample and the geometrical based method involves determining the area of a pixel and taking the ratio of total counts in the ventricle to those in the pixel containing the highest counts.

Emptying and Filling Rates

The ventricular emptying and filling rates can be estimated from the gradients of the activity-time curve: on the downward part, ie diastole to systole, and on the upward part, ie systole to diastole, respectively. The parameters are usually calculated from the maximal gradients, ie peak emptying rate PER and peak filling rate PFR, or at a particular point on the slopes, e.g. one third. The values are expressed in units of end diastolic volume (EDV) per second, where the counts in the left ventricle at end diastole are used to represent EDV. The times from the start of the downward or upward part of the curve to the point on the curve, where the values are estimated, e.g. time to peak emptying rate tPER and time to peak filling rate tPFR, are expressed in milliseconds. The normal values vary among institutions but are approximately: PER >2.0, PFR >2.5, tPER <240, tPFR <180.

Wall Motion

This assesses the motion of the endocardial wall and is evaluated visually by overlaying the ROIs created at the different parts of the cardiac cycle. Only 12–16 frames should be used, fewer than this will compromise temporal resolution and more will compromise count statistics. If a greater number have been acquired they should be summed.

Parametric Images

A number of parametric images can be generated by creating an activity-time curve for each pixel in the cardiac area, and these are best displayed using a colour coding. One displays regional ejection fraction or stroke volume, generated from activity-time curves in the ventricular area. The parameters are calculated as above. Another involves fitting a cosine curve to each activity-time curve constructed over the whole cardiac area, allowing a phase and amplitude value to be calculated for each. The phase demonstrates the regional initiation of the cardiac sequence, and appears as a wave front that sweeps through the heart. This is particularly useful for demonstrating sites of delayed emptying. The amplitude indicates the variation in counts between the peak of the cosine curve and the mean value and demonstrates the regional stroke volume. This is particularly useful for revealing areas of wall motion abnormalities. A histogram of the number of pixels demonstrating particular phases normally exhibits two peaks: in the atrial and ventricular phases, which are distorted in the case of cardiac abnormalities.

Regurgitation Analysis

This is used to quantify inappropriate flow through the aortic or mitral valves. Regions of interest are drawn over both the left and right ventricles, excluding as much of the other cardiac structures as possible. The differences between end diastolic and end systolic counts are calculated for each ventricle from their respective activity-time curves.

First Pass Study

In this investigation, the transit, of an activity bolus, through the thorax is evaluated. With $^{99}Tc^{m}$, a LEHS collimator and a wide PHA energy window of ±15% are used, to increase detection sensitivity. The limiting factor on acquisition quality is the count statistics rather than spatial resolution, and usually a 32×32 matrix is used to maximise pixel count density particularly in the end systolic frame, although a 64×64 matrix can be used if the count rate is high or in lower frame rate acquisitions. Because of the high count rates experienced, if large field of view cameras are used, the periphery should be shielded to reduce pulse pileup as much as possible. Activity-time curves generated in this study are 'saw toothed'; each peak and trough representing the end diastolic and end systolic parts of each cardiac cycle. A representative cycle is constructed from a number of these cycles and this is used in subsequent processing. The quality of this is affected by the bolus, count rate, cycles used, background subtraction and, possibly, patient motion. Two types of study are undertaken: ventricular and shunt.

Acquisition

The acquisition should be undertaken, such that when the bolus is in the right ventricle, the count rate over the whole field of view is at least 1.5×10^5 counts s^{-1}; suboptimal studies being likely when it is $<10^5$ counts s^{-1}. It is also suggested that background corrected end diastolic counts 2×10^3 counts s^{-1} are adequate (Esquerre and Coca 1995). An ECG signal is sometimes used to identify diastolic frames.

Left Ventricle

The frame duration should be varied to suit the heart rate, but a standard 25×10^{-3} s is usually used unless the heart rate is very high, >150 min^{-1}, or the diastolic function is of particular interest, in which case, a duration of $10–20 \times 10^{-3}$ s should be used. Such frame durations will involve acquiring up to 2000 frames. The administration must be very rapid, and the FWHM of the bolus in the superior vena cava, should be <1 s and, if possible, <0.5 s.

Right Ventricle

In comparison to the left ventricular study, the administered bolus, in this study, reaches the right ventricle without significant dispersion and the FWHM of the bolus in the superior vena cava should be much wider, at 2–3 s.

Processing

The data should not be processed until the quality of the administration bolus, in the superior vena cava (SVC), has been assured both qualitatively, by visual inspection, and quantitatively, by calculation of the FWHM. The latter uses an activity-time curve from an SVC ROI on a series of 1 s frames created from the original data, with the following quality descriptions: good <1 s, adequate 1–1.5 s and delayed >1.5 s. The activity-time curve should not show more than one peak, otherwise background may be oversubtracted during processing. The processing involves producing a representative cycle which is used in all parameter calculation, the sequence is:

1. an ROI is drawn around the ventricle of interest and an activity-time curve created
2. on this, the contiguous cardiac cycles to be used for creating a representative cycle are marked (only those whose end diastolic counts are at least 70% of the peak end diastolic counts)
3. a background ROI is drawn over the right lung, avoiding any cardiac structure
4. this is overlaid onto the image frame, immediately before the appearance of activity in the left ventricle
5. the counts in this are subtracted from the frames forming the chosen cardiac cycles, with the application of a washout factor because the counts are decreasing
6. the corrected end diastolic and end systolic frames are redisplayed, the initial ROI modified if necessary, and created
7. the chosen cycles, from the final activity-time curve, are concatenating into a representative cycle
8. the image frames in the representative cycle are usually time-smoothed

The background subtraction is a crucial step in the processing since variation in the background frame can substantially alter the calculated EF, volumes and apparent wall motion. The frame used to determine the background should be as close as possible to the one in which the bolus appears in the left ventricle, but should not itself show any activity in the left ventricle and no residual activity in the right ventricle. If a suitable frame is not available, background subtraction should not be undertaken.

Count Statistics

The adequacy of the acquisition count rate is evaluated by assessing the counts under the ventricular ROI on the end diastolic frame of the representative cycle. This is more accurate than checking the raw data because this also assesses that there are sufficient cycles for analysis and background has not been oversubtracted. The counts should be >2500 and preferably >4000, with high resolution wall motion images requiring >5000.

Quantitative Analysis

This includes ejection fraction, and systolic emptying and diastolic filling parameters. The analysis is performed using similar calculations, to those in the equilibrium study, applied to the final background corrected representative cycle. Left ventricular volume can be estimated using either count based or geometric methods. However, normal values vary depending on the type of processing used, especially the type of background correction, and the patient's position.

Parametric Images

These comprise regional ejection fraction, stroke volume, and amplitude and phase images. Their quality depends on the image statistics and whether there has been translational movement of the heart.

Display

The image frames comprising the final representative cycle should be displayed in cine mode, using colour coding because of the low pixel count density and the subtle changes from cardiac cavities to background. Spatial smoothing should be used if the count density is low. The image should be normalised to the peak activity in the ventricle to properly appreciate the count changes in the ventricles because this will avoid the swamping effect of possibly higher counts in the aorta or left atrium. This provides a qualitative assessment of regional wall motion. The parametric images should also be displayed with colour coding.

Shunt Study for Left to Right Intracardiac Shunting

In this study, the frame duration is not crucial since the analysis does not require a temporal resolution better than 100×10^{-3} s. Although the suggested 32×32 matrix size is suitable, the less demanding frame rate requirement also allows use of the 64×64 matrix. The administration should still be rapid, however, because the analysis assumes that the activity accumulation in the lungs and then its washout is monoexponential.

Processing

An ROI is constructed over the right lung rather than the left, to avoid activity in the left heart and great vessels, and a high count density activity-time curve is created. This represents the activity bolus transit through the lungs, and must have good statistics and not be contaminated by activity from other cardiac structures. The analysis identifies a second activity-time curve superimposed on the first; the second curve representing the fraction of the bolus undergoing a second recirculation through the lungs after left to right shunting in the heart, see Fig. III.2.9. Because the shunted fraction appears slightly later than the initial bolus, it can be separated from it. The first curve is identified by fitting a γ function, from the take-off point of the curve to the start of the second curve. Visual assessment of the closeness of the fit is very important and it is likely that the first or final point will need to be altered. The fitted curve is then subtracted from the raw data to reveal the shunt component. Another curve is then fitted to this and the areas calculated as: A_1 = the area under the first fitted curve and A_2 = the area under the second fitted curve. An example of the relative sizes of these are if half of the pulmonary venous blood flow underwent shunting, then the second curve would have an area half that of the first. The shunt ratio of pulmonary to systemic blood flow, $Q_p:Q_s$, is calculated as (Gelfand and Hannon 1988):

$$A_1/(A_1 - A_2) \qquad \text{(III.2.14)}$$

Gastrointestinal Tract

Much of the processing involved in the assessment of the GI tract is in term of motility, whereby a tracer is followed along the tract.

Oesophagus

A bolus of activity is administered and a dynamic study acquired. Transit is investigated by monitoring the disappearance of the activity by defining an ROI around the entire oesophagus. From the activity-time curve; the maximum count rate, C_m, and the

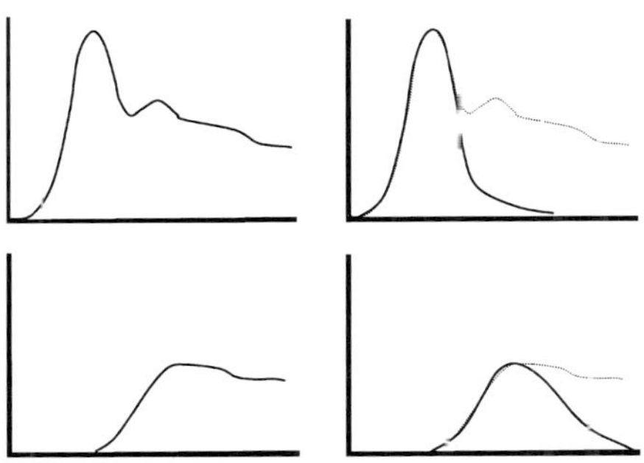

Fig. III.2.9. Shunt analysis. Top left: activity-time curve from ROI over right lung. Top right: γ function fitted to the first curve. Bottom left: fitted curve subtracted from first curve. Bottom right: γ function fitted to subtracted curve

count rate at subsequent times, C(t), are recorded. A measure of the transit time of the bolus is then made by plotting $[C_m - C(t)]/C_m$ against time. Peristalsis is monitored by creating a compressed image for the area in the above ROI. For each frame of the study, the counts in each row of pixels are summed to form a column one pixel wide and as long as the ROI. The columns of summed pixels are then concatenated in temporal sequence to form one image. The vertical position represents distance along the oesophagus, the horizontal position represents time during the study and the intensity represents amount of activity. Thus as the bolus moves along the oesophagus, blocks of colour representing the activity move down and to the right of the image. Any abnormalities in peristalsis are revealed by inappropriate movement of this activity. This type of image is also known as a space-time plot, see Fig. III.2.7.

Gastric Emptying

Regularly spaced sequential images are acquired, for set durations of ~1 min, in the anterior and posterior orientations; and the decrease in stomach activity is assessed using irregular ROIs. The counts have to be recorded in both orientations to compensate for the affects of attenuation which, if only the anterior were used, would cause an apparent increase in activity soon after administration, as distal movement in the stomach resulted in anterior as well as distal activity movement in the body. To correct for this, geometric means of the counts are calculated, and the clearance curves are generated from these. This study can also be undertaken as a dual radionuclide study whereby both the solid and the liquid phases can be assessed at the same investigation using $^{99}Tc^{m}$ and ^{111}In.

Lung

Lung investigation is a good example of the complexity of a dual radionuclide study. It is undertaken as ventilation, perfusion or ventilation–perfusion; the ventilation phase being undertaken with inhalation of radioactive gases or aerosols, and the perfusion phase using $^{99}Tc^{m}$ labelled to small particles and injected intravenously.

Ventilation

Ventilation imaging using the long lived gases ^{127}Xe and ^{133}Xe demands a series of respiratory manoeuvres, the first being inhalation to breath-hold, during which an image representing ventilation is acquired. This restricts acquisition to one orientation and duration to breath-hold capability; the latter resulting in poor quality images, especially with ^{133}Xe which has a energy, of 81 keV, too low for efficient imaging. Subsequent rebreathing in a closed ventilatory circuit, establishes equilibrium during which the image represents lung volume. Both imaging orientation and duration, with consequential improvement in image quality, can be extended during this period. The final phase, of washout over a number of respiratory cycles, allows a single orientation acquisition, which represents ventilatory turnover rate, the inverse of the mean transit time, at the end of which the image reveals any areas of lung which are trapping gas. With ^{133}Xe, the perfusion phase is undertaken following the ventilation image so that there is no downscatter into the 81 keV window. With ^{127}Xe, the perfusion can be undertaken before the ventilation study because the photopeak is sufficiently higher than that of $^{99}Tc^{m}$.

The alternative ventilation agent is the radioaerosol. This is a solution of $^{99}Tc^{m}$ labelled to a compound that will not be absorbed rapidly by the lung mucosa. It is introduced into the lungs as small droplets and behaves as a pseudogas, remaining airborne as long as laminar flow is maintained and being deposited when this turns turbulent, usually in the peripheral airways. Consistent imaging, of the peripheral airways, requires that the mass median diameter of the particles be maintained in a tightly controlled range, and there are a number of trapping baffles within the administration unit which remove particles too small or too large, so that the aerosol penetrates to the periphery rather than being deposited in the large early airways. In patients with bronchial disease, turbulent airflow occurs early in the bronchial tree, however, and results in deposition of the droplets centrally, apparent as accumulations of activity in the middle thorax. Once deposited the activity distribution is reasonably stable with a biological half-life of ~90 minutes. Thus the perfusion phase may be undertaken at about 2 hours after administration. When perfusion imaging is undertaken immediately after the radioaerosol imaging, the ratio of perfusion to ventilation count rate should be 5:1 or greater, to ensure that perfusion defects are not flooded by the ventilation activity (Forstrom et al. 1996).

With both long lived gases and radioaerosols, the patient is moved between the ventilation and perfusion imaging phases, which means that visual comparison, or quantitative analysis, of the two components is not exact. Using a short lived gas, e.g. the 13 s half-life $^{81}Kr^{m}$ eluted from a ^{81}Rb–$^{81}Kr^{m}$ generator, can avoid this. At normal respiratory rates, the equilibrium image represents regional ventilation rather than volume (Fazio and Jones 1975), because almost all of the radioactivity will be removed by decay during exhalation, and there will be no residual to diffuse into poorly ventilated lung volumes during the subsequent inspiration. This facilitates multiple ventilation acquisitions during relaxed tidal breathing. However, the image tends to a volume image as the respiratory rate increases, e.g. with hyperventilation and in neonates, because the rapid expirations allow radioactive gas to diffuse into poorly ventilated lung volumes. The generator is eluted, close to the camera, by a constant stream of moist air at a flow rate of ~3 l m^{-1}. The radiation exposure from the generator, in its transport shield, is very high and it must be surrounded in a secondary lead shield immediately on receipt and certainly during use; because, not only will it irradiate staff and patients, but it will also produce imaging artefacts. The eluted radioactive gas is supplied to a breathing mask via a capillary tube, to minimise the transit time. Rather than supplying the mask directly, the gas is fed into a wider tube as a reservoir, see Fig. III.2.10. This ensures a constant radioactive concentration during inspiration, which is important for obtaining an image of regional ventilation.

Ventilation–Perfusion

Using $^{81}Kr^{m}$, both phases of this investigation can be undertaken at the same time, which removes the possibility of patient movement and means that the images of each phase are in the same orientation and can be compared exactly. The most usual clinical referral is for investigation of suspected pulmonary embolism, in which a photopenic area on perfusion and normal ventilation is expected. At 190 keV the $^{81}Kr^{m}$ photons are sufficiently higher energy, than the 140 keV photons from $^{99}Tc^{m}$, for the

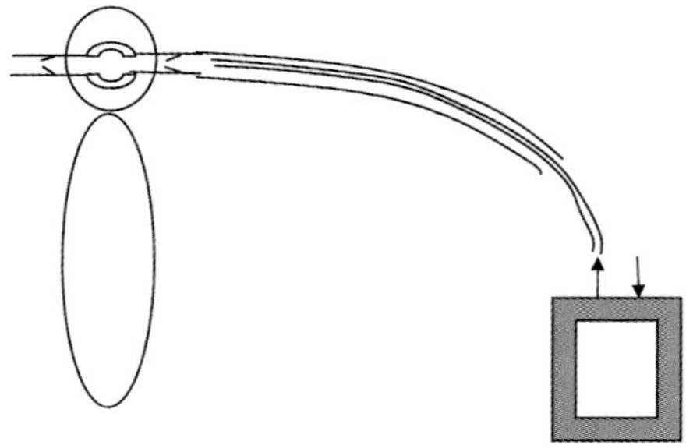

Fig. III.2.10. Capillary and reservoir tube in $^{81}Kr^{m}$ delivery system

ventilation image to be uncontaminated from the perfusion distribution. However, to obtain perfusion images uncontaminated with downscatter from the higher energy $^{81}Kr^{m}$ photons, the acquisition should be undertaken with the supply of gas interrupted. A more efficient use of camera time with better patient tolerance is to use a dual acquisition technique, with simultaneous acquisition of both ventilation and perfusion images, and the supply of gas only being interrupted during the time that the camera is being moved to another orientation. In this protocol, downscatter into the $^{99}Tc^{m}$ window could degrade the image quality and reduce the contrast of photopenic areas on the perfusion distribution. This can be demonstrated on phantom studies but does not affect clinical results if the ^{81}Rb–$^{81}Kr^{m}$ generator activity is <300 MBq (Hastings et al. 1995).

Renal

Renal imaging investigations comprise static studies of the parenchyma and dynamic studies of the perfusion, filtration and excretion, termed renography; and most of the current techniques were developed in the late 1970s and 1980s (Cosgriff 1998). There are subtle differences between renography used in native kidneys and that used in transplanted kidneys, but standardised methods have not yet been implemented for either (Prigent et al. 1999). Differences in calculated parameter values among hospitals have been identified as being most probably due to variations in study protocols and analysis programmes, which suggests the need for external quality assurance on a regular basis (Heikkinen 1999; Heikkinen et al. 2001; Tondeur et al. 2000); and has been implicated as a crucial factor in the declining popularity of deconvolution techniques in renography (Cosgriff 1998).

Native Kidneys

Parenchyma

Static renal investigation is used to calculate the relative function of each kidney, by comparing total background corrected counts obtained from ROIs constructed on the image. Despite a variation in techniques, the spread of values is reasonably narrow among institutions. Considering data only from the posterior projection may be inaccurate, if attenuation effects are not corrected, because of possible variations in depth between the two kidneys (Fleming et al. 1998). This is not necessary for normally located kidneys in children, however (Hervas et al. 2001), the mean difference in relative function measured using the geometric mean method and the non-depth corrected posterior view being estimated to be only 1.2% (Lythgoe et al. 1998) or 0.1±0.8% for images comprising 2.5×10^5 counts (Prigent et al. 1999). In other than this group, relative function is usually estimated by calculating the geometric means of the counts from both anterior and posterior images. In very poorly functioning kidneys, however, the anterior view is often not used because of its considerable attenuation effect and the difficulty of identifying the kidneys for accurate construction of the ROIs. Relative function is calculated from counts obtained from the posterior view only, but any error from variations in relative depth is unlikely to surpass that which would be introduced by the poor counting statistics of the anterior projection image.

If an estimation of the activity in each kidney is required rather than relative function, the ROI counts must be corrected for attenuation and accurate values of depth and linear attenuation coefficient, μ_l, are needed. Depth can be measured, as the distance from the centre of the kidney to the posterior surface with the distance measuring function of the camera (Prigent et al. 1999), or estimated using derived formulae (Lythgoe et al. 1998). Previously published equations are acceptable for predicting adult depth, but substantially underestimate paediatric depth. More accurate equations, suitable for both adults and paediatrics, were derived as: right kidney, $D = 16.778 \times w/h + 0.752$ and left kidney, $D = 16.825 \times w/h + 0.397$, where w is the body weight in kilograms and h the height in centimetres (Inoue et al. 2000). For a 20% PHA window, the value of μ_l is strongly dependent on the size of the ROI and on whether background subtraction was performed, varying from 0.119 cm^{-1}, for a loose ROI with no background subtraction, to 0.150 cm^{-1}, for a tight kidney ROI and background subtraction. Using values from 0.099 to 0.153 cm^{-1}, would cause a change of up to 30% in estimates of activity for a 6 cm deep kidney (Hindie et al. 1999).

Renography

Renography comprises perfusion and filtration–excretion phases, following administration of the activity as a bolus. Wide variations, in acquisition and processing methodology, exist (Cosgriff 1989; Houston et al. 2001), and consensus reports regarding the renogram have been compiled by O'Reilly et al. (1996), O'Reilly (2003) and Prigent et al. (1999).

Acquisition

Because the collimator cannot be changed between phases, the type is determined on the basis of the longer frame duration filtration–excretion phase, and is usually a LEGP, unless ^{123}I is used in which case a special collimator is recommended. Higher spatial resolution is not needed and reduced detection sensitivity would result in an unacceptably decreased count density or necessitate higher administration activities. A matrix of 128×128 should be used; except for paediatrics, in which cases a 64×64 matrix with an acquisition zoom of up to 2 is used. The perfusion phase usually extends over the first 60 s of acquisition, with a frame rate of 2 s. The filtration–excretion phase comprises 10 or 20 s duration frames. For basic renography, the peak count rate, after background subtraction, should be at least 200 counts s^{-1}, which corresponds to a statistical error of ~2% for a frame time of 20 s. This is sufficient to ensure that the renogram will require little, if any, smoothing prior to relative function estimation (Cosgriff et al. 1992). The random error, due to counting statistics, in the calculation of relative function is the same using both 10 or 20 s frame durations and is generally very low, at <3%, when the usual administration activities are used (Moonen and Jacobsson 1997).

For deconvolution analysis, the optimum frame rate is still a matter of discussion, but often a duration of 10 s is used. The required peak count rates are higher than in basic renography, however; for normal kidneys being 1000 counts s^{-1} in the whole kidney and 400 in the parenchyma (Cosgriff et al. 1992).

Processing

Perfusion phase

The perfusion can be assessed against an arterial standard, by visually comparing activity-time curves from the kidneys and aorta using ROIs constructed just inside their relevant borders. If the renal ROI extends beyond the kidney, the counts per pixel will be reduced progressively as the area increases. The calculation is less sensitive to area, however, if the ROI is smaller than the kidney, see Table III.2.2. Background subtraction is not undertaken because activity not well mixed before recirculation and such subtraction can give erroneous results, but the kidney curves are normalised to the area of the aorta, and hence comparison is with counts $pixel^{-1}$ values. Generally, renal blood flow quantitation in native kidneys is not undertaken because this requires higher injected activity without adding any useful clinical data. However, there is a technique for calculating a perfusion index based on the relationship with counts in the aorta which returns a value of renal blood flow as a percentage of cardiac output (Peters 1991).

Filtration–Excretion Phase

Methodological variations, including algorithm and site of background region, were influencing factors in the consistency of relative function estimation. The background subtraction method influenced time-to-peak, and curve smoothing influenced mean transit time (Houston et al. 2001). Analysis of the filtration phase requires ROI construction over each kidney, in this case just outside their borders, and a background

Table III.2.2. Kidneys with ROIs at border, larger and then smaller

ROI position	At border	Larger	Smaller
Counts in kidney	2000	2000	2000
Area of kidney (pixels)	200	200	200
Counts in ROI	2000	2000	1800[a]
Area of ROI (pixels)	200	300	150
True counts/pixel	10	10	10
Calculated counts/pixel	10	6.7	12
% error	0	–33	+20

[a] This is not as low as 1500 (i.e. 10 counts $pixel^{-1}$) because of the bulkiness of the kidney towards its centre, i.e. because of a reduction of tissue towards its border

area. The parameter of concern is the background subtracted total count in each kidney and, therefore, if the renal ROI is smaller than the kidney, the counts will be reduced progressively as the area decreases but the values will less sensitive to area if this is larger than the kidney, see Table III.2.3. Therefore, to ensure that all of the parenchyma is included, the ROI should be outside the kidney and may be fairly well clear of the border. The most usual method of constructing renal ROIs is by manual drawing but a number of semiautomatic techniques are also used. These include using a circular or elliptical region as an outer limit with thresholding to determine the renal isocontours, and a rectangular region enclosing the kidney and touching both poles. There is no clinically significant difference in relative function estimations when using rectangular instead of kidney shaped regions, even though the rectangular regions include variable amounts of highly vascular structures. If the ROIs are constructed manually, the renal pelvis should be included otherwise reproducibility suffers, and several one minute duration frames formed from the first few minutes of the study, before any of the tracer has left the kidney, should be used as templates.

Depth Correction

In the supine position, only 14% of adults and 7.5% of children exhibit a difference in kidney depth >1 cm, but this increases in other positions. A 1 cm difference would only result in a true relative function of 50:50 being calculated as 48:52 without depth correction, assuming a linear attenuation coefficient of 0.10 cm^{-1}. Thus for routine renography in the supine position, depth correction is not required (Cosgriff et al. 1992; O'Reilly et al. 1996; O'Reilly 2003), except in adults when there is marked asymmetry in kidney size, e.g. when the ROI area ratio is 1.5 or the kidney exhibits an unusual shape or location. It is, however, necessary to apply depth correction when calculating absolute function, in both adults and children (Cosgriff and Brown 1990).

When depth correction is indicated, the most convenient method is to use the technique detailed above in 'parenchyma', acquiring a lateral image at the end of the dynamic and measuring the distance from the centre of the kidney to the posterior surface with the distance measuring function of the camera, the most appropriate value of μ_l being 0.087 cm^{-1} (Inoue et al. 2000), although reported values vary from 0.10–0.14 cm^{-1} (Cosgriff and Brown 1990; Cosgriff et al. 1992). When using a dual head camera, the geometric mean method of simultaneous anterior and posterior acquisi-

Table III.2.3. Kidneys with ROIs at border, larger and then smaller

ROI position	At border	Larger	Smaller
Counts in kidney	10,000	10,000	10,000
Kidney counts in ROI	10,000	10,000	8,000[a]
Area of ROI (pixels)	200	300	150
Background counts in ROI	2,000	3,000	1,500
Kidney + bkd counts in ROI	12,000	13,000	9,500
Kidney counts in ROI (– bkd)	10,000	10,000	8,000
% error	0	0	–20

[a] This is not as low as 7500 (i.e. 50 counts/pixel) because of the bulkiness of the kidney towards its centre, i.e. because of a reduction of tissue towards its border

tions should not be used, because errors are introduced, into the anterior projection, by several factors, such as the difference in superposition of bone and bowel gas.

Background

The main purpose of background subtraction is to provide an accurate estimate of relative renal function (Cosgriff et al. 1992). It is a complicated procedure but is necessary for accurate analysis (Taylor et al. 1997), because background represents 50–80% of the counts in the kidneys during the 2–3 minute part of the study when relative function is usually calculated. It is, however, very difficult to recommend a standard background subtraction procedure (Blaufox et al. 1996), the positioning of the ROI being the major problem, because it is difficult to find an area which is completely representative of the background contribution to the kidney ROI. An inappropriate choice may give a vascular spike at the beginning of the renogram curve. A suprarenal ROI can include too much liver or spleen, in which case oversubtraction of the vascular contribution will result; conversely a sub-renal ROI will underestimate this vascular component. The background ROI should not be too close to the kidney to avoid scatter from renal activity and should not extend beyond the body. Different ROI shapes, such as rectangular, elliptical and surrounding the kidney outline, perform equally well; operator independent construction probably being the most reproducible for sequential studies (Taylor et al. 1997). Whatever method of construction and position is used, the ROI counts should be area normalised to the kidney ROI before subtraction.

Parameter Calculations

To allow a more precise interpretation of the data, facilitate comparison among patients and to enable improved follow-up comparisons; it is recommended that the renogram y-axis be expressed in terms of % injected dose (Cosgriff et al. 1992) and a number of parameters be calculated from the activity-time curves.

Relative Indices. The most commonly used index is the relative function which is expressed as the percentage contributions from the right and left kidneys, and is based on the relative counts between the curves, during the period 1–2 or 1–2.5 minutes before any tracer leaves the kidney, unless the peak of the curve has already been reached. Any activity, occurring before 1 minute, should not be considered because this contains a significant amount of non-renal vascular activity (Blaufox et al. 1996). There are several methods of comparing the relative counts; the two recommended are the integral and the Patlak-Rutland (Peters 1994) methods, although implementation of the latter in commercial systems is presently limited. In the integral method, the area under the background corrected renogram during the analysis period is used, and represents the cumulative uptake of activity in the kidneys. The Rutland-Patlak plot is a double correction method, which aims to eliminate the residual vascular activity remaining after background subtraction. It requires a cardiac ROI centred on the highest count rate region of the left ventricle, which should be kept constant for successive tests on the same patient. The activity-time data is replotted; the y-axis becoming the background corrected kidney counts divided by the cardiac counts; and the x-axis becoming the integral of the cardiac counts divided by the cardiac counts, which is an equivalent of time. The relative function parameter is then determined from the mean upslope the resultant curve. Normal values will eventually be available for the two recommended techniques from a large scale study in normal volunteers.

Two other analysis techniques are common. In the mean slope method, the mean slope of the background corrected renogram is determined. The normalised slope method attempts to correct for the decreasing tracer concentration in the blood, as does the Rutland-Patlak method and also requires a cardiac ROI. However, in this case, the data is only replotted on the ordinate (y-axis). This becomes the first derivative of the background corrected kidney counts divided by the corresponding heart counts. The relative function parameter is then determined from the mean upslope the resultant curve.

Absolute Indices. The relative function index does not indicate whether both the kidneys are functioning well or poorly, and a number of methods for quantifying renal function have been proposed and are considered to perform well in clinical practice but need to be validated on a large scale. These include (Taylor 1999):

- Time to Peak
  Time in seconds to reach peak count
- Residual Activity
  30 minute count as a percentage of peak count or count at 2–3 minutes
- Excretion Index
  Ratio of 1–2 or 1–2.5 to 20–21 minute counts
- Half Value Time
  Time in s from peak count to half peak
- Lasix Index
  Ratio of counts at lasix administration to counts 600 s later

Some involve taking into account the amount of activity administered and the sensitivity of the system by using a known amount of activity in a standard, such as:

- Accumulation Index
  1-2 or 1–2.5 minute counts corrected for injected activity

An estimate of glomerular function may be made by using imaging data only with no blood sample. The most widely used technique is the Gates method. Static images of the administration activity, before and after administration, are acquired at a set distance from the camera and a renogram study is acquired immediately after administration. Background corrected counts are obtained for the kidneys between 2–3 minutes after administration. These are corrected for depth and the renal uptake of the tracer calculated as a percentage of activity administered (obtained as counts). This is converted to GFR using a regression equation (Gates 1982; Dubovsky and Yester 1988):

$$\text{GFR} = \%\ \text{renal uptake} \times 9.81270 - 6.82519 \qquad \text{(III.2.15)}$$

In a similar manner, estimates of ERPF can be obtained using ^{123}I iodohippurate with counts obtained between 1–2 minutes after administration (Schlegel and Hamway 1976). Although these methods deliver good precision, methods using blood samples yield more accurate results (Dubovsky and Yester 1988).

Restoration. When the administered activity arrives at the kidneys, it is no longer a bolus but is dispersed. The shape of the activity-time curve depends on this, the input function, since it determines the delivery rate to the kidney, as well as on the way the kidneys handle the radiopharmaceutical, the retention function. The input function can be assessed using an activity-time curve from an ROI over the heart, since this occurs after the dispersive effect of the lungs. The retention function, provides the best assessment of kidney behaviour, being the curve that would be obtained if the activity were injected directly into the renal artery and, can be derived from the measured renogram by deconvolving it with the measured input function, thus removing the effects of bolus dispersion and recirculation. If background is not subtracted from the renogram before deconvolution, then a very early spike is seen, which represents the arriving bolus of activity. This is followed by a plateau, which represents the minimum transit time of tracer through the ROI, and then a falling phase which represents the variable transit times through the renal system, see Fig. III.2.11. From the retention function, both whole kidney, including the renal pelvis in the ROIs, and parenchymal, excluding the renal pelvis, transit times can be calculated, by dividing the area under the curve by the plateau height. These are, however, very susceptible to the effects of statistical noise on the activity-time curves, and usually require higher administration activities (Dubovsky et al. 1999) as indicated in 'acquisition' above.

Ureteric Motility

The dynamics of urine movement can be assessed using the same type of compressed image technique as used in oesophageal motility analysis.

Transplanted Kidneys

Again, administration is by bolus injection and the only difference from the native kidney acquisition is that the perfusion frame duration is 1 s or faster for the first 40–60 s (Dubovsky et al. 1995, 1999).

Perfusion Phase

The most widely used quantitative indices of perfusion are Hilson's perfusion index (Hilson et al. 1978; Hilson 1991) and Kirchner's kidney/aorta ratio (Kirchner et al. 1978), which relate renal blood flow in the graft to the blood flow in the iliac artery or abdominal aorta respectively. The renal ROI is constructed with the same constraints

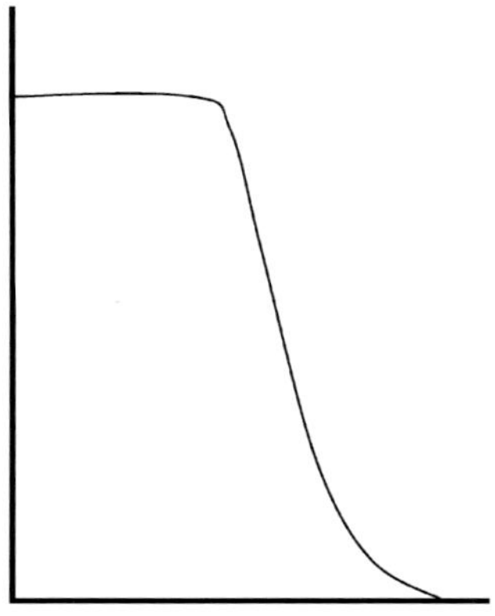

Fig. III.2.11. Retention function of activity in the kidney. The ordinate represents activity and the abscissa represents time

as in the native kidney perfusion analysis. The iliac ROI should be positioned a distance, equivalent to the length of the renal artery, distal to the anastomosis. Background should be subtracted from both activity-time curves using a relatively small ROI lateral, but not too close, to the kidney. The index is calculated from the start of perfusion in the iliac to its peak, as: 100 × area under iliac/area under kidney. With improved perfusion, the index reduces and vice versa, a very approximate limit of normal being 150%. Other indices of perfusion are: flow index, vascular transit time and blood flow as a percentage of cardiac output (Dubovsky et al. 1999). Additional values have also been obtained in cortical rather than whole kidney ROIs (Anaise et al. 1986).

Filtration–Excretion Phase

This is processed in the same way as for the native kidney, except that the position of the background ROI has to be altered, usually to a lateral area, because of the situation of the transplanted kidney in the abdominal pelvis.

Thyroid

Estimation of the administered activity accumulation in the thyroid is undertaken by comparing the counts in the thyroid with those from a known activity in a standard. This can be done using a LEGP (Becker et al. 1996b) but, often a pinhole collimator is used for imaging with an aperture 5 mm or less (Becker et al. 1996a), and this can be used to calculate the uptake. In the latter case, however, great care must be taken maintain the same separation from the collimator to both the thyroid and the phantom. A spacer is used for this, which sets the exact distance from the pinhole collimator to the skin. This is sprung so that undue pressure is not put on the patient if the collimator is moved slightly too close. Again background must be drawn and the calculation of thyroid uptake is the same as in the non-imaging section.

For imaging with $^{99}Tc^{m}$, an anterior image is acquired for $1–2 \times 10^5$ counts or 5 min. With ^{123}I, the corresponding parameters are generally $0.5–1 \times 10^5$ counts or 10 min (Becker et al. 1996a).

References

Anaise D, Oster ZH, Atkins ML, Arnold AN, Weis S, Waltzer WC, Rapaport FT (1986) Cortex perfusion index: a sensitive detector of acute rejection crisis in transplanted kidneys. J Nucl Med 27: 1697–1701

Becker D, Charles ND, Dworkin H, Hurley J, McDougall IR, Price D, Royal H, Sarkar S (1996a) Procedure guideline for thyroid scintigraphy: 1.0. Society of Nuclear Medicine. J Nucl Med 37: 1264–1266

Becker D, Charkes ND, Dworkin H, Hurley J, McDougall IR, Price D, Royal H, Sarkar S (1996b) Procedure guideline for thyroid uptake measurement: 1.0. Society of Nuclear Medicine. J Nucl Med 37:1266–1268

Bergmann H, Busemann-Sokole E, Horton PW (1995) Quality assurance and harmonisation of nuclear medicine investigations in Europe. Eur J Nucl Med 22:477–480

Blaufox MD, Aurell M, Bubeck B, Fommei E, Piepsz A, Russell C, Taylor A, Thomsen HS, Volterrani D (1996) Report of the radionuclides in nephrourology committee on renal clearance. J Nucl Med 37:1883–1890

Blokland JAK, Camps JAJ, Pauwels EKJ (1997) Aspects of performance assessment of whole body imaging systems. Eur J Nucl Med 24:1273–1283

Burger C, Berthold T (2000) Physical principles and practical aspects of clinical PET imaging. In: von Schulthness GK, Buck A, Engel-Bicik I, Steinert HC (eds) Clinical positron emission tomography. Correlation with morphological cross-sectional imaging. Lippincott Williams and Wilkins, Philadelphia

Chandler ST, Thomson WH (1996) Institution of Physics and Engineering in Medicine and Biology report 73 mathematical techniques in nuclear medicine. IPEMB, York

Cosgriff PS (1989) UK gamma camera renography survey. Nucl Med Commun 10:214

Cosgriff PS (ed) (1997) Cost B2 final report (EUR 17916 EN): Quality assurance of nuclear medicine software. Office for Official Publications of the European Communities, Luxembourg

Cosgriff PS (1998) Quality assurance in renography: a review. Nucl Med Commun 19:711–716

Cosgriff P, Brown H (1990) Influence of kidney depth on the renographic estimation of relative renal function. J Nucl Med 31:1576–1577

Cosgriff PS, Lawson RS, Nimmon CC (1992) Towards standardization in gamma camera renography. Nucl Med Commun 13:580–585

De Geeter F, Franken PR, Defrise M, Andries H, Saelens E, Bossuyt A (1996) Optimal collimator choice for sequential iodine-123 and technetium-99m imaging. Eur J Nucl Med 23:768–774

De Lima JJP (1996) Nuclear medicine and mathematics. Eur J Nucl Med 23:705–719

DePuey EG, Garcia EV (2001) Updated imaging guidelines for nuclear cardiology procedures, part 1. J Nucl Cardiol 8:G1–G58

Dubovsky EV, Yester MV (1988) Glomerular filtration rate, renal transit times, effective renal plasma flow, and differential renal function. In: Gelfand MJ, Thomas SR (eds) Effective use of computers in nuclear medicine. McGraw-Hill, New York, pp 348–383

Dubovsky EV, Russell CD, Erbas B (1995) Radionuclide evaluation of renal transplants. Semin Nucl Med XXV:49–59

Dubovsky EV, Russell CD, Biscof-Delaloye A, Bubeck B, Chaiwatanarat T, Hilson AJW, Rutland M, Oei HY, Sfakianakis GN, Taylor A Jr (1999) Report of the radionuclides in nephrourology committee for evaluation of transplanted kidney (review of techniques). Semin Nucl Med XXIX: 175–188

El Fakhri G, Moore SC, Kijewski MF (2002) Optimization of Ga-67 imaging for detection and estimation tasks: dependence of imaging performance on spectral acquisition parameters. Med Phys 29:1859–1866

Esquerre JP, Coca FJ (1995) First-pass radionuclide angiocardiography using a single-crystal gamma camera: are count statistics actually the limiting factor? J Nucl Med 36:1528–1530

Fazio F, Jones T (1975) Assessment of regional ventilation by continuous inhalation of radioactive krypton-81m. Br Med J 3:673–676

Fleming JS, Cosgriff PS, Houston AS, Jarritt PH, Skrypniuk JV, Whalley DR (1998) UK audit of relative renal function assessment of DMSA scintigraphy. Nucl Med Commun 19:989–997

Forstrom LA, Dunn WL, O'Connor MK, Decklever TD, Hardyman TJ, Howarth DM (1996) Technical pitfalls in image acquisition, processing and display. Semin Nucl Med XXVI:278–294

Garcia EV (ed) (1996) Imaging guidelines for nuclear cardiology procedures, part 1. J Nucl Cardiol 3:G1–G46

Gates GF (1982) Glomerular filtration rate: estimation from fractional renal accumulation of ^{99m}Tc-DTPA (stannous). Am J Roentgenol 138:565

Gelfand MJ, Hannon DW (1988) Left-to-right shunts. In: Gelfand MJ, Thomas SR (eds) Effective use of computers in nuclear medicine. McGraw-Hill, New York, pp 271–287

Green MV, Bacharach SL, Borer JS, Bonow RO (1991) A theoretical comparison of first-pass and gated equilibrium methods in the measurement of systolic left ventricular function. J Nucl Med 32: 1801–1807

Gustafsson A, Arlig A, Jacobsson L, Ljungberg M, Wikkelso C (2000) Dual-window scatter correction and energy window setting in cerebral blood flow SPECT: a Monte Carlo study. Phys Med Biol 45:3431–3440

Hastings DL, Jeans SP, Wall WH, Hall BJ, Miller DE (1995) The effect on diagnostic quality of using dual isotope imaging for ^{81}Kr^{m} ventilation and ^{99}Tc^{m}-MAA perfusion lung scanning. Nucl Med Commun 16:281–289

Heikkinen JO (1999) A dynamic phantom for radionuclide renography. Phys Med Biol 44:N39–N53

Heikkinen JO, Kuikka JT, Ahonen AK, Rautio PJ (2001) Quality of dynamic radionuclide renal imaging: multicentre evaluation using a functional renal phantom. Nucl Med Commun 22: 987–995

Hervas I, Marti JF, Gonzalez A, Ruiz JC, Alonso J, Bello P, Manzano F, Torres I, Mateo A (2001) Is the depth correction using the geometric mean really necessary in a ^{99}Tc^{m}-DMSA scan in the paediatric population? Nucl Med Commun 22:547–552

Hilson A (1991) The renal transplant perfusion index: where are we now? Eur J Nucl Med 18: 227–228

Hilson AJW, Maisey MN, Brown CB, Ogg CS, Bewick MS (1978) Dynamic renal transplant imaging with Tc-99m-DTPA(Sn) supplemented by a transplant perfusion index in the management of renal transplants. J Nucl Med 19:994–1000

Hindie E, Buvat I, Jeanguillaume C, Prigent A, Galle P (1999) Quantitation in planar renal scintigraphy: which mu value should be used? Eur J Nucl Med 26:1610–1613

Hines H, Kayayan R, Colsher J, Hashimoto D, Schubert R, Fernando J, Simcic V, Vernon P, Sinclair RL (1999) Recommendations for implementing SPECT instrumentation quality control. Nuclear Medicine Section – National Electrical Manufacturers Association (NEMA). Eur J Nucl Med 26: 527–532

Houston AS, White DR, Sampson WF, Macleod MA, Pilkington JB (1998) An assessment of two methods for generating automatic regions of interest. Nucl Med Commun 19:1005–1016

Houston AS, Whalley DR, Skrypniuk JV, Jarritt PH, Fleming JS, Cosgriff PS (2001) UK audit and analysis of quantitative parameters obtained from gamma camera renography. Nucl Med Commun 22:559–566

Hutton BF, Osiecki A (1998) Correction of partial volume effects in myocardial SPECT. J Nucl Cardiol 5:402–413

IAEA (1991) Tecdoc 602 quality control of nuclear medicine instruments. International Atomic Energy Agency, Vienna

Inoue Y, Yoshikawa K, Suzuki T, Katayama N, Yokoyama I, Kohsaka T, Tsukune Y, Ohtomo K (2000) Attenuation correction in evaluating renal function in children and adults by a camera-based method. J Nucl Med 41:823–829

IPSM (1992) Report 66 quality control of gamma cameras and associated computer systems. Institute of Physical Sciences in Medicine, London

Kelty NL, Cao Z-J, Holder LE (1997) Technical considerations for optimal orthopedic imaging. Semin Nucl Med XXVII:328–333

Kirchner PT, Goldman WH, Leuyman SB, Kiepler RF (1978) Clinical application of the kidney to aortic blood flow index (K/A) ratio. Contrib Nephrol 11:120–126

Kuyvenhoven JD, Busemann Sokole E, van Rijk P, Clausen M, Jamar F, Slegers G, Thierens H, Cosgriff P, Dierckx RA (2000a) European Association of Nuclear Medicine. A survey of guidelines in 27 EANM associated societies by the EANM Task Group on Quality Assurance and Standardisation. Eur J Nucl Med 27:BP31–BP44

Kuyvenhoven JD, Busemann Sokole E, van Rijk P, Dierckx RA (2000b) Expert groups, task groups and committees, national societies and the nuclear medicine community: towards "EANM protocols". Eur J Nucl Med 27:1437–1440

Lau YH, Hutton BF, Beekman FJ (2001) Choice of collimator for cardiac SPET when resolution compensation is included in iterative reconstruction. Eur J Nucl Med 28:39–47

Lawson RS (1999) Application of mathematical methods in dynamic nuclear medicine studies. Phys Med Biol 44:R57–R98

Lythgoe MF, Gradwell MJ, Evans K, Gordon I (1998) Estimation and relevance of depth correction in paediatric renal studies. Eur J Nucl Med 25:115–119

Macey DJ, DeNardo GL, DeNardo SJ (1999) Planar gamma camera quantitation of ^{123}I, ^{99m}Tc or ^{111}In in the liver and spleen of an abdominal phantom. Cancer Biother Radiopharm 14:299–306

Moonen M, Jacobsson L (1997) Effect of administered activity on precision in the assessment of renal function using gamma camera renography. Nucl Med Commun 18:346–351

Mullani NA (2000) Comparing diagnostic accuracy of γ camera coincidence systems and PET for detection of lung lesions. J Nucl Med 41:959–960

Norrgren K, Svegborn SL, Areberg J, Mattsson S (2003) Accuracy of the quantification of organ activity from planar gamma camera images. Cancer Biother Radiopharm 18:125–131

O'Connor MK (1996) Instrument- and computer-related problems and artifacts in nuclear medicine. Semin Nucl Med XXVI:256–277

O'Connor MK, Brown ML, Hung JC Hayostek RJ (1991) The art of bone scintigraphy – technical aspects. J Nucl Med 32:2332–2341

O'Reilly P (2003) Consensus Committee of the Society of Radionuclides in Nephrourology. Standardization of the renogram technique for investigating the dilated upper urinary tract and assessing the results of surgery. BJU Int 91:239–243

O'Reilly P, Aurell M, Britton K, Kletter K, Rosenthal L, Testa T. Consensus on diuresis renography for investigating the dilated upper urinary tract (1996) Radionuclides in Nephrourology Group. Consensus Committee on Diuresis Renography. J Nucl Med 37:1872–1876

Parker RP, Smith PHS, Taylor DM (1984) Basic science of nuclear medicine. Churchill Livingstone, Edinburgh

Peters AM (1991) Quantification of renal haemodynamics with radionuclides. Eur J Nucl Med 18: 274–286

Peters AM (1994) Graphical analysis of dynamic data: the Patlak-Rutland plot (editorial). Nucl Med Commun 15:669–672

Peters AM, Myers MJ (1998) Physiological measurements with radionuclides in clinical practice. Oxford University Press, Oxford
Port SC (1994) The first-pass ventricular function study: why now? J Nucl Med 35:1301–1302
Port SC (1999) Imaging guidelines for nuclear cardiology procedures, part 2. American Society of Nuclear Cardiology. J Nucl Cardiol 6:G53–G83
Prigent A, Cosgriff P, Gates GF, Granerus G, Fine EJ, Itoh K, Peters M, Piepsz A, Rehling M, Rutland M, Taylor A Jr (1999) Consensus report on quality control of quantitative measurements of renal function obtained from the renogram: international consensus committee from the scientific committee of radionuclides in nephrourology. Semin Nucl Med XXIX:146–159
Rosenthal MS, Cullom J, Hawkins W, Moore SC, Tsui BMW, Yester M (1995) Quantitative SPECT imaging: a review and recommendations by the focus committee of the Society of Nuclear Medicine Computer and Instrumentation Council. J Nucl Med 36:1489–1513
Rousset OG, Ma Y, Evans AC (1998) Correction for partial volume effects in PET: principle and validation. J Nucl Med 39:904–911
Schiepers C, Almasi JJ (1988) Equilibrium gated blood pool imaging at rest and during exercise. In: Gelfand MJ, Thomas SR (eds) Effective use of computers in nuclear medicine. McGraw-Hill, New York, pp 136–205
Schlegel JU and Hamway SA (1976) Individual renal plasma flow determination in 2 minutes. J Urol 126:282–285
Sharp PF, Chesser RB, Mallard JR (1982) The influence of picture element size on the quality of clinical radionuclide images. Phys Med Biol 27:913–926
Sorenson JA, Phelps ME (1987) Physics in nuclear medicine, 2nd edn. Grune and Stratton, New York
Starck SA, Carlsson S (1997) Digital filtering of bone scans: an ROC study. Nucl Med Commun 18: 98–104
Taylor A (1999) Radionuclide renography: a personal approach. Semin Nucl Med XXIX:102–127
Taylor A Jr, Thakore K, Folks R, Halkar R, Manatunga A (1997) Background subtraction in technetium-99m-MAG3 renography. J Nucl Med 38:74–79
Tindale WB, Barber DC (2000) Diagnostic impact: a challenge for quantitative nuclear medicine. Nucl Med Commun 21:217–219
Todd-Pokropek A (2002) Advances in computers and image processing with applications in nuclear medicine. Q J Nucl Med 46:62–69
Tondeur M, Melis K, de Sadeleer C, Verelst J, van Espen M-B, Ham H, Piepsz A (2000) Inter-observer reproducibility of relative ^{99}Tc^{m}-DMSA uptake. Nucl Med Commun 21:449–453
Wittry MD, Juni JE, Royal HD, Heller GV, Port SC (1997) Procedure guideline for equilibrium radionuclide ventriculography. Society of Nuclear Medicine. J Nucl Med 38:1658–1661
Zanzonico P, Heller S (2000) The intraoperative gamma probe: basic principles and choices available. Semin Nucl Med XXX:33–48
Zaidi H, Hasegawa B (2003) Determination of the attenuation map in emission tomography. J Nucl Med 44:291–315
Zito F, Gilardi MC, Magnani P, Fazio F (1996) Single-photon emission tomographic quantification in spherical objects: effects of object size and background. Eur J Nucl Med 23:263–271

CHAPTER III.3

Single Photon Tomographic Imaging

Abbreviations

2D	two dimension
3D	three dimension
CT	computed tomography
ERNA	equilibrium radionuclide angiography
FOV	field of view
FWHM	full width at half maximum
LEGP	low energy general purpose
LEHR	low energy high resolution
LVEF	left ventricular ejection fraction
MPI	myocardial perfusion imaging
PHA	pulse height analyser
ROI	region of interest
SPET	single photon emission tomography
TCR	transmission count rate to emission crosstalk count rate

Single Photon Emission Tomographic (SPET) imaging overcomes the loss of contrast suffered by planar images, which impairs the detectability of small lesions, particularly those which are deep lying and which exhibit reduced radionuclide accumulation. The reconstructed data can also be reoriented into, for example, coronal or sagittal sections, for better visualisation of the relative positions of activity distributions, which may help to localise the position of abnormalities more accurately. The technique also has the potential to quantify the regional distribution of activity, which allows representation of the activity distribution in units of MBq ml^{-1} rather than just counts per pixel, and thus better indicates organ function and radiation dosimetry (Rosenthal et al. 1995; Fleming and Alaamer 1996). A number of acquisition and processing factors are peculiar to SPET and, although software for single photon tomographic imaging was introduced in the mid 1980s, a range of imaging protocols are in use and are sometimes inappropriate, which demonstrates the on-going need to encourage the correct use of the instrumentation (Heikkinen et al. 1999). The image data is heavily and sophisticatedly processed. This includes attenuation correction and scatter compensation which can improve image resolution and contrast, significantly. It is possible to reconstruct images that are far from optimum. Artefacts, which would have been easily recognised in the raw images, can become camouflaged. It is thus imperative that there is a complete understanding of the processing techniques and their constraints, and an ability to verify that they have been implemented correctly (DePuey 1994; Rosenthal et al. 1995; IAEA 1991). Because of the gantry movement, collision sensors are mandatory but those at the collimator surface are often only triggered when a significant force is applied because the collimator must sometimes be in contact with the surface of the patient to minimise separation. It is therefore essential that the system not be left unattended when a patient is being imaged in case the patient moves or the system experiences a fault, either of which could result in injury to the patient or, less importantly, damage to the system.

Factors Determining Image Quality and Affecting Lesion Detectability

The performance of tomographic studies represents a huge leap in technical difficulty from planar studies, and lack of attention to these technical aspects will result in the generation of suboptimal studies and artefacts in the image data (O'Connor et al. 1991). Once the tomographic system is optimally adjusted, the quality of the transaxial images depends solely on the procedures used to acquire and process the data.

Acquisition Parameters

The following acquisition parameters have to be optimised: collimator, pixel size, count density, detector orientation, rotation orbit, rotation arc, projection number, energy window, physiological gating and, in the case of dynamic SPET, frame duration. Although post-processing may improve the results if data is acquired inappropriately, it will not produce data equivalent to that obtained under optimal conditions (Rosenthal et al. 1995).

Collimator

The type of collimator is the most important factor in determining detection efficiency and spatial resolution (Tsui et al. 1998). Normally, parallel hole collimators are used because they do not introduce spatial distortions and are consequently the easiest to use. Because spatial resolution deteriorates with collimator–patient separation and because a large separation is often experienced during camera rotation, a collimator with as high a spatial resolution as possible should be used to maintain as good a spatial resolution with depth as possible (Mueller et al. 1986; Rosenthal et al. 1995). The disadvantage of this is the deterioration experienced in detection sensitivity, and if a low count rate or a short acquisition duration is anticipated, a collimator with a lower spatial resolution and higher detection sensitivity should then be substituted. In multihead systems, higher resolution collimators should be used because the improvement in resolution more than compensates for the loss in sensitivity (Fahey et al. 1992). An alternative, sometimes used in myocardial perfusion studies, is to use ultra high resolution collimators with long bore holes, which helps to minimise the collimator to cardiac separation, or to set the relative orientation of the detectors to 76° rather than to 90°.

Another technique for maintaining spatial resolution with depth is the use of a fan beam collimator, which accomplishes this by focussing. These have been shown to detect small deep cold lesions better than parallel collimators (Li et al. 1994). Considerable care must be taken when acquiring data using a fan beam collimator because the magnification across the patient makes it difficult to ensure that the whole area of interest is kept with the acquisition field of view, at each projection angle. Otherwise truncation artefacts will be created in the reconstructed image. Consequently the use of the collimator is restricted to small organs and is often used for brain and cardiac studies (O'Connor 1996). An advantage of the magnification, however, is an improvement in detection efficiency (Rosenthal et al. 1995). The projection data is usually rebinned into parallel projections so that it is then compatible with standard reconstruction software (Bruyant 2002). Care must be taken during processing, however, because not treating the data as fan-beam or using the wrong value of focal length will lead to severe distortions in the reconstructed images (O'Connor 1996).

Pixel Size

The same data sampling requirements apply to tomographic acquisition as to planar imaging, in that according to the Nyquist theorem a pixel size no larger than half the planar system spatial resolution distance is required to accurately sample the projection data without degrading the image.

The projection image pixel size is the same as that within the transaxial slice, and also determines the transaxial slice thickness. For a surface system spatial resolution of 5 mm, a pixel size of 2.5 mm would be required which, for a 500 mm field of view (FOV), translates to an acquisition matrix of 200 × 200 when using a magnification factor of one. If a smaller pixel size is obtained by using a magnification higher than unity, the FOV will be reduced and care must be taken that important data is not excluded. The complete area of interest must remain within the FOV at every projection angle for artefact free reconstruction.

However, because reconstructed spatial resolution need only be optimised in the area of interest on the reconstructed transaxial slice, the required pixel size should be determined from the planar system spatial resolution at the closest point of the area of interest, rather than at the surface of the collimator or patient. It is generally better to err on the side of the pixel being too small rather than too large, because any aliasing experienced in the projection images because the pixel size is too large will be amplified during the processing to cause image artefacts (Rosenthal et al. 1995). There is, however, no advantage to having a pixel size significantly less than the system resolution of the camera at this point in the reconstructed transaxial slice. There is a significant disadvantage, however, since slices will be reconstructed from fewer acquired counts, because they are the thickness of a pixel, and will therefore have unnecessarily high noise. This suggests that, in the above example, a larger pixel size would be suitable and an acquisition matrix nearer to 128 × 128 might be appropriate. The deterioration in tomographic spatial resolution caused by the larger pixel size can be less than that due to increased statistical noise occasioned by smaller pixels. Therefore, the final choice of pixel size should be determined for each clinical application based on both planar system spatial resolution at the smallest collimator-area of interest separation and anticipated count rate.

Count Density

Because noise is amplified by the reconstruction process, it is more important than in the planar case to acquire as high a count density as possible, at each projection angle. Since the maximum activity that can be administered to the patient is determined on radiological protection principles, the determinant of count density is purely on the duration that the patient can remain comfortable and motionless during the complete acquisition and on the effective half-life of the radiopharmaceutical (Rosenthal et al. 1995). Since even slight movement can significantly degrade the reconstructed image contrast, this acquisition duration should be adequate but not excessive. For this reason, multiple head cameras, in which data are acquired on all heads simultaneously, should be used to improve acquisition count density rather than to shorten acquisition durations.

Detector Orientation

The detector must view the same volume during its entire rotation otherwise the data passed to the reconstruction process will be incomplete. The faces of the detectors must therefore be absolutely parallel to the axis of rotation, and this can be checked using a spirit level, positioned on the face of the collimator rather than on the back of the detector head.

Rotation Orbit

The simplest camera orbit around the patient is circular but this results in a varying separation between the collimator face and the surface of the patient, better laterally, and worsening in the anterior–posterior orientations. This causes a deterioration in planar spatial resolution at certain points in the acquisition arc, particularly in the anterior–posterior projections, with consequential loss of tomographic spatial resolution. This can be minimised by using a 180° rotation arc, which can often accommodate a smaller separation than would be required to describe a 360° arc, see Fig. III.3.1. It can also be minimised by reducing the separation between the collimator face and the patient using a non-circular orbit. Image resolution achieved with a non-circular orbit will be patient dependent but will always be equal to, or better than, that achieved with a conventional circular orbit (O'Connor et al. 1991; Rosenthal et al. 1995; Forstrom et al. 1996). However, non-circular orbit acquisitions have increased variation of spatial resolution due to increased variation in detector to source distance. Spatial distortion and imaging artefacts have been observed (Eisner et al. 1986; DePuey and Garcia 2001), caused by the variations in spatial resolution over the rotation (Knesaurek et al. 1989) and whose magnitude are correlated with increasing long–short axis ratios (Maniawski et al. 1991). These can be minimised, particularly in 180° studies, by implementing attenuation, resolution and scatter corrections (Knesaurek et al. 1989).

Four methods of defining the rotation orbit are available. The first is an elliptical orbit determined by setting the lateral limits of the body from which an ellipse is constructed geometrically. A more sophisticated approach is to perform a non-geometric orbit. This can be accomplished by defining the limit of the anterior body surface as well as the lateral aspects. The inferior body surface is constrained by the imaging table and should be known to the system. An alternative approach, used in myocardial perfusion imaging, is to define the position of the heart on a projection image in conjunction with identification of the anterior and the left lateral borders of the patient.

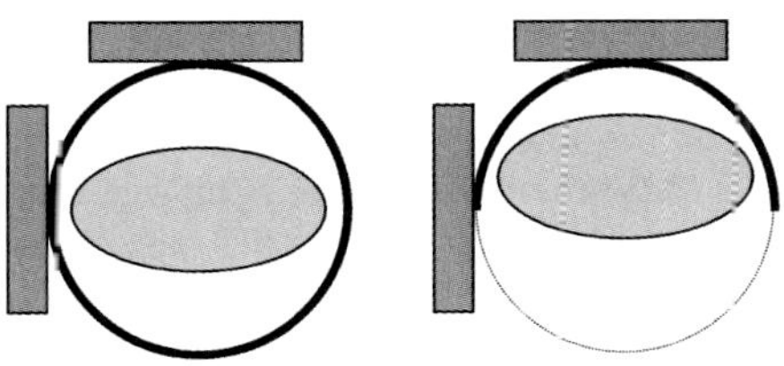

Fig. III.3.1. Patient–detector separations for 360° and 180° rotation arcs

The camera can then maintain a minimum separation and keep the heart within the field of view throughout the acquisition arc. Finally, true body contouring, in which the relative position of the gamma camera and the subject is continuously adjusted, without prior surface identification is accomplished by some form of proximity device located at the collimator face.

To accomplish these movements, the position of the camera, the imaging bed or both must be controlled by the acquisition computer.

Rotation Arc

Projection images should be acquired over a 360° rotation arc to facilitate optimal reconstructed spatial resolution over the whole transaxial slice. However, if the area of interest is located asymmetrically in the transaxial slice, for instance the heart, liver or lumbar spine, reconstructed spatial resolution need only be optimised in this area. A 180° rotation allows concentration of the imaging time over the sectors providing the best planar spatial resolution because of smallest separation and best count rate from smallest attenuation, provide significantly improved contrast (Eisner et al. 1986; Knesaurek et al. 1989). Disadvantages of the restricted acquisition arc can be the creation of image artefacts and a rapid loss of reconstructed spatial resolution in the more distant half of the transaxial slice, resulting from the absence of projection data along this arc (Garcia 1994). Triple head and fixed 180° double head systems always describe a 360° acquisition arc but do not always use data from the whole acquisition arc for reconstruction.

Data Acquisition Mode

Although step-and-shoot acquisition mode is the more usual, continuous acquisition is often used to improve counting statistics or shorten acquisition duration. A small amount of blurring is caused, particularly tangentially, and this increases across the transaxial field of view, proportionally to the distance from the axis of rotation (Bieszk and Hawman 1987). There is no significant loss of tomographic spatial resolution, when 128 projections are used over a 360° rotation, such that the angular difference between each projection is only 3°. The data acquired in continuous mode is rebinned into projections before analysis. A hybrid mode of continuous step and shoot acquisition involves acquisition during detector movement from one angle to the next, and results in increased counting statistics as compared to the standard step and shoot, whilst reducing most of the small amount of blurring associated with continuous acquisition (Cao et al. 1996; DePuey and Garcia 2001). A factor in favour of the continuous mode is when the ratio of movement to acquisition time, in the step-and-shoot option, is >10% (Rosenthal et al. 1995).

Projection Number

In the step and shoot acquisition, the linear distance between projections is a critical parameter in the quality of the reconstruction. For accurate and artefact free reconstruction, there must be enough projections to fill the Radon space. This means that the linear distance between projections should comply with the Nyquist theorem in the same way as pixel size, and must not be greater than half the tomographic resolution. The number of projections are, however, completely independent of the pixel size. If too few projections are acquired, the reconstructed images will be distorted by radial streaks and loss of contrast, the streaks being more pronounced towards the periphery. If too many projections are acquired, the projection images will comprise unnecessarily few counts. For a parallel hole collimator, the linear distance between projections is inversely proportional to the number of projections and proportional to the distance, along the projection line from the centre of rotation towards the collimator face, see Fig. III.3.2. Because it is adequate to optimise reconstructed spatial resolution only in the area of interest rather than over the whole transaxial slice, the linear separation need only be defined at the minimum radius of rotation that encloses the area of interest (Fig. III.3.3).

The number of projections are calculated by assuming that the linear length can be approximated by the arc length and by dividing this into the circumference of the circle enclosing the area of interest, which is $2\pi r$ where r is the minimum radius of rotation (Rosenthal et al. 1995; Forstrom et al. 1996; Hutton 1996). For the situation in which the tomographic spatial resolution at the periphery of the area of interest is 12 mm, the maximum linear separation of the projections at the periphery should be 6 mm. In this situation, the number of projections required is at least $2\pi r/6$. For radii of rotation of 15 cm, 10 cm and 5 cm, the required projection views are 157, 105 and 52 respectively. Only half of these views are required if the rotation arc is 180° rather than 360°. As with pixel size, there is no advantage in having a projection separation significantly less than the system resolution of the camera because of the increased noise at each projection angle. Again, the final choice of sample interval should be determined for each clinical application.

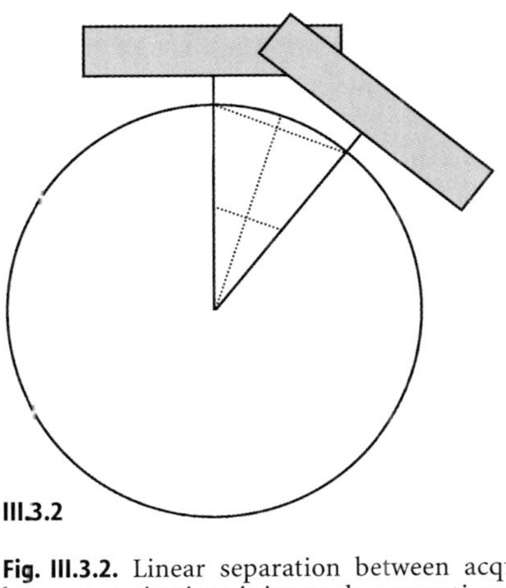

III.3.2

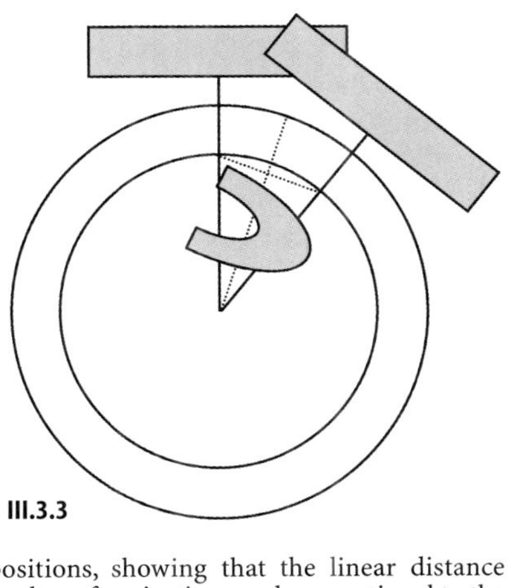

III.3.3

Fig. III.3.2. Linear separation between acquisition positions, showing that the linear distance between projections is inversely proportional to the number of projections and proportional to the patient–detector separation

Fig. III.3.3. The linear distance between projections, and therefore the number of projections required, is defined at the minimum radius of rotation that encloses the area of interest

Energy Window

Improvements in tomographic contrast can be achieved with a reduction in the contribution of scatter to the projection views, by matching the PHA window to the energy resolution of the camera, see 'pulse height analyser' in Chapter II.3.

Dynamic Acquisition

Dynamic acquisition is unusual but is possible, the maximum frame rate, that can be achieved, depending on the speed at which the camera can rotate. With 'slip-ring' technology, this may be 4 times per minute which, if the camera has 2 heads, gives a frame rate of 8 min^{-1}, fast enough for renography (Peters and Myers 1998) but with severe noise problems.

Patient Position

When a complete investigation, such as cardiac or cerebral perfusion, can involve two separate acquisitions some time apart, it is vital to maintain consistent patient positioning between the two studies. This is achieved by recording the gantry start angle and radius, and bed position on the first study and reproducing them on the second study. In a 180° acquisition rotation, the choice of prone or supine positioning can have a significant effect on image quality because of attenuation through the width of the table. Perpendicularly, most will attenuate only 5–10%, of $^{99}Tc^{m}$, and 8–12%, of ^{201}TI photons; but tangentially, the attenuation will be much more and depend on construction (O'Connor and Bothun 1995), see Fig. III.3.4.

Data Integrity

The imaged activity distribution must be stationary during the complete acquisition otherwise the data passed to the reconstruction process will be incomplete. A changing concentration will result in artefacts, e.g. bladder filling with radioactive urine during the acquisition which makes imaging of the pelvis difficult. An absence of activity redistribution and integrity of the acquired data must be confirmed by reviewing a cine display before reconstruction is started. If there is poor or missing data on one projection image only, this can usually be replaced by averaging adjacent frames. If more than one frame is poor, however, it will usually be necessary to re-acquire the data.

Motion Correction

The area being imaged must remain stationary relative to the axis of rotation of the camera during the complete acquisition otherwise tomographic spatial resolution will

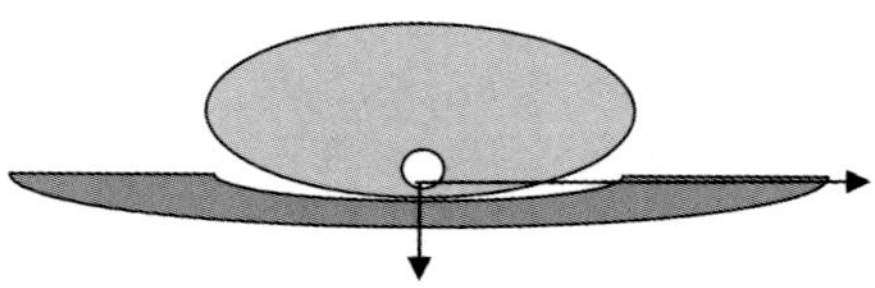

Fig. III.3.4. Imaging table attenuation. This is much higher tangentially than perpendicularly

be reduced and artefacts may be created. The latter have variable appearance and can be difficult to recognise. The acquisition data must be reviewed, for such an occurrence, before reconstruction by first inspecting the cine presentation and then the sinogram. The cine review will reveal patient motion along the axis of rotation, excessive background activity, and any truncation of the area of interest. It will also demonstrate variations in camera performance with angle by showing count variations between frames and axial shift in multi-detector systems (Hines et al. 1999a,b, 2000).

A sinogram is a concatenation of all the projection images into a single image volume. This volume is constructed by stacking each projection image vertically, in rotational sequence, to form a three-dimensional volume. Thus, there is a row for each angle at which data are acquired and the vertical axis corresponds to the angular position of the detector (Bruyant 2002). The horizontal axes represent the count locations on the detector. The position of each focus of activity should describe a sinusoidal variation from the top of the sinogram to the bottom, as the projection angle progresses. Movement will be revealed as dislocations in this pattern and bad acquisition projection as unduly high or low counts across the whole row (IPSM 1985). The sinogram will give similar information as cine review of planar images but with greater sensitivity for transverse patient motion which will show up as discontinuities, these will be less for vertical shifts (Hines et al. 1999a,b, 2000). A small shift can be corrected using a frame shift program using, for example: cross-correlation, diverging squares or two-dimensional fit (O'Connor et al. 1998) but a major shift can only be overcome by repeating the acquisition.

Processing Parameters

Transaxial images are created from projection images by some method of tomographic reconstruction. Significant differences, in quality of reconstruction, can be obtained by using different techniques and reconstruction parameters.

Filtered Back-Projection

The traditional reconstruction technique is filtered back-projection. This always involves the application of a ramp filter which cancels the erroneous activity values created throughout the reconstructed image by simple back-projection, leaving activity distributions only at the true positions. If the projection images were noise free and infinite in number, the ramp filter would give perfect reconstruction of the transaxial slices. Because, however, the projection images are affected by Poisson statistics and the number of projection images are limited, noise is present throughout the reconstructed image at all frequencies. The ramp filter will amplify the high frequencies and produce an extremely noisy reconstructed image, much worse than is usually seen in planar images. As a result, it is almost always necessary to apply a constraining or smoothing filter to the ramp filter to impose an upper limit, or cut-off, on the frequencies passed through to back-projection, to avoid this excessive amplification of noise at the upper frequencies. Increasing the cut-off retains more high frequencies, which makes the image noisier but maintains spatial resolution. Reducing the cut-off removes more low frequencies which makes the image smoother but degrades spatial resolution (Hansen 2002).

This cut-off frequency is usually determined in relation to the Nyquist frequency. This is the highest spatial frequency that the matrix can accurately sample and retaining frequencies higher than this provides no additional information in the image. The Nyquist frequency is the inverse of twice the pixel size. For instance, for a pixel size of 5 mm, the Nyquist frequency would be 1/10 mm^{-1}, 0.1 mm^{-1} or 1 cm^{-1}. Suppression of spatial frequencies below this would, theoretically, be expected to affect spatial resolution. However, it may be necessary to reduce the cut-off frequency depending on image noise, but it should be set as high as possible, consistent with acceptable image quality, to preserve as much information as possible (O'Connor et al. 1991). In practice, suppression of frequencies above half the Nyquist frequency does not seem to significantly impair spatial resolution. Suppression of lower frequencies does however degrade resolution.

Thus, the optimum cut-off frequency depends on the radionuclide being used, being higher for those such as $^{99}Tc^{m}$ for which the camera is optimised and lower for radionuclides which produce an image with poorer spatial resolution. It also depends on the size and extent of the defect to be detected (Gilland et al. 1988). Clinically, filter cut-off values ~0.25 $pixel^{-1}$ however lower values of 0.14 $pixel^{-1}$ have been reported to be superior (Sankaran et al. 2002). Care must be exercised when specifying the frequencies as proportion of the Nyquist frequency, particularly when the pixel size is not well matched to spatial resolution. As the pixel size is made smaller, the same value of cut-off frequency will attenuate less high frequency components. For instance, for a planar system spatial resolution of 10 mm on a 500 mm camera, the maximum pixel size which will satisfy the Nyquist sampling theorem is 5 mm. This can be achieved with a matrix of 64×64 and a zoom of 1.56. In this case, the Nyquist frequency is 1 cm^{-1}, and a cut-off at 0.5 of the Nyquist frequency would attenuate frequencies in the image below 0.5 cm^{-1}. If a higher zoom, such as 2.0, is used the pixel size will be 3.9 mm. In this case, the Nyquist frequency will be 1.28 cm^{-1}, and a cut-off at 0.5 of the Nyquist frequency would attenuate frequencies in the image below 0.64 cm^{-1}, resulting in a sharper and noisier image.

Filters

A number of filters are available, but all have one thing in common; a smooth transition at the cut-off point, because a sharp cut-off at the upper frequency limit would produce artefacts in the reconstructed image. The difference among the filters is their shape, i.e. the manner in which they allow the various frequencies in the projection images to pass through to reconstruction. A smooth filter begins to roll-off early and decreases slowly. It should be used with acquisitions which contain low counts, e.g. $<2 \times 10^6$ in total, and investigations which demand a lower spatial resolution. A sharp filter is one which rolls-off late and decreases rapidly. It should be used with acquisitions which contain high counts, e.g. $>3 \times 10^6$ in total, and investigations which demand a higher spatial resolution (Taylor 1994). These shapes are altered by parameters within each filter. Conventional filters are independent of the distribution of the counts in the image, and include: Butterworth, Hamming (Hann), Parzen and Shepp-Logan (see Chapter III.5). The latter incorporates the ramp function. Of these, the Butterworth and Hamming are the most commonly used.

The count density will, however, vary over different parts of the image and, therefore, the optimum filter shape will be different at different points in the image. Another, more complicated, set of filters attempts to correct for this by amplifying intermediate frequencies (Taylor 1994). These are the 'adaptive' or 'restoration' filters, the most common of which are Metz and Wiener. These also incorporate knowledge of the imaging capabilities of the γ camera and, as well as removing noise from the image, they avoid reducing the spatial resolution by not blurring the edges (Hansen 2002). They are, however, difficult to use because they can create artefactual activity distributions.

The choice of filter function is crucial for determining both the noise and spatial resolution in the final image, the optimum filter being one which smoothes enough to suppress noise but not enough to cause significant degradation in spatial resolution. Over smoothing can obscure genuine abnormalities, while undersmoothing can complicate interpretation of the images because of excessive noise The most appropriate filter shape is determined by both count density and gamma camera system resolution and the best criterion seems to be subjective evaluation of clinical studies. When count density in the acquired data is low, statistical uncertainties will be relatively high, and image noise can be reduced by suppressing more of the higher frequencies. Where the acquired count density is higher, less smoothing is needed. Default values are usually established on the basis of research studies but it may be considered better to alter these in certain circumstances. However, such adjustment can easily result in image artefacts and therefore should be resisted unless experimental evaluation, perhaps using phantom data, has been undertaken (DePuey and Garcia 2001). A restriction in filter types may help to standardise image processing approaches (van Laere et al. 2001b).

Butterworth

The flexibility and ease of design of Butterworth filters have made them the filters of choice in most nuclear medicine procedures (Gilland et al. 1988). It has a low frequency plateau which passes low frequencies unaltered but progressively attenuates higher frequencies. As well as a cut-off, it has an order which controls the slope of the transition. This defines how quickly the plateau changes to cut-off, the slope increasing as the order increases. Although a higher order produces a higher spatial resolution image, its effect is much less severe than the cut-off (Germano 2001; Hansen 2002).

Hamming (Hann)

This filter progressively attenuates frequencies and does not exhibit the low frequency plateau of the Butterworth filter. The parameter that determines its exact behaviour is the cut-off, or critical, frequency. Lower values mean more high frequencies are attenuated and the image is smoother. Higher values mean more high frequencies are passed and the image is sharper but noisier. The Hann filter differs from the Hamming only in the value of its constants.

When to Filter

Equivalent results are obtained by applying the ramp and smoothing filters before or after back-projection. However, a number of effects are possible. If each line of pixels in each projection image is filtered by a one dimensional filter before back-projection, the effect on the reconstructed transaxial slice is the same as applying a two-dimensional filter to the reconstructed slice after back-projection. It does not matter whether the ramp or the smoothing filter is applied first, or whether they are applied simultaneously. This is as far as the ramp can go because this filter only concerns the individual lines of pixels in the projection images or the individual reconstructed transaxial slices. The back-projection procedure is not implemented across transaxial slices and the ramp filter is not applied other than in the plane of the reconstruction. However, the smoothing can be extended across transaxial planes to create a better reconstruction. This can be achieved by applying a post-reconstruction axial filter or by applying a three dimensional filter to the reconstructed volume after back-projection using only the ramp filter. An equivalent effect on the reconstruction volume is achieved by applying a two-dimensional smoothing filter to the whole projection image before back-projection. The usual sequence, therefore, is to undertake the latter and then apply the ramp filter during back-projection.

Artefacts Due to Filtering

The combined effect of the ramp and smoothing filter can produce artefacts immediately adjacent to discrete areas of high activity in the reconstructed image. These artefacts appear as severely reduced activity which can obscure underlying structures in their proximity (Forstrom et al. 1996). Some of the areas which may be affected are: the heart because of activity in the liver, the spine because of activity in the kidney and the pelvis because of activity in the bladder. The severity of the effect depends on the behaviour of the smoothing filter and this provides a method of determining whether an area of reduced activity is artefactual or real, by raising the cut-off frequency which should minimise an artefact but have no effect on a real activity distribution except to increase the noise level.

Iterative Reconstruction

The alternative to filtered back-projection, only routinely feasible recently, is iterative reconstruction. The advantages of the technique are that noise is less pronounced and streak artefacts are reduced, particularly on low count density reconstructions (Zaidi and Hasegawa 2003). Also, corrections for non-uniform attenuation, scatter and resolution can be incorporated into the reconstruction. A number of iterations are required for the reconstructed image to converge to an acceptable presentation; with few iterations, the reconstructed image is blurred and little detail is apparent in it. With more iterations, the image becomes sharper but the noise level increases. This is because noisy reconstructed images may give projections that are very close to the measured noisy projections (Bruyant 2002).

Two methods of using this technique are common. The stopping criterion used clinically is to stop after a fixed number of iterations and not run into convergence (Zaidi and Hasegawa 2003). At this point, the image should look reasonably sharp but the noise should still be at a reasonable level. The alternative is to allow a larger number of iterations and then apply a smoothing filter to mitigate the effects of the increased noise level. If a filter is used, it should only be post-reconstruction (Port 1999).

Slice Thickness

The minimum thickness of each transaxial slice is the size of a pixel on the projection image. However, spatial resolution in the direction along the patient, i.e. between slices, depends on separation from the collimator face if a parallel hole collimator is used. If the projection views have been acquired using a very small pixel size, the transaxial slice thickness may be inappropriately narrow. There is no advantage to having a slice thickness significantly less than the system resolution of the camera, since the slices will be reconstructed from fewer acquired counts and will therefore have unnecessarily high statistical uncertainty. The transaxial slice thickness can be increased by combining two or more rows of pixels in the projection image before reconstruction. However, the transaxial image contrast will deteriorate with thicker transaxial slices due to the partial volume effect.

Partial Volume Effect

This occurs when a pixel in the projection image is not completely occupied by an activity distribution and causes an artefactual reduction in the counts contained in the pixel. The effect becomes particularly apparent with small activity distributions and with poor spatial resolution. The activity concentration is progressively over or under estimated, depending on the activity distribution. Structures <2 times the reconstructed spatial resolution of the system can appear to contain less activity than larger structures, thus suffering a loss of contrast and being more difficult to detect (Mullani 2000); the effect being more severe for cold rather than hot activity distributions (Kojima et al. 1989). The same effect occurs if the activity distribution is moving relative to the projection image and occupies a pixel only for part of the acquisition duration. For instance, part of the myocardial wall of the heart can appear to have an area of reduced perfusion.

Corrections Applied to Reconstructions

The reconstructed image must be corrected for effects such as attenuation and scatter. This may be performed before, during or after the reconstruction.

Attenuation

Two effects of attenuation can be observed on the reconstructed image, reduction in counts which increases with depth and spatial elongation of high activity areas in the direction of lower attenuation (Gillen et al. 1991; Zaidi and Hasegawa 2003). The former can be severe and, although the decrease in counts can be compensated, will lead to increased statistical noise, which cannot be recovered by attenuation correction (Turkington 2000). The spatial distortion is consistent with the operation of filtered back projection (Hansen and Kramer 2000), and its magnitude can be appreciated by considering the difference in relative dimensions laterally and antero-posteriorly. This may be 10 cm or more, and means that, for a source at the centre of the body, the count rate antero-posteriorly may be as much as 2.5 times more than laterally. This spatial distortion can adversely affect the detection of small objects adjacent to areas of increased activity (Turkington 2000), but is significantly less prominent when iterative reconstruction is used rather than filtered back projection (Hansen and Kramer 2000; Lappi et al. 2002). Two techniques are available for attenuation correction: analytical and transmission.

Analytical Methods

These assume that the attenuation is uniform throughout a transaxial slice and, consequently, work well in the brain and abdomen.

Linear Attenuation Coefficient, μ_l

They require a value of the linear attenuation coefficient, μ_l, of the tissue being imaged at the energy of the radiation being detected. The narrow beam value of μ_l is applicable when γ photons are attenuated by both absorption and Compton scattering, see Table III.3.1. In practice, some of the scattered photons are still detected within the photopeak because this has to be reasonably wide to maintain a reasonable detection efficiency, and for $^{99}Tc^m$ there is only a small difference in energy with relatively large angle of Compton scatter. For instance a photon scattered through 45° has an energy of 129 keV and will be accepted by a PHA 20% window centred on 140 keV. In this case, the detected counts will be higher than expected, particularly from shallow depths. Therefore, if scatter correction is not implemented, attenuation is not described by a single exponential but depends on detection geometry and PHA window. The effective values for soft tissue are: 0.124 cm^{-1} for $^{99}Tc^m$, 0.184 for ^{201}Tl and 0.110 for ^{111}In. For both ^{201}Tl and ^{111}In, μ_l decreases with increasing depth; in the former because of the increasing fraction of scattered radiation in the 40% PHA window used with this radionuclide, and in the latter because a change in mean photon energy with depth caused by the use of two PHA windows with this radionuclide (Starck and Carlsson 1997). However, if scatter correction is implemented, the original narrow beam value should be used.

Body Contour

These analytical methods require definition of the body contour which must be the edge of the body and not the edge of the organ, for instance it is the edge of the head not the edge of the brain. Several techniques have been used for determining body contours. One approach is to assume that the cross-section of the patient is elliptical. This ellipse can be defined by identifying the major axes of the ellipse using radioactive point sources placed at the lateral edge of the body and the same outline is used for all the transverse sections. Another approach is to use the radioactivity distribu-

Table III.3.1. Narrow beam linear attenuation coefficients (μ_l) cm^{-1}, used when γ photons are attenuated by both absorption and Compton scattering (King et al. 1995; Galt et al. 1999)

	Lungs	Soft tissue	Bone
^{201}Tl	0.063	0.191	0.429
$^{99}Tc^m$	0.051	0.153	0.286

tion throughout the transaxial slice and a threshold pixel count to define the edge of the body, which can be achieved using an image reconstructed with scattered radiation detected in a lower energy PHA acquisition window. A transmission study may be undertaken using a separate lower PHA energy window to detect radiation from a lower energy radioactive source placed on the side of the body, e.g. fixed to the detector gantry. Analysis of the transmitted radiation would indicate the body outline at all projection angles.

Transmission Methods

The analytical methods do not perform well when the attenuating medium is non-uniform. This is particularly extreme in the thorax because of the air in the lungs and so affects myocardial perfusion studies. Transmission methods overcome this limitation. In these methods, an attenuation map of the object is created. This is undertaken by taking multiple transmission scans from multiple projections in the same imaging session as the emission measurements (Almeida et al. 1998), ensuring that the patient is in the identical position for both. A map of the attenuation coefficients is created by reconstructing the transmission projections using filtered back-projection, after they have been corrected for crosstalk contamination from the emission photons. It is not possible to use this attenuation map when reconstruction the emission image with filtered back-projection but it can be utilised directly in iterative reconstruction (Galt et al. 1999); consequently most attenuation correction systems use an iterative algorithm (Port 1999). The transmission method, in theory, produces better results than the analytical method but requires additional set-up and scan time, radiation exposure to the patient, and the implementation of these methods has become contentious (Bailey and Meikle 2000).

The most common form of transmission attenuation system uses external radioactive sources, mostly ^{153}Gd, for which a 20% PHA window is centred on 100 keV to cover the 97 and 103 keV emitted photons (El Fakhri et al. 2000a). The first acquisition, in the sequence, is a blank transmission scan; undertaken without the patient in the system but using the same collimator that will be used in the patient investigation. As well as providing the baseline for calculation of attenuation correction factors, this also provides some quality control relating to the proper operation of the transmission sources. The projection images should not be uniform but should also not display any structural non-uniformities. The patient is then positioned, centrally in the field of view to avoid loss of transmission data by truncation; which is especially important when fan beam collimators are used.

The transmission and emission images can be acquired in three sequences: simultaneous, sequential, and interleaved sequential. Patient motion between the different phases is potentially a significant problem and even a small misalignment, of as little as 2 cm, can produce serious errors in corrected emission images (Matsunari et al. 1998) which, because they depend on the type of misregistration and may appear at different points in the reconstructed emission image, are difficult to identify as artefacts (Stone et al. 1998).

In the simultaneous mode, there can be no patient movement between the data sets and the total acquisition duration is just the emission duration. In the sequential mode, the acquisition duration is extended to the sum of the separate acquisitions and the potential for patient movement, between data sets, is increased. However, transmission contamination of the emission images is avoided because the transmission source is shielded during the emission acquisition and emission crosstalk into the transmission window can be measured during the emission acquisition. The interleaved sequential mode can only be undertaken with scanning line sources; emission and emission crosstalk into the transmission window being acquired from most of the field of view, while transmission data is acquired from a line moving according to the source position. The acquisition duration is similar to the sequential mode, but the potential for patient movement is less because there is no temporal separation between the emission and transmission acquisitions. In all three acquisition modes, since the transmission count rate usually exceeds the emission count rate, the total acquisition duration is determined mainly on the required emission counts (DePuey and Garcia 2001).

The count density of the transmission images affects the quality of the corrected reconstructed emission images and it is only at the higher count densities that the noise in the emission data remains the dominant factor. The spatial resolution of the transmission images also affects the quality of the corrected reconstructed emission images. The highest quality corrected emission images are obtained when the spatial resolution of the transmission images is similar to, or better than, that of the emission images; artefacts being created when it is almost perfect, or much worse than the emission images (Tsui et al. 1998).

Clinical Quality Control

As the transmission count rate decreases with decay of the source, increasingly noisy attenuation maps, and therefore corrected reconstructed emission images, are obtained. This is particularly evident with radionuclides like ^{153}Gd, which have a half-life of only 281 days, and exacerbated in large patients with greater attenuation. The clinical utility of a transmission source for a particular patient can be assessed before the SPET study, using the ratio of transmission count rate to emission crosstalk count rate in the transmission window, TCR, measured on a planar acquisition. As the transmission source becomes less effective, the emission crosstalk component will become larger and the TCR will decrease. It is suggested is that: if $0.8<\text{TCR}<4$, the transmission acquisition duration should be increased, and if $\text{TCR}<0.8$, the transmission acquisition should not be undertaken. Before being used to correct emission data, the reconstructed attenuation maps should be reviewed visually, and should not be used if any of three artefacts are evident: increased intensity at the body edge, indicating truncation of transmission data; decreased intensity in regions of known high attenuation, suggesting a high emission crosstalk into the transmission window; a generally noisy image, suggesting too few transmission counts (DePuey and Garcia 2001).

CT Systems

In the recently implemented CT system, a sequential transmission emission acquisition is undertaken with the transmission data being acquired first, over the same axial distance as the camera field of view, and the patient is then translated for emission acquisition (Patton 2000).

Segmentation

In order to reduce statistical errors in the attenuation maps, a process of segmentation has been introduced, which relies on there being only a limited number of different attenuating tissues types in the body, e.g. lung, soft tissue, bone (Galt et al. 1992; Pan et al. 1997). Segmentation, of the attenuation map, is implemented by associating a particular tissue type to each pixel, which restricts statistical noise by allowing only known values of μ_l to be included in the map (Burger and Berthold 2000). If this is not performed accurately, however additional errors will be passed through to the reconstruction (Delbeke and Sandler 2000).

Scatter

In a typical PHA window, Compton scattered photons comprise ~25–35% compared to unscattered photons for $^{99}Tc^m$ and ~95% for ^{201}Tl, depending on the amount of tissue through which the photons travel. This results in loss of contrast in the projection and reconstructed images, and artefacts in the reconstructed images. Its effects become more apparent when attenuation is corrected and it is the most complicated factor affecting SPET images because the scatter response function broadens with depth in the patient and is influenced by both the activity and the tissue density distributions (Tsui et al. 1998). Although, for most clinical studies, contrast enhancement can be achieved by display manipulation and the value of scatter correction is not proven, it can reduce the error in measurement of lesion size, improve image quality and contrast (O'Connor et al. 1995; El Fakhri et al. 2000b) and is essential if absolute quantification is required. Scatter correction improves image contrast and is essential if absolute quantification is required. However, for most clinical cases contrast enhancement can be achieved by display thresholding and the value of scatter correction is limited. If scatter correction is implemented and then analytical attenuation correction is undertaken, the value of the attenuation coefficient used in the correction must be changed from the lower effective value to the higher narrow beam value. This is because attenuation will no longer be reduced by Compton scattered photons being detected in the photopeak. Correcting transmission data for scatter has no effect on attenuation corrected reconstructed emission images, however (El Fakhri et al. 2000a).

Resolution Recovery

The usual deterioration in reconstructed spatial resolution with distance experienced with the use of parallel hole collimators can be reduced using resolution recovery or restoration filters. Complete resolution recovery is difficult to achieve, however, because it is normally accompanied by an increase in image noise. Correction for the depth-dependent collimator response improves spatial resolution and contrast, but reduces the signal to noise ratio (El Fakhri et al. 2000b). It also tends to slow iterative reconstruction, requiring many more iterations than normal.

Display

Because of the extended field of view of the camera, the projection data is acquired not for an individual transaxial slice but rather over a number of slices. The reconstructed image volume may thus be reoriented into sagittal, coronal (see Table III.3.2) or oblique cross sections to facilitate better anatomical localisation. Localisation remains difficult without morphological information, and implementation of combined SPET/CT, originally for attenuation correction, provides the potential for great diagnostic benefit, see 'Image fusion with CT' in Chapter III.4. Multiple slices in each of the above orientations can be a confusing way to appreciate the relative positions of activity distributions, and 3D visualisation techniques, of surface and volume rendering, can be used to display all the data in one image. Surface rendering generates a solid representation in which the surface is shaded to simulate light reflected toward the viewer or is delineated by a wire mesh. The interior voxels are usually hidden. However, the surface can be made transparent to allow internal objects to be seen. A surface is determined from each tomographic slice using count thresholding, edge detection etc. Volume rendering generates a semitransparent image, using a ray casting technique, in which all voxels contribute to the 3D rather than relying on a surface representation. The voxel intensities along each ray are projected to determine the final pixel intensity in the 3D image, which avoids the intermediate step of deriving a surface. The transparency allows objects to be viewed through other objects (Milo 1996). Using these techniques, the whole dataset can be viewed from different angles by cine display. The volumes of lesions can be calculated, from such stacked data, by summing the pixels determined as being in the lesion by thresholding, 2D edge detection or manual delineation of edges (Hillel and Hastings 1993). Because of the partial volume effect, these methods fail for small source volumes, which will therefore be overestimated (IPSM 1985).

Clinical Studies

Abdomen

If low energy radionuclides are used, a high resolution collimator should be installed, at the expense of detection sensitivity. Pixel sizes should be defined for the area of interest and are usually based on a 128×128 matrix for $^{99}Tc^{m}$ and a 64×64 matrix for other radionuclides. Because the patient position during acquisition is relaxed and stable, acquisition durations of up to 40 minutes can be achieved, which provides reasonable count density. However, in the lower abdomen and pelvis a short imaging time

Table III.3.2. Data orientation in transaxial, sagittal and coronal presentations. Where: A anterior, P posterior, R right, L left, S superior, I inferior

Transaxial			Sagittal			Coronal		
	A			S			S	
R		L	P		A	R		L
	P			I			I	

is used to avoid the bladder filling with activity. The detectors in dual head systems are oriented at 180°, and a full 360° acquisition arc is almost always used, often with a non-circular rotation since otherwise, a large radius of rotation would be unavoidable because of the body dimensions. Much less frequently, when the area of interest is towards a particular body surface, a 180° arc is used. The number of projections should be defined for the area of interest and usually involve 3° increments for $^{99}Tc^{m}$ and 6° increments for other radionuclides. One pixel thick slices are usually reconstructed. Analytical attenuation correction using the Chang method works well because attenuation is reasonably uniform and the edge of the body can be easily identified. It is not usually necessary to re-orientate the reconstructed data.

Bone

Bone SPET is the second most frequently performed tomographic study in routine nuclear medicine (Sarikaya et al. 2001). Because the patient position during acquisition is relaxed and stable, acquisition durations of up to 40 min can be achieved (Kelty et al. 1997), which provides reasonable count density. However, in the lower abdomen and pelvis a short imaging time is used to avoid the bladder filling with activity. The low energy collimator type used is ultra-high or high resolution at the expense of detection sensitivity (O'Connor et al. 1991; Sarikaya et al. 2001). A 128 × 128 matrix is used; except in the skull, for which a 64 × 64 matrix with zoom of 1.5–2 is preferred. The detectors in dual head systems are oriented at 180° and a rotation arc of 360° is usually used. The number of projections should be defined on the area of interest, ~70 and ~110 being required for complete angular sampling of the head and pelvis. In practice, 60–70 views are adequate for most clinical studies but increasing to 120–130 (Kelty et al. 1997) may produce better definition of structures such as the ribs. Bone images contain high frequency information and filters such as the Butterworth, are better at preserving this than a simple Hann filter (O'Connor et al. 1991), although iterative reconstruction provides better image quality (Blocklet et al. 1999; Sarikaya et al. 2001). When there is high activity in the bladder, digital filtering of the projection data can significantly improve the quality of the reconstruction in the hip area (O'Connor and Kelly 1990). Oblique angle presentation of the pelvis and hips improves lesion localisation (Gates 1996), which is often aided by 3D image analysis (Sarikaya et al. 2001).

Spine

A 180° rotation arc, with ~40 projections, over the posterior surface of the body is used when imaging the lumbar spine. In this investigation, the patient should be positioned in the prone position to reduce the curvature of the spine, minimise collimator to patient separation and eliminate the high attenuation of photons passing through the full width of the imaging table (O'Connor et al. 1991), see Fig. III.3.5. Analytical atten-

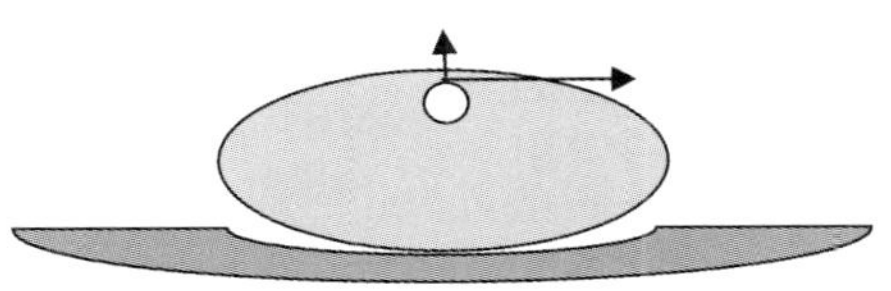

Fig. III.3.5. Minimising the effect of imaging table attenuation by prone positioning

uation correction using the Chang method works well because attenuation is reasonably uniform and the edge of the body can be easily identified. One pixel thick slices are reconstructed. Re-orientation of the reconstructed images into oblique axes is undertaken so that individual transaxial slices do not cut through different vertebrae (Gates 1996). A 360° rotation is used in the thoracic spine because, here, the vertebral bodies are a significant distance from the posterior surface of the body and will be poorly reconstructed if a restricted rotation is used (O'Connor et al. 1991).

Breast

For breast imaging using $^{99}Tc^{m}$, a high resolution collimator is usually used, with a 360° rotation of 3° increments and 30–35 s per increment. SPET with iterative reconstruction provides a significant improvement in the detection of small-sized breast tumours compared with planar images. A 128 × 128 matrix is preferable to a 64 × 64 matrix (Garin et al. 2001).

Cerebral Perfusion

A recent study has shown that there is a need for external quality assurance of clinical brain perfusion SPET investigations (Heikkinen et al. 1998). The radionuclides used are ^{201}Tl, $^{99}Tc^{m}$ and ^{123}I. Because of the small size of important cerebral structures, the low energy parallel hole collimators used for $^{99}Tc^{m}$ and ^{123}I are high or ultra high resolution (Serrano et al. 2002) at the expense of detection sensitivity. A general purpose collimator is not suitable. A fan beam is preferable to parallel hole because it provides better spatial resolution with depth and higher detection sensitivity (Juni et al. 1998; Vines and Ichise 1999; Catafau 2001; Tatsch et al. 2002), important because of the many small structures where subtle activity variations may be present. For $^{99}Tc^{m}$ the use of a narrower, asymmetric PHA window, of 17% centred on 142 keV, improved the contrast but at the expense of an increase in statistical noise due to fewer counts (Gustafsson et al. 2000). For ^{201}Tl, the high resolution or general purpose collimator is used. It is possible to achieve a small radius of rotation by, for instance, imaging at the edge of the camera field of view to avoid the shoulders of the patient. Consequently, reconstructed spatial resolution can be as good as 10 mm at the edge of the brain. A pixel size of <5 mm is therefore demanded, which can be achieved with an acquisition matrix of 128 × 128 or 64 × 64 with zoom (Serrano et al. 2002). Because the patient position during acquisition is relaxed and stable, particularly if a good head restraint is used, acquisition durations of up to 40 min can be achieved which provides a good count density. However, the risk of patient motion during such long acquisitions must be considered. At least 5×10^{6} counts should be acquired (Tatsch et al. 2002). However, relatively poor spatial resolution is usually obtained, in adults, when a single head system is used. This is even more pronounced in children because of the small volume of the head and, in the neonate, partial volume effects can have obvious consequences on the images (Piepsz 1995). It may be useful to temporally segment the acquisition to avoid complete loss of data due to patient motion (Juni et al. 1998).

A full 360° acquisition arc is always used, usually with a circular rotation. The detectors in dual head systems are oriented at 180° and although the radius of rotation is likely to be 15 cm, the radius of the brain is more likely to be 10 cm. With a reconstructed spatial resolution of 10 mm, the required linear separation of the acquisition projections at the periphery of the brain will be 5 mm. The number of projections must therefore be greater than $2\pi \times 100/5=126$, which indicates that 128 should be used, or 3° increments. Continuous data acquisition rather than step and shoot will probably save ~10% of the acquisition duration, and the quality of the acquisition will not be compromised because of the fine angular separation.

One pixel thick slices are usually reconstructed by filtered back-projection using a Butterworth filter with an order of 5–10 and a cut-off frequency of 0.2–0.9 cm^{-1} or the Shepp-Logan filter (Serrano et al. 2002). The Butterworth cut-off frequency, in cm^{-1}, for an order 7 can be optimised to the average number of counts per pixel: 0.4 for 50–250, 0.45 for 250–500, 0.5 for >500 (Stelter et al. 2001). Resolution recovery or spatially varying filters should be used with caution, as they may produce artefacts. The image quality can be improved using iterative reconstruction, however (Kauppinen et al. 2000). Attenuation correction should always be undertaken (Juni et al. 1998). Analytical correction using the Chang method, with $\mu_l = 0.09$–$0.12\ cm^{-1}$, is usually used because attenuation is reasonably uniform and the edge of the brain can be easily identified (van Laere et al. 2001a), there is little loss of accuracy in making this assumption (Iida et al. 1998; Tatsch et al. 2002). There is no significant improvement when transmission attenuation correction is used (van Laere et al. 2001a). Scatter is normally neglected but correction, e.g. using two energy windows (Hashimoto et al. 1997), does improve contrast (Gustafsson et al. 2000) and is required for quantitative analyses (Iida et al. 1998; Stodilka et al. 1998).

The reconstructed data is always re-oriented so that the final images are presented in a similar plane regardless of the patient orientation during acquisition. The initial translation is to correct for any rotation of the head across the body, this is done by creating slices parallel to the sagittal axis. Following this, oblique slices are created parallel to either the fronto-occipital or fronto-cerebellar line or parallel to the longitudinal edge of the temporal lobe. The latter orientation is designed to better differentiate between the medial and lateral aspects of the temporal lobes (Catafau 2001). Coronal and sagittal displays should be used (Juni et al. 1998) with a consistent display normalisation, e.g. to maximum pixel count. Discontinuous colour maps should be used with caution because they can accentuate asymmetries between the hemispheres (Catafau 2001).

Cardiac

This is the most frequently performed SPET investigation in routine nuclear medicine (Sarikaya et al. 2001) and comprehensive guidelines have been published by DePuey and Garcia (2001) and Port (1999).

Myocardial Perfusion

Two radionuclides are used for myocardial perfusion imaging, SPET MPI, $^{99}Tc^{m}$ and ^{201}Tl. A higher count density can be achieved with $^{99}Tc^{m}$ because higher activities can be administered and the emission energy is more suitable for detection by current cameras. Two types of study are undertaken: non-gated and gated. Inadequate acquisition has been documented in 6% of studies and inadequate processing in 10% of studies (Prvulovich et al. 1998).

Acquisition

Non-Gated and Gated

Usually, the patient is imaged twice; once at rest and once after exercise. With $^{99}Tc^{m}$, if both studies are performed on the same day, the first study has to be undertaken using a smaller activity administration so that higher activity in the subsequent study can 'swamp' the residual activity from the first administration.

In these studies, extracardiac activity can have complex effects on reconstruction, which depends on the amount of extracardiac activity and pattern of attenuation (Heller et al. 1997). There is no reliable correction for such artefacts, although they are less prominent with iterative rather than filtered back projection reconstruction. They can often be eliminated by repeating the acquisition after the activity level in the adjacent structure has decreased. Imaging should not be started too soon after exercise because, during recovery from strenuous exercise, the heart creeps upwards from its lower position during exercise to its normal position. Acquisition during this 'creep' causes blurring of the image and artefactual perfusion defects in the myocardium (Britten et al. 1998).

The reconstructed image quality depends on the count density in the projection images. Because artefactual perfusion defects may be created as a result of poor counting statistics, the peak myocardial pixel count, in an anterior planar projection, should >100 for a ^{201}Tl study and >200 for a $^{99}Tc^{m}$ study. Imaging tables, which are not classified as low attenuation, can slightly decrease the apparent myocardial count distribution in the septum with high attenuation tables and prone positioning also causing apparent reduction in the antero-lateral wall (O'Connor and Bothun 1995).

Collimator

A high spatial resolution collimator should be used with $^{99}Tc^{m}$ (Prvulovich et al. 1998; Strauss et al. 1998), and with a dual head system, in which the detectors are set at 90°, an ultra-high resolution collimator with long bore holes is sometimes used to reduce the collimator–cardiac spatial separation as much as possible (Rajabi et al. 2002). However, improved reconstruction may be possible using a general purpose collimator in combination with distance-dependent spatial resolution recovery (Lau et al. 2001). The lower resolution general purpose collimator should be used with ^{201}Tl, because of the reduced acquisition count rate.

PHA Energy Window

A 20% symmetric PHA window is usually used for $^{99}Tc^{m}$, but this may be reduced on cameras with better energy resolution, resulting in decreased scatter and improved image resolution but, possibly, longer acquisition durations. The window settings for ^{201}Tl should be 30–35%.

Pixel Size

Considering the reduced spatial resolution at depth, the low activity administration in the one day protocol and the partial volume effect caused by the beating heart, a pixel size 6.4±0.2 mm is recommended for both $^{99}Tc^{m}$ and ^{201}Tl. This is achieved using a matrix of 64 × 64 (Rajabi et al. 2002) with zoom, of 1.2–1.5, appropriate to the field of view size. Such magnification does make ensuring that the whole heart remains in the field of view at every projection difficult, however, unless cardiac tracking is available, in which the reduced field of view is moved around the whole field of view to maintain the heart near its centre.

Acquisition Duration
The patient position during acquisition is not relaxed and acquisition durations have to be limited. For the low administration of the one day protocol, the duration has to be as long as possible, usually 30 min. The high administrations of the one or two day protocols can manage with a shorter duration, such as 20–25 min (Strauss et al. 1998), although 10–13 min can be used when using $^{99}Tc^{m}$ with a dual detector system. Multi-rotation acquisitions of five 4.5 min rotations have been used, with motion correction, to reduce the problems associated with upward creep of the heart following exercise (Britten et al. 1998).

Orbit
For single and dual head systems, a 180° rotation arc (Bural et al. 2002) over the left anterior sector is preferred instead of a complete 360° rotation. The latter usually results in better uniformity of normal myocardium due to more consistent spatial resolution and may provide more accurate quantitative information; and a 180° orbit may have significant erroneous inhomogeneity and overestimate defect size, especially when the heart is not at the centre of the orbit, which is common (Liu et al. 2002). However, avoiding acquisition data from the right and posterior sectors reduces noise caused by attenuation and degraded spatial resolution due to the increased separation from the detector, and is particularly true for the lower energy ^{201}Tl, defect contrast being improved (O'Connor et al. 2000) by up to 10% (Nakajima et al. 1998). A circular orbit is the most popular and, although the radius of rotation of the camera might be as high as 25 cm, the circle enclosing the area of interest is more restricted, particularly if a 180° rotation is used instead of 360°. Elliptical orbits can be used to reduce degradation in spatial resolution by minimising the imaging separation. However, image artefacts are significantly more severe than those from circular acquisitions and, because of the significant difference in images reconstructed from circular and elliptic acquisitions, standardised normal files acquired from circular acquisitions should not be used for comparison with patient data acquired from elliptic acquisitions (Abufadel et al. 2001).

Projections
Both continuous step and shoot and continuous modes are preferred to step and shoot because of the improvement in counting statistics. The number of projections required when using $^{99}Tc^{m}$ with a LEHR collimator should be at least 64 over an acquisition arc of 180°, in order to avoid any degradation of spatial resolution, but 32 projections are sufficient in the relatively low spatial resolution study comprising ^{201}Tl and a LEGP collimator, again over an acquisition arc of 180° (Strauss et al. 1998).

Attenuation Correction
Artefacts resulting from variable photon attenuation in the thorax are the most significant factors limiting interpretation (King et al. 1995; Galt et al. 1999; El Fakhri et al.

1999). Analytical attenuation correction does not work well because attenuation is non-uniform throughout the thorax. Transmission attenuation is possible, however, if the detectors on a dual head system are oriented at 90°, and restores activity distribution uniformity in the myocardium (Heller et al. 1997; El Fakhri et al. 2000b), resulting in images with no differences in uniformity between men and women (Araujo et al. 2000), between 180° and 360° rotations, and between 90° dual-detector and 120° triple-detector systems (LaCroix et al. 1998).

Care has to be exercised when implementing transmission attenuation in different activity studies, e.g. in the one day protocol. Down-scatter results in underestimation of attenuation in simultaneous emission and transmission and, if not corrected, differences between high and low activity studies may result in interpretation errors (Almquist et al. 1999). In scanning line source systems used with ^{201}Tl studies, the down-scatter from the 100 keV ^{153}Gd photons to the emission window is sufficiently low that it can be ignored but the electronic transmission window can cause a degradation in image quality because of the consequential slight decrease in emission counting rate (Bailey and Meikle 2000).

Although the general concepts of transmission attenuation are common among systems and have been shown to work, there are many different implementations. These must be tested using standard phantom and investigation protocols before they are used in routine investigations (Wackers 1999; Bailey and Meikle 2000; Darcourt and Buvant 2000; Hendel et al. 2002; Wackers 2002).

Gated Only

Attenuation correction will not remove the effect of diaphragmatic respiratory motion, which may cause artefactual reduction in relative count density in both the anterior wall and the inferior wall relative to the lateral wall, and which suggests a possible need for respiratory gating (Pitman et al. 2002).

Images of myocardial perfusion at different parts of the cardiac cycle can be reconstructed by gating the acquisition at each projection using the R wave of the ECG, which must be reliable and consistent. It is recommended that both stress and rest SPET perfusion studies be undertaken as gated acquisitions, as long as the count density is adequate, which is of particular concern in the low activity administrations; left ventricular ejection fractions calculated using ^{201}Tl being inadequate for clinical use (Wright et al. 2000). Typically, such gated studies are acquired over the same duration as ungated studies, and a dual head system has a significant advantage because the increased sensitivity compensates in part for the short acquisition time per cardiac cycle section. With this sensitivity advantage, gating of the study into 8 or 16 equal time segments over the cardiac cycle is feasible with dual head systems, but usually 8 images are acquired at each projection over equal time periods of the cardiac cycle so that each has ~1/8 of the counts of the complete study (Rubio et al. 2002; Hyun et al. 2002; Schaefer et al. 2002; Dede and Narin 2002; Petersen et al. 2002; Muylle et al. 2002; Dziuk et al. 2002), because 16 segments result in too low a count density. The same considerations, regarding R–R acceptance window, apply as in the planar blood pool acquisition. It may also be necessary to wait a little longer after exercise before commencing imaging, however, because the function of the heart will be changing.

A combination of gating and attenuation correction has been shown to provide the highest diagnostic accuracy (Links et al. 2002). The individual time frames are usually reconstructed independently, but the noise in the final images can be dramatically reduced by temporal filtering among them (Vanhove et al. 2002).

Collimator

In gated studies, although many institutions prefer using a high resolution collimator, the recommended collimator for $^{99}Tc^m$ is the general purpose, in order to maximise detection sensitivity.

Acquisition Duration

These vary widely, from as little as 10 min using a triple head system (Schaefer et al. 2002) to 20–30 min (Rubio et al. 2002; Hyun et al. 2002; Cantinho et al. 2002) and over 40 min (Pena et al. 2002; Harada et al. 2002). ^{201}Tl acquisition durations can be established on the basis of the count rate in a preliminary 4 min planar study, to ensure at least 5×10^5 background subtracted myocardial counts.

Processing

Non-Gated and Gated

Compensation for detector resolution and scatter improves detection sensitivity compared to only compensation for attenuation (Sankaran et al. 2002) but is implicated in artefact generation (Harel et al. 2001).

Patient Motion

The most important post-acquisition quality control procedure is the display of the raw projection data in cine mode and the sinogram, in order to detect patient and heart motion which is believed to affect 10–20% of studies (Germano 2001). Artefacts in the reconstructed images that may be caused by uncorrected motion are: reduced activity in opposite sections of the myocardium, usually the anterior and inferior walls, streaks from the myocardium or a focus of increased activity in a part of the myocardium (DePuey 1994). Motion of less than 1 pixel should not produce clinically significant artefacts. Significant artefacts are possible from at least 2 (Germano 2001) and more likely 3 pixel motion (Forstrom et al. 1996), from at least 6.5 mm and more likely 13 mm motion with quantitative abnormalities from axial movement being more frequent than from lateral movement and when the movement occurred at the beginning or the end of the acquisition (Cooper et al. 1992). With double head acquisition, the largest defects were created when the motion occurred during the middle of the acquisition. Motion correction significantly improved lesion detectability (Matsumoto et al. 2001).

When motion exceeds 1 pixel, if available, a validated motion correction program may be applied before reconstruction. In general, however, such programs have not been extensively validated. Manual correction by an experienced technologist was as accurate as an automatic 2D fit method, but was more accurate than cross-correlation and diverging squares automatic methods (O'Connor et al. 1998). Useful correction must correct axial motion to within 6.5 mm and lateral movement to within 13 mm (Forstrom et al. 1996). Otherwise, the acquisition should be repeated if $^{99}Tc^m$ is used; repeat imaging with ^{201}Tl being complicated because of redistribution of the radiopharmaceutical.

Filters

Both filtered back-projection and iterative reconstruction give similar results, although iterative reconstruction can reduce the count-loss artefact in inferior and posterior walls (Bai et al. 2001). In filtered back-projection many filters are in use. With $^{99}Tc^{m}$ these include: Hamming with cut-off frequency 0.43–0.5 or 0.46–0.5 cm^{-1}; Butterworth of order 3–9 and cut-off frequency 0.3–0.5 cm^{-1} or order 5–9 and cut-off frequency 0.35–0.5 cm^{-1}; Metz with FWHM 3–5 mm and order 3–10 or FWHM 3.5–4.5 mm and order 8–9.5; Wiener with 2.5–5 mm and order 3.5–4 (Bural et al. 2002; Rajabi et al. 2002). However, filters used for stress studies may differ from those for rest studies, because of the difference in count density, with lower cut-off frequencies being used in the low administration one day acquisition (Ohnishi et al. 1997) and the low count density gated study.

Reorientation

So that the final images are presented in a similar plane, regardless of the heart orientation, it has become common practice that the reconstructed data set be reoriented to coincide with the axes of the heart, parallel slices being created along three oblique axes: vertical long axis, horizontal long axis and short axis planes (Germano 2001). These must be consistent between the stress and rest studies, to enable accurate comparison. The display of these reoriented slices must also be consistent: the vertical long axis starting at the septum and progressing to the lateral wall; the horizontal long axis starting at the inferior wall and progressing to the anterior wall; and the short axis starting at the apex and progressing to the base.

Polar Map

Quantitative analysis can be useful to supplement visual interpretation. Most quantitative analyses require that the tomographic slices be displayed in a polar map format. This is created by concatenating the short axis slices, starting with the apex at the centre and ending with the base at the periphery. The data transferred from each short axis image to the polar map is the maximum count, between the endocardial and epicardial surfaces, along a series of rays from the centre of the left ventricle. The individual short axis data sets are normalised to one another. The polar map is formatted to a standard size, this facilitates comparison with a plot constructed from normal studies. The study plot is normalised to the normal plot using the highest count pixel of the heart under investigation, assuming this to represent the most normal part of the myocardium. Because of differences in body shape, however, separate normal plots are required for men and women. Once compared to the normal plots, a series of polar map images can be created to represent defect extent, severity and also the amount of reversibility between stress and rest studies.

3D Display

Reconstructed data may be presented in 3D format, in either static or cine mode, which can help to appreciate the location and extent of perfusion abnormalities.

Gated Only

Gated SPET can help to determine if an area of apparently reduced perfusion is real or caused by an attenuation artefact. An area which demonstrates reduced activity at both stress and rest but is not accompanied by a wall motion abnormality is suggestive of an artefact. In a gated study, the projection images for each cardiac time interval are used to reconstruct transaxial slices at a particular point in the cardiac cycle. Thus, series of SPET images of the heart are obtained at several points from diastole to systole to diastole. All the non-gated quality assurance procedures are applicable to gated acquisition, the most important of which is to review the summed gated projection data in cine mode, which should reveal gating errors due to arrhythmias, as a variation in image count density among projections. Also, a heart rate histogram should be reviewed to assess R–R consistency.

Filters

For $^{99}Tc^{m}$, processing by filtered back-projection usually uses a Butterworth filter with order 4–5 and cut-off frequency 0.16–0.6 cm^{-1} (Gonzalez et al. 2002; Muylle et al. 2002; Wright et al. 2002). A lower cut-off frequency being used for 16 segments because of increased statistical noise. Smoother reconstruction filters lead to calculation of lower left ventricular volumes and higher ejection fractions but, even if the optimal filter is used, the uncertainties are still too great to be clinically acceptable (Wright et al. 2002).

Interpretation

The behaviour of the ventricular wall motion can be determined from a cine playback of a three-dimensional volume rendered image. The left ventricular volume and ejection fraction can be calculated from ROIs constructed at the inner threshold of the myocardium at different parts of the cardiac cycle. Although the agreement between gated SPET and ERNA appear sufficient for routine evaluation of LVEF, gated SPET significantly over- or underestimated ejection fraction compared with ERNA and the latter should be used when precise measurements are required (Manrique et al. 1999; Ford et al. 2001). The biggest limitations to accuracy are adjacent extracardiac activity and >50% decrease in activity uptake in perfusion defects (Achtert et al. 1998). An 8 frame acquisition, underestimated end diastolic volume, overestimated end systolic volume, and therefore underestimated left ventricular ejection fraction, compared with a 16 frame acquisition (Kubo et al. 2000).

The thickening of the myocardial wall from diastole to systole can be assessed using the apparent increase in counts in the wall during the cardiac cycle rather than the physical appearance of the wall edge which can be misleading (Shen et al. 1999). This is a consequence of the partial volume effect which reduces as the wall thickens, causing the counts to increase roughly in proportion to the increase in thickness. However, the variation in count density with wall thickness is not linear, and will underestimate small volumes.

Blood Pool

Acquisition

The most appropriate detector configuration is a dual head system with the heads oriented at 90° and loaded with a LEHR collimator; 64 projections, 32 per head, being acquired during a 180° rotation over a duration of 16 minutes. If the R–R interval is very irregular the number of projections can be halved whilst maintaining the total acquisition duration. With single head systems, a LEGP collimator can be used and 32 projections acquired, but over 32 minutes duration. The pixel size should ideally be 4–5 mm, using a matrix of 64×64 with zoom appropriate to the field of view size. The R–R acceptance window should be wider than used in planar acquisition, at ±25–35%, because poor counting statistics cannot be tolerated by the reconstruction, creating streak artefacts. 16 frames per R–R cycle are considered optimal, left ventricular ejection fractions decreasing at 8 frames per cycle and counting statistics not allowing more frames per cycle. If only systolic parameters and ejection fraction are required, these 16 frames can be acquired over only the first part of the R–R interval, but accurate assessment of diastolic parameters will not be possible.

Processing

When 16 frames per cardiac cycle are acquired and reconstruction is by filtered back-projection, a Butterworth filter of order 5–7 and cut-off frequency of 0.25–0.35 cm^{-1} can be used, with the cut-off being reduced as noise increases (Muylle et al. 2002). Summed short axis oblique slices are reoriented from the transaxial images, and ventricular function parameters are extracted in a manner similar to planar processing, except that background subtraction is not required. Left ventricular ejection fractions calculated from this data will be 7–10% higher than calculated from planar data because of the complete exclusion of left atrial activity, and will be similar to values calculated from first pass data. Similarly, determination of right ventricular ejection fraction is possible because there is no overlap of activity in other cardiac structures. Regional wall motion can be assessed using a 3D display in cine mode.

Thorax

For detection and localisation of 1 cm spherical lesions in ^{67}Ga thoracic SPET reconstructed using filtered back-projection with Butterworth filter order 5 and Chang analytical attenuation correction, the optimum cut-off frequency was 0.32–0.47 cm^{-1} (Wells et al. 1999). However, detection and localisation can be improved using iterative reconstruction (Wells et al. 2000).

References

Abufadel A, Eisner RL, Schafer RW (2001) Differences due to collimator blurring in cardiac images with use of circular and elliptic camera orbits. J Nucl Cardiol 8:458–465

Achtert AD, King MA, Dahlberg ST, Pretorius PH, LaCroix KJ, Tsui BM (1998) An investigation of the estimation of ejection fractions and cardiac volumes by a quantitative gated SPECT software package in simulated gated SPECT images. J Nucl Cardiol 5:144–152

Almeida P, Bendriem B, de Dreuille O, Peltier A, Perrot C, Brulon V (1998) Dosimetry of transmission measurements in nuclear medicine: a study using anthropomorphic phantoms and thermoluminescent dosimeters. Eur J Nucl Med 25:1435–1441

Almquist H, Arheden H, Arvidsson AH, Pahlm O, Palmer J (1999) Clinical implication of down-scatter in attenuation-corrected myocardial SPECT. J Nucl Cardiol 6:406–411

Araujo LI, Jimenez-Hoyuela JM, McClellan JR, Lin E, Viggiano J, Alavi A (2000) Improved uniformity in tomographic myocardial perfusion imaging with attenuation correction and enhanced acquisition and processing. J Nucl Med 41:1139–1144

Bai J, Hashimoto J, Suzuki T, Nakahara T, Kubo A, Iwanaga S, Mitamura H, Ogawa S (2001) Comparison of image reconstruction algorithms in myocardial perfusion scintigraphy. Ann Nucl Med 15:79–83

Bailey DL, Meikle SR (2000) Does attenuation correction work? J Nucl Med 41:960–961

Bieszk JA, Hawman EG (1987) Evaluation of SPECT angular sampling effects: continuous versus step-and-shoot acquisition. J Nucl Med 28:1308–1314

Blocklet D, Seret A, Popa N, Schoutens A (1999) Maximum-likelihood reconstruction with ordered subsets in bone SPECT. J Nucl Med 40:1978–1984

Britten AJ, Jamali F, Gane JN, Joseph AE (1998) Motion detection and correction using multi-rotation 180 degrees single-photon emission tomography for thallium myocardial imaging. Eur J Nucl Med 25:1524–1530

Bruyant PP (2002) Analytic and iterative reconstruction algorithms in SPECT. J Nucl Med 43: 1343–1358

Bural G, Ceri IO, Erkilic M, Tercan E, Saka O (2002) The change of cardiac axis in exercise and rest ^{99m}Tc MIBI scintigraphy: it's effect on image quality during SPECT reconstruction. Eur J Nucl Med 29:S177

Burger C, Berthold T (2000) Physical principles and practical aspects of clinical PET imaging. In: von Schulthness GK, Buck A, Engel-Bicik I, Steinert HC (eds) Clinical positron emission tomography. Correlation with morphological cross-sectional imaging. Lippincott Williams and Wilkins, Philadelphia

Cantinho GC, Pena HP, Monteiro JM, Daniel AD, Chester HC, Godinho FG (2002) Normal functional parameters of left ventricle in perfusion gated SPECT. Eur J Nucl Med 29:S206

Cao Z, Holder LE, Chen CC (1996) Optimal number of views in 360 degrees SPECT imaging. J Nucl Med 37:1740–1744

Catafau AM (2001) Brain SPECT in clinical practice, part I perfusion. J Nucl Med 42:259–271

Cooper JA, Neumann PH, McCandless BK (1992) Effect of patient motion on tomographic myocardial perfusion imaging. J Nucl Med 33:1566–1571

Darcourt J, Buvat I (2000) Does attenuation correction work? J Nucl Med 41:961

Dede F, Narin Y (2002) Comparison of left ventricular function at rest and post-stress Tc-99m MIBI gated SPECT in normal and ischaemic myocardium. Eur J Nucl Med 29:S210

Delbeke D, Sandler MP (2000) The role of hybrid cameras in oncology. Semin Nucl Med XXX: 268–280

DePuey EG (1994) How to detect and avoid myocardial perfusion SPECT artifacts. J Nucl Med 35: 699–702

DePuey EG, Garcia EV (2001) Updated imaging guidelines for nuclear cardiology procedures, part 1. American Society of Nuclear Cardiology. J Nucl Cardiol 8:G1–G58

Dziuk M, Canizales A, Britton KE, Pietrzykowski J, Cholewa M (2002) The comparison of adenosine exercise Tc-99m tetrafosmin with treadmill MIBI gated SPECT. Eur J Nucl Med 29:S211

Eisner RL, Nowak DJ, Pettigrew R, Fajman W (1986) Fundamentals of 180° acquisition and reconstruction in SPECT imaging. J Nucl Med 27:1717–1728

El Fakhri GN, Buvat I, Pelegrini M, Benali H, Almeida P, Bendriem B, Todd-Pokropek A, Di Paola R (1999) Respective roles of scatter, attenuation, depth-dependent collimator response and finite spatial resolution in cardiac single-photon emission tomography quantitation: a Monte Carlo study. Eur J Nucl Med 26:437–446

El Fakhri G, Buvat I, Almeida P, Bendriem B, Todd-Pokropek A, Benali H (2000a) Should scatter be corrected in both transmission and emission data for accurate quantitation in cardiac SPET? Eur J Nucl Med 27:1356–1364

El Fakhri G, Buvat I, Benali H, Todd-Pokropek A, Di Paola R (2000b) Relative impact of scatter, collimator response, attenuation, and finite spatial resolution corrections in cardiac SPECT. J Nucl Med 41:1400–1408

Fahey FH, Harkness BA, Keyes JW Jr, Madsen MT, Battisti C, Zito V (1992) Sensitivity, resolution and image quality with a multi-head SPECT camera. J Nucl Med 33:1859–1863

Fleming JS, Alaamer AS (1996) The influence of collimator characteristics on quantitation in SPECT. J Nucl Med 37:1832–1835

Ford PV, Chatziioannou SN, Moore WH, Dhekne RD (2001) Overestimation of the LVEF by quantitative gated SPECT in simulated left ventricles. J Nucl Med 42:454–459

Forstrom LA, Dunn WL, O'Connor MK, Decklever TD, Hardyman TJ, Howarth DM (1996) Technical pitfalls in image acquisition, processing and display. Semin Nucl Med XXVI:278–294

Galt JR, Cullom SJ, Garcia EV (1992) SPECT quantification: a simplified method of scatter and attenuation correction for cardiac imaging. J Nucl Med 33:2232–2237

Galt JR, Cullom SJ, Garcia EV (1999) Attenuation and scatter compensation in myocardial perfusion SPECT. Semin Nucl Med XXIX:204–220

Garcia EV (1994) Quantitative myocardial perfusion single-photon emission computed tomographic imaging: quo vadis? (Where do we go from here?) J Nucl Cardiol 1:83–93

Garin E, Devillers A, Girault S, Laffont S, Schill O, Bernard AM, Moisan A, Bourguet P (2001) Scintimammography: better detection of small-sized lesions with tomographic than planar images, a phantom study. Nucl Med Commun 22:1045–1054

Gates GF (1996) Oblique angle bone SPECT imaging of the lumbar spine, pelvis, and hips. An anatomic study. Clin Nucl Med 21:359–362

Germano G (2001) Technical aspects of myocardial SPECT imaging. J Nucl Med 42:1499–1507

Gilland DR, Tsui BMW, McCartney WH, Perry JR, Berg J (1988) Determination of the optimum filter function for SPECT imaging. J Nucl Med 29:643–650

Gillen GJ, Gilmore B, Elliott AT (1991) An investigation of the magnitude and causes of count loss artifacts in SPECT imaging. J Nucl Med 32:1771–1776

Gonzalez MJ, Ricart Y, Martin-Comin J, Pallares C, Beltran P, Gomez-Hospital J, Cequier A, Esplugas E, Ramos M (2002) LVEF, motility index and LV volumes. Gated-SPECT versus ventriculography performed within 24 hours. Eur J Nucl Med 29:S209

Gustafsson A, Arlig A, Jacobsson L, Ljungberg M, Wikkelso C (2000) Dual-window scatter correction and energy window setting in cerebral blood flow SPECT: a Monte Carlo study. Phys Med Biol 45:3431–3440

Hansen CL (2002) Digital image processing for clinicians, part II: filtering. J Nucl Cardiol 9:429–437

Hansen CL, Kramer M (2000) Attenuation smear: a 'paradoxical' increase in counts due to attenuation artifact. Int J Card Imaging 16:455–460

Harada M, Shimizu A, Murata M, Kubo M, Mitani R, Dairaku Y, Matsuzaki M (2002) Simultaneous evaluation of systolic and diastolic regional functions with quadruple contours display method using ECG-gated myocardial SPECT. Eur J Nucl Med 29:S211

Harel F, Genin R, Daou D, Lebtahi R, Delahaye N, Helal BO, Le Guludec D, Faraggi M (2001) Clinical impact of combination of scatter, attenuation correction, and depth-dependent resolution recovery for ^{201}Tl studies. J Nucl Med 42:1451–1456

Hashimoto J, Kubo A, Ogawa K, Amano T, Fukuuchi Y, Motomura N, Ichihara T (1997) Scatter and attenuation correction in technetium-99m brain SPECT. J Nucl Med 38:157–162

Heikkinen J, Kuikka JT, Ahonen A, Rautio P (1998) Quality of brain perfusion single-photon emission tomography images: multicentre evaluation using an anatomically accurate three-dimensional phantom. Eur J Nucl Med 25:1415–1422

Heikkinen J, Ahonen A, Kuikka JT, Rautio P (1999) Quality of myocardial perfusion single-photon emission tomography imaging: multicentre evaluation with a cardiac phantom. Eur J Nucl Med 26:1289–1297

Heller EN, DeMan P, Liu YH, Dione DP, Zubal IG, Wackers FJ, Sinusas AJ (1997) Extracardiac activity complicates quantitative cardiac SPECT imaging using a simultaneous transmission-emission approach. J Nucl Med 38:1882–1890

Hendel RC, Corbett JR, Cullom SJ, DePuey EG, Garcia EV, Bateman TM (2002) The value and practice of attenuation correction for myocardial perfusion SPECT imaging: a joint position statement from the American Society of Nuclear Cardiology and the Society of Nuclear Medicine. J Nucl Cardiol 9:135–143 and J Nucl Med 43:273–280

Hillel PG, Hastings DL (1993) A three-dimensional second-derivative surface-detection algorithm for volume determination on SPECT images. Phys Med Biol 38:583–600

Hines H, Kayayan R, Colsher J, Hashimoto D, Schubert R, Fernando J, Simcic V, Vernon P, Sinclair RL (1999a) Recommendations for implementing SPECT instrumentation quality control. Nuclear Medicine Section – National Electrical Manufacturers Association (NEMA). Eur J Nucl Med 26:527–532

Hines H, Kayayan R, Colsher J, Hashimoto D, Schubert R, Fernando J, Simcic V, Vernon P, Sinclair RL (1999b) National Electrical Manufacturers Association recommendations for implementing SPECT instrumentation quality control. J Nucl Med Technol 27:67–72

Hines H, Kayayan R, Colsher J, Hashimoto D, Schubert R, Fernando J, Simcic V, Vernon P, Sinclair RL (2000) National Electrical Manufacturers Association recommendations for implementing SPECT instrumentation quality control. J Nucl Med 41:383–389

Hutton B (1996) Angular sampling necessary for clinical SPECT. J Nucl Med 37:1915–1916

Hyun IY, Kim DH, Seo JK, Kwan J, Park KS, Choe W-S, Lee WH (2002) Normal parameters of left ventricular volume and ejection fraction measured by gated myocardial perfusion SPECT: comparison of Tc-99m MIBI and Tl-201. Eur J Nucl Med 29:S205

IAEA (1991) Tecdoc 602 quality control of nuclear medicine instruments. International Atomic Energy Agency, Vienna

Iida H, Narita Y, Kado H, Kashikura A, Sugawara S, Shoji Y, Kinoshita T, Ogawa T, Eberl S (1998) Effects of scatter and attenuation correction on quantitative assessment of regional cerebral blood flow with SPECT. J Nucl Med 39:181–189

IPSM (1985) Report 44 an introduction to emission computed tomography. Institute of Physical Sciences in Medicine, London

Juni JE, Waxman AD, Devous MD Sr, Tikofsky RS, Ichise M, Van Heertum RL, Holman BL, Carretta RF, Chen CC (1998) Procedure guideline for brain perfusion SPECT using technetium-99m radiopharmaceuticals. Society of Nuclear Medicine. J Nucl Med 39:923–926

Kauppinen T, Koskinen MO, Alenius S, Vanninen E, Kuikka JT (2000) Improvement of brain perfusion SPET using iterative reconstruction with scatter and non-uniform attenuation correction. Eur J Nucl Med 27:1380–1386

Kelty NL, Cao Z-J, Holder LE (1997) Technical considerations for optimal orthopedic imaging. Semin Nucl Med XXVII:328–333

King MA, Tsui BMW, Pan T-S (1995) Attenuation compensation for cardiac single-photon emission computed tomographic imaging, part 1. Impact of attenuation and methods of estimating attenuation maps. J Nucl Cardiol 2:513–524

Knesaurek K, King MA, Glick SJ, Penney BC (1989) Investigation of causes of geometric distortion in 180° and 360° angular sampling in SPECT. J Nucl Med 30:1666–1675

Kojima A, Matsumoto M, Takahashi M, Hirota Y, Yoshida H (1989) Effect of spatial resolution on SPECT quantification values. J Nucl Med 30:508–514

Kubo N, Morita K, Katoh C, Shiga T, Konno M, Tsukamoto E, Morita Y, Tamaki N (2000) A new dynamic myocardial phantom for the assessment of left ventricular function by gated single-photon emission tomography. Eur J Nucl Med 27:1525–1530

LaCroix KJ, Tsui BM, Hasegawa BH (1998) A comparison of 180 degrees and 360 degrees acquisition for attenuation-compensated thallium-201 SPECT images. J Nucl Med 39:562–574

Lappi S, Lazzari S, Sarti G, Pieri P (2002) Assessment of geometrical distortion and activity distribution after attenuation correction: a SPECT phantom study. J Nucl Cardiol 9:508–514

Lau YH, Hutton BF, Beekman FJ (2001) Choice of collimator for cardiac SPET when resolution compensation is included in iterative reconstruction. Eur J Nucl Med 28:39–47

Li J, Jaszczack RJ, Turkington TG, Metz CE, Gilland DR, Greer KL, Coleman RE (1994) An evaluation of lesion detectability with cone-beam, fanbeam and parallel-beam collimation in SPECT by continuous ROC study. J Nucl Med 35:135–140

Links JM, DePuey EG, Taillefer R, Becker LC (2002) Attenuation correction and gating synergistically improve the diagnostic accuracy of myocardial perfusion SPECT. J Nucl Cardiol 9:183–187

Liu YH, Lam PT, Sinusas AJ, Wackers FJ (2002) Differential effect of 180 degrees and 360 degrees acquisition orbits on the accuracy of SPECT imaging: quantitative evaluation in phantoms. J Nucl Med 43:1115–1124

Maniawski PJ, Morgan HT, Wackers FJT (1991) Orbit-related variation in spatial resolution as a source of artifactual defects in thallium-201 SPECT. J Nucl Med 32:871–875

Manrique A, Faraggi M, Vera P, Vilain D, Lebtahi R, Cribier A, Le Guludec D (1999) ^{201}Tl and ^{99m}Tc-MIBI gated SPECT in patients with large perfusion defects and left ventricular dysfunction: comparison with equilibrium radionuclide angiography. J Nucl Med 40:805–809

Matsumoto N, Berman DS, Kavanagh PB, Gerlach J, Hayes SW, Lewin HC, Friedman JD, Germano G (2001) J Nucl Med 42:687–694

Matsunari I, Boning G, Ziegler SI, Kosa I, Nekolla SG, Ficaro EP, Schwaiger M (1998) Effects of misalignment between transmission and emission scans on attenuation-corrected cardiac SPECT. J Nucl Med 39:411–416

Milo T (1996) Spect image display techniques. In: Henkin RE, Boles MA, Dillehay GL, Halama JR, Karesh SM, Wagner RH, Zimmer AM (eds) Nuclear medicine. Mosby, St Louis, pp 254–259

Mueller SP, Polak JF, Kijewski MF, Holman BL (1986) Collimator selection for SPECT brain imaging: the advantage of high resolution. J Nucl Med 27:1729–1738

Mullani NA (2000) Comparing diagnostic accuracy of γ camera coincidence systems and PET for detection of lung lesions. J Nucl Med 41:959–960

Muylle K, Vanhove C, Maenhout A, Franken PR (2002) Comparison of 180-degree and 360-degree data acquisition in gated myocardial perfusion and in gated blood pool tomography. Eur J Nucl Med 29:S210

Nakajima K, Taki J, Yamamoto W, Michigishi T, Tonami N (1998) Effect of 360 degrees and 180 degrees rotation SPET acquisitions on myocardial polar map: comparison of ^{201}Tl-, ^{99}Tc^{m}- and ^{123}I-labelled radiopharmaceuticals. Nucl Med Commun 19:315–325

O'Connor MK (1996) Instrument- and computer-related problems and artifacts in nuclear medicine. Semin Nucl Med XXVI:256–277

O'Connor MK, Bothun ED (1995) Effects of tomographic table attenuation on prone and supine cardiac imaging. J Nucl Med 36:1102–1106

O'Connor MK, Kelly BJ (1990) Evaluation of techniques for the elimination of "hot" bladder artifacts in SPECT of the pelvis. J Nucl Med 31:1872–1875

O'Connor MK, Brown ML, Hung JC, Hayostek RJ (1991) The art of bone scintigraphy – technical aspects. J Nucl Med 32:2332–2341

O'Connor MK, Caiati C, Christian TF, Gibbons RJ (1995) Effects of scatter correction on the measurement of infarct size from SPECT cardiac phantom studies. J Nucl Med 36:2080–2086

O'Connor MK, Kanal KM, Gebhard MW, Rossman PJ (1998) Comparison of four motion correction techniques in SPECT imaging of the heart: a cardiac phantom study. J Nucl Med 39:2027–2034

O'Connor MK, Leong LK, Gibbons RJ (2000) Assessment of infarct size and severity by quantitative myocardial SPECT: results from a multicenter study using a cardiac phantom. J Nucl Med 41: 1383–1390

Ohnishi H, Ota T, Takada M, Kida T, Noma K, Matsuo S, Masuda K, Yamamoto I, Morita R (1997) Two optimal prefilter cutoff frequencies needed for SPECT images of myocardial perfusion in a one-day protocol. J Nucl Med Technol 25: 256–260

Pan TS, King MA, Der-Shan L, Dahlberg ST, Villegas BJ (1997) Estimation of attenuation maps from scatter and photopeak window single photon emission computed tomographic images of technetium 99m-labeled sestamibi. J Nucl Cardiol 4: 42–51

Patton JA (2000) Instrumentation for coincidence imaging with multihead scintillation cameras. Semin Nucl Med XXX: 239–254

Pena HP, Cantinho GC, Veiga AV, Marona DM, Gomes PG, Godinho FG (2002) Perfusion gated SPECT – ejection fraction variability, as a function of the software. Eur J Nucl Med 29: S205

Peters AM, Myers MJ (1998) Physiological measurements with radionuclides in clinical practice. Oxford University Press, Oxford

Petersen CL, Kjoer A, Hviid AM (2002) Substantial difference in left ventricular architecture and perfusion can occur despite unchanged stroke volume. Eur J Nucl Med 29: S210

Piepsz A (1995) Recent advances in pediatric nuclear medicine. Semin Nucl Med XXV: 165–182

Pitman AG, Kalff V, van Every B, Risa B, Barnden LR, Kelly MJ (2002) Effect of mechanically simulated diaphragmatic respiratory motion on myocardial SPECT processed with and without attenuation correction. J Nucl Med 43: 1259–1267

Port SC (1999) Imaging guidelines for nuclear cardiology procedures, part 2. American Society of Nuclear Cardiology. J Nucl Cardiol 6: G53–G84

Prvulovich EM, Jarritt PH, Vivian GC, Clarke SE, Pennell DJ, Underwood SR (1998) Quality assurance in myocardial perfusion tomography: a collaborative BNCS/BNMS audit programme. British Nuclear Cardiology Society/British Nuclear Medicine Society. Nucl Med Commun 19: 831–838

Rajabi H, Bitarafun AR, Yaghoobi N, Firouzabady H, Rustgou F (2002) Filter selection for Tc99m-sestamibi myocardial perfusion SPECT imaging. Eur J Nucl Med 29: S207

Rosenthal MS, Cullom J, Hawkins W, Moore SC, Tsui BMW, Yester M (1995) Quantitative SPECT imaging: a review and recommendations by the focus committee of the Society of Nuclear Medicine Computer and Instrumentation Council. J Nucl Med 36: 1489–1513

Rubio AR, Garcia-Burillo A, Gonzalez-Gonzalez JM, Oller G, Canela T, Richart JA, Aguade S, Roca I, Castell J (2002) Interstudy repeatability of gated-SPECT quantitative parameters. Eur J Nucl Med 29: S208

Sankaran S, Frey EC, Gilland KL, Tsui BMW (2002) Optimum compensation method and filter cutoff frequency in myocardial SPECT: a human observer study. J Nucl Med 43: 432–438

Sarikaya I, Sarikaya A, Holder LE (2001) The role of single photon emission computed tomography in bone imaging. Semin Nucl Med XXXI: 3–16

Schaefer WM, Namdar T, Koch K-C, Block S, Nowak B, Buell U (2002) Validation of left ventricular ejection fraction (LVEF) by gated-99mTc-tetrofosmin-SPECT (g-SPECT) in routine clinical practice. Eur J Nucl Med 29: S207

Serrano M, Silva C, Serena A, Nogueiras JM, Vale I, Outomuro J, Campos L (2002) A simplified visual method for quantification of brain perfusion SPECT. Eur J Nucl Med 29: S194

Shen MY, Liu YH, Sinusas AJ, Fetterman R, Bruni W, Drozhinin OE, Zaret BL, Wackers FJ (1999) Quantification of regional myocardial wall thickening on electrocardiogram-gated SPECT imaging. J Nucl Cardiol 6: 583–595

Starck SA, Carlsson S (1997) The determination of the effective attenuation coefficient from effective organ depth and modulation transfer function in gamma camera imaging. Phys Med Biol 42: 1957–1964

Stelter P, Junik R, Krzyminiewski R, Gembicki M, Sowinski J (2001) Semiquantitative analysis of SPECT images using $^{99}Tc^{m}$-HMPAO in the treatment of brain perfusion after the attenuation correction by the Chang method and the application of the Butterworth filter. Nucl Med Commun 22: 857–865

Stodilka RZ, Kemp BJ, Msaki P, Prato FS, Nicholson RL (1998) The relative contributions of scatter and attenuation corrections toward improved brain SPECT quantification. Phys Med Biol 43: 2991–3008

Stone CD, McCormick JW, Gilland DR, Greer KL, Coleman RE, Jaszczak RJ (1998) Effect of registration errors between transmission and emission scans on a SPECT system using sequential scanning. J Nucl Med 39: 365–373

Strauss HW, Miller DD, Wittry MD, Cerqueira MD, Garcia EV, Iskandrian AS, Schelbert HR, Wackers FJ (1998) Procedure guideline for myocardial perfusion imaging. Society of Nuclear Medicine. J Nucl Med 39: 918–923

Tatsch K, Asenbaum S, Bartenstein P, Catafau A, Halldin C, Pilowsky LS, Pupi A (2002) European Association of Nuclear Medicine procedure guidelines for brain perfusion SPET using (99m)Tc-labelled radiopharmaceuticals. Eur J Nucl Med 29: BP36–BP42

Taylor D (1994) Filter choice for reconstruction tomography. Nucl Med Commun 15: 857–859

Tsui BM, Frey EC, LaCroix KJ, Lalush DS, McCartney WH, King MA, Gullberg GT (1998) Quantitative myocardial perfusion SPECT. J Nucl Cardiol 5: 507–522

Turkington TG (2000) Attenuation correction in hybrid positron emission tomography. Semin Nucl Med XXX: 255–267

Van Laere K, Koole M, Versijpt J, Dierckx R (2001a) Non-uniform versus uniform attenuation correction in brain perfusion SPET of healthy volunteers. Eur J Nucl Med 28: 90–98

Van Laere K, Koole M, Versijpt J, Vandenberghe S, Brans B, D'Asseler Y, De Winter O, Kalmar A, Dierckx R (2001b) Transfer of normal ^{99m}Tc-ECD brain SPET databases between different gamma cameras. Eur J Nucl Med 28: 435–449

Vanhove C, Franken PR, Defrise M, Deconinck F, Bossuyt A (2002) Reconstruction of gated myocardial perfusion SPET incorporating temporal information during iterative reconstruction. Eur J Nucl Med 29: 465–472

Vines DC, Ichise M (1999) Evaluation of differential magnification during brain SPECT acquisition. J Nucl Med Technol 27: 198–203

Wackers FJT (1999) Attenuation correction, or the emperor's new clothes (editorial). J Nucl Med 40: 1310–1312

Wackers FJ (2002) Should SPET attenuation correction be more widely employed in routine clinical practice? Against. Eur J Nucl Med 29: 412–415

Wells RG, Simkin PH, Judy PF, King MA, Pretorius H, Gifford HC (1999) Effect of filtering on the detection and localization of small Ga-67 lesions in thoracic single photon emission computed tomography images. Med Phys 26: 1382–1388

Wells RG, King MA, Simkin PH, Judy PF, Brill AB, Gifford HC, Licho R, Pretorius PH, Schneider PB, Seldin DW (2000) Comparing filtered backprojection and ordered-subsets expectation maximization for small-lesion detection and localization in ^{67}Ga SPECT. J Nucl Med 41: 1391–1399

Wright GA, McDade M, Keeble W, Martin W, Hutton I (2000) Quantitative SPECT myocardial perfusion imaging with ^{201}Tl: an assessment of the limitations. Nucl Med Commun 21: 1147–1151

Wright GA, McDade M, Martin W, Hutton I (2002) Quantitative gated SPECT: the effect of reconstruction filter on calculated left ventricular ejection fractions and volumes. Phys Med Biol 47: N99–N105

Zaidi H, Hasegawa B (2003) Determination of the attenuation map in emission tomography. J Nucl Med 44: 291–315

Dual Photon Imaging

Abbreviations

2D two dimensions
3D three dimensions
C-OS-EM coincidence-list-ordered sets expectation-maximisation
CT computed tomography
DPET dual photon emission tomography
FDG fluorodeoxyglucose
FOV field of view
GSO gadolinium oxyorthosilicate
LSO lutetium oxyorthosilicate
MAC mean or maximum activity concentration
MRI magnetic resonance imaging
OS-EM ordered subset estimation maximisation

PHA	pulse height analyser
PMT	photomultiplier tube
ROI	region of interest
SPET	single photon emission tomography
SUR	standardised uptake ratio
SUV	standardised uptake value

Dual photon imaging provides an additional perspective to functional imaging; by utilising the radioisotopes of the low atomic number elements: carbon, nitrogen and oxygen, which are particularly important physiologically (Marsden and Sutcliffe-Goulden 2000). The only radionuclides, of these, suitable for imaging are the β^+ emitters: ^{11}C, ^{13}N and ^{15}O. It is another β^+ emitting radionuclide ^{18}F, however, which has driven the development of this modality; by exploiting its ability to substitute for hydrogen in many biologically active molecules without significantly altering their behaviour in the body. The short half-lives, of the majority of radionuclides, has been one of the main barriers to the widespread clinical use of DPET, necessitating a medical production cyclotron in close proximity and rapid radiopharmaceutical production, which introduces complexity and high cost. The popularity of ^{18}F is mainly due to its relatively long half-life which allows transportation from production to reasonably distant imaging sites. Development of new single photon radiopharmaceuticals has enabled SPET to imitate the physiological functionality, but there is presently no single photon equivalent of ^{18}FDG, which is used in ~98% of clinical DPET investigations. Accurate attenuation correction in DPET enables the count density, throughout the reconstructed images, to be made directly proportional to the local radioactivity concentration, which allows absolute quantification of radiopharmaceutical accumulation and values of physiological parameters to be expressed directly, e.g. blood flow in millilitres per second per gram or metabolic rate in moles per second per gram. However, most studies involve only images (Marsden and Sutcliffe-Goulden 2000; Cherry and Phelps 1996). Of particular interest is the very significant recent introduction of on-board or in-line x-ray CT which, not only appears to be the best technique for undertaking attenuation correction but also, provides the very important 'functional anatomic mapping'. This continues the progress in multi-modality imaging and image fusion, which have become areas of major interest since departments have tended to become 'all digital'. Much of the physics perspective of DPET is similar to that of SPET. There are a number of differences, however, including clinical use of dedicated units and γ cameras; acquisition and processing considerations; and specific clinical procedures.

Factors Determining Image Quality and Affecting Lesion Detectability

A number of factors affect the image quality and therefore the detectability of lesions. Several of these are common to both dedicated systems and coincidence γ cameras.

Acquisition Considerations

Environment

Because DPET systems are operated without the absorptive collimators used in single photon imaging, they are extremely sensitive to extraneous radiation and, in a busy nuclear medicine department, substantial wall shielding, to protect them, may be required. Such extraneous radiation may originate from radioactive contamination or high radiation exposure levels from sources or patients administered radioactive materials. The immediate environment must be well controlled and should be confirmed to be suitable before imaging is started. This is likely to be more difficult when coincidence γ cameras are used because, for some of the time, they are likely to be operated in single photon mode with collimators installed, in which configuration their sensitivity to extraneous radiation will be low. When used in DPET mode, a previously suitable environment may no longer be acceptable and, unless corrected, may cause artefacts in the reconstructed images. For single photon γ cameras, with lighter shielding than coincidence systems, sources of high energy annihilation photons may cause imaging artefacts, which may be difficult to detect if the exposure is transient. This is a particular problem in tomographic acquisition.

Acquisition Mode

A decision has to be made whether to acquire data in 2D or 3D mode. 3D acquisitions have been used mainly for brain studies (Al Sugair and Coleman 1998; Burger and Berthold 2000). The scatter fraction depends on the scattering tissue and on the thickness of tissue between the detectors, and is therefore higher during abdominal imaging, ~50%, than brain imaging, ~30% (Zaidi 2001). Most imaging of the chest and abdomen is, therefore, undertaken in 2D mode (Al Sugair and Coleman 1998), in order to reduce the effect of scatter from activity outside the field of view and also because the use of the 3D methodology in these areas has not yet been validated. The preferred mode for quantitative acquisition is also 2D, because this is considered the 'gold standard', the accuracy in estimating absolute activity in a region approaching 5% (Burger and Berthold 2000) and the relative accuracy within an image around 1–2%. These figures depend on all necessary corrections being applied appropriately, however. If they are not, the accuracy will be reduced. The range of activity administrations used for 2D acquisition with a dedicated DPET system is 200–750 MBq (Budinger 1998; Flanagan et al. 1998; Schelbert et al. 1998), reduced to ~1/2 for 3D acquisitions because of the increased sensitivity of the 3D mode necessitating a reduction in the maximum activity that can be within the field of view (Patton and Turkington 1999). These figures are again reduced when coincidence γ cameras are used to avoid saturation consequent on their inability to deal with the very high count rates involved. Collimated γ cameras, with their reduced sensitivity however, require activities of the order used in 2D dedicated DPET. The problem of high count rates is exacerbated when using radionuclides with particularly short half lives. With these, there is a limited time during which data can be acquired, thus high activities must be administered and high count rates must be tolerated by the system in order to achieve adequate counting statistics (Budinger 1998). The majority of work in DPET involves imaging the whole body, which requires several adjacent scans to be performed over the entire volume of interest. For this, the control software must accurately adjust the table position and combine the resulting images into one data set during the reconstruction. A rarely used acquisition type is the gated scan, which allows synchronisation of the acquisition with the cardiac cycle (Burger and Berthold 2000). Acquisitions can also be performed in dynamic mode, mainly in dedicated systems since there is a complete ring of detectors and rotation of a detector is not required. In these cases up to 3700 MBq may be administered to achieve statistical uncertainty of <20%, with multiple injections sometimes being used (Budinger 1998).

Attenuation Correction

Attenuation correction is a particular consideration in DPET, not only being essential for quantitation but also producing a more realistic radioactivity distribution (Turkington 2000) and improving lesion to background contrast (Coleman et al. 1999). Reconstruction artefacts otherwise resulting, include regional activity non-uniformities caused by artefactually increased or decreased activity, and spatial distortion of intense activity distributions with apparent elongation along the lines of lower attenuation (Coleman et al. 1999; Patton and Turkington 1999; Delbeke and Sandler 2000).

Another artefact is an edge effect along the body surface, where apparently increased activity is demonstrated if attenuation correction is not undertaken (Wahl 1999). The enhancement does not follow the body contour, but forms a convexity over it and any concavities, e.g. between the legs, will appear more shallow. This artefact originates from annihilations at the body surface. Annihilation photons emitted tangential to the body surface will not suffer attenuation; whereas in any other direction at least one will suffer substantial attenuation, see Fig. III.4.1. More radiation is detected along this projection line and the activity source appears elongated along a tangent to the surface. This effect is not seen for activities even a few cm below the surface because, in this geometry, there is no direction in which attenuation is avoided. It is thus possible to recognise that the emission image has not been attenuation corrected: the lungs appear to have activity in them, the skin shows increased activity and any areas of focal accumulation show spatial distortion.

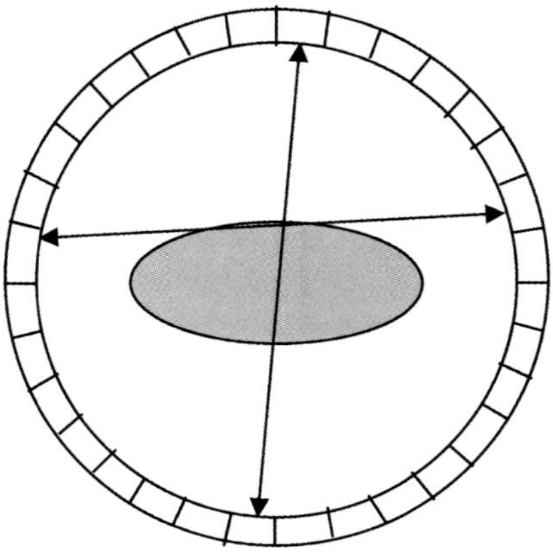

Fig. III.4.1. Edge effect attenuation artifact. Annihilation photons emitted tangentially to the body surface will not suffer attenuation; whereas in any other direction at least one will suffer substantial attenuation

Analytical correction, using a body contour and a uniform attenuation map, can be used where the attenuation is reasonably homogeneous, i.e. brain and parts of the abdomen. This shortens the study duration, by removing the need for a transmission scan; and avoids introducing more noise to the reconstructed image from the statistical variations in the transmission image. Analytical correction is less accurate than measured attenuation correction, however, because of the assumption of uniform attenuation and is most appropriate for qualitative or semiquantitative interpretation (Delbeke and Sandler 2000). Analytical correction cannot be used in studies involving the chest and parts of the abdomen because of the non-uniformity in attenuation (Patton and Turkington 1999; Burger and Berthold 2000) and, for these areas, transmission attenuation correction should be undertaken.

Severe artefacts can be introduced into the reconstructed image, however, by an incorrect attenuation map (Burger and Berthold 2000) and, therefore, the duration of the transmission acquisition must be long enough to adequately reduce the statistical fluctuations in the transmission image, which propagate through to the final image and introduce errors (Delbeke and Sandler 2000). This usually means that the transmission duration should be similar to that of the emission acquisition (Burger and Berthold 2000), which adds significantly to the length of the study. The quality of the reconstructed image is also determined by the spatial co-registration of the transmission and emission scans, which is compromised by patient movement, more likely with long acquisition durations (Flanagan et al. 1998; Burger and Berthold 2000; Patton 2000). The emission acquisition represents summed images over many breathing cycles but, in order to optimise image co-registration when CT transmission attenuation correction is used, normal expiration breath holding during the short CT acquisition is recommended (Goerres et al. 2002; Beyer et al. 2003). Another factor, that affects the co-registration, is if repositioning of the patient in the gantry, is required between the transmission and emission acquisitions (Delbeke and Sandler 2000). Patient movement between emission and transmission acquisitions can be detected, on the attenuation corrected emission scans, by under or over correction in large linear areas. An unacceptably short transmission scan will cause a speckled appearance, on attenuation corrected emission images, as statistical variations in the transmission map propagate through the reconstruction process.

To obtain the best attenuation data, the transmission acquisition should be undertaken before radionuclide administration. This avoids contamination of the transmission data by photons from the emission radionuclide. Until the middle 1990s, this was the transmission procedure that had to be undertaken (Almeida et al. 1998). The disadvantage of the technique was an increase in patient imaging duration. Since the delay, from administration to emission acquisition could ~1 hour; whatever the duration of the transmission acquisition, it increased the total study duration by >1 hour. The next best alternative is the analytical correction, when this is appropriate. The final alternative is a short post administration transmission scan and hybrid correction involving a segmented transmission map. This technique reduces the possibility of patient movement between the two acquisitions, and also shortens the duration of the total study (Flanagan et al. 1998). With the availability of post-administration transmission scanning protocols, attenuation correction now requires only a moderate increase in scanning time (Nuyts et al. 2002b).

Acquisition Sequence

The acquisition sequence comprises up to five acquisitions. A blank acquisition is undertaken to determine detector efficiency and to provide the baseline against which an attenuation correction transmission scan is compared. It is performed using a transmission source without a patient or phantom in the system (Budinger 1998; Burger and Berthold 2000; Turkington 2000). The blank scan is usually acquired for a longer duration than the transmission scan to reduce statistical noise being propagated into the reconstructed image (Cherry and Phelps 1996; Bailey 1998). If a transmission acquisition is undertaken before the emission acquisition, the activity is not administered until after the transmission acquisition (Turkington 2000). If the emission acquisition is undertaken at the same time as or before the transmission acquisition, a cross-talk acquisition can be undertaken, to determine the emission to transmission cross-talk contamination. If a transmission acquisition is required, this is undertaken without patient movement between this and the emission acquisition (Budinger 1998). Attenuation correction factors are calculated for each detector pair by taking the $\log_e$ ratio of the blank and transmission sinograms (Burger and Berthold 2000; Patton 2000). A calibration acquisition is undertaken to reconstruct images in absolute units of MBq cm^{-3}, it is necessary to calibrate the PET scanner against a standard source. This is commonly undertaken by imaging a uniform cylinder filled with a known activity, e.g. 100 MBq, of $^{68}Ge/^{68}Ga$ dissolved in a gel (Budinger 1998) or ^{18}F, often using analytical attenuation correction (Lodge et al. 1998).

Processing Considerations

The main problem for accurate reconstruction is patient motion but once this is avoided, a number of other factors have also to be taken into consideration. Before reconstruction, each projection ray undergoes a number of corrections for: detector efficiency, attenuation if required, randoms (Budinger 1998), and sensitivity profile. The emission scan can be reconstructed using filtered back projection with, typically, a Butterworth or Hamming smoothing filter of various parameters. Iterative reconstruction has improved the reconstructed image quality (Patton 2000; Hsu 2002). It has also improved lesion detection, by improving the signal to noise ratio particularly in low activity regions, since lesion detection is dependent on the image noise as well as on the contrast difference between lesion and background. The signal to noise ratio depends, however, on the number of iterations used (Nuyts et al. 2002b); 5 or 6 iterations and 8 subsets being used for coincidence γ camera cardiac imaging and 2 iterations and 8 subsets for dedicated system cardiac imaging (Nuyts et al. 2002a; de Sutter et al. 2000). The noise distribution over images reconstructed using filtered back-projection is reasonably homogeneous, causing the signal to noise ratio to deteriorate in areas of low activity. This is not the case with images reconstructed using iterative techniques; in which the noise varies over the image, being lower in low activity regions and resulting in improved signal to noise ratios. These improvements averaged 160% in low activity areas such as the lungs but only 25% in high activity areas such as the liver, and even produced a deterioration in very high activity areas (Bengel et al. 1997; Riddell et al. 2001). After reconstruction to transaxials the multiple volumes are combined to give a single set of transaxial slices before reformatting to coronals and sagittals.

Quantitative Analysis

Although DPET imaging can provide qualitative, semiquantitative and quantitative information; clinical studies are generally analysed using qualitative and semiquantitative techniques, quantitative methods being too invasive and time consuming for routine work. Semiquantitative measurements provide an objective method for differentiating lesions (Delbeke and Sandler 2000) but no definite diagnostic advantage over qualitative analysis. They are most useful for longitudinal comparison studies, such as for determination of tumour response to therapy (Flanagan et al. 1998).

Standard Uptake Values

The semiquantitative analysis produces an index which has been named in various ways but the most popular is the standardised uptake value (SUV) (Flanagan et al. 1998), another is the standardised uptake ratio (SUR). This is a representation of the accumulation of activity in an area of interest compared to the average distribution throughout the body. An SUV of one, describes a uniform activity concentration throughout the body. Greater than one, represents areas of increased uptake and less than one, areas of decreased uptake. A threshold of 2.5 is usually used to establish abnormal accumulations (Bar Shalom et al. 2000), and values up to 20 will appear in the brain, bladder and kidneys and also in malignant tumours. It is calculated by dividing the decay corrected activity, per unit volume, in an area of interest by the administered activity and patient body weight or lean body mass (Flanagan et al. 1998; Al Sugair and Coleman 1998). It has been suggested that tumour response is best assessed using these values (Bedigian et al. 1998).

$$\mathrm{SUV} = \mathrm{MAC}/(\mathrm{A} * \mathrm{W}) \qquad (\mathrm{III.4.1})$$

where
MAC is the mean or maximum activity concentration ($\mathrm{MBq\ cm^{-3}}$)
A is the administered activity (MBq)
W is the body weight (g)

Although SUVs are widely used, opinions differ as to their usefulness (Keyes 1995), because many factors affect their calculation (Engel-Bicik 2000). For instance, if any administration is extravasated, the SUV estimate will be too low since the administered activity part of the equation will be too high. The values of the SUVs depend on whether the mean or maximum MAC is used. The former varying with the method of defining the region and the latter considering only one point in the area of interest (Lee et al. 2000). They also change with reconstruction technique, and with the number of iterations and subsets used in iterative reconstruction (Ramos et al. 2001; Ivanovic and Falen 2002). Accurate values have been obtained using OS-EM reconstruction and hybrid attenuation correction (Visvikis et al. 2001), and similar values were obtained using simultaneous emission transmission scans and separate scans (Inoue et al. 1999). Similar uptake values have been obtained using CT and ^{68}Ge derived attenuation corrections (Kamel et al. 2002). SUVs can also be calculated from coincidence γ camera data for lesions with diameters >1.5 cm, using singles count rate-related calibration factors, but correction for partial volume and recovery effects is needed to improve the agreement with SUVs determined using dedicated DPET (Zimny et al. 2001).

Mathematical Models

One of the unique features of DPET is to quantitate the concentration of radiopharmaceutical in MBq cm^{-3}. By combining this with a tracer kinetic model and a dynamic sequence of quantitative images, the absolute rate of the biological process along a particular biochemical pathway can be estimated (Al Sugair and Coleman 1998; Delbeke and Sandler 2000; Engel-Bicik 2000).

Image Fusion with CT

Initially developed as a method for transmission attenuation correction, combined DPET/CT has recently caused a perturbation in diagnostic nuclear medicine by providing the possibility of routinely available 'functional anatomic mapping'. The outstanding potential diagnostic benefit of combining anatomical detail with functional information has long been recognised, and was originally tackled using software co-registration techniques. Even with all the efforts invested in these, routine applications did not emerge. The reasons included problems experienced in image transfer among modalities and in co-registration of the images. The procedures for the latter are difficult and only give approximate results, particularly in mobile areas of the body such as the abdomen. DPET/CT overcomes these problems; not only achieving a higher accuracy of image fusion, but doing this routinely and avoiding the need for separate attenuation correction scans. The acquisition sequence is that the CT scan is undertaken, and then the patient is automatically repositioned to the start position for DPET acquisition (Shreve 2000; Turkington 2000; Ell and von Schulthness 2002; Townsend and Beyer 2002).

Clinical Use of the Detection Systems

Image Display

Most visual diagnosis is undertaken using a rotating greyscale 3D maximum intensity projection.

Dedicated Systems

Before an acquisition is undertaken, it is necessary to correct for any variation in detection efficiency in the individual detector elements. This is done using the blank scan which produces a set of sinogram images, from which the correction maps are calculated and in which dark stripes reveal any defective detector blocks (Burger and Berthold 2000). Emission static acquisition durations typically range from 5 to 20 min for the body and from 15 to 30 min for the brain, with dynamic acquisitions ranging from 20 to 45 min; the static acquisitions being undertaken over either a single site, e.g. brain and heart, or stacked sequentially over the whole-body, as in oncology (Dahlbom et al. 1992; von Schulthess et al. 2000). Recently introduced GSO and LSO systems can sustain higher activities in the FOV and can acquire at high coincidence rates, allowing attenuation-corrected whole-body studies to be completed in ~30 min. Joining the images presents a problem, however, because of the reduced detection sensitivity at the periphery of the field of view, more severe in 3D acquisitions than in 2D, which can cause a 'zipper' effect, see Fig. III.4.2. This is apparent, on coronal and sagittal projections, as regular spaced lines of reduced activity across the body. The problem can be overcome by overlapping the axial fields of view. This may only require 2 or 3 slices, or ~4 mm, in 2D mode but it may require 35–50% of the FOV in 3D mode (Tarantola et al. 2003).

Separate transmission acquisitions can take from 8 to 20 minutes but, with segmentation of the attenuation maps, this can be reduced to 2–5 minutes depending on the source age (Burger and Berthold 2000; Khorsand et al. 2002) and it is possible to undertake emission and transmission acquisitions simultaneously. There remains an unsettled issue as to whether transmission correction should always be used or whether it can be avoided for some acquisitions; for instance, undertaking whole body imaging without attenuation correction and then performing attenuation correction only for selected areas. When the long separate transmission acquisitions are undertaken, the total acquisition duration is doubled which, for a whole-body investigation with 10 or more bed positions, is unmanageable. Although the lesion to background ratio is higher with attenuation correction (Delbeke and Sandler 2000), and the uncorrected images contain artefacts and lesion shape may be altered, lesions are not missed. Uncorrected images may be preferred, when quantitative or semiquantitative assessment is not necessary, particularly in whole-body oncology studies; because if the attenuation corrected images are reconstructed from noisy emission and transmission data, or a long total acquisition duration encourages patient movement between the emission and transmission acquisitions, the reconstructed image quality may be worse than non-attenuation corrected images, especially when iterative reconstruction is not used. There is less of an argument against undertaking transmission attenuation correction when the short duration hybrid protocol is available. If transmission acquisition is undertaken, images should be reviewed both with, and without, attenuation correction applied (Flanagan et al. 1998; Nuyts et al. 2002b; von Schulthess et al. 2000).

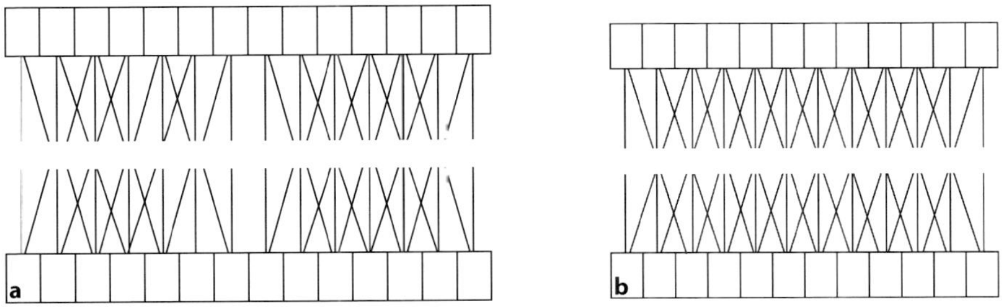

Fig. III.4.2. Zipper effect. (**a**) when images are stacked to form a whole-body acquisition, the reduced detection sensitivity at the periphery of each field of view, caused by the reduced numbers of projection lines, is reinforced. (**b**) this is avoided by slightly overlapping the fields

Collimated Gamma Cameras

Clinical applications are generally limited to investigations in which the lesions to be detected are large and a high uptake of ^{18}FDG is expected. The longer scanning times possible with this radionuclide and the high contrast images obtained from the high uptake, compensate for the poor spatial resolution and detection sensitivity of these systems (Jarritt and Acton 1996; Budinger 1998), which are regarded as being mainly suitable for assessing myocardial viability (Martin et al. 1995; Chen et al. 1997; Kuge and Katoh 2001; Blockland et al. 2002). Oncological applications are not reliable as tumours smaller than ~2–3 cm cannot be detected (Martin et al. 1995; Blockland et al. 2002).

Coincidence Gamma Cameras

These provide good diagnostic information when used with ^{18}FDG, whose longer half-life allows adequate count statistics to be acquired from longer acquisition durations, using activity concentrations that do not saturate the detectors (Budinger 1998). The system is operated with the collimators removed, which makes the crystal very sensitive to changes in temperature and also to physical damage, with dual head systems set at a relative orientation of 180°. In general, the count rate capability is not limited by the efficiency of detection but by the singles capability of each detector; therefore, the distance between detectors should be set to the maximum possible. The effect of activity outside the FOV primarily originates from an increase in the detected random coincidences and appears to be more severe in torso imaging as a result of the brain uptake (IPEM 2003). Unlike a full-ring dedicated system, a coincidence γ camera does not acquire data over the whole area of interest simultaneously but rather samples it at various detector angles. The duration of the acquisition rotation is determined on the basis that, to avoid significant reconstruction artefacts, the radioactive distribution must remain constant throughout the acquisition. This will only be the case if the rotation duration does not exceed 10% of the effective half-life of the radiopharmaceutical. Thus for an unchanging ^{18}F distribution, a total acquisition should be undertaken using multiple rotations, each of duration <11 min, and preferably ~3 min (Patton 2000). For longer rotation durations, decay correction will be necessary, and for very short half-life radionuclides, it would be necessary to use a constant infusion technique to enable a steady state condition to be established (Jarritt and Acton 1996).

The blank scan takes about one minute. In that the transmission source has just to expose the crystals, a static image is acquired on a 128×128 matrix. Emission acquisition durations are typically 20–30 minutes, with the data being acquired in list mode (Schelbert et al. 1998; Patton and Turkington 1999; Patton 2000) with a PHA window set at 511 keV ± 10% (Schelbert et al. 1998). Typical step and shoot acquisitions use 3° increments into a 64×64 matrix (Schelbert et al. 1998). Variation of the angular increment showed no consequences in image quality, an increment of 6° being sufficient for high resolution images with no under-sampling effects such as those occurring in dedicated systems (Kunze et al. 2000).

In this mode, even visual interpretation should be undertaken on images after attenuation correction because of the reconstruction artefacts, which would otherwise result (Patton and Turkington 1999; Delbeke and Sandler 2000). Attenuation correction improves image clarity; reducing the artefactually high apparent activity in the lungs; restoring lesion shape, and providing an accurate body contour with appropriate intensity. Signal-to-noise ratio and lesion contrast are also better (Coleman et al. 1999; Chan et al. 2001). If ^{137}Cs is used as the transmission source, care must be taken regarding the activity administered to the patient because of the added count rate from the transmission source which can lead to high count rate artefacts. An additional photopeak count rate of 5×10^4 counts s^{-1} from the ^{137}Cs photons could mean another 2×10^5 counts s^{-1} in the total energy spectrum (Turkington 2000). When using a CT for the transmission acquisition, although the spatial resolution is of the order of 1 mm, the transmission data is reconstructed using filtered back projection to create attenuation maps on a 128×128 array (pixel size 4.2 mm), to correspond to the array size of the reconstructed emission scan (Patton 2000).

When switching back to single photon mode: PHA peak and window, and uniformity must be tested to ensure stability. This is because the operating conditions of the PMTs may have been changed and may require some time to stabilise (IPEM 2003).

Comparison

The better performance characteristics of dedicated systems predicts much better clinical results than obtained using coincidence γ cameras, with typical detection efficiency of the coincidence γ cameras being <3% of a modern dedicated system per axial cm of the field of view (Delbeke and Sandler 2000; Mullani 2000). Therefore image noise is much higher in coincidence γ cameras (Mullani 2000) and they exhibit a lower lesion detectability (Coleman et al. 1999; Weber et al. 1999; Delbeke and Sandler 2000) for small differences in contrast (Mullani 2000). An advantage of coincidence γ cameras is in imaging the whole body. Since the field of view is much larger than that of a dedicated system, it requires only as few as two table positions to cover the whole body, even on a tall patient, whereas a dedicated system requires up to ten (Cherry and Phelps 1996; von Schulthess et al. 2000; Al Sugair and Coleman 1998).

Oncology

Although the lesion to background ratio is higher, with dedicated systems, than with γ cameras (Weber et al. 1999); the coincidence γ camera, with thick NaI(Tl) crystals, has clinical potential (Delbeke and Sandler 2000; Inoue et al. 2001) for detecting lesions larger than ~1.5 cm (Shreve et al. 1998; Boren et al. 1999; Zimny et al. 1999a,b; Bar-Shalom et al. 2000; Ak et al. 2001; Haslinghuis-Bajan et al. 2002) which show increased tracer accumulation (Landoni et al. 1999). Above this size, both collimated and coincidence γ cameras detect 80–96% of lesions found by dedicated systems. Below this size, only a coincidence γ cameras with a 15.9 mm crystal, and using C-OS-EM reconstruction with attenuation correction detects ~60%, the others achieving only 25–36%. Small lesions may be easier to detect in lung and breast than in abdomen, because of the low attenuation and background activity in the former, and the opposite in the latter (Budinger 1998; Delbeke and Sandler 2000; Mullani 2000; Ak et al. 2001) and coincidence γ cameras appear to have similar detection sensitivity, to dedicated systems, for lesions >2 cm (Weber et al. 1999; Yutani et al. 1999). In the lung, coincidence γ cameras and dedicated systems provided similar visual detection sensitivity but the lesion to background contrast was lower in the coincidence γ camera images. The latter was improved with attenuation correction (Tatsumi et al. 1999). Transmission attenuation correction markedly improves image quality and the detection of small lesions, and advances the diagnostic performance of coincidence γ cameras closer to that of dedicated systems (Zimny et al. 1999a,b).

Cardiology

Because images of the heart do not require high spatial resolution, a dedicated system is not mandatory and a coincidence γ camera system is adequate (Gemmell et al. 2002; Dilsizian et al. 2001), and can reliably demonstrate the extent of viable and scarred myocardium (Matsunari et al. 2001). Even collimated cameras have been found to be clinically effective for the detection of viable myocardium (Budinger 1998). They are reported as being superior to coincidence γ cameras used without attenuation correction, because of the marked effect of attenuation on coincidence γ camera images (Bax and Wijns 1999; Hasegawa et al. 1999). Coincidence γ cameras without attenuation correction produce technically inferior and clinically inaccurate information regarding viability when compared with dedicated systems using attenuation correction (Dilsizian et al. 2001).

Neurology

Collimated γ cameras are not suitable for brain imaging because of their poor spatial resolution. Even with ultra high energy fan beam collimators, which improved both spatial resolution and sensitivity, diagnostic quality images cannot be obtained because of the acquisition duration, of 40–60 min, required (Delbeke and Sandler 2000).

Clinical Studies

Oncology

By far the most important clinical application is the detection and staging of tumours using ^{18}FDG. In brain imaging, a 3D acquisition is undertaken for 20 min, with transmission attenuation correction, and reconstruction is undertaken on a 128×128 matrix using a non-iterative method (Engel-Bicik 2000). For single body acquisitions on dedicated systems, emission acquisition typically aims to collect 5–15×10^6 counts and a long transmission acquisition ~1.25×10^8 counts. Reconstruction is on a matrix of 128×128 (Schelbert et al. 1998). Whole-body 2D imaging on dedicated systems, involving sequential adjacent acquisitions over as much of the body as required, contains ~10^6 counts per plane (Al Sugair and Coleman 1998; Bengel et al. 1997; Budinger 1998; Steinert et al. 2000a,b). The sequentially recorded sets of images for each bed position are corrected for radionuclide decay, rearranged into a stack of transaxial images and resliced into a set of coronal whole-body images (Schelbert et al. 1998).

Although attenuation corrected and uncorrected images were equally sensitive for detecting lung (Farquhar et al. 1999) and whole body lesions (Imran et al. 1998), attenuation correction is encouraged (Bedigian et al. 1998), especially when small centrally located lesions are a possibility (Mullani 2000). Using this, the diagnostic yield of coincidence γ cameras may improve (Ak et al. 2001). Imaging of the chest without attenuation correction is not reliable (Blockland et al. 2002).

Neurology

Typically, dedicated systems are used; coincidence γ cameras not yet being considered established (Bartenstein et al. 2002). In these, the range of adult administration activities is 300–600 MBq for 2D acquisition and 125–250 MBq for 3D acquisition.

In dedicated systems, a one minute acquisition is undertaken for positioning and then either a 2D or a 3D acquisition is used (Arigoni et al. 2000; Buck and Wieser 2000); 3D being recommended for children to reduce the radiation dose. If 3D acquisition is undertaken, scatter correction is mandatory. Typically, acquisition is undertaken over 10–20 min, to collect $50–200\times10^6$ counts. To reduce the possibility of patient movement compromising the whole study, this can be undertaken in 6×5 min acquisitions, and the sinograms, with no signs of movement, added before reconstruction (Bartenstein et al. 2002).

Attenuation correction is mandatory, with calculated attenuation correction being most commonly used because this is the most straightforward and least time-consuming, giving results that are adequate for routine clinical applications. The necessity of transmission attenuation is presently a matter of debate (Arigoni and Buck 2000; Buck and Wieser 2000; von Schulthess et al. 2000). Transmission attenuation correction is usually undertaken using segmentation techniques. The different attenuation correction techniques result in systematic differences in the reconstructed images, however. Emission acquisition is usually over 10–20 min to acquire at least 100×10^6 counts. Reconstruction is usually on a matrix of 128×128 to give a pixel size of 2–4 mm and is routinely undertaken by filtered back projection using either Hamming or Shepp-Logan filters; although at clinical count levels, a post reconstruction Metz filter is reported to produce images of better quality, control noise more effectively and enhance contrast when compared to Hann filter (Varga et al. 1997). Image realignment can be undertaken prospectively using a head holder or retrospectively, for instance with MRI, using software (Bartenstein et al. 2002).

Image analysis is either visual or quantitative, display of coronal and sagittal images being mandatory and 3D display possibly helpful. Quantitative analysis is voxel based using ROIs, for instance, created on MRI and superimposed on the DPET images; and then using statistical parametric mapping, stereotactic normalisation, or smoothing on a voxel by voxel basis.

Cardiology

A dedicated system must have an axial field of view large enough, at least 14–18 cm, to image the heart in one acquisition. Coincidence γ camera systems are used with an axial acceptance angle of 12° (de Sutter et al. 2000). Most routine investigations are performed without cardiac or respiratory gating, and the cyclical motion degrades the object spatial resolution considerably. Several dynamic acquisition investigations can be performed, including: myocardial perfusion using $^{13}NH_3$, ^{82}Rb, or $H_2{}^{15}O$, fatty acid metabolism using ^{11}C palmitate or acetate, and various receptor studies, but the most common is ^{18}FDG for myocardial viability. For most studies, 300–370 mCi of ^{18}FDG are administered and a single static acquisition is undertaken (Bergmann 1998). Reconstruction is on a matrix of 128×128 (Hasegawa et al. 1999; de Sutter et al. 2000) and transmission attenuation correction should always be undertaken, even if only qualitative analysis is performed (von Schulthess et al. 2000); misleading results may be produced if attenuation correction is not used (de Sutter et al. 2000).

References

Ak I, Blokland JAK, Pauwels EKJ, Stokkel MPM (2001) The clinical value of F-18-FDG detection with a dual-head coincidence camera: a review. Eur J Nucl Med 28:763–778

Al Sugair A, Coleman RE (1998) Applications of PET in lung cancer. Semin Nucl Med XXVIII: 303–319

Almeida P, Bendriem B, de Dreuille O, Peltier A, Perrot C, Brulon V (1998) Dosimetry of transmission measurements in nuclear medicine: a study using anthropomorphic phantoms and thermoluminescent dosimeters. Eur J Nucl Med 25:1435–1441

Arigoni M, Buck A (2000) PET imaging in dementias. In: von Schulthness GK, Buck A, Engel-Bicik I, Steinert H Ch (eds) Clinical positron emission tomography. Correlation with morphological cross-sectional imaging. Lippincott Williams and Wilkins, Philadelphia

Bailey DL (1998) Transmission scanning in emission tomography. Eur J Nucl Med 25:774–787

Bar-Shalom R, Valdivia AY, Blaufox MD (2000) PET imaging in oncology. Semin Nucl Med XXX: 150–185

Bartenstein P, Asenbaum S, Catafau A, Halldin C, Pilowski L, Pupi A, Tatsch (2002) European Association of Nuclear Medicine procedure guidelines for brain imaging using [^{18}F]FDG. Eur J Nucl Med 29:B43–B48

Bax JJ, Wijns W (1999) Fluorodeoxyglucose imaging to assess myocardial viability: PET, SPECT or gamma camera coincidence imaging? J Nucl Med 40:1893–1895

Bedigian M, Benard F, Smith R, Karp J, Alavi A (1998) Whole-body positron emission tomography for oncology imaging using singles transmission scanning with segmentation and ordered-subsets-expectation maximization (OS-EM) reconstruction. Eur J Nucl Med 25:659–661

Bengel FM, Ziegler SI, Avril N, Weber W, Laubenbacher C, Schwaiger M (1997) Whole-body positron emission tomography in clinical oncology: comparison between attenuation-corrected and uncorrected images. Eur J Nucl Med 24:1091–1098

Bergmann SR (1998) Cardiac positron emission tomography. Semin Nucl Med XXVIII:320–340

Beyer T, Antoch G, Blodgett T, Freudenberg LF, Akhurst T, Mueller S (2003) Dual-modality PET/CT imaging: the effect of respiratory motion on combined image quality in clinical oncology. Eur J Nucl Med 30:588–596

Blokland JA, Trindev P, Stokkel MP, Pauwels EK (2002) Positron emission tomography: a technical introduction for clinicians. Eur J Radiol 44:70–75

Boren EL Jr, Delbeke D, Patton JA, Sandler MP (1999) Comparison of FDG PET and positron coincidence detection imaging using a dual-head gamma camera with 5/8-inch NaI(Tl) crystals in patients with suspected body malignancies. Eur J Nucl Med 26:379–387

Buck A, Wieser H-G (2000) PET imaging in epilepsy. In: von Schulthness GK, Buck A, Engel-Bicik I, Steinert H Ch (eds) Clinical positron emission tomography. Correlation with morphological cross-sectional imaging. Lippincott Williams and Wilkins, Philadelphia

Budinger TF (1998) PET Instrumentation: what are the limits? Semin Nucl Med XXVIII:247–267

Burger C, Berthold T (2000) Physical principles and practical aspects of clinical PET imaging. In: von Schulthness GK, Buck A, Engel-Bicik I, Steinert H Ch (eds) Clinical positron emission tomography. Correlation with morphological cross-sectional imaging. Lippincott Williams and Wilkins, Philadelphia

Chan W-L, Freund J, Pocock NA, Szeto E, Chan F, Sorensen BJ, McBride B (2001) Coincidence detection FDG PET in the management of oncological patients: attenuation correction versus non-attenuation correction. Nucl Med Commun 22:1185–1192

Chen EQ, MacIntyre WJ, Go RT, Brunken RC, Saha GB, Wong C-Y O, Neumann DR, Cook SA, Khandekar SP (1997) Myocardial viability studies using fluorine-18-FDG SPECT: a comparison with fluorine-18-FDG PET. J Nucl Med 38:582–586

Cherry SR, Phelps ME (1996) Positron emission tomography: methods and instrumentation. In: Sandler MP, Patton JA, Coleman RE, Gottschalk A, Wackers FJT, Hoffer PB (eds) Diagnostic nuclear medicine, 3rd edition. Williams and Wilkins, Baltimore, pp139–159

Coleman RE, Laymon CM, Turkington TG (1999) FDG imaging of lung nodules: a phantom study comparing SPECT, camera-based PET, and dedicated PET. Radiology 210:823–828

Dahlbom M, Hoffman EJ, Ho CK, Schiepers C, Rosenqvist G, Hawkins RA, Phelps ME (1992) Whole-body positron emission tomography, part I. Methods and performance characteristics. J Nucl Med 33:1191–1199

Delbeke D, Sandler MP (2000) The role of hybrid cameras in oncology. Semin Nucl Med XXX: 268–280

De Sutter J, de Winter F, van de Wiele C, De Bondt P, D'Asseler Y, Dierckx R (2000) Cardiac fluorine-18 fluorodeoxyglucose imaging using a dual-head gamma camera with coincidence detection: a clinical pilot study. Eur J Nucl Med 27:676–685

Dilsizian V, Bacharach SL, Khin MM, Smith MF (2001) Fluorine-18-deoxyglucose SPECT and coincidence imaging for myocardial viability: clinical and technologic issues. J Nucl Cardiol 8:75–88

Ell PJ, von Schulthness GK (2002) PET/CT: a new road map. Eur J Nucl Med 29:719–720

Engel-Bicik I (2000) Brain tumours and inflammation. In: von Schulthness GK, Buck A, Engel-Bicik I, Steinert H Ch (eds) Clinical positron emission tomography. Correlation with morphological cross-sectional imaging. Lippincott Williams and Wilkins, Philadelphia

Farquhar TH, Llacer J, Hoh CK, Czernin J, Gambhir SS, Seltzer MA, Silverman DHS, Qi J, Hsu C, Hoffman EJ (1999) ROC and localization ROC analyses of lesion detection in whole-body FDG PET: effects of acquisition mode, attenuation correction and reconstruction algorithm. J Nucl Med 40:2043–2052

Flanagan FL, Dehdashti F, Siegel BA (1998) PET in breast cancer. Semin Nucl Med XXVIII:290–302

Gemmell HG, McKiddie F, Davidson J, Chilcott F, Welch A, Egred M (2002) The use of a dual-headed gamma camera to image myocardial perfusion using $^{13}NH_3$ – comparison with full-ring PET. Eur J Nucl Med 29:S176

Goerres GW, Kamel E, Heidelberg T-NH, Schwitter MR, Burger C, von Schulthness GK (2002) PET-CT image co-registration in the thorax: influence of respiration. Eur J Nucl Med 29:351–360

Hasegawa S, Uehara T, Yamaguchi H, Fujino K, Kusuoka H, Hori M, Nishimura T (1999) Validity of F-18-fluorodeoxyglucose imaging with a dual-head coincidence gamma camera for detection of myocardial viability. J Nucl Med 40:1884–1892

Haslinghuis-Bajan LM, Hooft L, van Lingen A, van Tulder M, Deville W, Mijnhout GS, Teule GJ, Hoekstra OS (2002) Rapid evaluation of FDG imaging alternatives using head-to-head comparisons of full ring and gamma camera based PET scanners – a systematic review. Nuklearmedizin 41:208–213

Hsu CH (2002) A study of lesion contrast recovery for iterative PET image reconstructions versus filtered backprojection using an anthropomorphic thoracic phantom. Comput Med Imaging Graph 26:119–127

Imran MB, Kubota K, Yamada S, Fukuda H, Yamada K, Fujiwara T, Itoh M (1998) Lesion-to-background ratio in nonattenuation corrected whole-body FDG-PET images. J Nucl Med 39: 1219–1223

Inoue T, Oriuchi N, Koyama K, Ichikawa A, Tomiyoshi K, Sato N, Matsubara K, Suzuki H, Aoki J, Endo K (2001) Usefulness of dual-head coincidence gamma camera with thick NaI crystals for nuclear oncology: comparison with dedicated PET camera and conventional gamma camera with thin NaI crystals. Ann Nucl Med 15:141–148

IPEM (2003) Report 86 quality control of gamma camera systems. Institute of Physics and Engineering in Medicine, London

Ivanovic M, Falen SW (2002) Effects of PET acquisition and image reconstruction parameters on standard uptake values. Eur J Nucl Med 29:S108

Jarritt PH, Acton PD (1996) PET imaging using gamma camera systems: a review. Nucl Med Commun 17:758–766

Kamel E, Hany TF, Burger C, Treyer V, Lonn AHR, von Schulthness GK, Buck A (2002) CT vs ^{68}Ge attenuation correction in a combined PET/CT system: evaluation of the effect of lowering the CT tube current. Eur J Nucl Med 29:346–350

Keyes JW Jr (1995) SUV: standard uptake or silly useless value? J Nucl Med 36:1836–1839

Khorsand A, Garf S, Pirich CH, Zettinig G, Eidherr H, Kletter K, Sochor H, Maurer G, Porenta G (2002) Assessment of myocardial perfusion by dynamic NH_3-PET imaging: comparison of two quantitative methods. Eur J Nucl Med 29:S176

Kuge NTY, Katoh C (2001) Current status and future of metabolic cardiac imaging. Nucl Med Commun 22:847–850

Kunze W-D, Baehre M, Richter E (2000) PET with a dual-head coincidence camera: spatial resolution, scatter fraction and sensitivity. J Nucl Med 41:1067–1074

Landoni C, Gianolli L, Lucignani G, Magnani P, Savi A, Travaini L, Gilardi MC, Fazio F (1999) Comparison of dual-head coincidence PET versus ring PET in tumor patients. J Nucl Med 40: 1617–1622

Lee JR, Madsen MT, Bushnel D, Menda Y (2000) A threshold method to improve standardized uptake value reproducibility. Nucl Med Commun 21:685–690
Lodge MA, Badawi RD, Marsden PK (1998) A clinical evaluation of the quantitative accuracy of simultaneous emission/transmission scanning in whole-body positron emission tomography. Eur J Nucl Med 25:417–423
Marsden P, Sutcliffe-Goulden J (2000) Principles and technology of PET scanning. Nucl Med Commun 21:221–224
Martin W, Delbeke D, Patton J, Hendrix B, Weinfeld Z, Ohana I, Kessler R, Sandler M (1995) FDG-SPECT: correlation with FDG-PET. J Nucl Med 36:988–995
Matsunari I, Yoneyama T, Kanayama S, Matsudaira M, Nakajima K, Taki J, Nekolla SG, Tonami N, Hisada K (2001) Phantom studies for estimation of defect size on cardiac ^{18}F SPECT and PET: implications for myocardial viability assessment. J Nucl Med 42:1579–1585
Mullani NA (2000) Comparing diagnostic accuracy of γ camera coincidence systems and PET for detection of lung lesions. J Nucl Med 41:959–960
Nuyts J, Mortelmans L, van de Werf F, Djian J, Sambuceti G, Schwaiger M, Touboul P, Maes A (2002a) Cardiac phantom measurement validating the methodology for a cardiac multi-centre trial with positron emission tomography. Eur J Nucl Med 29:1588–1593
Nuyts J, Stroobants S, Dupont P, Vleugels S, Flamen P, Mortelmans L (2002b) Reducing loss of image quality because of the attenuation artifact in uncorrected PET whole-body images. J Nucl Med 43:1054–1062
Patton JA (2000) Instrumentation for coincidence imaging with multihead scintillation cameras. Semin Nucl Med XXX:239–254
Patton JA, Turkington TG (1999) Coincidence imaging with a dual-head scintillation camera. J Nucl Med 40:432–441
Ramos CD, Erdi YE, Gonen M, Riedel E, Yeung HWD, Macapiniac HA, Chisin R, Larson SM (2001) FDG-PET standardized uptake values in normal anatomical structures using iterative reconstruction segmented attenuation correction and filtered back-projection. Eur J Nucl Med 28:155–164
Riddell C, Carson RE, Carrasquillo JA, Libutti SK, Danforth DN, Whatley M, Bacharach SL (2001) Noise reduction in oncology FDG PET images by iterative reconstruction: a quantitative assessment. J Nucl Med 42:1316–1623
Schelbert HR, Hoh CK, Royal HD, Brown M, Dahlbom MN, Dehdashti F, Wahl RL (1998) Procedure guideline for tumor imaging using fluorine-18-FDG. Society of Nuclear Medicine. J Nucl Med 39:1302–1305
Shreve P, Steventon R, Deters E, Kison P, Gross M, Wahl M (1998) Oncological diagnosis with 2-[fluorine-18]fluoro-2-deoxy-D-glucose imaging: dual head coincidence gamma camera versus positron emission tomographic scanner. Radiology 207:431–437
Shreve PD (2000) Adding structure to function. J Nucl Med 41:1380–1382
Steinert H Ch, Kacl G, von Schulthness GK (2000a) Head and neck cancer and thyroid cancer. In: von Schulthness GK, Buck A, Engel-Bicik I, Steinert HC (eds) Clinical positron emission tomography. Correlation with morphological cross-sectional imaging. Lippincott Williams and Wilkins, Philadelphia
Steinert H Ch, Kubik-Huch R, von Schulthness GK (2000b) Breast carcinoma. In: von Schulthness GK, Buck A, Engel-Bicik I, Steinert HC (eds) Clinical positron emission tomography. Correlation with morphological cross-sectional imaging. Lippincott Williams and Wilkins, Philadelphia
Tarantola G, Zito F, Gerundini P (2003) PET instrumentation and reconstruction algorithms in whole-body applications. J Nucl Med 44:756–769
Tatsumi M, Yutani K, Watanabe Y, Miyoshi S, Tomiyama N, Johkoh T, Kusuoka H, Nakamura H, Nishimura T (1999) Feasibility of fluorodeoxyglucose dual-head gamma camera imaging in the evaluation of lung cancer: comparison with FDG PET. J Nucl Med 40:566–573
Townsend DW, Beyer T (2002) A combined PET/CT scanner: the path to true image fusion. Br J Radiol 75:24–30
Turkington TG (2000) Attenuation correction in hybrid positron emission tomography. Semin Nucl Med XXX:255–267
Varga J, Bettinardi V, Gilardi MC, Riddell C, Castiglioni I, Rizzo G, Fazio F (1997) Evaluation of pre- and post-reconstruction count-dependent Metz filters for brain PET studies. Med Phys 24:1431–1440
Visvikis D, Cheze-LeRest C, Costa DC, Bomanji J, Gacinovic S, Ell PJ (2001) Influence of OSEM and segmented attenuation correction in the calculation of standardised uptake values for [18F]FDG PET. Eur J Nucl Med 28:1326–1335
Von Schulthness GK, Kacl G, Stumpe DM (2000) The normal PET scan. In: von Schulthness GK, Buck A, Engel-Bicik I, Steinert HC (eds) Clinical positron emission tomography. Correlation with morphological cross-sectional imaging. Lippincott Williams and Wilkins, Philadelphia
Wahl RL (1999) To AC or not to AC: that is the question. J Nucl Med 40:2025–2028

Weber WA, Neverve J, Sklarek J, Ziegler SI, Bartenstein P, King B, Treumann T, Enterrottacher A, Krapf M, Haussinger K, Lichte H, Prauer H, Thetter O, Schwaiger M (1999) Imaging of lung cancer with fluorine-18 fluorodeoxyglucose: comparison of a dual-head gamma camera in coincidence mode with a full-ring positron emission tomography system. Eur J Nucl Med 26:388–395

Yutani K, Tatsumi M, Shiba E, Kusuoka H, Nishimura T (1999) Comparison of dual-head coincidence gamma camera FDG imaging with FDG PET in detection of breast cancer and axillary node metastasis. J Nucl Med 40:1003–1008

Zaidi H (2001) Scatter modelling and correction strategies in fully 3-D PET. Nucl Med Commun 22: 1181–1184

Zimny M, Kaiser H, Cremerius U, Reinartz P, Shreckenberger M, Sabri O, Buell U (1999a) Dual-head gamma camera 2-[fluorine-18]-fluoro-2-deoxy-D-glucose positron emission tomography in oncological patients: effects of non-uniform attenuation correction on lesion detection. Eur J Nucl Med 26:818–823

Zimny M, Kaiser HJ, Cremerius U, Sabri O, Schreckenberger M, Reinartz P, Bull U (1999b) F-18-FDG positron imaging in oncological patients: gamma camera coincidence detection versus dedicated PET. Nuklearmedizin 38:108–114

Zimny M, Kaiser HJ, Wildberger J, Nowak B, Cremerius U, Sabri O, Buell U (2001) Analysis of FDG uptake with hybrid PET using standardised uptake values. Eur J Nucl Med 28:586–592

CHAPTER III.5

Image Processing and Analysis

Peter J. Riley

Abbreviations

IA	image analysis
IP	image processing
CT	computed tomography
MTF	modulation transfer function
MRI	magnetic resonance imaging
NIH	National Institutes of Health
ROI	region of interest
TF	transfer function
SNR	signal-to-noise ratio
SPET	single photon emission tomography

Diagnostic images in nuclear medicine are displayed as digital images on workstations and are then typically sent to film recorders for hardcopy and subsequent analysis on film viewers. This has been the traditional technique of reviewing other medical images for x-ray, CT, MRI etc. Film viewing is an art and requires good control over the ambient light levels for eye adaptation leading to optimal visual acuity. However, there is an increasing trend to review images on workstations, where it is possible to apply a variety of techniques to enhance image details to aid in the diagnosis. Control of ambient light for eye adaptation is still important when reviewing digital images. Nuclear medicine was one of the first disciplines to routinely rely upon digital image review and to employ image processing for image enhancement and to undertake quantitative image analysis for dynamic studies.

Recall from previous chapters that a digital image is composed of individual picture elements, or PIXELS, which may take on a variety of brightness levels determined by the image depth, i.e. the amount of memory allocated to hold the value of the pixel brightness level. Hence, an image depth of 8 bits allows for 256 grey levels, whilst 12 bits allows for 4096 grey levels. Nuclear medicine images usually are displayed with 8 bit depth, whilst digital x-ray images are typically 12 bits deep.

The size of the digital image on the workstation monitor is given in pixels as length × height. Nuclear medicine images are typically 128×128 pixels whilst a chest x-ray may be 2000×2500 pixels. The nuclear medicine image with an 8 bit depth, i.e. 128×128×8, will require 16 kB of memory for storage, whilst the chest x-ray, 2000×2500×16 (without compression), will require 10 MB. The video RAM required to display these images must be correspondingly as large and the monitors must have a screen size at least as large to view the full image. Diagnostic quality monitors for chest film viewing are very large, and consequently expensive, and must comply with radiology standards for contrast range and luminosity. Nuclear medicine image monitors should also be of diagnostic quality, but these requirements are usually met with a good generic monitor.

In the following sections we shall look at the techniques employed in Image Processing and in Image Analysis.

Image Processing

Image Processing (IP) refers to the production of new digital images in which certain types of information have been enhanced to overcome limitations of normal visual acuity.

Transfer Functions and Colour Maps

The discussion to date has referred to greyscale images, in which the brightness of a pixel has been scaled to correspond to the concentration of radioactivity in the corresponding volume element, or VOXEL, in the patient. However, the human eye has a complex sensitivity which is not linear but logarithmic from the bright to the dark areas of an image. This sensitivity is very fine, but depends upon proper viewing conditions and eye adaptation. Simple IP can convert a conventional greyscale image into a colour image where subtle gradations can be readily enhanced and thus reduce the dependency upon the observer's visual acuity. The Transfer Function (TF) is a simple representation of this process. Figure III.5.1 shows the digital image transfer function and Fig. III.5.2 shows the spectrum transfer function.

The diagrams show the range of pixel values in the input image, 0–255, and the corresponding assignment of a colour to that pixel value. Figure III.5.1 shows the TF for a greyscale image, where pixel values of 0 are displayed as black, pixel values of 255 are displayed as white, and the intermediate values are shown as greys. Figure III.5.2 shows the TF for a colour image, where pixel values of 0 are displayed as purple and pixel values of 255 are displayed as red: the colour range runs from purple, blue, cyan,

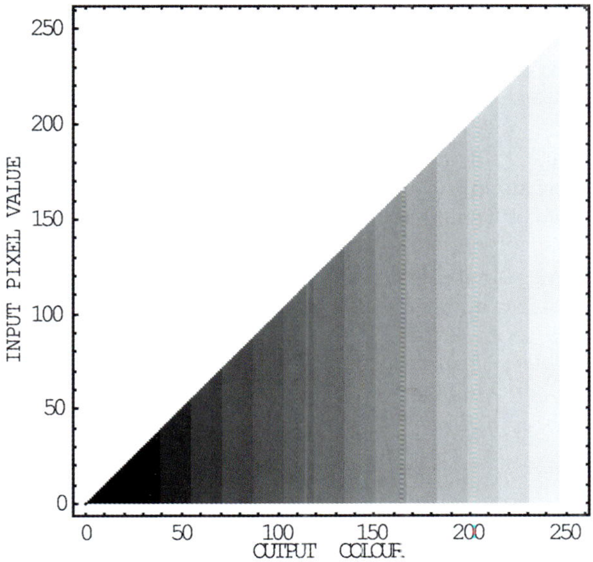

Fig. III.5.1. Digital image transfer function

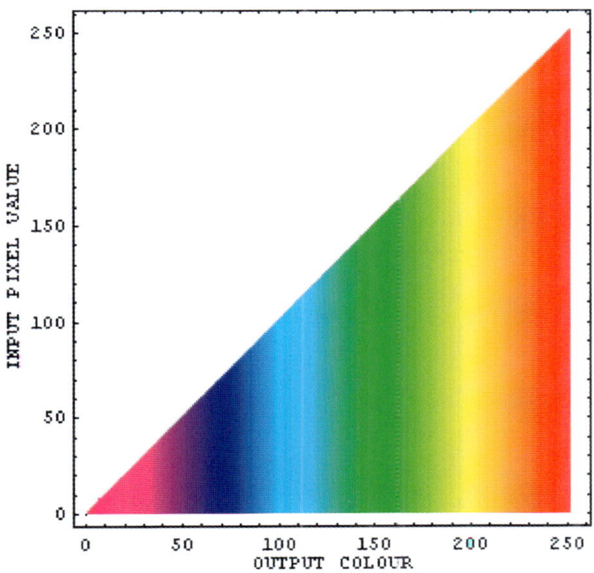

Fig. III.5.2. Spectrum transfer function

green, yellow, orange to red. From Fig. III.5.2, it can be seen that an input pixel value of 75 would appear blue in the output, whilst a value of 150 would be green. This particular colour scale is referred to as a spectrum map. Colour scales commonly encountered in nuclear medicine are:

1. Spectrum: purple, blue, cyan, green, yellow, orange, red. Notice that each colour has a variable range in the Spectrum scheme e.g. green is much broader than yellow.

2. National Institute of Health (NIH): this is the same as the Spectrum but the range of each colour has been made more nearly equal.
3. Blackbody: which ranges from dark red to red to yellow to white.
4. Flow: cyan, blue, black, red, yellow, white.

Various manufacturers also have their own proprietary colour mappings. Hence, it is important to know which scheme an image is displayed in for correct interpretation. Usually, the image is accompanied by a colour scale at the side or bottom of the image to establish this reference.

The TFs considered above have a simple linear relationship between the input and the output images. Other types of transforms are also possible:
1. Logarithmic: this TF is convex-up and will expand the lighter levels.
2. Exponential: this TF is concave-down and will expand the darker levels
3. Sigmoid: this TF is S-shaped and combines the features of exponential and logarithmic functions.

Window Levels

A technique employed in most medical image review stations is image windowing. In this case the TF sets all pixel values below a window level to be all black, and uses the display's full grey level range to display only those pixel values within a window width, all pixel values above the window width are set to white. Figure III.5.3 demonstrates that an image window is defined by the window level and the window width. An example of an image with a level 0 and width 256 is shown in Fig. III.5.4, whereas Fig. III.5.5 shows an image with level 50 and width 100, which gives much more contrast in certain tissue regions.

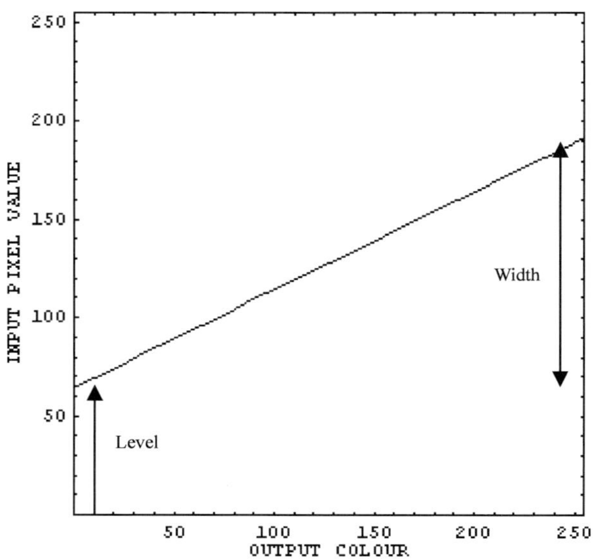

Fig. III.5.3. Example of an image window being defined by the window level and the window width. In the diagram the level is 64 and the width is 128. All pixel values of 64 and below are displayed as black, and all pixel values above 192 are displayed as white. The pixel values between 64 and 192 use the full grey level range to enhance contrast within this range

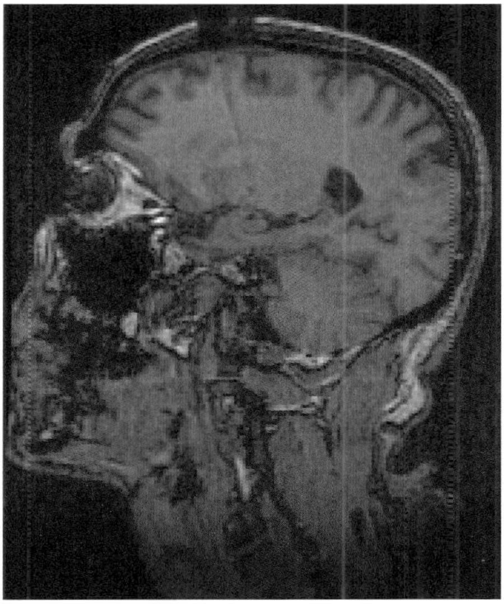

Fig. III.5.4. An image with a level of 0 and width 256

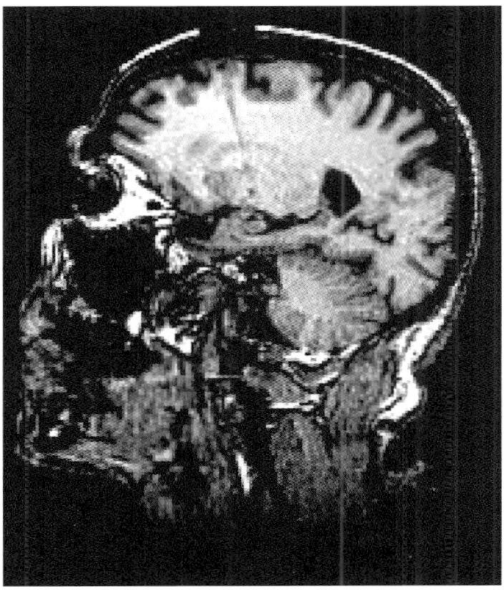

Fig. III.5.5. An image with level 50 and width 100, giving much more contrast in certain tissue regions than shown in Fig. III.5.4

Image Histogram

The distribution of pixel values in an image can be conveniently displayed as an histogram, see Fig. III.5.6. The image in Fig. III.5.6 does not have much contrast, its histogram shows that most of the image grey levels are compressed into the darker areas. Figure III.5.7 shows the same image following the application of an algorithm which redistributes the pixel data to expand the content more equally across all available grey levels; this technique is Histogram Equalisation.

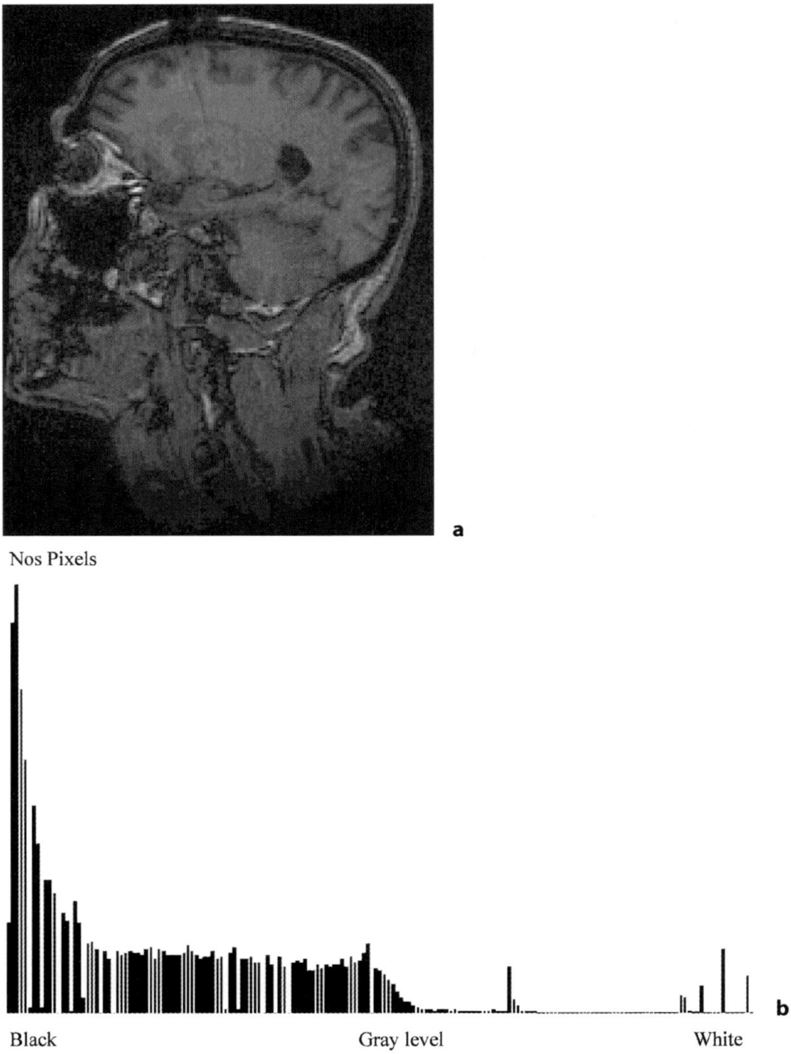

Fig. III.5.6. The histogram (**b**) shows the distribution of the number of pixels with a particular grey level in the image (**a**)

The histogram can be used in an ROI to quantitate tissue characteristics. In a CT image the pixel values correspond to x-ray attenuation indices, in a nuclear medicine image they correspond to activity and uptake indices. The relative areas under regions of the histogram give the percentage occurrence of these tissue characteristics.

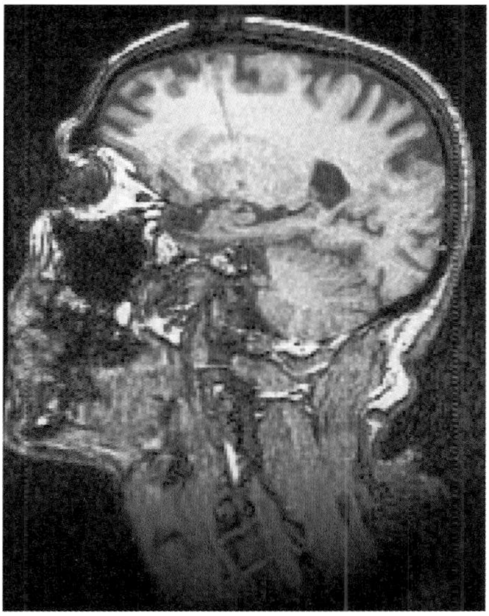

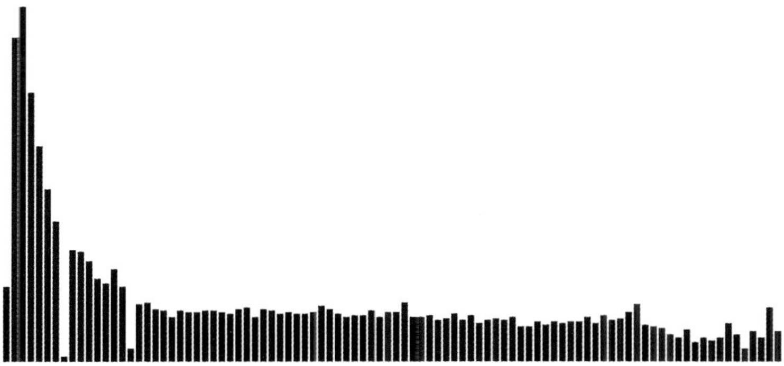

Fig. III.5.7. The same image as in Fig. III.5.6 following a Histogram Equalisation algorithm. Again (**a**) shows the image and (**b**) the histogram

Image Processing Kernels

Image characteristics may be enhanced by the application of convolution kernels. These are processing matrices which are applied to each pixel in an image and which change the pixel value in a manner depending upon the neighbouring pixel values, see Table III.5.1. In this case, the value of the pixel "e" will be recalculated as:

$$e = ar + bs + ct + du + ev + fw + gx + hy + iz \qquad \text{(III.5.1)}$$

For example, consider the central pixel in a small region of an image, represented in Table III.5.2 which is acted upon by the kernel shown in Table III.5.3. This will result in the central pixel, originally of value 1, returning a new value of 1, i.e. there is no change.

Table III.5.1. Image array to which a convolution kernel is applied

Image array			Kernel		
a	b	c	r	s	t
d	e	f	u	v	w
g	h	i	x	y	z

Table III.5.2. Pixel values in a small region of an image

1	1	1
1	1	1
1	1	1

Table III.5.3. Convolution kernel acting on pixel values shown in Table III.5.2

−1	−1	−1
−1	9	−1
−1	−1	−1

Consider the next case of an image array shown in Table III.5.4 acted upon by the same kernel as above. This will result in the central pixel, originally of value 2, returning a new value of 8. A similar outcome would be obtained for a longer central column, shown in Table III.5.5 which will result in the values shown in Table III.5.6 after rescaling negatives to 0; note that the border pixels have not been processed.

Table III.5.4. Pixel values in a small region of an image

1	2	1
1	2	1
1	2	1

Table III.5.5. Pixel values in a larger region of an image

1	1	2	1	1
1	1	2	1	1
1	1	2	1	1
1	1	2	1	1
1	1	2	1	1

Table III.5.6. Results of convolution kernel shown in Table III.5.3 acting on pixel values shown in Table III.5.5

1	1	2	1	1
1	0	8	0	1
1	0	8	0	1
1	0	8	0	1
1	1	2	1	1

In the original image the column of pixel values of 2 will appear as a faint vertical line against a background of pixel values of 1. The effect of this particular kernel is to increase the brightness of these edge pixels to a pixel value of 8 against a background of pixel values of 0–2. This is an Edge Enhancement kernel of order 3×3.

Convolution kernels are typically found as 3×3, 5×5 and 7×7 matrices. Most IP software will come with a predefined library of processing kernels and will also allow the user to define their own kernels for customised processing. The following are common kernels found in IP libraries:

1. Edge Enhancement: kernels which show the original image but which enhance all edges in the image, whatever the angle of the edge within the image. There are numerous types of kernels which will range from subtle to overt edge enhancement in the image.
2. Edge Direction: kernels which will enhance edges running in a particular direction within the image, e.g. verticals, or horizontals, or diagonals.
3. Laplacians: kernels which suppress the underlying image but extract the edges in an image, resulting in an outline image. May also be used as edge direction extractors.
4. Smoothing: kernels which soften an image to varying degrees, with a loss of resolution.
5. Noise Reduction: kernels which use averaging to despeckle an image, but with some loss of detail.

Image Arithmetic

Images may be added and subtracted together to enhance content.

Noise Reduction

In interventional procedures using x-ray fluoroscopy the patient may be irradiated for lengthy periods, so patient dose reduction techniques are required. If the dose rate is decreased the image becomes noisy due to quantum mottle and there is a danger of losing diagnostic detail. The signal strength in an image can be characterised by the signal-to-noise ratio (SNR). The noise in an image goes as the square-root of the signal (the number of x-rays per voxel). Consider an image with signal S, noise S and SNR = S. Now add two such images with a resultant signal = 2S, noise = (2S), and SNR = 1.41S. There is an overall increase in the perceived SNR.

This technique can be used to add several frames of noisy video which are then displayed at a slower frame rate e.g. instead of 20 frames s^{-1} display 4 integrated frames at 5 frames s^{-1} and with twice the SNR. For many procedures the decreased frame rate is not detectable.

Image Subtraction

Another technique, originally used in interventional fluoroscopy, results in a clear image of the vascular structure (digital subtraction angiography). An initial masking image is obtained of the ROI, then a second image is acquired following the injection of contrast medium into the vascular structure, and finally the digitally subtracted image of the contrast–mask is displayed to show just the vascularity without the overlying tissue. As there may be patient movement between these acquisitions there is often a pixel-shift mechanism employed to obtain the maximum subtraction effect.

In nuclear medicine SPET studies of the brain, a baseline image set may be acquired to act as a mask image and a subsequent action image acquired following some action on the part of the patient. This action may be a specific type of task, motor, verbal or cognitive, or it may be inadvertent as with epileptic seizure. The difference image (action–mask) will highlight those regions of the brain which have been activated during the action. Similar techniques are applied in heart studies and in functional Magnetic Resonance Imaging.

Similarly, in nuclear medicine tomographic acquisitions there is a simultaneous acquisition of on-peak and off-peak images, the latter representing the scatter contribution. The scatter image is automatically subtracted from the on-peak image prior to reconstruction to enhance detail in the image.

Unsharp Masking

This technique combines both a convolution kernel and image subtraction. The image is first convolved with a Low Pass Filter kernel to obtain an image containing the low spatial resolution components. This filtered image is then subtracted from the original image to leave the high spatial resolution components, resulting in sharpening of detail in the image.

Image Analysis

In addition to enhancing image characteristics with IP, there are various quantitative techniques which can be applied to medical images with Image Analysis (IA).

Dynamic Studies

Nuclear medicine was the first medical imaging modality to acquire quantitative functional information of imaged organs.

One of the most common dynamic studies is the renogram, in which a time series of successive images are acquired and analysed to give a time-varying graph of the passage of radioactivity through an organ. The simplest technique assumes that patient movement is negligible and the operator uses a lightpen to outline the organ margin on a reference image; this outline mask is then propagated through all the other images and the total activity within the defined ROI is calculated for each image.

The limitations on the accuracy of this type of analysis are twofold:

1. the definition of the organ margin for the ROI is operator dependent.
2. there is an underlying assumption that patient movement is negligible, allowing the use of a single outline to be defined as a mask for all the images.

Fortunately, neither of these assumptions lead to critical errors in the majority of patient studies. However, both of these problems could be addressed with the application of some artificial intelligence automation. An IP algorithm for the detection of the organ margin could be used on each image to give a standardised ROI and to adjust for any patient movement.

Curve Fitting

Quantitative analysis of the activity-time curves can be obtained by fitting ideal equations to the empiric data. For example, the excretion curve on the renogram is often approximated by an exponential and the corresponding excretion half-life may be quoted as an index of the excretion time.

In this case the raw data for the excretion curve, the set of (y,t) values where y is the activity at time t, is first converted to (z, t) where $z = \log(y)$. A linear regression fit to this data is then undertaken to obtain a linear equation:

$$z = st + i \tag{III.5.2}$$

where s is the slope of the line and i the z intercept

From this equation the corresponding exponential equation is:

$$y = A \exp(-st) \tag{III.5.3}$$

where $A = \exp(i)$

A convenient measure of the excretion rate may be given by the half-life, which is $t_{1/2} = 0.693/s$.

There are other functions which one may wish to fit to the raw data. For example, the blood curve may better be modelled with a double exponential of the form:

$$y = A \exp(-at) + B \exp(-bt) \tag{III.5.4}$$

or the overall uptake and clearance curve is sometimes modelled as a Gaussian:

$$y = 1/\sqrt{(2\pi)}\,\sigma\,[\exp - (t-\mu)^2/(2\sigma^2)] \tag{III.5.5}$$

In such cases there is no linear-fit algorithm available as in the preceding example. It is then necessary to undertake a recursive fit to a model equation, where initial "guestimates" of the unknown variables are used in the model equation. The recursive process then attempts to minimise the root-mean-square error to refine those values in subsequent reiterations, until a predefined error level has been reached (e.g. 0.0001).

Various model functions should be available on the nuclear medicine computer system for the user to undertake non-linear recursive data fitting.

Data Smoothing

Sometimes there is too much noise on the data to be able to undertake non-linear analysis of the data, e.g. equation fitting. The noise can crash the recursive process. It then becomes necessary to firstly smooth the dataset by employing filters.

The simplest smoothing algorithm is the Linear Filter, which takes the weighted average of the data around a point and replaces the original data for that point with the newly calculated average. For example, the (1, 2, 1) linear filter replaces the point y_n with the value

$$(1\,y_{n-1} + 2\,y_n + 1\,y_{n+1})/4 \tag{III.5.6}$$

This is a 3-point moving average.

Likewise, a (1, 2, 4, 2, 1) filter replaces y_n with

$$(1\,y_{n-2}+2\,y_{n-1}+4\,y_n+2\,y_{n+1}+1\,y_{n+2})/10 \qquad \text{(III.5.7)}$$

to give a 5-point moving average.

Other types of filter are available which act upon the spatial-frequency spectrum of the dataset, whether the data is from a physiological curve or from an image. The Butterworth filter is often employed:

$$A(f)=\sqrt{[1/(1+|f/f_c|^{2n})]} \qquad \text{(III.5.8)}$$

This gives an attenuation factor, A, at each frequency f, where f_c is the critical frequency and n is the order; both f_c and n are design parameters which are specified by the user to change the characteristics of the attenuation curve (see Fig. III.5.8). As the curves show, the low frequency components undergo little attenuation (multiplicative factors close to 1.0), whereas the higher frequency components undergo greater attenuation as the frequency increases. The critical frequency f_c corresponds to the frequency at which the attenuation has the value of 0.707. The order of the curve n determines the width of the transition zone, becoming narrower as n become larger.

There are a number of such filters which are used in SPET prior to reconstruction. The noise in a projected image is exacerbated by the application of the ramp filter which preferentially multiplies the high frequency "white" noise in each projection. The ramp filter is modified by using a multiplicative filter, such as a Butterworth multiplied by the ramp.

Other attenuation filters include:

Hamming	$A(f)=0.54+0.46\cos(\pi f/f_c)$	(III.5.9)
Hann	$A(f)=0.5+0.5\cos(\pi f/f_c)$	(III.5.10)
Shepp–Logan	$A(f)=\sin(\pi f/f_c)/(\pi f/2f_c)$	(III.5.11)

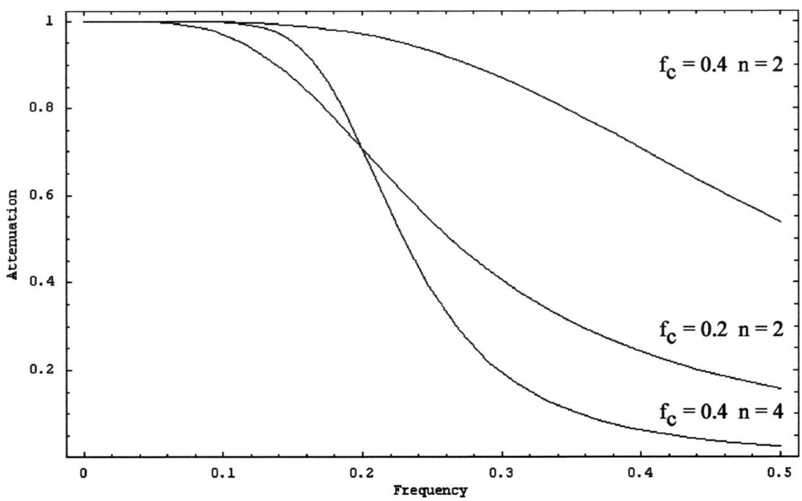

Fig. III.5.8. A family of Butterworth filters employed in frequency-space to attenuate high frequency noise in datasets

With these filters there is only the one adjustable factor, f_c. All of these filters demonstrate a smooth roll-off of the higher spatial frequencies to minimise aliasing artefacts which appears as a rippling effect in the filtered image.

Image Quality

Image analysis can also provide information on the quality of the imaging system. The most comprehensive information is obtained from the Modulation Transfer Function (MTF). The modulation is a measure of the contrast; the object modulation m_o is the inherent object contrast, and the image modulation m_i is a measure of the observable contrast. The Modulation Transfer is just the ratio of m_i/m_o and is a measure of the effectiveness of the imaging system in transferring its input information to appear as an output. All images can be broken down into spatial frequency components and each frequency will have a Modulation Transfer value. The total frequency response of the imaging system gives the MTF.

These concepts can be illustrated with a series of simplified "objects", the Line Pair test objects with various spatial frequencies.

Figure III.5.9 shows the object modulation for two Line Pair test objects with spatial frequencies of 1 lp cm^{-1} and 2 lp cm^{-1}. These show a sinusoidal variation of a radionuclide activity ranging from a minimum A_{min} to a maximum A_{max}. The corresponding inherent object modulation is:

$$m_o = (A_{max} - A_{min})/(A_{max} + A_{min}) \qquad \text{(III.5.12)}$$

Similarly, as shown in Fig. III.5.10, the image modulation is determined from indirect measurement of activity from the corresponding image densities I_{min} and I_{max}, so the image modulation is:

$$m_i = (I_{max} - I_{min})/(I_{max} + I_{min}) \qquad \text{(III.5.13)}$$

As a sample calculation, the m_o of the 1 lp cm^{-1} object with $A_{max} = 1$ and $A_{min} = 0.08$ (normalised) is therefore 0.85. The corresponding m_i for that object with $I_{max} = 0.31$ and $I_{min} = 0.1$ is 0.51. Therefore, the Modulation Transfer at 1 lp cm^{-1} is 0.51/0.85=0.6 (note that the image densities are normalised to the density for a flood field i.e. at 0 lp cm^{-1}).

A similar analysis for the 2 lp cm^{-1} object gives the same m_o, the I_{max} and I_{min} are 0.12 and 0.10, to give an $m_i = 0.085$, and the modulation transfer at 2 lp $cm^{-1} = 0.1$.

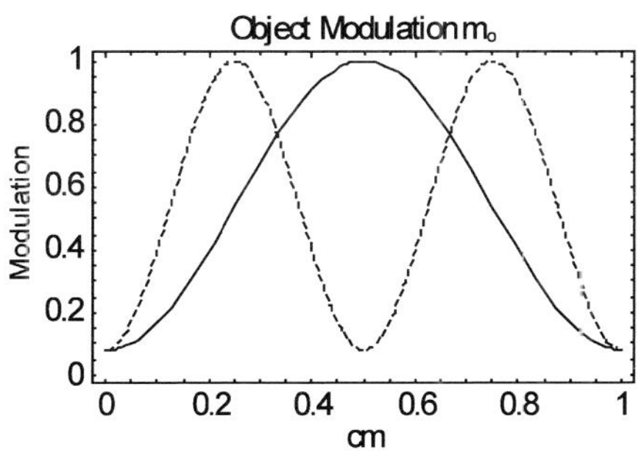

Fig. III.5.9. The object modulation for line pair test objects of 1 lp cm^{-1} and 2 lp cm^{-1} (dashed)

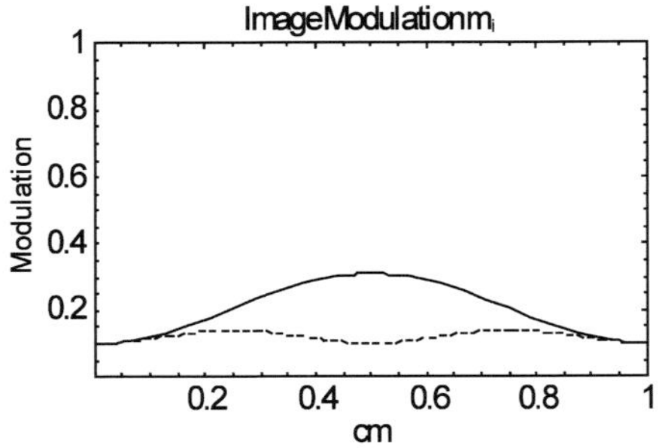

Fig. III.5.10. The image modulation for line pair test objects of 1 lp cm^{-1} and 2 lp cm^{-1} (dashed)

The graph, shown in Fig. III.5.11, displays the complete MTF for all such spatial frequency test objects.

In practice, Line Pair test objects are not used for such a direct measurement of the MTF. Instead, a thin line-source of activity is imaged and a Fourier Transform of the image gives the Line Spread Function, which contains all the spatial frequency information necessary for the MTF analysis.

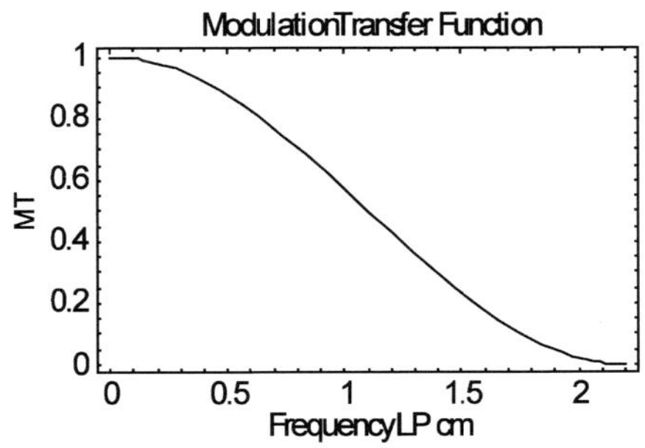

Fig. III.5.11. The corresponding modulation transfer function for all such spatial frequency test objects

PART IV

Quality Assurance

Quality Assurance

Abbreviations

ICRP	International Commission on Radiological Protection
QA	quality assurance
sd	standard deviation

The concept of quality assurance is an essential component in the radiation protection regimen recommended by the ICRP (ICRP 1991) in that the data obtained from a clinical investigation must provide the maximum diagnostic information for the minimum appropriate radiation exposure, and the likelihood of technical failure during an investigation must be reduced to an appropriately low level. This requires that each department develop and implement quality assurance protocols with which they are comfortable; dependent on the complexity of the investigation equipment and investigations in use, the historical behaviour of the equipment, the quality assurance equipment and staff expertise available, and any national regulations in force. These should be designed to provide sufficient information without seriously reducing clinical investigation time (IPSM 1992). Surveys have found that an encouraging degree of harmonisation exists (Bergmann et al. 1995) although variations do occur (Marshall and Wells 1999).

In a busy department, the quality assurance regimen is the one in which it is easiest to 'cut corners' because it is possible to reassuringly convince oneself that the quality of the investigations are not being compromised; if this is done the patient is, without doubt, being 'short changed'. A well designed and implemented QA programme is the foundation upon which all other work in the department is based and depends. Although many failures will be detected clinically and all data must be scrutinised for such occurrences (Hines et al. 1999a,b, 2000), activity distributions and failures are so unpredictable that many will not be evident, especially those involving slow and subtle changes, which will eventually result in errors in interpretation (IPSM 1992). Many recommendations are provided by a number of organisations and, when the protocols are being developed, the original reference should be consulted for a complete description of the procedure. These should then be adapted to suit the department (Daube-Witherspoon et al. 2002) and, thereafter, the protocol should be performed consistently and accurately, to allow slow and subtle changes to be detected.

Operational tests must be undertaken on radiopharmaceuticals before they are used clinically. The programme for equipment is more involved, however, and starts with a detailed specification. This is essential, not only to achieve the optimal match of operating characteristics and clinical requirements, but also to provide parameters against which to test for acceptance (AAPM 1987). For something as complicated as the modern γ camera–computer system, this is not a simple task but it is aided by published advice (Lunt at al. 1985; Tindale 1995; Wells and Buxton-Thomas 1995) including coincidence γ camera systems (British Nuclear Medicine Society website, IPEM 2004). Acceptance tests are mandatory to ensure that the equipment is installed and operating to the level specified (Hines et al. 1999a,b, 2000) and are critical in ensuring long term high quality performance. These require all appropriate documentation to be available and start with a physical inspection (IAEA 1991). Simultaneously with the acceptance tests, reference tests should be undertaken. These reflect operating conditions under clinical conditions and provide images and numerical results against which to test the ongoing performance of the equipment. Routine tests are reference tests which are undertaken at various regular intervals (IPSM 1992) to detect any slow and subtle deterioration in function that would be likely to affect the clinical efficacy of the equipment in the future. Their aim is also to reduce the downtime of the equipment, to a minimum, by finding and correcting any fault before failure. The final group of tests are the operational checks, which are undertaken each day before the start of clinical work.

Both images and numerical measurements must be recorded, to enable the results of routine testing to be compared with those obtained at reference (IPSM 1992) and, subsequently, when the system was operating optimally. Limits of acceptability should be defined on the basis of a statistical analysis of these results, so that random variations can be accommodated (Young et al. 1990). This is so that when these limits are exceeded consistent action can be implemented, such as the performance of additional tests or maintenance. Maintenance and faults must also be recorded. A sudden change in a test value could be detected if its value exceeded a 2 × sd variation from the mean historical value. In such a case, the measurement should be repeated because an erroneous value in a second such measurement would be extremely unlikely and then immediate attention should be considered. More difficult changes to detect are slow alterations over a long period; such a gradual degradation suggesting that the system may require maintenance in the near future (Hines et al. 1999a,b, 2000). For this, a trend analysis of recent sequential values is required. Detection is even more difficult when the only abnormality is an increase in the variation of the measurement. To detect this, the variance in the most recent set of measurements must be compared with those in a previous set

References

AAPM (1987) American Association of Physicists in Medicine report 22 rotating scintillation camera SPECT acceptance testing and quality control. American Institute of Physics, New York

Bergmann H, Busemann-Sokole E, Horton PW (1995) Quality assurance and harmonisation of nuclear medicine investigations in Europe. Eur J Nucl Med 22:477–480

Daube-Witherspoon ME, Karp JS, Casey ME, DiFilippo FP, Hines H, Muehllehner G, Simcic V, Stearns CW, Adam LE, Kohlmyer S, Sossi V (2002) PET performance measurements using the NEMA NU 2-2001 standard. J Nucl Med 43:1398–1409

Hines H, Kayayan R, Colsher J, Hashimoto D, Schubert R, Fernando J, Simcic V, Vernon P, Sinclair RL (1999a) Recommendations for implementing SPECT instrumentation quality control. Nuclear Medicine Section – National Electrical Manufacturers Association (NEMA). Eur J Nucl Med 26: 527–532

Hines H, Kayayan R, Colsher J, Hashimoto D, Schubert R, Fernando J, Simcic V, Vernon P, Sinclair RL (1999b) National Electrical Manufacturers Association recommendations for implementing SPECT instrumentation quality control. J Nucl Med Technol 27:67–72

Hines H, Kayayan R, Colsher J, Hashimoto D, Schubert R, Fernando J, Simcic V, Vernon P, Sinclair RL (2000) National Electrical Manufacturers Association recommendations for implementing SPECT instrumentation quality control. J Nucl Med 41:383–389

IAEA (1991) Tecdoc 602 quality control of nuclear medicine instruments. International Atomic Energy Agency, Vienna

ICRP (1991) International Commission on Radiological Protection publication 60, 1990 recommendations of the International Commission on Radiological Protection. Pergamon, Oxford

IPSM (1992) Report 66 quality control of gamma cameras and associated computer systems. Institute of Physical Sciences in Medicine. London

IPEM (2004) The basics of gamma camera PET. Institute of Physics and Engineering in Medicine, London

Lunt MJ, Davies MD, Kenyon NG (1985) A camera specification for tendering purposes. Nucl Med Commun 6:493–496

Marshall DSC, Wells CP (1999) South Thames quality standard for nuclear medicine equipment. Nucl Med Commun 20:789–798

Tindale WB (1995) Specifying dual-detector gamma cameras and associated computer systems. Nucl Med Commun 16:534–538

Wells CP, Buxton-Thomas M (1995) Gamma camera purchasing. Nucl Med Commun 16:168–185

Young KC, Kouris K, Awdeh M, Abdel-Dayem HM (1990) Reproducibility and action levels for gamma camera uniformity. Nucl Med Commun 11:95–101

Radiopharmaceutical

Abbreviations

ITLC instant thin layer chromatography
R_f relative front

Quality assurance of radiopharmaceuticals is mandatory on any reconstituted within the institution. National regulations will dictate the pharmacological sterility checks that need to be undertaken on both the product and the facilities. Because of the rapid administration of these products following manufacture, pharmaceutical quality control tests before administration are difficult. However, a number of such tests can be undertaken. These include: pyrogen testing and particle size estimation when appropriate. Facility tests include microbiological safety and performance of the work cabinets and microbiological safety of the rooms themselves (UKRG 2001). National regulations will also determine the exact physical and chemical tests that must be undertaken.

In manufactured radiopharmaceuticals involving radiolabelling with $^{99}Tc^m$, the first quality tests involve the eluate taken from the ^{99}Mo–$^{99}Tc^m$ generator. Following manufacture, quality assurance can then be undertaken on the product radiopharmaceuticals.

Generator

The expected elution yield should be calculated and compared with that actually obtained as an indication of whether or not the generator is working properly. The expected elution yield, which should be 80–90%, is calculated from the nominal ^{99}Mo activity and reference date multiplied by: the elution efficiency and the factor 0.964 to take into account equilibrium and decay scheme considerations (see '^{99}Mo–$^{99}Tc^m$, activity profile' in Chapter I.2).

Radionuclidic Purity

Ideally the eluate should contain only the desired radionuclide. A contaminating radionuclide does not add to diagnostic information but increases the radiation dose to the patient and may degrade the image quality, if its photon energy is similar to or greater than the useful radionuclide. Radionuclidic purity is defined as the ratio, in %, of the activity of the required radionuclide to the total activity of the eluate. For the ^{99}Mo–^{99}Tc^{m} generator, the expected contaminant is ^{99}Mo escaping from the column primarily because of damage in transit. The limits of acceptable contamination may be prescribed by national regulations. When this is not the case, a good rule of thumb is to keep the radiation dose to the patient from the contaminant to <10% of that from the desired radionuclide. Because it has a longer half-life and the relative contamination will increase with time, this may necessitate imposing a limit on the useable time of the eluate.

The contamination level of ^{99}Mo should be determined, in at least the first elution after any movement of the generator (UKRG 2001), by shielding the eluate in 6 mm of lead and measuring the resultant activity in a dose calibrator. The activity of ^{99}Tc^{m} will be attenuated by a factor of 10^{-5} whereas the activity of ^{99}Mo will only be attenuated by ~0.5, because of its higher energy γ photons, of 740 and 778 keV. This is then compared with the activity of ^{99}Tc^{m} measured without the shield. Although ^{99}Mo will also be detected; because the ^{99}Tc^{m} radioactivity in the eluate is at least 1000 times that of ^{99}Mo, the error in the estimation of the ^{99}Tc^{m} activity is small and can be neglected. Typically, the ^{99}Mo content is 3.8×10^{-4}% (Marengo et al. 1999).

Chemical Purity

A radiopharmaceutical should contain only the desired chemical, and chemical purity is an indication of the level of unwanted chemical species in the eluate. Al^{3+} may be present, from the large amounts of alumina in the column. Testing for this is undertaken most simply using colour indicator papers and a standard solution of Al^{3+}. The eluate contains less than a particular amount when the intensity of the colour it produces is less than that produced by a standard solution. Typically, the Al^{3+} content is <10 ppm (Marengo et al. 1999). Similarly, pH measurement, using narrow range pH paper, should be undertaken (UKRG 2001).

Radiopharmaceutical

Because ^{99}Tc^{m} may form several products during labelling with a pharmaceutical, it is important to determine the proportion of the ^{99}Tc^{m} that is in the required chemical form. This proportion may change during storage after manufacture and before administration.

Radiochemical Purity

Radiochemical purity is an indication of the percentage of the $^{99}Tc^m$ which is present as the desired radiopharmaceutical. It is normally high, approaching 97%. Depending on national regulations, radiolabelling quality must be evaluated on, at least, every new batch and also if imaging suggests a radiopharmaceutical problem. Such a quality control programme helps to eliminate many sources of labelling failure (Decristoforo et al. 2000; UKRG 2001).

To determine the radiochemical purity, the various radiochemical entities in the radiopharmaceutical must be separated. This is most commonly achieved by chromatography and paper or instant thin layer chromatography (ITLC), e.g. silica gel, are the usual methods because they are rapid, simple and inexpensive. This involves an absorbent strip (stationary phase) and a solvent (mobile phase), which is used in a closed container so that the atmosphere becomes saturated with the solvent vapour. Fresh solvent must be used for each test and the chromatography strips must be handled carefully by the edges or with forceps otherwise separation artefacts may be produced.

A few μl of the radiopharmaceutical are placed near the bottom (origin) of the chromatographic strip, over as small an area as possible and without the applicator, e.g. needle, touching the surface of the strip. Before the spot dries, to avoid any oxidation of the radiopharmaceutical to free pertechnetate, the strip should be placed in the solvent. The origin must not be immersed otherwise the separation may be poor. The solvent moves along the chromatographic strip by adsorption and capillary action, which separates the radiochemical species, the most soluble moving fastest. When the solvent front has moved an adequate distance along the strip, the strip is removed from the solvent, the exact position of the solvent front is marked with pencil and the strip is allowed to dry.

The identity of each radiochemical species can be determined by measuring the distance that it has moved along the strip. Expressed as a proportion of the distance to the solvent front, this is known as the relative front (R_f). It is characteristic of each radiochemical species and should be constant for a particular combination of stationary and mobile phases. The relative activity of each species can be determined by measuring the distribution of activity along the length of the strip, relative to the total activity. The radiochemical purity of $^{99}Tc^m$ is often determined using two different stationary and mobile phases. One combination determines the percentage of hydrolysed/reduced technetium (HR%) present and the other combination determines the percentage free pertechnetate (F%) present.

The radiochemical purity is then calculated as:

$$100 - HR\% - F\% \quad \text{(IV.2.1)}$$

References

Decristoforo C, Siller R, Chen F, Riccabona G (2000) Radiochemical purity of routinely prepared $^{99}Tc^{m}$ radiopharmaceuticals: a retrospective study. Nucl Med Commun 21:349–354

Marengo M, Aprile C, Bagnara C, Bolzati C, Bonada C, Candini G, Casati R, Civollani S, Colombo FR, Compagnone G, del Dottore F, Di Guglielmo E, Ferretti PP, Lazzari S, Minoia C, Pancaldi D, Ronchi A, Sanita di Toppi G, Saponaro R, Torregiani T, Uccelli L, Vecchi F, Piffanelli A (1999) Quality control of $^{99}Mo/^{99}Tc^{m}$ generators: results of a survey of the Radiopharmacy Working Group of the Italian Association of Nuclear Medicine (AIMN). Nucl Med Commun 20:1077–1084

UKRG (2001) Report of a joint working party: the UK Radiopharmacy Group and the NHS Pharmaceutical Quality Control Committee. Quality assurance of radiopharmaceuticals. Nucl Med Commun 22:909–916

Non-Imaging

Abbreviations

FWHM	full width at half maximum
MCA	multichannel analyser
PHA	pulse height analyser
PMT	photomultiplier tube
QA	quality assurance

Monitors

The tests for both contamination monitor and exposure meter are similar. They aim to check for obvious physical damage, battery failure and the response to a known activity. Operational checks should check the probe for damage, the state of the battery; and the measured count rate or exposure rate, for a known activity standard source, against expected count rate for contamination and exposure rate for the exposure meter. Routine tests should involve calibration of the scale against known activities and exposure rates respectively; and often have to be undertaken by authorised institutions.

Dose Calibrators

Published recommendations regarding the quality assurance of dose calibrators include: IAEA (1991), IPSM (1992), IEC (1992, 1994). QA measurements should be within the manufacturer specifications for the unit and should generally not be more than 10% different to an expected reading. A number of certified standard sources are required to undertake these tests. These are 'vial' shaped and usually in the activity range 4–40 MBq. ^{57}Co is essential because its emission energy is close to that of ^{99}Tc^{m}. Other sources are used to cover a wide energy range and include: ^{133}Ba, ^{137}Cs, ^{60}Co and ^{129}I.

Operational Tests

The position of the syringe holding part of the dipper should be confirmed to be at the usual height. The background activity should be measured, using the most usual calibration factor or one with a higher amplification. If the indicated activity is higher than anticipated, the dipper and well insert should be removed and, if contaminated, cleaned. Any residual activity must then be subtracted from measurements until it is non-significant. If this and environmental contamination is not the cause, the more involved routine tests should be undertaken.

Constancy Check

The constancy of activity measurement, using the ^{57}Co source, should then be assessed for each of the electrometer factors in routine use, by correcting for decay and comparing to the values at acceptance. During routine use, this constancy should be continually assessed by comparing the measured activity with that anticipated in, for instance, ^{99}Tc^{m} generator eluate, reconstituted ^{99}Tc^{m} kits, commercial preparations and patient administrations. Any unexplained measurement should be further evaluated using the more involved routine tests.

Routine Tests

Manufacturer specific checks such as zero drift, and battery or electronic status should be undertaken at the recommended intervals.

Electronic Linearity

The electrometer amplifier output should be inversely linear to the radionuclide factors. This is checked by plotting the reciprocal of the activity measurements, obtained in the constancy check, against the radionuclide factors.

Accuracy

The accuracy for measuring $^{99}Tc^{m}$ is assessed, approximately, by comparing the measured activity of the ^{57}Co standard source (A) with its certified activity corrected for decay (A_c).

$$\text{accuracy} = 100 * (A - A_c)/A_c\ (\%) \qquad \text{(IV.3.1)}$$

The accuracy for other radionuclides is then assessed over a wide energy range, using the other standard sources.

Geometric Variation

The measurement accuracy will alter with both source volume and container; and should be assessed by comparing the measured activity of each combination (A) with a standard source (A_s), such as $^{99}Tc^{m}$ in a volume of 10 ml in a glass elution vial. The severity of any discrepancies also depends on the radionuclide, being more severe, with low energy γ emitters such as ^{123}I and β^- emitters, than with $^{99}Tc^{m}$. A correction factor should be implemented for any combination causing errors >5%.

$$\text{correction factor} = A_s/A \qquad \text{(IV.3.2)}$$

When multiple calibrators are in use, the correction factors should be calculated against one unit taken to be the 'reference'.

Precision

Precision may alter over repeated measurements. This is assessed over a wide activity range, e.g. using $^{99}Tc^{m}$ sources of 10–50,000 MBq, and over a wide energy range using the standard sources. Each is measured ten times and removed from the calibrator between measurements.

$$\text{precision} = 100 * (A_s - A_m)/A_m\ (\%) \qquad \text{(IV.3.3)}$$

where A_m is the mean of the 10 measurements.

Count Rate Linearity

The measurement accuracy should be independent of the count rate. This is assessed over a wide range of $^{99}Tc^{m}$ activities, with the sources contained in vials of ~20 ml and the high activity being the highest available.

Method 1

The high activity is measured at regular intervals until the activity falls to a few MBq.

Method 2

Different volumes, eg 0.1–10 ml, of a high concentration solution are dispensed into the same size vials, and then made up to a standard volume. The activities of the sources are measured at regular intervals until the activity, in the lowest, falls to a few MBq.

Analysis

At activities <100 MBq, the true activity of the source should = the measured activity. Higher values of true activity can then be calculated by extrapolation, allowing the deviation to be assessed.

Scintillation Counters

These comprise well counters and automatic sample counters and published recommendations regarding their quality assurance include: Todd et al. (1980), IAEA (1991), Fleming et al. (2002). Measurements are made using a long half-life source that matches the size of the well, with a sufficiently low activity (~4 kBq) that deadtime is not significant. A suitable radionuclide is ^{137}Cs, although ^{129}I is better for systems designed primarily to measure low energy emitters such as ^{125}I.

Operational Tests

The high voltage applied to the PMT should be confirmed to be at the expected value. With the check source installed, the photopeak should be displayed on the multichannel analyser, MCA, display and confirmed to be in the expected channel. If necessary, the amplifier gain should be adjusted to position the photopeak in the correct channel, and recorded. Rapid fluctuations may be caused by an unstable power supply, by temperature changes or by electronic faults. Slow drift may indicate deterioration of the crystal or PMT. Without the check source installed, background counts should be acquired, at the photopeak of the check source, for a set duration. If this is higher than anticipated, any removable sample tube holders should be checked for contamination and cleaned, if necessary. If this or environmental contamination is not the cause, the more involved routine tests should be undertaken.

Routine Tests

Energy

Calibration

The photopeaks should be confirmed to occur at the correct position on the multichannel analyser by acquiring photopeak spectra, with at least 2500 counts in the peak channel, for all radionuclides in routine use.

Resolution

From the calibration photopeak data, the energy resolution is calculated for each radionuclide.

$$\mathrm{FWHM} = 100 * \mathrm{FWHM}/\mathrm{E_p} \qquad \text{(IV.3.4)}$$

where E_p is the energy of the radionuclide photopeak in keV

A sudden deterioration suggests crystal damage. A slow change suggests PMT problems or yellowing of the crystal from ingress of moisture.

Linearity

From the calibration photopeak data, the recorded photopeak energies are plotted against the known photopeak energy of each radionuclide. Non-linearity suggests a problem in the amplifier and a zero offset suggests a problem in the PHA.

Sensitivity

With the standard source installed, the usual PHA window is centred on the photopeak and counts acquired for a set duration, sufficient for at least 10^4. The count rate per kBq after background subtraction and decay correction is calculated. Values less than anticipated, suggest incorrect energy calibration or degraded energy resolution. In automatic sample changing counts, another factor may be the mechanical system because incorrect installation of the source will result in a reduced detection sensitivity.

Efficiency

The efficiency can then be calculated:

$$E\,(\%) = 100 * (R_s - R_b)/(A_s * e) \qquad \text{(IV.3.5)}$$

where
R_s is the count rate of the source in counts per second
R_b is the count rate of the background in counts per second
A_s is the activity of source in becquerels
e is the the frequency of emission as a fraction of disintegrations

Geometric Variation

The geometric variation in efficiency should be evaluated by counting a series of samples of varying volume, and the optimal volume determined.

Precision (χ^2)

The reproducibility is assessed by undertaking 10 sequential acquisitions, using the standard source and a PHA window centred on the photopeak, for a constant duration such that at least 10^4 counts are obtained. Background subtraction is not necessary. In automatic sample counters, the source should not be moved between measurements because this may introduce an error related to source positioning. The χ^2 statistic is calculated:

$$\chi^2 = \Sigma\,(C_i - C_m)^2/C_m \qquad \text{(IV.3.6)}$$

where
C_i are individual counts
C_m is the mean of the counts

A χ^2 value which is too low (<3.3 for 10 measurements) or too high (>16.9) indicates too little or too much variation in the measurements, respectively, and suggests a problem. The low value possibly originating from a high count rate and a high value from electrical noise.

Count Rate Linearity

The measurement accuracy should be consistent at both low and high count rates, and its variation should be assessed using a range of $^{99}Tc^{m}$ activities.

Method 1

An initial activity, eg ~200 kBq, is measured at regular intervals until the activity falls to a few kBq.

Method 2

Different volumes, eg 0.05–5 ml, of a concentration of ~200 kBq/ml are dispensed into counting tubes, and then made up to a standard volume. The activities of the sources are measured at regular intervals until the activity, in the lowest, falls to a few kBq.

Analysis

Background subtraction is not necessary. At activities <10 kBq, the true activity of the source should = the measured activity. Higher values of true activity can then be calculated by extrapolation, allowing the deviation to be assessed.

Probes

Surface

The mechanical stability of the unit, including the collimator and probe mountings should be checked. Then the QA of the detector and electronics are undertaken as for scintillation counters. However, the standard sources are usually certified 400 kBq discs or rectangles, and the activity linearity source should be in the MBq range. In use, these are positioned in a mounting at the front of the detector, on the axis and at a consistent distance from the crystal. The error limits are usually larger than with the counters because of positioning uncertainties.

Intraoperative

Operational Tests

These units do not usually provide an energy spectrum display and therefore the positioning of the PHA window on the photopeak cannot be checked visually. An operational check is therefore to measure the constancy using a standard source (Zanzonico and Heller 2000).

Routine Tests

Extensive quality assurance protocols have not yet been developed (Zanzonico and Heller 2000) but a number of tests have been described. The sources used in these are those administered clinically or similar with more convenient longer half-lives.

Absolute and angular sensitivity measurements have been made in air and at 5 cm depth in water, with values varying by factors of over 30 among different designs. Spatial resolution measurements have been made at 5 cm depth in water, and it is suggested that spatial resolution allowing localisation to within 1 cm is probably sufficient. Shielding adequacy, count rate linearity and energy resolution also form part of regular quality assurance (Tiourina et al. 1998; Perkins and Britten 1999). Practical evaluation using clinically relevant phantom studies also forms an important part of the assessment, an example is determination of the mean error between an actual lesion and the estimated location (Zanzonico and Heller 2000). Another is prediction of the minimum separation between the injection site and lymph node required to allow the node to be identified in the presence of high injection site activity; the best probes allowing localisation 2–3 cm closer to the injection site than the poorest performing probes (Britten 1999).

References

Britten AJ (1999) A method to evaluate intra-operative gamma probes for sentinel lymph node localisation. Eur J Nucl Med 26:76–83

Fleming J, Williams N, Skrypniuk J, Thorley P, Rose M, Driver I (2002) Report 85 radioactive sample counting – principles and practice. Institute of Physics and Engineering in Medicine, York

IAEA (1991) Tecdoc 602 quality control of nuclear medicine instruments. International Atomic Energy Agency, Vienna

IEC (1992) International Electrotechnical Commission publication IEC61145 calibration and usage of ionization chamber systems for assay of radionuclides. IEC, Geneva

IEC (1994) International Electrotechnical Commission publication IEC1303 medical electrical equipment: particular methods for declaring the performance of radionuclide calibrators. IEC, Geneva

IPSM (1992) Report 65 protocol for establishing and maintaining the calibration of medical radionuclide calibrators and their quality control. Institute of Physical Sciences in Medicine, York

Perkins AC, Britten AJ (1999) Specification and performance of intra-operative gamma probes for sentinel node detection. Nucl Med Commun 20:309–315

Tiourina T, Arends B, Huysmans D, Rutten H, Lemaire B, Muller S (1998) Evaluation of surgical gamma probes for radio-guided sentinel node localisation. Eur J Nucl Med 25:1224–1231

Todd JH, Short MD, Bessent RG, Hawkins LA, Leach KG, Ottewell D, Morgan DW (1980) HPA topic group report 33 an introduction to automatic radioactive sample counters. Hospital Physicists' Association, London

Zanzonico P, Heller S (2000) The intraoperative gamma probe: basic principles and choices available. Semin Nucl Med XXX:33–48

CHAPTER IV.4

Single Photon Planar Imaging

Abbreviations

CFOV central field of view
COR centre of rotation
ECG electrocardiogram
FOV field of view
FWHM full width at half maximum
FWTM full width at tenth maximum
IAEA International Atomic Energy Agency
IEC International Electrotechnical Commission
LE low energy
LEGP low energy general purpose
LFOV large field of view
LSF line spread function
MCA multichannel analyser
MTF modulation transfer function
NEMA National Electrical Manufacturers' Association
PHA pulse height analyser
PMT photomultiplier tube
PSF point spread function
QA quality assurance
ROI region of interest
SFOV small field of view
SPET single photon emission tomography
UFOV useful field of view

Quality assurance measurements on γ cameras are complicated; and variations among institutions have highlighted the need for a standard implementation (Marshall and Wells 1999). The measurements have to satisfy two vital demands. The first is to provide a pre-purchase comparison of the performance of different cameras and, in post-purchase acceptance tests, to determine whether the user specifications have been achieved. The second is to assess the acceptability of a particular camera, for clinical imaging, over the duration of its use; and to detect any deterioration in performance early enough to allow preventive servicing, before failure, in order to reduce clinical down-time and the possibility of interrupted patient investigations. Many comprehensive recommendations have been published, which can be categorised into manufacturer oriented and user oriented. The manufacturer oriented tests developed by the National Electrical Manufacturers' Association, NEMA (Hines et al. 1999a,b, 2000; NEMA 1994, 2001), are usually intrinsic and undertaken in air whereas those developed by the International Electrotechnical Commission, IEC (IEC 1992, 1998), are undertaken in the presence of scatter and therefore more closely simulate the clinical performance of the camera. The user oriented tests are described in: AAPM (1982, 1987); IAEA (1991); IPSM (1992); Ficken and McCartney (1994); Waddington et al. (1995); IPEM (2003). At acceptance, data should be compared with the manufacturer's factory or installation data and at routine testing, they should be compared with reference or previous data (IAEA 1991). Generally, numerical test measurements should be within 10% of the anticipated value. The IAEA have also compiled a quality control atlas, which contains examples that will help to gain an understanding of the interpretation of quality control tests and how to recognise artefacts. The atlas also contains images from clinical investigations, in which there was unsuspected malfunctioning or suboptimal use of the system, or operator error (IAEA 2003).

Specification

Appreciation of the user oriented protocols is required to construct specification documents for γ cameras. This is a difficult procedure for these instruments (IPSM 1992) but advice has been published (Lunt 1985; Wells and Buxton-Thomas 1995) and is regularly updated, for instance to include dual-detectors (Tindale 1995) and coincidence γ cameras, with a standard format document, on the British Nuclear Medicine Society website, allowing a ready comparison of different systems.

Installation

The power supply to the camera head, and particularly the high-voltage supply to the PMTs, must be as stable as possible. Any interruption will require a period of stabilisation, after re-application of the power, before the camera can be used for imaging. A rule of thumb is that this should be twice the power-out period, up to a maximum of 24 hours. The electronics are delicate and must be protected from electrical spikes which may occur at such power-outs, which may necessitate installing external electrical protection. It is essential to accommodate the manufacturer's environmental specifications, particularly with regard to acceptable temperature and humidity ranges and rates of change, and protection from magnetic fields. Long term reliability and performance quality, and therefore user confidence and liking, depends on thorough acceptance testing, to ensure that the purchase specifications have been achieved. It must be undertaken immediately after installation and before the camera is put into clinical use; a camera not performing optimally at this stage has a high likelihood of never doing so (IAEA 1991). It is at this stage that the reference tests are undertaken and the reference values and images established.

Acquisition Conditions

The acquisition conditions during QA measurements are extremely important, both to obtain reliable data and also to maintain the camera in a useable condition. Manufacturer's instructions should be followed, eg it may be necessary to use a lead mask when making intrinsic measurements on some cameras. Unless otherwise specified in a particular test, the acquisitions should be undertaken with the following conditions:

1. the PHA should be peaked before each measurement.
2. the PHA window should be that used clinically, which is usually 20% but this may be reduced to as low as 15% on some systems with very good energy resolution.
3. the count rate should be $<2\times10^4$ counts s^{-1}, and this should be less than will result in a 10% count rate loss.
4. there should be no acquisition zoom.
5. the position of the gantry and bed should be recorded such that the acquisition geometry may be reproduced between different investigations.
6. for multiple-headed cameras, measurements should be undertaken on each individual detector head and the values should be similar.
7. to reduce the possibility of contaminating the detector, it should be covered with a plastic sheet.
8. Particular care must be taken when measurements are taken with the collimator removed, because the crystal is very vulnerable to physical damage and the camera is very sensitive to extraneous radiation.

Sources

The performance of the camera will often vary with photon energy therefore, if possible, should evaluated using the radionuclide most appropriate to the test which will usually be $^{99}Tc^{m}$. The best point sources are made using the smallest volume in a narrow plastic, rather than glass, syringe. A plastic blind hub should be used, rather than a needle which can cause both hot and cold artefacts (O'Connor 1996). This type of source should be positioned at the centre of the field of view, at a consistent distance of at least 5 UFOVs from the crystal.

When point sources are not appropriate and extended sources of uniform intensity must be used, these can be fillable sources of various radionuclides or solid sources of ^{57}Co. Their diameter should be at least 3 cm greater than the UFOV and, for consistency, they should be positioned in the same orientation, relative to the detector. The uniformity of the source should be checked; on delivery for solid sources and immediately before use for fillable sources. This is done using a 10^6 count acquisition and, if there are any obvious non-uniformities on the image or a difference when compared to previous or standard images (Busemann-Sokole et al. 1996), a similar but 90° rotated acquisition. A non-uniformity which rotates attributes the problem to the source rather than to the camera.

Fillable flood sources are either made of plastic or glass. A problem with plastic sources is that the walls of the source distort outwards under pressure of the liquid. These should, therefore, be filled horizontally to reduce this effect. Both types should be filled and mixed well at least one hour before use to allow even dispersion of the radionuclide. There must be no obvious signs of bubbles or solid matter, including microorganisms. A major problem with this type of source is the radiation exposure to the person making it.

Solid flood sources have the disadvantages of cost and decay but the advantages are a constant uniformity and reduced radiation exposure. They are usually made of ^{57}Co, in which case a check for ^{56}Co and ^{58}Co contamination should be made at delivery. These shorter-lived, 70–80 days, contaminants have high-energy photon emissions, >500 keV, that can create image artefacts and cause errors in the measurement of detection sensitivity (Busemann-Sokole et al. 1996; O'Connor 1996). If problems exist, they can be reduced by letting the source age for a few months before use or by separating it from the detector as much as possible during measurements.

Phantoms

Quadrant bar or orthogonal hole phantoms provide linear arrangements of different separations in lead, which should be matched to the intrinsic spatial resolution of the camera. The spatial resolution FWHM is 1.75 times the width of the smallest resolved bars or the minimum inter-hole spacing. The phantoms should be positioned as closely as possible to the crystal face and carefully aligned in the x and y directions. Measurements are usually repeated with the phantoms rotated in 90° increments and then inverted and the rotation repeated, resulting in 8 acquisitions.

Operational Checks

Daily and whenever they are changed, collimators should be checked for dents and scratches. With good collimator holders and changers the chances of damage during loading are small but collimators are delicate, even though they tend to be covered with a protective material. If damage is apparent, the uniformity should be checked by acquiring a system flood image. The detector head mountings should be checked for freedom from mechanical defects, the head and gantry movements for reliability and safety, and the emergency stops and collision avoidance sensors for operation.

PHA Energy Calibration

This checks that the photopeak positions are as anticipated.
Acquisition mode: MCA
Photopeak: radionuclides in use

This is usually done using an MCA display and by identifying the peaks in keV; but if the camera uses an automatic PHA peaking, the sources used should not be patients because the PHA windows could be centred over the energies of the scattered photons. The PHA positions may not match the known photon energies exactly because cameras tend to be only adjusted to the $^{99}Tc^{m}$ energy and not to the energies of other radionuclides. What cannot be accepted are short term changes in photopeak position because camera performance could change within a patient investigation; slow changes can be accommodated, if checked regularly. The shape of the photopeak is also important; one showing apparently more scatter than anticipated may indicate erroneous correction files in some advanced system.

Background Count Rate

This test checks for contamination, particularly on the crystal which is very important since a small activity can significantly degrade intrinsic uniformity and cause artefacts on clinical images. The test is also very sensitive for revealing noise in the detector electronics.

Photopeak: $^{99}Tc^{m}$
Matrix: 128
Acquisition: 100 s

The acquisition is undertaken with the detector facing downwards, with or without a collimator loaded. The latter does not check for collimator contamination, however, which is easier to detect in the intrinsic test than on a flood acquisition. The obtained count rate should be low, the usual level being determined historically over a number of acquisitions. The image should be visually inspected, and should be uniform with no count groupings. An abnormal result necessitates the detector being re-oriented to check for local environmental contamination.

Uniformity and Sensitivity

Uniformity is the most important indication of γ camera function because it is affected by most faults (O'Connor 1996), and can be assessed either extrinsically or intrinsically.

Matrix: 64 or 128
Acquisition counts: $1–3\times10^{6}$ for SFOV, $2–5\times10^{6}$ for LFOV

The acquisition time should be recorded and the sensitivity, as counts s^{-1} MBq^{-1}, calculated.

Extrinsic

The advantage of this test is that is does not require removal of the collimators and in a dual head system can be undertaken for both heads simultaneously by bringing them close together. It checks both the detector and collimator uniformity and will reveal any sizable collimator contamination or damage (O'Connor 1996).

Radionuclide: ^{57}Co
Source: sheet
Activity: 70 MBq for SFOV to 750 MBq for LFOV (depending on collimator sensitivity)

Intrinsic

This checks the uniformity of only the detector and not the collimator.
Radionuclide: $^{99}Tc^{m}$
Source: point
Activity: ~40 MBq

For first few months after installation, the routine test with quantitation should be undertaken, instead of this, because there is likely to be a drift in the PMTs and therefore uniformity, and this needs to be monitored regularly to determine when the correction files need updating.

Evaluation

The image should be inspected visually (Ficken and McCartney 1994), on both the display and hardcopy, and compared to previous acquisitions. Any non-uniformities indicate a problem, including even subtle changes such as slightly reduced counts at the periphery.

Routine Tests

These tests are presented in order of operational convenience.

Physical Inspection

A thorough physical inspection should be undertaken, involving both mechanical and electrical considerations. Gantry movements are now very sophisticated; all movement combinations and safety devices must be checked for expected operation within allowed safety limits. Electrical safety should include annual measurements of earth continuity, insulation and leakage current (IPSM 1992).

PHA Energy Calibration

This is performed as in the operational tests but for a number of radionuclides of various energies, such as $^{99}Tc^{m}$, ^{123}I, ^{67}Ga, ^{131}I. It is intended to assess the energy linearity of the MCA, consistency in which is essential in setting clinical PHA photopeaks.

Shield Leakage

The detector head must shield against extraneous radiation, so that the detector only records photons incident on the front of the crystal.

Radionuclide:	$^{99}Tc^{m}$ or radionuclide with highest γ energy in clinical use
Source:	point source or a partially shielded vial
Activity:	4–40 MBq such that the count rate in front of the collimator is $<10^4$ counts s^{-1}
Collimator:	appropriate to radionuclide energy
Matrix:	64
Acquisition:	100 s a time, such that 10^4 counts at the collimator

A background and then a reference acquisition, with the source 100 mm above the collimator and on its central axis, are undertaken. Test acquisitions are then made at 12 sites around the detector head shielding, especially at sites of joints in shielding and cable exit points. After background correction, the counts at each test site are compared to those at the reference site. Any design weaknesses, eg flat collimators, should be particularly well investigated.

Acquisition Orientation Settings

The orientation of an image is important for defining the anatomical positioning of areas of interest. This can be manipulated by the acquisition system and this test confirms proper operation.

Radionuclide:	$^{99}Tc^{m}$
Source:	point
Activity:	~40 MBq
Matrix:	64
Acquisition:	10 s

A static image is acquired, for each of the acquisition orientation settings. The anticipated orientation of the acquired images should be then be confirmed.

Intrinsic Tests

Energy Resolution

This test assesses the ability of the detector to discriminate between scattered and unscattered radiation. An MCA spectrum is acquired for $^{99}Tc^{m}$ and the photopeak of the ^{57}Co spectrum is then established to calibrate the energy scale. It is possible to undertake this measurement intrinsically or extrinsically. However, most recommendations suggest the former.

Radionuclides:	$^{99}Tc^{m}$ and ^{57}Co
Source:	point
Activity:	~4 MBq
Photopeak:	spectrum display scanned in 1 keV increments

Matrix: 64
Acquisition: ~10 s, such that the counts in the peak channel 10^4

The energy resolution is calculated using equation II.1.1 presented in Chapter II.1. The major factors that degrade energy resolution are poor alignment of the PMT gains, failure of one or more PMTs, defects in or deterioration of the crystal, separation of the PMT-light guide assembly from the crystal and high count rates (IAEA 1991).

Uniformity

This is a simple test of basic camera function.

Acquisition

Radionuclide: $^{99}Tc^{m}$
Source: point
Activity: 10–20 MBq
Matrix: 64 for SFOV, 128 for LFOV
Acquisition counts: 20×10^6 for SFOV, 30×10^6 for LFOV (at least 10^4 counts in the central pixel)
Count rate: as usual and also at 7.5×10^4 counts s^{-1}

At acceptance because of expected drift, this test should be undertaken <1 week after any on-site corrections have been acquired.

Analysis

Visual evaluation should be undertaken at both normal and high contrast settings. A number of quantitative indices can then be calculated, sometimes after 9-point smoothing of the image:

1. mean pixel count
2. standard deviation
3. covariance of pixel counts (= 100 × sd/mean pixel count)
4. integral non-uniformity
5. differential non-uniformity
6. percentage of pixels containing counts within:
   2.5%, 5%, 7.5%, 10%, 12.5% and 15% of the mean count

Evaluation

The image should be visually inspected for variations in density, particularly for any areas that clearly stand out or any areas indicating an out-of-tune PMT. If any of the uniformity values are worse than anticipated, the fundamental detector performance should be evaluated by turning off the flood corrections and acquiring an uncorrected flood image. The major factors that degrade intrinsic uniformity are poor alignment of the PMT gains, failure of one or more PMTs, spatial non-linearities, defects in or deterioration of the crystal, separation of the PMT-light guide assembly from the crystal, incorrect setting of the position or width of the PHA window and high count rates (IAEA 1991).

Narrowed and Off-Centred PHA Windows

Intrinsic uniformity depends on both the energy of the radionuclide and the PHA window. Off-peak imaging can detect crystal hydration defects which appear initially as white spots, turning yellow later. At first, the spots are thin and their total area is small, but slowly they can affect the whole scintillator. The optical property of the hydrated crystal is different from the not hydrated one; the hydrated spots absorbing and scattering light in a different way, causing a loss in the number of detectable photons from scintillation events. The result is a non-homogeneous light response for given gamma photon energy, with a varying geometrical distribution in the crystal (Keszthelyi-Landori 1986).

Because of the much reduced sensitivity, the acquisition is shortened.

Acquisition: 15×10^6 counts (at least 5×10^3 in the central pixel)
PHA window: narrow, 15%

This window width is often used clinically and uniformity should be maintained at that measured for a 20% window. Any deterioration suggests that the on-site acquired corrections need updating. Otherwise, the coupling between the PMT-light guide assembly and crystal may not be intact.

PHA window: very narrow, 10%

The image may reveal the PMT positions and the uniformity may be worse than above but any non-uniformities should be symmetrically distributed over the FOV.

PHA window: very narrow, 10% with a –5% and then a +5% shift

Evaluation need only be by visual inspection which should reveal the PMT positions but, again, any non-uniformities should be symmetrically distributed over the FOV.

Radionuclides Other than $^{99}Tc^m$

Cameras are optimised for $^{99}Tc^m$, and other radionuclides may experience some degradation in uniformity.

Radionuclide: low energy such as ^{201}Tl and high energy such as ^{131}I
Acquisition: 30×10^6 counts

The uniformity indices should not be significantly different from those at 140 keV.

Spatial Linearity

This is often undertaken at the same time as spatial resolution because the same test pattern, geometry and acquisition parameters are used.

Radionuclide: $^{99}Tc^m$

Qualitative

The advantage of this test is that the evaluation can be undertaken over the entire crystal.

Phantom: quadrant bar/orthogonal hole
Source: point
Activity: 40–100 MBq
Matrix: maximum
Acquisition counts: $1–5 \times 10^6$

The images should be evaluated visually and should be linear over the entire FOV without any local distortion. If there are any changes from reference images, the intrinsic spatial resolution should be tested.

Quantitative

This uses a 3 mm thick lead mask with 1 mm wide parallel slits, which is positioned immediately adjacent to the crystal. Two acquisitions are undertaken, with the slits aligned along each axis of the camera.

Source: point
Activity: 750 MBq
Count rate: usual and at 7.5×10^4 counts s^{-1}
Matrix: maximum
Zoom: such that, pixel size ≤0.1 FHWM across the slits
Acquisition: 5×10^6 counts such that peak channel counts 10^3

Count profiles are constructed across the slit images, and a two dimensional array of peak locations is generated. From these integral and differential spatial linearity are calculated. The integral linearity is the variation between adjacent pairs of line source images, expressed as the maximum displacement for both CFOV and UFOV. The differential linearity measures the maximum 'ripple' in adjacent segments of any one line image, expressed as the standard deviation averaged in each direction for both CFOV and UFOV. Accuracy should be 0.1 mm. The major factor that degrades spatial distortion is a drift of individual PMTs, however, any of those listed as affecting flood-field uniformity can be responsible.

Spatial Resolution

This test assesses the amount of blurring in the image, using a line source whose width is much smaller than the camera's spatial resolution, because the width on the image then represents the spatial resolution of the camera.

Radionuclide: $^{99}Tc^m$

Intrinsic spatial resolution degradation may be caused by electronic faults, misalignment of PMT gains, crystal problems and high count rates.

Qualitative

This is good for detecting regional variations in spatial resolution over the whole field of view.

Phantom:	quadrant bar or orthogonal hole
Source:	point
Activity:	20–40 MBq
Matrix:	512
Acquisition counts:	6×10^6

Comparison with previous images should not reveal any loss of clarity. FWHM can be estimated from the widths of the smallest resolved separations, as described in 'sources'.

Quantitative

Quantitative estimates are given by the FWHM, FWTM or modulation transfer function, MTF, (see 'Image quality' in Chapter III.5) of the line spread function, LSF, or the point spread function, PSF.

Line Source

The difficulty with this test is constructing and using the phantom. It is a very heavy object and must be placed as close to the crystal surface as possible. The weight of the phantom must therefore be supported by the edges of the detector shield and not by the crystal itself. It needs to provide two line sources of <0.5 mm internal diameter and 50 mm long, eg butterfly line. These are fitted into a lead block of thickness 50 mm and separated by, eg 50 mm. Slits in the lead block, 0.5 mm wide, transmit radiation from the sources to the crystal. To avoid scattered radiation, the crystal adjacent to the source needs to be shielded with 3 mm thick lead. Acquisition is initially undertaken with the line sources oriented along the x- and then the y-axes, with one as close as possible to the centre of the FOV because this is where the minimum spatial distortion should occur. The test should then be repeated for 4 other positions over the FOV.

Activity:	~40–70 MBq in each line
Matrix and zoom:	such that pixel size $\leq 0.1 \times$ FWHM
Acquisition:	$6\text{–}10 \times 10^6$ counts such that counts in each peak channel 10^4

To confirm phantom alignment, narrow count profiles are constructed at the top, middle and bottom of the line image. The peaks should occur in the same pixel row or column, otherwise the phantom should be realigned and the acquisition repeated. Count profiles, up to 3 pixels wide, should then be constructed across the line images. For each profile, FWHM and FWTM are calculated by linear interpolation between adjacent pixels. These are converted to mm by calibrating the pixel size using the known line separation and the values for the two lines are averaged.

Point Source

This uses the data acquired for the quantitative analysis of spatial linearity. FWHM and FWTM are calculated and averaged over both the CFOV and the UFOV.

Sensitivity

The acquisition time should be recorded and the sensitivity, as counts s^{-1} MBq^{-1}, calculated.

Radionuclide: $^{99}Tc^{m}$
Source: point
Activity: ~40 MBq
Matrix: 128
Acquisition counts: $1–3\times10^{6}$ for SFOV, $2–5\times10^{6}$ for LFOV

Degradation in intrinsic plane sensitivity can be caused by poor spatial non-linearity, high count rates, incorrect PHA settings, PMTs gain misalignment or failure, crystal deterioration, and coupling problems between the PMT-light guide assembly and crystal (IAEA 1991).

Multiple Window Spatial Registration

This evaluates the error for estimating the positions of different energy photons from the same source. Since it is energy dependent, it is measured for the radionuclides used clinically and thus ^{67}Ga is generally used.

Radionuclide: ^{67}Ga
Photopeaks: if only 2 available, use highest and lowest
Activity: 40–80 MBq
Count rate: $<10^{4}$ counts s^{-1} in any PHA window
Matrix: maximum

Qualitative

Phantom: quadrant bar or orthogonal hole
Source: point
Acquisition counts: 5×10^{6}

Images are acquired using individual PHAs and then all PHAs. The latter should not reveal any deterioration compared to the former.

Quantitative

This requires a shielded source such that a small aperture transmits the radiation. This is placed on the crystal itself, so care has to be taken.

Source: 2 ml in 10 ml vial in lead shield with a 3–5 mm diameter circular aperture
height 25 mm, wall thickness 6 mm
Zoom: maximum such that pixel size ≤2.5 mm
Counts: 10^{5}, such that peak pixel counts 10^{4}
Source positions: centre of the FOV, and up to 8 other positions along the x and y axes

Images are acquired using individual PHAs and all together, for each source position. Count profiles are constructed through the images in both the x and y directions, the variation in the peak positions are determined.

System Tests

Collimator

The collimators should be inspected for dents, scratches or distortion and, if discovered, should be subjected to system uniformity tests. A small dent in the collimator will result in an apparent cold spot. Low energy collimators, and in particular foil construction, are susceptible to damage during delivery or in use as mechanical stress or inadequate bonding may lead to a slight separation of the foils which appears as a line of increased activity (O'Connor 1996). Each collimator must also be loaded onto the camera head and the collimator mounting mechanism checked for alignment and proper working.

Hole Angulation

For accurate imaging, the holes of a parallel hole collimator should be parallel, and for the non-slant type they should be perpendicular to the crystal face when loaded. This is particularly important when the camera is used for SPET, when angulation errors of 1° can seriously degrade the reconstructed spatial resolution (Eckholt and Bergmann 2000), introducing a 3.5 mm error at a separation of 20 cm (Busemann-Sokole 1987). The most sensitive tests for hole angulation are the centre of rotation, COR, offset at different radii of rotation, see Chapter IV.5, and the following collimator hole alignment tests (Babicheva et al. 1997) which can be undertaken using either a multi-source phantom or a point source.

Radionuclide: $^{99}Tc^{m}$
Matrix: maximum
Acquisition: 5×10^6 counts

Multi-Source Phantom

The phantom is a Perspex sheet slightly larger than the collimators. In it, one mm diameter holes are drilled at 5 cm spacings, creating a matrix of point sources. This sheet can be located in one of two slots cut in the side supports exactly 20 cm apart, at 10 cm and 30 cm from the crystal face. The holes should be exactly perpendicularly aligned when the sheet is moved between slots, within the limits of manufacture and use (Busemann-Sokole 1987). With the holes loaded with activity, acquisitions are undertaken at 10 cm and 30 cm separations; the duration of the second acquisition being adjusted to equate the count statistics with the first. This phantom can be simulated using a point source and an accurate computer-controlled imaging table; the measurement accuracy for absolute angulation errors being better than 0.32°, and regional variations in channel tilt being detected with an accuracy better than 0.16° (Eckholt and Bergmann 2000).

Sources: multiple point
Activity: highest radioactive concentration available

Count profiles are constructed through the point images and any peak shift, d, between the 10 cm and 30 cm images determined. This should be minimal when the collimator holes are perfectly orthogonal with the detector plane (Busemann-Sokole 1987).

Point Source

The simplest approach, however, is to monitor the appearance of a point source at an extremely large source–camera separation, of 4–9 m. In this geometry, the image of the point source fills the FOV and any septal deviations, well below the level which will produce detectable effects in SPET, are revealed as changes in count density (Malmin et al. 1990).
Source: point
Activity: 200–300 MBq

Acquisitions are undertaken with the source at the centre of the field of view and then half-way to the edge of the collimator, along each axis. Abnormalities appear as a distorted image shape, or linear steaks or a mottled appearance on the image (Malmin et al. 1990).

Sensitivity, Parallel Hole

This test assesses the proportion of γ photons absorbed by the collimator. It uses a thin flat phantom with thin walls, so that self attenuation is as low as possible, with a large area at least 15 cm diameter, to reduce the effect of any system non-uniformity.
Radionuclide: ^{99}Tc^{m} for LE and ME collimators
^{131}I for HE collimators
Source: at 10 cm above each collimator
Activity: ~40 MBq, accurately known
Matrix: 64
Acquisition: 100 s, background 300 s

Source and background acquisitions are undertaken. For each collimator, the sensitivity, in counts s^{-1} MBq^{-1}, and the relative sensitivity, compared to the LEGP collimator, are then calculated.

Uniformity

Radionuclide: ^{99}Tc^{m} or ^{57}Co
Source: sheet, on or 10 cm above the collimator
Activity: 100–400 MBq
Matrix: maximum
Acquisition: 30×10^6 counts

Both the raw images and the images corrected using the intrinsic flood are visually evaluated. When a pinhole collimator is tested, the uniformity will vary across the FOV but should be symmetrical. Any non-uniformity, not apparent on the intrinsic image, indicatives a collimator problem. Quantitative assessment is undertaken as for the intrinsic test, and values should increase by 2% (IAEA 1991).

Spatial Linearity

This tests the spatial linearity of both the collimator and the camera. It is undertaken in the same way as the intrinsic test, except that a sheet source rather than a point source is used.

Radionuclide:	$^{99}Tc^{m}$ or ^{57}Co
Source:	sheet, on the phantom
Activity:	200–400 MBq
Matrix:	512
Acquisition:	10^{6} counts

If any deterioration is detected, the intrinsic spatial resolution should be re-evaluated.

Spatial Resolution

This tests the spatial resolution of both the collimator and the camera. Scatter is incorporated to give clinically meaningful results.

Matrix:	maximum

Qualitative

This assesses the system spatial resolution over the whole FOV.

Phantom:	quadrant bar or orthogonal hole on collimator, and at 5 and 10 cm in air and Perspex
Radionuclide:	$^{99}Tc^{m}$ or ^{57}Co
Source:	sheet, on phantom
Activity:	200–500 MBq
Acquisition:	$1–6 \times 10^{6}$ counts

Evaluation is undertaken in the same way as for the intrinsic test. Sometimes, Moire interference patterns are created, these may be reduced by rotating the phantom. Any local degradation of spatial resolution, compared to the intrinsic image, indicates a collimator problem.

Quantitative

The phantom comprises parallel straight lines of internal diameter <0.5 mm and length >UFOV, fixed on a Perspex sheet of thickness <1 cm, and separated by 6 cm to avoid overlap on the scatter images.

Radionuclide: to suit the collimators being tested
Sources: 2–6 lines
Activity: 40–80 MBq in each line
Acquisition: $1–6 \times 10^6$ counts, such that peak channel counts 10^4
Zoom: maximum such that pixel size $\leq 0.1 \times$ FWHM

The sources are positioned, equally spaced around the central axis and aligned along the x and then the y axis. Acquisitions are undertaken on the collimator and then separated by 5–15 cm in air and with varying amounts of Perspex. The analysis is undertaken in the same manner as for the intrinsic test.

Total Performance

This gives an impression of the overall clinical performance of the system.
Radionuclide: $^{99}Tc^m$
Source: thyroid or liver slice phantom
Activity: 7 MBq in thyroid and 70 MBq in liver slice phantom
Collimator: LE with highest spatial resolution
Matrix: pixel size appropriate to system spatial resolution
Acquisition: 4×10^5 counts

The images should be visually assessed in comparison with reference images.

Whole-Body Tests

Tests for whole-body performance have been described, in detail, by Blockland et al. (1997).

Uniformity

This test will detect scan speed and electronic timing errors, which result in periodic count loss (O'Connor 1996).
Radionuclide: $^{99}Tc^m$ or ^{57}Co
Source: on the lower collimator
Activity: sheet 100–400 MBq or point ~20 MBq
Matrix: maximum
Scan speed: various
Scan length: maximum

With the sheet source, the whole-body images should appear uniform on visual inspection. With both the sheet and point sources, count profiles created along the scan direction should not show variations greater than statistically expected. This test can also be undertaken using several sources positioned at known distances along the longitudinal axis of the bed. In this case, the longitudinal count profile must be normalised to the activity of each source.

Exposure Time Corrections

When corrections are applied to the data from the qualitative uniformity test, a count profile created across the scan direction should be flat.

Spatial Resolution

This evaluates the effect of the mechanical motion.

Qualitative

Phantom:	quadrant bar or orthogonal hole at 45° to the scan direction, with and without Perspex
Radionuclide:	^{57}Co
Source:	sheet, on the phantom
Activity:	100–400 MBq
Matrix:	maximum
Scan speed:	as used clinically

Whole-body and static acquisitions are undertaken, for the same counts, with the camera as close as possible to the phantom. Both images should appear similar on visual evaluation.

Quantitative

Radionuclide:	^{99}Tc^{m}
Source:	2 lines separated by 10 cm, on the imaging table internal diameter ≤1 mm length = the width of scanned area
Activity:	~200 MBq in each line
Matrix:	pixel size 0.25 × FWHM system spatial resolution
Scan speed:	as used clinically

With a camera–source separation of 10 cm, acquisitions are undertaken with the sources positioned along and then across the table, at the middle of the table length and then at 1/2 FOV from each end. 25–30 mm wide count profiles across the lines are created and the FWHM and FWTM calculated and converted to mm using the known line spacing.

Linearity

Visual assessment of spatial linearity is afforded using a test similar to that of qualitative spatial resolution but with the phantom oriented parallel or perpendicular to the direction of motion. This will, however, reveal only coarse non-linearities and lacks reproducibility.

Image Size

The actual scan length and width, used in the uniformity test, are measured and then compared to the distances calculated from the pixels in the image and the known pixel size.

Multiple Scan Alignment

If more than one scan is required to image a patient, the uniformity test with profiles across the scan direction will reveal any gap or overlap in the alignment. The qualitative spatial resolution test will provide visual information on local alignment of the scans, which may be different at each end of the scan. Acquisitions should, therefore, be undertaken over the whole scan length.

Maximum Count Rate

This is a rapid test which can be undertaken if the count rate performance is not required.

Radionuclide:	$^{99}Tc^{m}$
Source:	point
Activity:	~4 MBq
Matrix:	64
Acquisition:	20 s

The count rate is determined with the source suspended centrally at a distance from the crystal face. This is repeated as the source is moved closer to the camera, until the count rate peaks and declines.

Count Rate Performance

The count-rate linearity and high count-rate performance are characterised by the count rate for a 20% count loss and the maximum count rate, and are measured both with and without scatter. For a multiple detector system, the measurements should be undertaken on each detector alone and then simultaneously on all detectors.

Radionuclide:	$^{99}Tc^{m}$
Matrix:	64
Acquisition:	10 s or 10^5 counts, whichever is longer

Intrinsic

Without Scatter

Decaying Source Method

This takes 2 days to complete.

Source: small vial, in an open lead shield of depth 5 cm walls and floor 2.5 cm thick open end covered with 6 mm of copper
Activity: ~80 MBq
i.e. more than required for the count rate to peak

A background acquisition is undertaken for 600 s, then the source is positioned 1.5 m above the crystal on the central axis. Acquisitions are repeated whenever the count rate falls by 10^4 counts s^{-1} and until it is $<4\times10^3$. The count, start time and acquisition duration are recorded, thus allowing estimation of the observed count rate. The input count rate is estimated by extrapolating from $<10^4$ counts s^{-1}, which allows the maximum count rate and the observed count rate at 20% count loss to be calculated. A quicker but more complicated method involves a similar source and shield but with multiple copper sheets which can be removed in sequence. The attenuation of each must be accurately known.

Multiple Source Method

Source: 4–6 points
Activity: various, eg 20–250 MBq, accurately measured

Regular acquisitions are undertaken during decay, for various combinations of the sources, with particular attention being paid to the low count rates in the linear region of the curve. Count rate measurements are decay corrected and the analysis is then similar to the decaying source method.

With Scatter

A more realistic assessment of the clinical situation can be obtained using a distributed source and scattering material, such as the IEC phantom (IEC 1992). The phantom can either be loaded with a high activity and acquisitions undertaken as it decays, or it can be loaded with increasing activity and acquisitions undertaken with each increase.

System

The last test can also be undertaken with the source positioned on the collimator.

ECG-Gated Acquisition

This study assesses the consistency of ECG triggering which can degrade the temporal resolution of gated investigations.
Radionuclide: $^{99}Tc^{m}$
Source: point
Activity: ~100 MBq
Matrix: 64
Acquisition: routine gated study

A study is acquired using a volunteer, or a simulator, to produce the ECG signal, and an activity-time curve generated from an ROI created over the point image. The variation in counts should not be more than anticipated statistically.

References

AAPM (1982) American Association of Physicists in Medicine report 9 computer aided scintillation camera acceptance testing. American Institute of Physics, New York

AAPM (1987) American Association of Physicists in Medicine report 22 rotating scintillation camera SPECT acceptance testing and quality control. American Institute of Physics, New York

Babicheva R, Bennie N, Collins L, Gruenewald S (1997) Parallel hole collimator acceptance tests for SPECT and planar studies. Australas Phys Eng Sci Med 20:242–247

Blokland JAK, Camps JAJ, Pauwels EKJ (1997) Aspects of performance assessment of whole body imaging systems. Eur J Nucl Med 24:1273–1283

Busemann-Sokole E (1987) Measurement of collimator hole angulation and camera head tilt for slant and parallel hole collimators used in SPECT. J Nucl Med 28:1592–1598

Buseman-Sokole E, Heckenberg A, Bergmann H (1996) Influence of high-energy photons from cobalt-57 flood sources on scintillation camera uniformity images. Eur J Nucl Med 23:437–442

Eckholt M, Bergmann H (2000) Angulation errors in parallel-hole and fanbeam collimators: computer controlled quality control and acceptance testing procedure. J Nucl Med 41:548–555

Ficken V, McCartney W (1994) SPECT quality control: a program recommended by the American College of Nuclear Physicians and the ACNP corporate committee. J Nucl Med Technol 22: 205–210

Hines H, Kayayan R, Colsher J, Hashimoto D, Schubert R, Fernando J, Simcic V, Vernon P, Sinclair RL (1999a) Recommendations for implementing SPECT instrumentation quality control. Nuclear Medicine Section – National Electrical Manufacturers Association (NEMA). Eur J Nucl Med 26: 527–532

Hines H, Kayayan R, Colsher J, Hashimoto D, Schubert R, Fernando J, Simcic V, Vernon P, Sinclair RL (1999b) National Electrical Manufacturers Association recommendations for implementing SPECT instrumentation quality control. J Nucl Med Technol 27:67–72

Hines H, Kayayan R, Colsher J, Hashimoto D, Schubert R, Fernando J, Simcic V, Vernon P, Sinclair RL (2000) National Electrical Manufacturers Association recommendations for implementing SPECT instrumentation quality control. J Nucl Med 41:383–389

IAEA (1991) Tecdoc 602 quality control of nuclear medicine instruments. International Atomic Energy Agency, Vienna

IAEA (2003) Quality control atlas for scintillation camera systems. International Atomic Energy Agency, Vienna

IEC (1992) International Electrotechnical Commission publication IEC 60789. characteristics and test conditions of radionuclide imaging devices: Anger type gamma cameras. IEC, Geneva

IEC (1998) International Electrotechnical Commission publication IEC 61675-3 characteristics and test conditions of radionuclide imaging devices. Part 3: gamma camera based wholebody imaging systems. IEC, Geneva

IPEM (2003) Report 86 quality control of gamma camera systems. Institute of Physics and Engineering in Medicine, London

IPSM (1985) Report 44 an introduction to emission computed tomography. Institute of Physical Sciences in Medicine, London

IPSM (1992) Report 66 quality control of gamma cameras and associated computer systems. Institute of Physical Sciences in Medicine. London

Keszthelyi-Landori S (1986) NaI(Tl) camera crystals: imaging capabilities of hydrated regions on the crystal surface. Radiol 158:823–826

Malmin R, Stanley P, Guth WR (1990) Collimator angulation error and its effect on SPECT. J Nucl Med 31:655–659

Marshall DS, Wells CP (1999) South Thames quality standard for nuclear medicine equipment. South Thames Region Nuclear Medicine Specialty Service Committee. Nucl Med Commun 20: 789–798

NEMA (1994) Standards publication NU1 performance measurements of scintillation cameras. National Electrical Manufacturers' Association, Washington

NEMA (2001) Standards publication NU1 performance measurements of scintillation cameras. National Electrical Manufacturers' Association, Washington DC

O'Connor MK (1996) Instrument- and computer-related problems and artifacts in nuclear medicine. Semin Nucl Med XXVI:256–277

Tindale WB (1995) Specifying dual-detector gamma cameras and associated computer systems. Nucl Med Commun 16:534–538

Waddington WA, Clarke GA, Barnes KJ, Gillen GJ, Elliot AT, Short MD (1995) A reappraisal of current methods for the assessment of planar gamma camera performance. Nucl Med Commun 16:186–195

Wells CP, Buxton-Thomas M (1995) Gamma camera purchasing. Nucl Med Commun 16:168–185)

Single Photon Tomographic Imaging

Abbreviations

FOV field of view
FWHM full width at half maximum
PHA pulse height analyser
QA quality assurance
ROI region of interest
SPET single photon emission tomography
UFOV useful field of view

These tests are not undertaken daily but only as part of regular more complicated testing; the frequency depending partly on the stability of the system. For consistency in the QA programme, these tests must be taken using identical acquisition and processing parameters. The position and orientation of the sources are crucial, as are the gantry parameters which should be recorded from the digital displays, if available. Performance requirements for tomography are more stringent than for planar imaging; and this applies particularly to uniformity, energy resolution and spatial resolution, although mechanical stability must also be considered. Many parameters may not be characterised by a single value, varying for example with: position in an image, radius of rotation, and reconstruction technique (IPSM 1992). Comprehensive recommendations have been published by: AAPM (1987), IAEA (1991), Ficken and McCartney (1994), Graham et al. (1995), IEC (1998, 2001), IPSM (1985, 1992), Hines et al. (1999a,b, 2000), NEMA (1994, 2001) and IPEM (2003).

A number of tests are now described; in order of performance convenience and the acquisition conditions, for other than the system tests, are those previously used for planar measurements which are:
Source: $^{99}Tc^{m}$
Photopeak: peak, e.g. 140 keV
PHA window: clinically used, e.g. 20%
Zoom: none

Gantry

Physical and Mechanical Inspection

The mechanics must be well adjusted for the system to be capable of acquiring acceptable SPET data.

Imaging Bed

The attenuation of the bed should be 5–10% (Kelty et al. 1997) and it can be assessed using transmission imaging.
Source: point
Activity: ~60 MBq
Matrix: 128
Acquisition: 100 s

A background acquisition is undertaken. Images are then acquired with the source suspended above the detector and then with the bed within the field of view. The attenuation is calculated using the attenuation equation I.3.23 introduced in Chapter I.3. Visual evaluation of transmission images, acquired using a flood source, will reveal the position of any metal parts in the bed.

The long axis of the imaging bed should be reasonably concentric with the system axis of rotation. It can be assessed by setting the detectors to the maximum radius of rotation and then measuring the distance from the bed to the head at 90° and then at 270°. The measurements should be repeated with the bed at the extremes of its travel.

Rotation, Constancy of Speed

Gantry rotation may be affected by vibration and inappropriate motion during non-circular rotations, and should be assessed over various orbits. For single heads and double head systems not set at 180°, gantry rotation may be slower when the heads are ascending rather than when they are descending, especially when heavy collimators are loaded.

Source:	none
Matrix:	64
Rotation:	360°
Views:	32
Acquisition:	5 s per view
Radius of rotation:	as wide as possible

Variations in rotation speed should be excluded by timing the rotation from 0° to 180° and from 180° to 360°, in both directions, when the heaviest collimators are loaded.

Detector Head Position

All of the digital indications of gantry and head positions should be checked using physical measurements.

Gantry Angle

For the detector faces to be parallel to the axis of rotation, the gantry must be vertical i.e. minimal tilt error. This is tested with a spirit level, oriented along the axis of rotation and positioned on the face of the detector, when the head is at 0° and then at 180° for different radial positions. If detector tilt is indicated digitally, the readings should be zero for the above combinations. Tilt error will also be revealed during centre of rotation analysis.

Intrinsic Tests

Matrix Offset

This checks that the acquisition matrices are centred on the fields of view.

Radionuclide: $^{99}Tc^{m}$
Source: as in Chapter IV.4 quantitative multiple window spatial registration
Activity: ~40 MBq
Acquisition counts: 10^5
Matrix: maximum
Zoom: none

The shielded source is positioned, accurately to within 1 mm, on the crystal at the centre of the field of view. An acquisition is undertaken and count profiles are constructed along the x and y axes. The centroids should be at 256.

Pixel Size

The pixel size must be accurately known, during reconstruction, for all combinations of matrix sizes and zoom factors and in both x and y directions, e.g. for analytical attenuation correction.

Radionuclide: $^{99}Tc^{m}$
Source: as in matrix offset
Activity: ~40 MBq
Acquisition: 5×10^4 counts for each source in the FOV
Matrix: all possible
Zooms: all possible

Images are acquired for 4 source positions: along the x^+, x^-, y^+ and y^- axes at ~5 cm from the edge of the field of view, known to within 1 mm. Count profiles are constructed through each point image and the separation of the sources calculated. The pixel size is then calculated by comparing the source separation in mm and pixels.

System Tests

Tomographic Uniformity and Sensitivity Variation with Orientation

This test will reveal uniformity or sensitivity variation with detector orientation.

Radionuclide: ^{57}Co or $^{99}Tc^{m}$
Source: sheet, attached to the collimator
Activity: 100–400 MBq
Head geometry: usual clinical or 180°
Matrix: 64 or 128
Tomographic rotation: 360°
Projections: finest angular sampling

Acquisition: constant duration per projection ≥300 counts $pixel^{-1}$
or 10^6 counts in each projection image

Changes in uniformity, with rotation, will be apparent on visual inspection of the raw data, during cine mode display (O'Connor 1996). Variations in sensitivity will be apparent in the sinogram as reduced intensity, which will correspond to areas of reduced counts in the reconstructed images.
Planar acquisition: 5–30 × 10^6 counts at detector orientations: 0°, 90°, 180° and 270°

Each image is visually evaluated for structured non-uniformities. Quantitative evaluation involves the calculation, for each image, of the: UFOV integral uniformity deviation from the mean of all images; and the sensitivity deviation from the mean of all images, obtained from the total counts per view corrected for decay. Both deviations should be <±1%.

System Tests Involving Tomographic Reconstruction

These tests must be undertaken with each collimator used in SPET.

Sources

A number of sources are required.

Point

There are two methods to construct a point source: as described in 'Sources' Chapter IV.4 or, preferably, in a thin walled glass capillary tube. In the latter, a high concentration radioactive solution is introduced but with a maximum source length of <2 mm.

Sheet Source

This is the same as described in 'Sources' Chapter IV.4, except that a scattering medium needs to be added to simulate acquisition in the clinical situation.

Uniformity Phantom

This is a hollow cylinder, of diameter ~20 cm and length >10 cm, which can be filled with a radioactive solution. It forms one section of the total performance phantom.

Spatial Resolution Phantom

This is a cylinder of uniform attenuation into which point or axially oriented line sources can be introduced, e.g. uniformity phantom.

Total Performance Phantom

This is a hollow cylinder, which can be filled with a radioactive solution, and into which Perspex shapes can be introduced. It usually comprises sections of: uniformity, linearity and various sizes of spheres or rods to create cold areas.

Centre of Rotation

The reconstruction process assumes that the centre of the image corresponds exactly with the physical axis of rotation. This tests determines the offset, across the axis of rotation, that must be applied to accomplish this.

Radionuclide:	$^{99}Tc^{m}$
Source:	point, on the imaging bed $<\pm 2$ cm of the centre of the FOV along the axis of rotation activity beyond the edge, i.e. not attenuated
Activity:	~40 MBq
Head orientation:	clinically used, 90° and 180° for variable angle cameras
Matrix:	128
Rotation:	360°
Orbit:	circular
Radius of rotation:	15–25 cm
Projections:	32
Acquisition:	$\geq 1-2 \times 10^4$ counts per projection

Acquisitions are undertaken with the point source offset towards the edge of the trans-axial FOV, because this will provide more information than a source on the axis of rotation. Displacement, of the source position, is plotted against angle of rotation: across and along the axis of rotation.

The variation in displacement along the axis of rotation should not show any recognisable pattern, and deviations >0.5 pixels from the average offset indicate a problem with head tilt, collimator hole angulation or gantry alignment. The variation in displacement across the axis of rotation should describe a sinusoidal pattern. From this, the centre of rotation offset should be calculated to within ±1.0 mm. It is preferable that it coincide with the centre of the image matrix because this yields the largest field of view and reduces the effect of changing matrix size. Therefore, the maximum offset from the axis of rotation should be <6 mm.

As a Function of Axial Position Along the Axis of Rotation

When a collimator is undergoing acceptance testing, the above test should be undertaken over the whole FOV of the collimator. This includes: acquisition positions on the axis of rotation, displaced by 10 cm across the axis of rotation, at the edge of the trans-axial FOV, and at 2 cm increments along the axis of rotation. The various offset values should not differ by more than 0.5 pixels (Cerqueira et al. 1988), 1.5 mm (IPEM 2003) or 2 mm (O'Connor 1996).

As a Function of Radius of Rotation

The hole angulation integrity of a collimator can be assessed by estimating the centre of rotation offset at various radii of rotation, the difference should be <0.25 pixels between radii of 15 cm and 25 cm (Busemann-Sokole 1987).

Tomographic Spatial Resolution, in Air

This tests the whole process of tomographic acquisition and reconstruction, including: acquisition characteristics such as gantry vibration, centre of rotation accuracy and reconstruction software. Any errors will be revealed as a degradation in tomographic spatial resolution compared with planar values. There are two problems with this technique. The pixel size used for tomography does not allow many pixels within the FWHM, although it is possible to achieve reasonable accuracy. Particularly with filtered back-projection reconstruction, tomographic spatial resolution will vary with position in the transaxial image, therefore, the source must be positioned on the axis of rotation.

Radionuclide:	$^{99}Tc^{m}$
Source:	point or line, on the imaging bed activity beyond the edge, i.e. not attenuated
Activity:	~40 MBq
Tomographic matrix:	128
Zoom:	yes
Rotation:	360°
Orbit:	circular
Radius of rotation:	as small as possible
Projections:	128
Acquisition:	10^4–10^5 counts per projection
Planar	
Head orientation:	0°
Matrix:	maximum
Zoom:	yes
Acquisition:	5×10^5 counts
Collimator–source:	as radius of rotation

Acquisitions are undertaken, for the source at the centre of the field of view and then displaced, in both directions, along the axis of rotation. Reconstruction uses only a ramp filter and no attenuation correction. The reconstructed point should be round and not elongated any direction. FWHM values calculated from count profiles, both across and along the axis of rotation, should be similar to those from the planar image.

This technique can also be used to investigate the characteristics of fan-beam collimators (O'Connor 1996) and the relative alignment of the detectors, along the axis of rotation, in multi-detector systems; the latter being undertaken by reconstructing the data from the heads separately and then combined.

Slice Thickness, at the Centre of the FoV

This evaluates the spatial resolution along the axis of rotation.

Radionuclide:	$^{99}Tc^m$
Source:	line, on the imaging bed at the centre of the FOV across the axis of rotation

Acquisition and reconstruction is undertaken as for the tomographic spatial resolution in air. The transaxial slice, in which the point source appears hottest, is located and the maximum pixel count and its x, y coordinates recorded. For these coordinates, the pixel count in adjacent slices is also recorded, a count profile of the point source along axis of rotation is generated and the FWHM calculated. This should be similar to the tomographic resolution determined by the tomographic resolution in air. This test can also be undertaken with scatter using the spatial resolution phantom.

Tomographic Spatial Resolution, with Scatter

This test indicates the spatial resolution which can be achieved clinically.

Phantom:	spatial resolution phantom at the centre of the FOV along the axis of rotation oriented along the axis of rotation
Radionuclide:	$^{99}Tc^m$
Activity:	~40 MBq
Matrix:	usual clinical
Projections:	usual clinical
Acquisition:	10^4–10^5 counts per projection
Rotation:	360°
Orbit:	usual clinical
Radius of rotation:	usual clinical, and larger

Acquisitions are undertaken with the source on the axis of rotation and then 5 cm offset from it. Reconstruction is undertaken using a ramp filter and all corrections. The FWHM values, for horizontal and vertical count profiles created across the reconstructed distributions, should be similar.

Linearity of Activity Estimation

This determines the accuracy of the activity levels across the reconstructed image.

Radionuclide:	$^{99}Tc^m$
Sources:	5 similar vials with accurately known activities placed across the imaging plane
Matrix:	128
Zoom:	none
Rotation:	360°
Orbit:	circular
Projections:	128
Acquisition:	$\geq 10^5$ counts per projection

The counts obtained from ROIs over the reconstructed sources should be linearly related to the known activities.

Total Performance

This assesses the performance of the whole system, under conditions similar to those used clinically.

Acquisition

Radionuclide:	$^{99}Tc^{m}$
Source:	total performance phantom at the centre of the FOV along the axis of rotation oriented along the axis of rotation
Activity:	200–700 MBq
Count rate:	$10 \pm 2 \times 10^{3}$ counts s^{-1}
Tomography	
Collimators:	all in clinical use
Matrix:	128
Zoom:	pixel size compatible with system spatial resolution
Rotation:	360°
Orbit:	circular and non-circular
Radius of rotation:	15 cm
Projections:	maximum
Acquisition:	$\geq 5 \times 10^{5}$ counts per projection
Planar	
Matrix:	128
Separation:	15 cm
Acquisition:	10^{6} counts

The acquisition time, T_{aq}, should be recorded for the planar acquisition. Tomographic reconstruction should be undertaken for all slice orientations (transaxial, sagittal, coronal) and, at acceptance, this should be repeated for all combinations of the tomographic acquisition parameters, e.g. projection number, matrix etc.

Measurements

These comprise assessments of: analytical attenuation correction, tomographic uniformity, sensitivity, sensitivity to scatter, image contrast and linearity. On all except analytical attenuation correction, the reconstruction should be performed using several smoothing filters, and both with and without uniformity correction.

Analytical Attenuation Correction

Reconstruction is undertaken using only a ramp filter and without attenuation correction. The linear attenuation coefficient is estimated from the counts at the centre and at 2 cm from the edge of the uniform section of the reconstructed phantom. These are obtained using five pixel wide count profiles constructed on a central transaxial image, in both the horizontal and vertical directions.

$$m = \ln(\text{edge counts}/\text{centre counts})/\text{radius cm}^{-1} \quad \text{(IV.5.1)}$$

The data is then reconstructed with attenuation correction, using both the calculated and default values. The accuracy of the body contour is assessed by superimposing it on all the reconstructed images. The accuracy of the attenuation correction is assessed by recreating the count profiles. These should be flat. If not, the pixel calibration should be checked.

Tomographic Uniformity

Four pixel wide transaxial slices are reconstructed using a ramp filter, attenuation correction and scatter correction, if available; both without and with system flood correction. Qualitative evaluation is undertaken by visually assessing each transaxial image, apart from the peripheral 2 pixels. Quantitative evaluation is difficult (Madsen 1997). One method is to estimate the contrast of any apparent ring artefact by constructing a 5 pixel wide profile through the centre of the image:

$$\text{contrast} = (C_R - C_B)/C_B \quad \text{(IV.5.2)}$$

where

C_R is the counts at the ring, peak or trough

C_B is the average counts in the uniform background, immediately on either side of the ring

Other methods involve estimating radial and tangential uniformity from a polar representation of each reconstructed image (Hughes 2001), and estimating uniformity values in annular rings concentric with the centre of rotation (Madsen 1997).

Tomographic Sensitivity to Scatter

Reconstruction is undertaken around the largest sphere or rod in the phantom.

$$\text{sensitivity to scatter} = C_S/(C_B - C_S) \quad \text{(IV.5.3)}$$

where

C_S is the counts in ROI over sphere or rod

C_B is the counts in equivalent background ROI adjacent to sphere or rod

Tomographic Image Contrast

Using the same reconstruction from the previous test, this evaluation uses smaller ROIs to analyse a number of spheres or rods of various sizes, which should be situated in the same plane of the phantom.

$$\text{contrast} = (C_B - C_S)/C_B \quad \text{(IV.5.4)}$$

Although tomographic contrast quantifies the ability of the system to detect small lesions, its value depends on the size of the sphere or rod. It is also difficult to obtain consistent results for small sizes. Therefore, the shape and the number of detected cold areas should also be recorded.

Tomographic Linearity

This is assessed by visually inspecting the linearity part of the phantom.

Transmission Attenuation Correction

The position of the transmission PHA window in relation to the must be checked (DePuey and Garcia 2001). This test assesses the adequacy of the transmission source activity.

Radionuclide:	$^{99}Tc^{m}$
Source:	uniformity phantom, oriented along the axis of rotation
Activity:	100–200 MBq
Collimator:	usual clinical
Matrix:	64
Rotation:	360°
Orbit:	circular
Radius of rotation:	usual clinical
Projections:	64
Acquisition:	≥600 s

Transmission and emission acquisitions are undertaken, and the attenuation maps reconstructed. The calculated mean attenuation coefficient, obtained by creating an ROI over 90% of a 5 cm slice constructed from the summed central slices, should be within the expected range for the transmission radionuclide. If the mean attenuation coefficient is within the expected range, the correction process should be evaluated by reconstructing the emission data without and with attenuation correction. Visual comparison of these images should show that the attenuation corrected images are more uniform (DePuey and Garcia 2001).

References

AAPM (1987) American Association of Physicists in Medicine report 22 rotating scintillation camera SPECT acceptance testing and quality control. American Institute of Physics, New York

Busemann-Sokole E (1987) Measurement of collimator hole angulation and camera head tilt for slant and parallel hole collimators used in SPECT. J Nucl Med 28:1592–1598

Cerqueira MD, Matsuoka D, Ritchie JL, Harp GD (1988) The influence of collimators on SPECT centre of rotation measurements: artefact generation and acceptance testing. J Nucl Med 29: 1393–1397

DePuey EG, Garcia EV (2001) Updated imaging guidelines for nuclear cardiology procedures, part 1. American Society of Nuclear Cardiology. J Nucl Cardiol 8:G1–G58

Ficken V, McCartney W (1994) SPECT quality control: a program recommended by the American College of Nuclear Physicians and the ACNP corporate committee. J Nucl Med Technol 22: 205–210

Graham LS, Fahey FH, Madsen MT, van Aswegen A, Yester MV (1995) Quantitation of SPECT performance: report of task group 4, nuclear medicine committee. Med Phys 22:401–409

Hines H, Kayayan R, Colsher J, Hashimoto D, Schubert R, Fernando J, Simcic V, Vernon P, Sinclair RL (1999a) Recommendations for implementing SPECT instrumentation quality control. Nuclear Medicine Section – National Electrical Manufacturers Association (NEMA). Eur J Nucl Med 26: 527–532

Hines H, Kayayan R, Colsher J, Hashimoto D, Schubert R, Fernando J, Simcic V, Vernon P, Sinclair RL (1999b) National Electrical Manufacturers Association recommendations for implementing SPECT instrumentation quality control. J Nucl Med Technol 27:67–72

Hines H, Kayayan R, Colsher J, Hashimoto D, Schubert R, Fernando J, Simcic V, Vernon P, Sinclair RL (2000) National Electrical Manufacturers Association recommendations for implementing SPECT instrumentation quality control. J Nucl Med 41:383–389

Hughes A (2001) Quantification of reconstructed SPECT non-uniformity. Nucl Med Commun 22: 703–712

IAEA (1991) Tecdoc 602 quality control of nuclear medicine instruments. International Atomic Energy Agency, Vienna

IEC (1998) International Electrotechnical Commission publication IEC 61675-2 characteristics and test conditions of radionuclide imaging devices. Part 2: single photon emission computed tomography. IEC, Geneva

IEC (2001) International Electrotechnical Commission publication IEC 61948-2 nuclear medicine instrumentation – routine tests, part 2. Scintillation cameras and single photon emission computed tomography imaging. IEC, Geneva

IPEM (2003) Report 86 quality control of gamma camera systems. Institute of Physics and Engineering in Medicine, London

IPSM (1985) Report 44 an introduction to emission computed tomography. Institute of Physical Sciences in Medicine, London

IPSM (1992) Report 66 quality control of gamma cameras and associated computer systems. Institute of Physical Sciences in Medicine. London

Kelty NL, Cao Z-J, Holder LE (1997) Technical considerations for optimal orthopedic imaging. Semin Nucl Med XXVII:328–333

Madsen MT (1997) A method for quantifying SPECT uniformity. Med Phys 24:1696–1700

NEMA (1994) Standards publication NU1 performance measurements of scintillation cameras. National Electrical Manufacturers' Association, Washington DC

NEMA (2001) Standards publication NU1 performance measurements of scintillation cameras. National Electrical Manufacturers' Association, Washington

O'Connor MK (1996) Instrument- and computer-related problems and artifacts in nuclear medicine. Semin Nucl Med XXVI:256–277

Dual Photon Imaging

Abbreviations

2D	two dimensions
3D	three dimensions
CFOV	central field of view
DPET	dual photon emission tomography
FWHM	full width at half maximum
FWTM	full width at tenth maximum
IEC	International Electrotechnical Commission
LSF	line spread function
NEC	noise equivalent count rate
NEMA	National Electrical Manufacturers' Association
PHA	pulse height analyser
ROI	region of interest

Operational Tests

When transmission attenuation correction is available, a blank scan should be performed. This will form the baseline for attenuation correction but will also reveal transmission source problems and, in a dedicated system, defective detector blocks (Buchert et al. 1999).

Routine Tests

Two approaches, to routine testing, have developed since 1990: the Society of Nuclear Medicine system (Karp et al. 1991), which was further developed by NEMA (NEMA 1994), and the European system (IEC 1998). Both considered only dedicated DPET, but the European system aimed to simulate the clinical situation as closely as possible whilst the NEMA system aimed to produce basic measurements. The NEMA system has been revised (NEMA 2001; Daube-Witherspoon et al. 2002): to increase the clinical relevance of the measurements, to make the tests simpler and less time-consuming, and to include consideration of 3D imaging and coincidence gamma cameras. Differences between the two systems remain, however, which reflects the different philosophies of the two committees (Adam et al. 1997). The revised NEMA system is designed primarily to test equipment used for whole-body acquisitions. For equipment designed primarily for brain imaging, the original NEMA system tests of scatter fraction and count rate should be used (NEMA 1994).

Acquisition and Processing Conditions

The acquisition parameters should be those that are used in typical patient studies. Included in this are: energy window, axial acceptance angle, coincidence time window and slice thickness. However, when spatial resolution is measured, an increased sampling is used. Reconstruction should also be similar to that used in a typical patient study, except in some tests which demand a ramp filter, only, to allow comparison among systems. ^{18}F is generally used because other radionuclides suffer very short half-lives or longer β^+ ranges.

NEMA System

Sources

The activity used should not cause the percent dead-time losses to exceed 5% or the random coincidence rate to exceed 5% of the total event rate. A point source should be made using a glass capillary tube of inside diameter <1 mm and the length of the activity should also be <1 mm.

Intrinsic

Spatial Resolution

This is measured in the radial and tangential directions within the transaxial slice and axially over a number of slices. A point source is imaged in air, at 6 locations. There are two axial positions: the centre of the axial field of view and offset by a quarter of the field of view. It has been suggested that, in ring systems, two extra axial positions be included, each offset 1/2 a slice separation, to help localise the source to the centre of a slice (Daube-Witherspoon et al. 2002).

For each of the axial positions, there are three transaxial positions: offset from the axis vertically by 1 cm and 10 cm and horizontally by 10 cm. At least 10^5 counts are acquired and the pixel size should be such that there are at least 3 pixels in the FWHM of the count profile through the reconstructed point. Reconstruction is by filtered back projection using a ramp filter with cut-off at the Nyquist frequency. Count profiles, of width ~2 times FWHM, are constructed through each reconstructed point in each of three directions: horizontally and vertically in the transaxial plane, and axially. From these, FWHM and FWTM are calculated. Thus for each imaged point, values of the radial (not at 1 cm vertical offset), tangential and axial resolution are calculated. These are averaged over the two axial positions. The results from this test are not those expected clinically, but a system with good intrinsic spatial resolution in air should also produce good spatial resolution in the clinical situation.

Count Rate Losses and Randoms

Most clinical investigations are undertaken at count rates high enough that coincidence events are lost because of dead-time and increased random coincidence rates. An appreciation of the behaviour of the system under clinical scanning conditions is afforded by the count rate performance (losses of events due to dead-time and randoms) as a function of activity. The phantom used is a 20 cm diameter 70 cm long solid polyethylene cylinder with an axial line source offset vertically downwards 4.5 cm from the axis. This has an inside diameter of 3.2 mm and is filled with a known activity, high enough that the peak true rate and peak NEC rate can be measured. Sequential acquisitions of at least 5×10^5 prompt counts, at intervals <half the $T_{1/2}$, are under-

taken until the randoms and dead-time losses are both <1% of the true coincidences rate. The total acquired counts in each acquisition are calculated from the sinogram of each transaxial slice, which is defined by a circle of diameter 24 cm.

The artefactual coincidences (scattered and random) are actually calculated by determining the true coincidences i.e. artefactual events = total events–true events, and these are assumed to be present only within ±2 cm of the source. This is done by analysing the sinogram profile for each projection angle, which describes the decrease in counts with distance from the source. The number of artefactual events within ±2 cm of the source, is calculated by averaging the counts at the ±2 cm positions. The number of true events is then calculated by subtracting this number from the total events within ±2 cm of the source. The contribution of each component (scattered, R_s, and random, R_r) to the artefactual coincidence rate can be calculated by considering the total (R_t) and true (R_{tr}) coincidence rate, and the intrinsic scatter fraction (SF) calculated in the following test.

$$R_r = R_t - R_{tr}/(1 - SF) \tag{IV.6.1}$$

$$R_s = R_{tr}\, SF/(1 - SF) \tag{IV.6.2}$$

Noise Equivalent Count Rate (NEC)

The NEC rate (R_n) is:

$$R_n = R_{tr}^{\,2} / R_t \tag{IV.6.3}$$

All of these coincidence rates are plotted against effective activity concentration, calculated from the source activity and phantom volume. From these, the peak R_t and R_n, and corresponding activity concentrations can be determined.

Scatter Fraction

The scatter fraction is an indication of the sensitivity of the system to scatter, and a low value is desirable, since all correction techniques add noise to the image.

$$\text{scatter fraction} = \text{scattered events/total events} \tag{IV.6.4}$$

when the measurement is made at a count rate low enough to ensure that random coincidences, dead-time effects and pileup are negligible.

Thus, the data used in this calculation is the low activity acquisition from the 'count rate losses and randoms' measurement. For each transaxial slice, the total events are calculated from the complete sinogram. The scattered events are actually calculated by determining the unscattered events i.e. scattered events = total events–unscattered events, and these are assumed to be present only within ±2 cm of the source. This is done by analysing the sinogram profile for each projection angle, which describes the decrease in counts with distance from the source. The number of scattered events within ±2 cm of the source, is calculated by averaging the counts at the ±2 cm positions. The number of unscattered events is then calculated by subtracting this number from the total events within ±2 cm of the source.

Sensitivity

This is the ratio of detected coincidences to activity, given as counts s^{-1} MBq^{-1}. The source is a 70 cm long plastic tube containing activity, which is low enough that the counting losses are <1% and the random rate <5% of the true coincidences rate. Acquisitions, sufficient to ensure at least 10^4 true coincidences per transaxial slice, are taken with the source suspended axially at the centre of the transaxial field of view and then offset by 10 cm. To ensure annihilation, the source must be surrounded with sufficient material to provide the electrons. This introduces attenuation effects, which have to be determined. This is done by taking 4 measurements using different surrounding metal sleeves and extrapolating the results to the situation when there is no surrounding material.

Accuracy of Corrections

Count Losses and Randoms

The accuracy of the corrections for dead-time losses and random coincidences is determined using the data acquired during the test of 'count rate losses and randoms'. This is reconstructed with dead-time losses and randoms corrections applied. For each transaxial slice, a circular 18 cm diameter ROI is defined over the centre of the phantom, rather than the centre of the source. R_{tr} is calculated as in the 'count rate losses and randoms' test. The residual error in corrected true coincidence rate, as a % of the expected true coincidence rate, is:

$$\Delta R = 100\,(R_{tr}/R_e - 1) \qquad \text{(IV.6.5)}$$

where R_e is the expected true coincidence rate obtained, by extrapolation assuming a linear relation, from values calculated at low effective activity concentrations. The maximum and minimum values of ΔR, over all slices, are plotted against the effective activity concentration.

Image Quality

This tests produces images which simulate those obtained in a whole-body study with hot and cold lesions. The phantom is that of a torso (IEC 1998), with a warm background at an activity concentration of 370 MBq/70 kg which is typical of patient studies. Non-uniform attenuation is simulated with lung equivalent material and the lesions with spheres. The hot spheres have inner diameters of 1.0, 1.3, 1.7 and 2.2 cm. The cold spheres have inner diameters of 2.8 and 3.7 cm. Two acquisitions are undertaken; with the activity concentration in the hot spheres equal to 8 and 4 times background. The 70 cm line source is filled with activity at the background concentration and placed adjacent to the torso to simulate activity outside the field of view. The acquisition protocol simulates a 100 cm whole body acquisition in a duration of 60 minutes, and is repeated 3 times.

Reconstruction is as used in a typical patient study, and the quality of the reconstructed image is quantitated using ROIs. A transaxial slice, which is centred on the spheres, is used. Circular ROIs with diameters equal to the spheres are drawn over the spheres. Background ROIs are also constructed, 12 on the central slice and 12 on each of slices ±1 and ±2 cm from the central slice. At each background ROI position, the ROIs of each size are drawn concentrically. For each sphere size, the background variability and %contrast are calculated. The background variability is determined as the coefficient of variation of the mean counts. The %contrast is calculated as:

$$100\,(C_S/C_B-1)/(A_S/A_B-1) \qquad \text{(IV.6.6)}$$

where
average counts
  C_S in the sphere ROI
  C_B in the background ROIs, of size equal to relevant sphere size
activity concentration
  A_S in the sphere
  A_B in the background

Attenuation and Scatter Correction

For each transaxial slice, 3 cm diameter circular ROI is constructed over the lung equivalent material, and 12 similar ROIs are constructed in the background area. The residual error in corrected true coincidence events after attenuation and scatter correction is:

$$\Delta C_L = 100\, C_L/C_B \qquad \text{(IV.6.7)}$$

where C_L is the average counts in the lung ROI

IEC System

The application of this system has been described by Adam et al. (1997).

Intrinsic

Spatial Resolution

The transaxial spatial resolution is measured, in air, using activity contained in a stainless steel needle, with inside diameter ~1 mm and wall thick enough to ensure annihilation. This is oriented axially and positioned in the centre of the field of view. Acquisitions are undertaken, in three different transaxial positions for each of 7 radial distances: 0, 1, 5, 10, 15, 20, 23 cm. Reconstruction, using a pixel size of ~0.5 mm, is by filtered back projection and ramp filter, and FWHM and FWTM values are calculated from count profiles constructed through the reconstructed points in the radial and tangential directions. In 3D mode, axial spatial resolution is measured similarly, with the source oriented transaxially. Acquisitions are undertaken with the source positioned at the centre of the axial field of view, and at axial offsets of ±2, ±4, ±6, ±7 cm, for each of 6 radial distances: 0, 5, 10, 15, 20, 23 cm. In 2D mode, the axial slice width is measured by moving the source axially in 0.5 mm increments. At each position, acquisitions are undertaken in three different transaxial positions for each of the 6 radial distances. For analysis, count profiles are constructed in the axial direction.

Recovery Coefficient

This is measured using activity contained in 6 spheres, of diameters 1, 1.3, 1.7, 2.2, 2.8, 3.7 cm, positioned in the same transaxial plane in a water filled cylinder, of diameter 22.6 cm. Acquisitions are undertaken with the spheres positioned centrally in the transaxial field of view and at the centre of the axial field of view, and then offset axially by 2 and 4 cm. Reconstruction uses a pixel size of 1.3 mm and involves analytical attenuation correction with $\mu = 0.096\ \text{cm}^{-1}$. 5 mm diameter ROIs are constructed centred on each sphere and the counts (C) determined. For each sphere the recovery coefficient is:

$$R_c = C/C_{37} \qquad \text{(IV.6.8)}$$

where C_{37} is the counts in the ROI over the 3.7 cm diameter sphere in which recovery is assumed to be 100%

Count Rate Losses and Randoms

These measurements are undertaken with three different phantom configurations which approximately realistic activity distributions by simulating the head, thorax and abdomen:

1. head: a lucite cylinder of diameter 20 cm containing a uniform activity concentration,
2. thorax: the IEC body phantom (IEC 1998) enclosing the empty phantom from (1) containing a 13 cm long 2.1 cm diameter rod source of activity,
3. thorax: as configuration (2) with phantom (1) containing water.

For each configuration, acquisitions are undertaken at regular intervals, with the phantom at the centre of the field of view, for at least 11 radionuclide half-lives. Reconstruction is undertaken without corrections and, for the head configuration, the transaxial field of view is limited to a circle of 24 cm diameter.

The total coincidence rate (R_t) and the random coincidence rate (R_r) are determined. The true coincidence rate (with scatter):

$$R_{trs} = R_t - R_r \qquad \text{(IV.6.9)}$$

The unscattered true coincidence rate:

$$R_{tr} = R_{trs}\,(1 - SF) \qquad \text{(IV.6.10)}$$

Again, R_e is the expected true coincidence rate obtained, by extrapolation assuming a linear relation, from values calculated at low effective activity concentrations.
The percent dead-time:

$$\%\,DT = 100\,(1 - R_{trs}/R_e) \qquad \text{(IV.6.11)}$$

Noise Equivalent Count Rate (NEC)

The NEC rate (R_n) is:

$$R_n = R_{trs}\,2/(R_t + R_r) \qquad \text{(IV.6.12)}$$

Scatter Fraction

The $FWHM_{10}$ is calculated, from the spatial resolution test, as the average of all the radial and tangential FWHMs at a radial distance of 10 cm. The scatter fraction is measured using a line source in a water filled cylinder of diameter 20 cm. Activity should be low enough to ensure that random coincidences, dead-time effects and pileup are negligible (Karp et al. 1991). Acquisitions are undertaken with the source at the centre of the phantom, and then at radial offsets of 4.5 and 9.0 cm. Reconstructions allow a field of view of 24 cm diameter. For each transaxial slice, the total events are calculated from the complete sinogram. The scattered events are actually calculated by determining the unscattered events i.e. scattered events = total events–unscattered events, and these are assumed to be present only within ±4 times $FWHM_{10}$ of the source. This is done by analysing the sinogram profile for each projection angle, which describes the decrease in counts with distance from the source. The number of scattered events within ±4 times $FWHM_{10}$ of the source, is calculated by averaging the counts at the ±4 times $FWHM_{10}$ positions. The number of unscattered events is then calculated by subtracting this number from the total events within ±4 times $FWHM_{10}$ of the source.

$$\text{scatter fraction} = \text{scattered events}/\text{total events} \quad \text{(IV.6.13)}$$

Sensitivity

This is calculated using the low activity, head configuration, data from the 'count rate losses and randoms' test. For each transaxial slice, the total counts (C_t) in the sinogram was determined and the slice sensitivity (S_s) for unscattered events calculated as:

$$S_s = C_t\,(1 - SF)/(T_a\,A) \quad \text{(IV.6.14)}$$

where
SF is the scatter fraction for the slice
T_a is the acquisition duration
A is the activity concentration

Accuracy of Corrections

Attenuation Correction

Three 5 cm diameter cylinders are inserted into the head phantom from the 'count rate losses and randoms' test; these are filled with:
- air to simulate lung
- water to simulate soft tissue,
- Teflon to simulate bone

The activity in the phantom is such that dead-time losses are <5%. Acquisitions are undertaken over a 10-hour period with the phantom 2.5 cm offset. Reconstruction is undertaken with a pixel size of 2 mm and all corrections applied, including scatter and attenuation. 3 cm diameter ROIs are constructed over the three inserts and 9 similar ROIs over the background area. For each insert, the %attenuation error is:

$$\%\,AE = 100\;C_I/C_B \quad \text{(IV.6.15)}$$

where
C_I is the count density in the insert ROI
C_B is the average count density in the background ROIs

The non-uniformity of the attenuation correction is assessed by determining the deviations, relative to C_B, of the maximum and minimum count densities of the background ROIs.

Scatter

This is assessed for the cardiac configuration using the IEC body phantom with arms (IEC 1998) which encloses the head phantom from the 'count rate losses and randoms' test, into which is obliquely mounted the IEC cardiac insert. The phantom is then loaded with:
- ventricle and arms – water
- myocardium – high activity
- lungs – air
- thorax walls – low activity concentration = 0.12 * myocardium

The phantom is positioned at the centre of the axial and transaxial field of view. A 60-minute acquisition is undertaken, when the dead-time losses are <5%. Following at least 10 half-lives, a 30-minute transmission acquisition is undertaken. Reconstruction, with a pixel size of 4.2 mm, is undertaken with all corrections applied but, with and without attenuation correction. For each slice involving the heart, ROIs are constructed over the ventricle and myocardium and count densities, C_v and C_m, determined. The relative scatter fraction is calculated as:

$$\%\ SF = 100\ C_v/C_m \tag{IV.6.16}$$

for attenuation correction and no attenuation correction.

Gamma Camera Systems

Although the new NEMA standard includes consideration of coincidence γ cameras, a number of tests, specific to the devices have been published (Fleming et al. 2000; IPEM 2003). In these, measurements should be undertaken for all dual photon collimators in use.

Operational Tests

Using a point source, the PHA peak and width, detection sensitivity and uniformity should be assessed. If transmission attenuation correction is to be used, a blank scan must be undertaken.

Intrinsic

Energy Resolution

A low activity, e.g. 1–2 MBq, source is used, such that dead-time effects will be minimal and the random fraction will be <2%. It is placed centrally in the field of view along the axis of rotation and an acquisition, using standard parameters to achieve at least 10^4 counts in the peak energy channel, is undertaken. The FWHM of the photopeak is calculated.

Spatial Resolution

A low activity, e.g. 1–2 MBq, point or line source is used, such that dead-time effects will be minimal and the random fraction will be <2%. It is placed centrally in the field of view along the axis of rotation and acquisitions, using standard parameters to achieve at least 10^6 counts, are undertaken on axis and then with 10 cm transaxial displacement. Reconstruction is undertaken using the small pixel size available and implementing all corrections. The FWHM is calculated from count profiles created through the reconstructed source in radial, tangential and axial directions. For the latter, if a line source is used, a separate acquisition is required with the source perpendicular to the axis of rotation.

Count Rate Losses and Randoms

The source used is either a point or the cylindrical phantom, positioned at the centre of the field of view. In both cases, the activity of ^{18}F must be higher than that required to achieve the peak count rate, e.g. 100 MBq. Measurements of the total coincidence count rate and the random coincidence count rate, i.e. singles rate, are taken at regular intervals during decay. This is continued until the linear part of the count rate curve is reached and repeated for all collimators in use. The measured rates are plotted against expected count rate, extrapolated from the low activity results and percent dead-time, peak coincidences and singles rates with respective activities, and activity for 50% coincidence dead-time are calculated.

Maximum Noise Equivalent Count Rate (NEC)

Various formulae have been described depending on the assumptions made about the noise added by the corrections. This formula corresponds to noiseless, or non-existent, corrections (Fleming et al. 2000):

$$\mathrm{NEC} = [R_t (1 - SF_l)^2]/[(1 + R_{rts}) \{(1 - R_{rts})/(1 - SF) + k_p \times R_{rts}\}] \qquad \text{(IV.6.17)}$$

An alternative is (IPEM 2003):

$$\mathrm{NEC} = R_t/[1 + (R_s/R_t) + (2 k_p R_r / R_t)] \qquad \text{(IV.6.18)}$$

where

$R_{rts} = R_r/(R_{tr}+R_s)$

k_p is the fractional volume of the field of view occupied by the phantom

SF_l is the scatter fraction evaluated from the line spread function (LSF) across the whole field of view

SF scatter fraction calculated only from the section of the LSF within the projection of the phantom

Scatter Fraction

Acquisitions, using the NEMA or IEC scatter fraction phantom geometry, are undertaken using a PHA window of 30% and at least 30 head angles. The scatter fraction is determined from the line spread function of the line source or from five ROIs created on at least 20 projections.

Sensitivity

In order to keep the randoms fraction and dead-time to a minimum, the activity used is such to ensure that the camera is operating on the linear part of the activity count rate curve, e.g. 1–2 MBq. With the source located at the centre of the field of view, acquisitions of a few minutes duration are performed such that the number of events in each scan $\sim 10^5$. The number of coincidence events s^{-1} are recorded. Using a point source, the sensitivity is calculated as counts s^{-1} MBq^{-1}. Using the NEMA or IEC scatter fraction phantom geometry, the sensitivity for a cylinder is calculated as counts s^{-1} MBq^{-1} and counts s^{-1} MBq^{-1} ml^{-1}.

Accuracy of Corrections

Attenuation Correction

A blank acquisition is undertaken and then the IEC phantom, in the 'attenuation correction' test geometry, is positioned 5 cm off centre transaxially and axially. A number of acquisitions are undertaken with various corrections implemented and at two activity levels; a high level similar to that used in the 'count rate losses and randoms' test and a low level such that the randoms fraction <2%. 10^7 counts should be acquired for both the transmission and emission acquisitions. Counts under the ROIs specified in the 'attenuation correction' test should be plotted against axial distance for each acquisition and reconstruction combination. Contrast indices are calculated as $1-(C_i/C_b)$ and the reconstructed uniformity as the variation in background values within and between transaxial slices.

Total Performance

The practical performance is assessed by imaging a total performance phantom loaded with ^{18}F, to give approximately the maximum NEC for the phantom, and imaged for 27 min. The quality of the images is assessed visually by selecting 9 mm thick slices from each section of the phantom.

References

Adam L-E, Zaers J, Ostertag H, Trojan H, Bellemann ME, Brix G (1997) Performance evaluation of the whole-body PET scanner ECAT EXACT HR following the IEC standard. IEEE Trans Nucl Sci 44:1172–1179

Buchert R, Bohuslavizki KH, Mester J, Clausen M (1999) Quality assurance in PET: evaluation of the clinical relevance of detector defects. J Nucl Med 40:1657–1665

Daube-Witherspoon ME, Karp JS, Casey ME, DiFilippo FP, Hines H, Muehllehner G, Simcic V, Stearns CW, Adam LE, Kohlmyer S, Sossi V (2002) PET performance measurements using the NEMA NU 2-2001 standard. J Nucl Med 43:1398–1409

Fleming JS, Goatman KA, Julyan PJ, Boivin CM, Wilson MJ, Barber RW, Bird NJ, Fryer TD (2000) A comparison of performance of three gamma camera systems for positron emission tomography. Nucl Med Commun 21:1095–1102

IEC (1998) International Electrotechnical Commission publication IEC 61675-1 characteristics and test conditions of radionuclide imaging devices, part 1. Positron emission tomography. IEC, Geneva

IPEM (2003) Report 86 quality control of gamma camera systems. Institute of Physics and Engineering in Medicine, London

Karp JS, Daube-Witherspoon ME, Hoffman EJ, Lewellen TK, Links JM, Wong W-H, Hichwa RD, Casey ME, Colsher JG, Hitchens RE, Muehllehner G, Stoub EW (1991) Performance standards in positron emission tomography. J Nucl Med 32:2342–2350

NEMA (1994) Standards publication NU 2 performance measurements of positron emission tomographs. National Electrical Manufacturers' Association, Washington

NEMA (2001) Standards publication NU 2 performance measurements of positron emission tomographs. National Electrical Manufacturers' Association, Washington

Image Presentation

Abbreviations

SMPTE Society of Motion Picture and Television Engineers

Since more interpretation is being undertaken from the display, this has taken on a greater importance than previously. A well adjusted monitor enhances interpretation sensitivity and consistency, by allowing accurate comparison with previous investigations and reference images. Hardcopy, however, remains a very important part of the diagnostic process, with the tendency being towards digital imagers. These are generally stable and any artefacts caused by them are easily identified because they are not present on the display monitor.

Display

Operational Checks

The screen should be checked for scratches and then the uniformity image, from the planar imaging operational quality assurance, should be visually inspected for blurring, spatial distortion or electrical interference.

Routine Checks

At acceptance of the whole system or new display, the most appropriate translation tables to be used for processing and interpretation of the various clinical investigations, should be established. Routine checks, to maintain consistency, are then conducted using a test pattern, such as that designed by the Society of Motion Picture and Television Engineers SMPTE (Gray et al. 1985), which should be available on most systems. Using this, the ambient light conditions should be adjusted to find the best combinations for viewing the different monochrome and colour tables.

Brightness and Contrast

The brightness and contrast settings on the monitor should be adjusted so that the full range of grey scales or colours, in the pattern, can be seen.

Grey Scale Linearity

A section of the pattern displays contrast changes from 0% to 100% in several increments. Each step should be uniform and show a different shade of grey or colour.

Spatial Linearity

The grid pattern should be straight and all the squares should be of the same size.

Spatial Resolution

Parts of the pattern contain fine details at both high and low contrast, which are sensitive to image noise. They should all be clearly visible with no evidence of streaking, smearing or banding.

Hardcopy

Operational Tests

The print should be checked for streaks and then the uniformity image, from the planar imaging operational quality assurance, should be visually inspected for blurring, spatial distortion or density changes which are not evident on the display.

Routine Tests

At acceptance of the whole system or new hardcopy unit, the most appropriate translation tables to be used for the various clinical investigations and users, should be established. This will involve a number of parameter settings, which may be different for each user. Careful record keeping is therefore essential including a representative SMPTE type test image for each acceptable combination, with density values being measured for the monochrome transmission images. Routine checks, to maintain consistency, are then conducted using the SMPTE type test pattern according to the quality parameters discussed in 'display' and also by comparison with the reference images.

Hardcopy artefacts can also be caused by film processing rather than by image creation. If wet processing is used, the process can be very variable and must be tested using a film exposed by a sensitometer. Regular monitoring of developer and fixer temperature, and cycle duration should also be undertaken. Independent sensitometry tests are not possible on dry processors but these are much less variable, because automatic quality control systems are incorporated and are usually initiated at power on, film refill or after a long idle time (Lu et al. 1999).

References

Gray JE, Lisk KG, Haddick DH, Harshbarger JH, Oosterhof A, Schwenker R (1985) Test pattern for video displays and hard-copy cameras. Radiol 154:519–527

Lu ZF, Nickoloff EL, Terilli T (1999) Monthly monitoring program on Dryview laser imager: one year experience on five Imation units. Med Phys 26:1817–1821

Computer

Abbreviations

ADC	analogue to digital converter
COST	Cooperation in the Field of Scientific and Technical Research
QA	quality assurance
ROI	region of interest

Dedicated computers are interfaced to almost all γ cameras, and these significantly affect the performance of the whole system. Methods of quality assurance are not as well developed as they are for the detectors and separating the two components can be difficult. Two aspects are involved; the camera–computer interface (IAEA 1991) and software (IPSM 1992; IPEM 2003).

Camera–Computer Interface

The main factors which must be evaluated are the performance of the analogue to digital converters, ADCs, and the timing of acquisitions.

ADCs

These can degrade uniformity by imposing lines on the image if their increments are irregular or a gradual change in intensity across the image and linearity distortion if their increments vary regularly. Evaluation of a static acquisition will reveal such artefacts.

Linearity

Integral linearity is the absolute difference between the actual position of the photon interaction and the position indicated by the ADC. It depends on the accuracy of converting the analogue position signal into a digital location and assessment requires precise quantitative measurements. More severe variations are likely in differential, rather than integral, non-linearity. This is the error in allocating an actual photon interaction position to an ADC 'bin', because their sizes are not constant, and it imposes vertical or horizontal stripes on the images.

Operational Evaluation

Visual inspection of the daily flood image, using high contrast, will reveal differential linearity problems as horizontal or vertical stripes.

Routine Evaluation

Radionuclide: $^{99}Tc^{m}$
Source: point
Activity: 100 MBq
Matrix: maximum
Acquisition counts: 10^8

An intrinsic image is acquired and 10 pixel wide profiles created, in both the x- and the y-directions. Integral non-linearity is estimated from the slope, which should be zero. The differential non-linearity is the largest difference between an actual pixel count and that expected from the slope.

Uniformity

Operational Evaluation

The daily flood image should be inspected for structural non-uniformities that do not rotate with changes in camera orientation settings.

Orientation

To check that the image is oriented correctly, a point source should be placed in one quadrant and the known position compared to the computer image. This should be repeated for all quadrants and orientation settings.

Acquisition Timing

This involves the accuracy of static acquisition durations and dynamic acquisition frame rates. It is assessed by stopwatch timing and comparison of acquired counts.

Static

Radionuclide: $^{99}Tc^{m}$
Source: point
Activity: ~7 MBq (to achieve ~5×10^{3} counts s^{-1})
~60 MBq (to achieve ~40×10^{3} counts s^{-1})
Matrix: 64
Acquisition duration: 100 s

Timed intrinsic acquisition durations, at both count rates, should confirm the termination limit chosen.

Dynamic

Data losses can occur within and between frames, and may be significant at high frame and count rates.
Radionuclide: $^{99}Tc^{m}$
Source: point
Activity: ~30 MBq (to achieve ~20×10^{3} counts s^{-1})
Matrix: 64
Acquisition duration: 20 s

The total count, C_s, is calculated from a static acquisition of duration, T_s. A dynamic study is acquired, with the shortest frame time possible, e.g. 100 frames of 0.2 s. At acceptance, all combinations of matrices and zooms should also be considered. The counts in each frame, C_f, are determined and a χ^2 test performed to confirm that the variation is reasonable. The apparent frame times, T_f, are calculated from $T_f = T_s \times C_f/C_s$ and the time lost per frame estimated by comparison with the mean. This should be minimal. The apparent total acquisition duration is estimated by summing these T_f values. This should equal the expected duration but, if time is being lost between frames, will exceed it.

Count Rate Performance

The computer may degrade the count rate performance of the system at high count rates, but should not do so in the clinical range of <20% count loss. This is assessed by comparing the count rates indicated by the camera and the computer during the planar count rate performance test.

Software

Software errors can cause very serious errors and results should always be treated with caution, particularly if they appear unreasonable. Comprehensive testing, of complete software packages, is very difficult but the evaluation of small segments, using appropriate test data, is possible (IAEA 1991). To develop the means by which the general quality of nuclear medicine software could be improved, COSTB2 (1988–1996), was undertaken under the auspices of COST, European Cooperation in the Field of Scientific and Technical Research, as part of the European Commission research and development programme (Cosgriff 1997). Software quality assurance comprises checks of utility software using test images and specific clinical application programs using clinical data (IPSM 1992).

Utility Software

These tests use simulated data, and should be performed when a new computer system, or utility software upgrade, is installed.

Low Level Image Processing

Image Display

The programs relating to image display can be tested by generating a 'ramp' image, I_1, such that the pixel count is constant in each column and increases across the image. Utility programs should then be used to extract information from this image and to alter its appearance. This could include: finding the maximum pixel count, displaying the image using different colour scales, varying the upper and lower display thresholds, and interpolating the data.

Image Quantification

The ROI and count profile functions can be tested by generating a uniform image over half of the field of view. The counts under various sizes and shapes of ROI, and the count profiles for various widths can then be evaluated for different positions on the image.

Image Arithmetic

Source: sheet
Frames: 64
Matrix: all
Acquisition: e.g. 5×10^5 and 10^6 counts

The subtraction, addition, multiplication and division functions can be tested, by acquiring uniformity images on different matrices. These can be manipulated and the counts from ROIs, covering most of the images, evaluated.

Activity–Time Curve Generation and Arithmetic

This accuracy of activity–time curve generation can be tested by acquiring a dynamic study, of a uniformity source, with the maximum number of frame duration phases.

Source: sheet
Frames: 64
Matrix: 64×64
Phases: maximum (e.g. duration 5, 10, 15, 20 s)

Several frames are acquired in each phase with the frame duration increasing during the study. The accuracy of curve generation can be assessed by constructing activity–time curves from ROIs covering approximately half of the image, and checking that the counts per frame vary as expected. The accuracy of curve addition and subtraction can be assessed by evaluating the counts in manipulated activity–time curves generated from two unequal sized ROIs totalling the previous one.

Complex Image Processing

More complex image processing functions are difficult to test but can be evaluated using software phantoms, which should include a range of examples to establish the limits of the algorithms. COST B2 revealed the need for such testing, particularly in SPET reconstruction. The same technique produced variations, in pixel values and signal-to-noise ratios, when implemented on different equipment or using different software versions; probably due to differences in the filter functions. Problems associated with centre-of-rotation correction proved to be less of a problem, but raw tomograms could not necessarily be reconstructed on different equipment. A similar need exists for testing that QA software produces correct results, varying results being produced, from the same data, by different systems (IAEA 1991); and that links to other systems and archiving routines work with no data loss.

Clinical Applications Programs

Software for processing clinical investigations has expanded rapidly, both in terms of availability and complexity, since the 1970s. Problems have been apparent since the early 1980s; for example, with interhospital variability in which poor quality software was been cited as a cause. Since COST B2 identified the level of software quality assurance in nuclear medicine, in 1988, as rudimentary, there has been an increasing awareness that concerted action is required. A solution proposed, to overcome the lack of standardisation in processing, is the development of phantom data, covering a range of normal and pathologic conditions, that can be used to comprehensively evaluate the capabilities and limitations of the software (Britton and Busemann Sokole 1990). This is particularly important as it becomes accepted that reproducible results within an institution are no longer sufficient but must be comparable with those in other institutions.

Types of Software

There are two types of clinical applications software: in-house written and commercially supplied. The quality of in-house software should be controlled in three areas; it should be designed with a good clinical understanding of the requirements and a knowledge of the limitations; the algorithms should be clear; and the production techniques should be good enough to ensure that the software code is generated without errors. The complete program must then be validated, however, to ensure that the clinical requirements are fulfilled and the limitations are avoided. This should be undertaken using real data and, if possible, simulated data because it is very difficult to include all eventualities using only the former, and carefully constructed simulated data permits controlled evaluation over a wide range of conditions (Collins et al. 1998). Commercial software should not be regarded as infallible and the results must always be scrutinised for inconsistencies and inaccuracies.

Testing

Three types of data are used for testing clinical software. The first comprises physical phantoms which are used to verify that the acquisition phase is accurate. The subsequent two are simulated and independently acquired patient data which are used to bypass the acquisition phase and verify that the processing phase is performing correctly.

Simulated data is generated mathematically and its advantages are that it can be well defined and its results are predictable. Comparison of the theoretical results with those actually obtained from the software indicate the latter's accuracy. Its limitations can be determined by degrading the data, e.g. by adding noise. A number of mathematical phantoms are now available including cardiac (Pretorius et al. 1997), brain (Collins et al. 1998) and hybrid dynamic renal (Cosgriff 1997). The latter is a synthetic study, reconstructed from various components of clinical data (Samal and Bergmann 1998). A more advanced dynamic renal phantom has been developed (Houston et al. 1999).

Patient data, whose results can be validated, should be archived for each software package. These should include clinically normal, extremely abnormal and borderline examples which will exercise the operational range of the software (Cosgriff 1997). Commercially supplied software would be expected to include such examples. As well as internal quality control, this mechanism also facilitates external quality control by distribution of identical clinical data among several departments; a procedure which would not have been possible, until recently, because of file incompatibilities (Cosgriff 1998).

There are two types of testing; the initial acceptance test confirming that the software is working correctly and that the users are operating it correctly; and the on-going routine evaluation confirming that this has not changed.

Acceptance

This quality control must be undertaken before software is used clinically. It involves familiarisation with: the documentation, algorithms and assumptions of the software; the clinical acquisition procedures; and the requirements and limitations of the processing, analysis and any normal data base results.

New software, software upgrades or other software modifications should be exposed to testing, using an appropriate combination of the above data, before being introduced clinically. This is particularly important if 'normal' database comparisons are used. Once the accuracy has been established, the reproducibility should be evaluated. For automatic and semiautomatic processing routines, this can be achieved by shifting the data slightly or by increasing the noise.

For operator dependent processing, the reproducibility for each user should be established by processing the archived patient data. The complete set should be processed in one session and repeated on several occasions. Intra and inter operator reproducibility of quantitative results can be evaluated using Bland–Altman plots (Bland and Altman 1986).

Routine

Periodically, the archived patient data should be reprocessed, by each user, to ensure that a change in processing technique has not occurred over time and that the level of reproducibility has not deteriorated. Again, this is particularly important if 'normal' database comparisons are used.

It may become necessary to introduce a change to the clinical procedure, in either the acquisition or the processing phases, for example to pixel size or reconstruction filter parameters. Before a change is implemented clinically, the effects of the change should be evaluated using simulated and archived patient data.

References

Bland MJ, Altman DG (1986) Statistical methods for assessing agreement between two methods of clinical measurement. Lancet 1:307–310

Britton KE, Busemann Sokole E (1990) Quality assurance in nuclear medicine software and 'COST'. Nucl Med Commun 11:334–338

Collins DL, Zijdenbos AP, Kollokian V, Sled JG, Kabani NJ, Holmes CJ, Evans AC (1998) Design and construction of a realistic digital brain phantom. IEEE Trans Med Imaging 17:463–468

Cosgriff PS (ed) (1997) Cost B2 final report (EUR 17916 EN): Quality assurance of nuclear medicine software. Office for Official Publications of the European Communities, Luxembourg

Cosgriff PS (1998) Quality assurance in renography: a review. Nucl Med Commun 19:711–716

Houston AS, Sampson WF, Jose RM, Boyce JF (1999) A control systems approach for the simulation of renal dynamic software phantoms for nuclear medicine. Phys Med Biol 44:401–411

IPEM (2003) Report 86 quality control of gamma camera systems. Institute of Physics and Engineering in Medicine, London

IPSM (1992) Report 66 quality control of gamma cameras and associated computer systems. Institute of Physical Sciences in Medicine. London

Pretorius PH, Xia W, King MA, Tsui BM, Pan TS, Villegas BJ (1997) Evaluation of right and left ventricular volume and ejection fraction using a mathematical cardiac torso phantom. J Nucl Med 38:1528–1535

Samal M, Bergmann H (1998) Hybrid phantoms for testing the measurement of regional dynamics in dynamic renal scintigraphy. Nucl Med Commun 19:161–171

PART V

Appendices

Radionuclide Decay Characteristics

Nuclide	Symbol	Half-life	Decay mode	$E_{\beta max}$	E_{photon}
Tritium	^{3}H	12.4 years	β^-	186 (100)	–
Carbon-11	^{11}C	20.4 months	β^+, γ	+970 (100)	511 (200)
Nitrogen-13	^{13}N	10.0 months	β^+, γ	+1200 (100)	511 (200)
Carbon-14	^{14}C	5730 years	β^-	156 (100)	–
Oxygen-15	^{15}O	122.2 s	β^+, γ	+1740 (100)	511 (200)
Fluorine-18	^{18}F	109.8 months	β^+, γ	+650 (100)	511 (200)
Sodium-22	^{22}Na	2.6 years	EC, β^+, γ	+546 (90)	1275 (100) 511 (180)
Sodium-24	^{24}Na	15.0 h	β^-, γ	1392 (100)	2754 (100) 1369 (100)
Calcium-45	^{45}Ca	163 days	β^-	257 (100)	–
Calcium-47	^{47}Ca	4.5 days	β^-, γ	1990 (18) 690 (82)	1297 (75)
Chromium-51	^{51}Cr	27.7 days	EC, γ	–	320 (10)
Iron-52	^{52}Fe	8.3 h	EC, β^+, γ		169 (99) 511 (115)
Iron-55	^{55}Fe	2.7 years	EC, γ	–	[6–7 (~28)]
Cobalt-56	^{56}Co	78.8 days	EC, β^+, γ	1500 (19)	1238 (67) 847 (100) 511 (40)
Cobalt-57	^{57}Co	270.9 days	EC, γ	–	122 (86)
Cobalt-58	^{58}Co	70.8 days	EC, β^+, γ	+ 475 (15)	811 (100) 511 (30)
Iron-59	^{59}Fe	44.5 days	β^-, γ	467 (53) 274 (46)	1292 (44) 1099 (56)
Cobalt-60	^{60}Co	5.3 years	β^-, γ	318 (100)	1332 (100) 1173 (100)
Gallium-67	^{67}Ga	78.3 h	EC, γ	–	300 (17) 185 (21) 93 (38)
Gallium-68	^{68}Ga	68.0 months	EC, β^+, γ	+1900 (88)	511 (178)
Germanium-68	^{68}Ge	288 days	EC, γ	–	9 (39)
Rubidium-81	^{81}Rb	4.6 h	EC, β^+, γ	+ (28)	511 (63) 446 (23)
Krypton-81m	$^{81}Kr^{m}$	13 s	γ	–	190 (100)

Nuclide	Symbol	Half-life	Decay mode	$E_{\beta max}$	E_{photon}
Krypton-81	^{81}Kr	2 10^5 years	EC, γ	–	276 (2)
Molybdenum-99	^{99}Mo	66.0 h	β^-, γ	1232 (82) 454 (17)	778 (4) 740 (12) 181 (6)
Technetium-99m	$^{99}Tc^m$	6.0 h	γ	–	141 (89)
Technetium-99	^{99}Tc	2 10^5 years	β^-	293 (100)	–
Indium-111	^{111}In	2.8 days	EC, γ	–	245 (94) 171 (91)
Iodine-123	^{123}I	13.2 h	EC, γ	–	159 (83) 27 (86)
Iodine-125	^{125}I	60.1 days	EC, γ	–	36 (7) [27–32 (138)]
Xenon-127	^{127}Xe	36.4 days	EC, γ	–	375 (17) 203 (68) 172 (26)
Iodine-129	^{129}I	1.6 10^7 years	β^-, γ	150 (100)	[30–35 (~69)]
Iodine-131	^{131}I	8.0 days	β^-, γ	606 (89) 334 (7)	637 (7) 365 (81) 284 (6)
Xenon-133	^{133}Xe	5.2 days	β^-, γ		81 (37) [30–36 (~46)]
Caesium-137	^{137}Cs	30 years	β^-, γ	173 (95)	662 (95)
Gadolinium-153	^{153}Gd	242 days	EC, γ	–	103 (21) 97 (30)
Thallium-201	^{201}Tl	73.1 h	EC, γ	–	167 (10) [68–82 (~95)]

Adapted from ICRP (1983) with permission from Elsevier.

β^- beta emission
β^+ positron emission
EC electron capture
$E_{\beta max}$ maximum β energy in keV, + represents β^+
E_{photon} main photon emissions in keV
( ) frequency of emission as a percentage of disintegrations
[ ] x-rays

References

ICRP (1983) International Commission on Radiological Protection publication 38 radionuclide transformations, energy and intensity of emissions, vol 11–13. Pergamon, Oxford

Units

Table V.2.1. SI unit prefixes

Multiplier	Prefix	Symbol
10^{-3}	milli	m
10^{-6}	micro	µ
10^{-9}	nano	n
10^{-12}	pico	p
10^{3}	kilo	k
10^{6}	mega	M
10^{9}	giga	G
10^{12}	tera	T

Table V.2.2. SI units and non-SI unit equivalents

SI unit	Non-SI unit
37 kBq	1 µCi
37 MBq	1 mCi
37 GBq	1 Ci
1 Gy	100 rad
1 Sv	100 rem
1 J	10^{7} erg
1.6×10^{-19} J	1 eV
2.58×10^{-4} C kg^{-1} of air	1 R

Bq becquerel, *Gy* gray, *Sv* sievert, *J* joule, *C* coulomb, *Ci* curie, *rad* rad, *rem* rem, *erg* erg, *eV* electronvolt, *R* roentgen

Collimator Details

Abbreviations

FWHM	full width at half maximum
HE	high energy
LE	low energy
LSF	line spread function
ME	medium energy
PSF	point spread function
SPET	single photon emission tomography
UHE	ultra high energy
VHE	very high energy

Parallel Hole Collimators

Parallel hole collimators are ~25–80 mm thick and contain $3–9 \times 10^4$ holes. Originally the holes were circular but are now usually hexagonal, in low and medium energy collimators, and constructed by assembling strips of lead foil. High energy collimators still have round holes, although square or triangular holes are sometimes used in high sensitivity collimators. These tend to be constructed using a casting technique (IPSM 1992). The latter has been improved recently and, since the uniformity of cast collimators is generally better than that of foil, these are now recommended in low and medium energy collimators, especially for SPET (Moore et al. 1992).

Septa

The lead between the holes is called the septum, and the holes are organised so as to maximise the exposure of the crystal for a particular septal thickness. A hexagonal arrangement provides the most efficient packing (Muehllehner et al. 1976), giving the highest sensitivity for a particular spatial resolution (IPSM 1992). This allows hexagonal, square or triangular holes to be formed with septa having a uniform thickness. Round holes, however, necessitate a varying thickness which reduces sensitivity. The minimum septal thickness is determined by the energy of the photons to be used. It must be such that only a small fraction of the incident photons will penetrate it and cross from one hole to another, otherwise an inaccurate image will be projected onto the crystal.

Because photon attenuation is an exponential process, the septa cannot be thick enough to absorb all off-axis photons and ~5% penetration along the shortest septal path is usually accepted. The septal thickness (t) and the shortest path length (p) are geometrically related by (Sorensen and Phelps 1987; Moore et al. 1992) (see Fig. V.3.1):

$$t = 2\,dp/(l - p) \qquad \text{(V.3.1)}$$

where
l is the length of the hole
d is its diameter

The limit of 5% penetration along the shortest path length can be imposed, on this equation, by considering the transmission in terms of the linear attenuation coefficient of the lead for the γ photons (μ):

$$\text{transmission} = e^{-\mu p} \leq 0.05$$

$$e^{-\mu p} \leq e^{-3}$$

$$\text{thus,}\quad t \geq 6\,d/(\mu l - 3) \qquad \text{(V.3.2)}$$

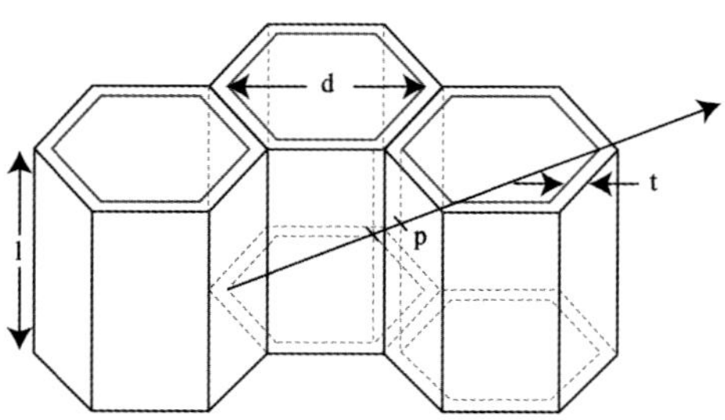

Fig. V.3.1. Penetration of collimator septa, showing relationship between the septal thickness (t) and the shortest path length (p)

Energy Range

In order to maintain the sensitivity of the collimator, t must be as thin as possible consistent with the above constraints. However because, in the energy range used in nuclear medicine, the value of μ for lead depends strongly on the photon energy, t is determined by the energy of the photons for which the collimator will be used. This is why collimators are categorised as low (LE), medium (ME), high (HE) or very high energy (VHE), based on the maximum photon energy for which the septal thickness is adequate. These thicknesses range from 0.18 mm in LE to >2 mm in UHE (IPSM 1992; Tsui et al. 1996) and, particularly in the LE category, result in the collimator being mechanically fragile and susceptible to physical damage if mishandled.

Spatial Resolution

For circular holes, the spatial resolution behaves as

$$R_p \approx d + d\,r/l \qquad \text{(V.3.3)}$$

where R_p is the spatial resolution defined as the FWHM of the point spread function or line spread function, measured at the rear face of the collimator, and r is the distance from the rear face of the collimator to source

Sensitivity

The sensitivity is only of the order of 0.1% and is defined as the fraction of photons, which are transmitted by the collimator, from those emitted by a point source. It is defined with only air, and no other attenuating material, between the source and the collimator, and is related solely to hole geometry and packing. The sensitivity behaves as (Moore et al. 1992; Tsui et al. 1996):

$$S_p \approx k\,d^4/[l^2\,(d+t)^2] \qquad \text{(V.3.4)}$$

where k is determined by the hole shape, being ~0.06 for round, ~0.07 for hexagonal and ~0.08 for square holes.

Thus short wide holes give the best sensitivity. Sensitivity does not depend on the source to collimator separation, if the measurements are made in air and over the whole collimator face and there is no significant septal penetration. Two geometrical effects cancel out. Individual hole sensitivity decreases as $1/r^2$ but the number of holes involved in the transmission increases as r^2. This is demonstrated in the behaviour of the PSF and LSF. As r increases, the amplitudes of the functions decrease but their widths increase, the areas under the curves remaining unchanged.

The requirements for better sensitivity and resolution work against each other. Because S_p varies as $\sim d^2/l^2$ and R_p varies as $\sim d/l$:

$$S_p \approx R_p^{\,2} \qquad \text{(V.3.5)}$$

Thus, an improvement in resolution is accompanied by a much reduced sensitivity.

In the above equation, (l) is actually the effective hole length. This is slightly less than the physical length (l_p) due to septal penetration by the γ photons being detected

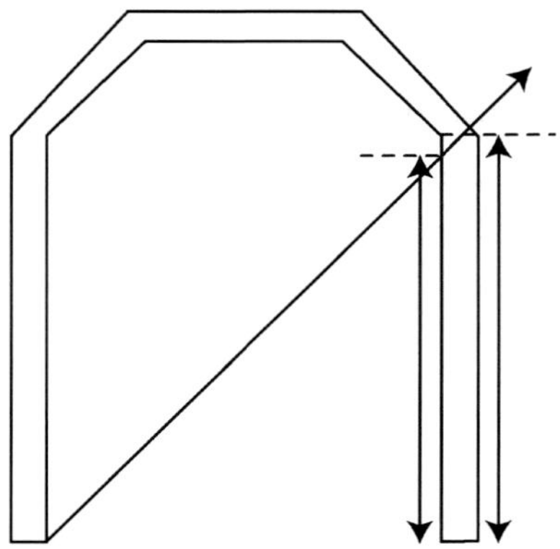

Fig. V.3.2. The effective hole length is slightly less than the physical length due to septal penetration by the γ photons

(see Fig. V.3.2). The relationship depends on the energy of the γ photons according to a consideration of linear attenuation coefficient:

$$l = l_p - 2/\mu \qquad \text{(V.3.6)}$$

Non-Parallel Hole Collimators

These comprise converging, diverging and pinhole types and the independence of sensitivity with source to collimator separation remains true only for uniform flood sources, which extend over the complete field of view. Also, in all cases, magnification varies with separation, causing spatial distortion.

The size of the field of view (F_m) varies with the magnification or minification (M) achieved, according to:

$$F_m = F/M \qquad \text{(V.3.7)}$$

where F is the full field of view of the camera.

Converging

The holes in this type of collimator are orientated towards a convergence point in front of the collimator, usually at a distance (f) of 40–50 cm from the front face. It is used to produce a magnified image of a small object. The magnification obtained is:

$$\begin{aligned} M &= \text{crystal to convergence point}/ \\ &\quad (\text{crystal to convergence point} - \text{front of collimator to object}) \qquad \text{(V.3.8)} \\ &= (f+w)/(f+w-r) \end{aligned}$$

where w is the thickness of the collimator.

Thus, the image magnification and orientation depend on the object position relative to the convergence point, see Table V.3.1.

Table V.3.1. Image magnification and orientation of a converging collimator depend on the object position relative to the convergence point

Object position	Image Size	Image orientation
r<f	Magnified	Non-inverted
f<r<2f	Magnified	Inverted
r>2f	Minified	Inverted

where
r is front of collimator to object
f is front of collimator to convergence point

Fanbeam

A variant of the converging collimator is the fanbeam, in which the holes are parallel in the axis along the patient but converge in the axis across the patient. It is being used in SPET to improve the spatial resolution with depth in the patient. The magnification obtained in the transverse axis is the same as for the converging collimator.

Diverging

The holes in this type of collimator are orientated away from a convergence point behind the collimator, usually at a distance (f) of 40–50 cm from the front face. It is used to produce a minified image of a large object. The minification obtained is:

$$\begin{aligned} M &= \text{distance from crystal to convergence point}/\\ &\quad(\text{distance from convergence point to object})\\ &= (f-w)/(f+r) \end{aligned} \qquad (V.3.9)$$

In usual operation, a diverging collimator decreases the size of the object by ~30%. The necessity for this type of collimator has decreased as field of view sizes have increased.

Fishtail

A variant of the diverging collimator is the fishtail, in which the holes are parallel in the axis along the patient but diverge in the axis across the patient. It was used in the older whole body scanning cameras to increase the lateral the field of view, to allow the full width of the patient to be imaged in one scan. The minification obtained in the transverse axis is the same as the diverging collimator. The requirement for this type of collimator has reduced with the introduction of rectangular large field of view cameras, which can image the whole transverse axis of the body using parallel hole collimators and avoid the introduction of spatial distortions.

Pinhole

This collimator has single small hole at the apex of a tungsten or other heavy metal cone, usually of length (f) 20–25 cm. It is used to produce a magnified inverted image of a small object. The magnification obtained is:

$$M = \text{distance from crystal to apex of collimator}/(\text{distance from collimator to object}) = f/r \quad (V.3.10)$$

The image magnification depends on the object position relative to the physical length of the collimator. When $r < f$, the image size is magnified and when $r > f$, it is minified.

Spatial Resolution and Sensitivity

The equations defining spatial resolution and sensitivity are more complicated than for parallel hole collimators. The effect of magnification or minification on system spatial resolution includes a modification of the intrinsic spatial resolution (Moore et al. 1992; Connolly et al. 1999):

$$R_s = \sqrt{[R_{np}^2 + (R_i^2/M^2)]} \quad (V.3.11)$$

where

R_s is the system spatial resolution
R_{np} is the spatial resolution of the collimator
R_i is the intrinsic spatial resolution

For R_{np}, a number of additional parameters have also to be considered:

d hole diameter, at the collimator face nearer to the crystal
θ orientation of the hole axes from the perpendicular to the face of the crystal
f focal length of the collimator

Converging or Fanbeam

$$R_c \approx [R_p/\cos(\theta)]\,[1 - w/(2\,(f + w))] \quad (V.3.12)$$

$$S_c \approx S_p \cos^2(\theta)\, f^2/(f - r)^2 \quad (V.3.13)$$

Thus spatial resolution and sensitivity vary across the field of view with the best being at the centre, where $\theta = 0$ and $\cos(\theta) = 1$. Also for a point source, sensitivity increases towards the convergence point, where photons are transmitted through all of the holes. Since magnification increases with source to collimator separation, spatial resolution will deteriorate less rapidly with depth in the patient than with the parallel hole collimator. Thus, both the spatial resolution and the detection sensitivity can be improved through a converging or fanbeam collimator.

Diverging

$$R_d \approx [R_p/\cos(\theta)]\,[1 + w/2f] \quad (V.3.14)$$

$$S_d \approx S_p \cos^2(\theta)\,(f + w)^2/(f + w + r)^2 \quad (V.3.15)$$

Pinhole

$$R_{pi} \approx d_{pi} + d_{pi}\, r/l \tag{V.3.16}$$

where d_{pi} = the aperture diameter.

$$S_{pi} \approx d_{pi} \cos^3(\beta)/16\, r^2 \tag{V.3.17}$$

where β is the orientation of the source from the perpendicular to the centre of the crystal.

For a point source, sensitivity decreases severely with source to collimator separation and positioning towards the periphery of the field of view. A hole diameter of 4–6 mm is considered optimum but both the spatial resolution and the sensitivity can be altered by changing the size of the hole using various inserts. A smaller hole will give increased spatial resolution and decreased sensitivity whereas one with a larger aperture will give decreased spatial resolution and increased sensitivity.

References

Connolly LP, Treves ST, Davis RT, Zimmerman RE (1999) Pediatric applications of pinhole magnification imaging. J Nucl Med 40:1896–1901

IPSM (1992) Report 66 quality control of gamma cameras and associated computer systems. Institute of Physical Sciences in Medicine. London

Moore SC, Kouris K, Cullum I (1992) Collimator design for single photon emission tomography. Eur J Nucl Med 19:138–150

Muehllehner G, Dudek J, Meyer R (1976) The influence of hole shape on collimator performance. Phys Med Biol 21:242–250

Sorenson JA, Phelps ME (1987) Physics in nuclear medicine, 2nd edn. Grune and Stratton, New York

Tsui BMW, Gunter DL, Beck RN, Patton J (1996) Physics of collimator design. In: Sandler MP, Patton JA, Coleman RE, Gottschalk A, Wackers FJT, Hoffer PB (eds) Diagnostic nuclear medicine, 3rd edn. Williams and Wilkins, Baltimore, pp 67–79

GFR Single Sample Equations

Abbreviations and Nomenclature

BSA	body surface area
t	sample time, in minutes, following administration
Q_0	injected activity in counts per minute
C(t)	plasma sample activity concentration at time t in counts per minute

The following are the details of single sample models, which can be used for the calculation of GFR.

Models Derived Specifically for Children

Groth (1984) at t=120

referenced to biexponential values

$$-\ln(C(t)\,ECV/Q_0)\,ECV/(t\,g(t)) \qquad (V.4.1)$$

where
$ECV = 5867\ BSA^{1.1792}$
$g(t) = 1.01\,e^{-0.00011t} + 0.538\,e^{-0.0178t}$

Groth and Aasted (1984) at t = 120

referenced to biexponential values

$$[72.295 \ln(t) - 425.41]\, y - 553.124 \ln(t) + 3236.76 \tag{V.4.2}$$

where
$y = \ln[C(t)\,(BSA/Q_0)\,10^7]$

Tauxe et al. (1987) at t = 90

referenced to biexponential values

$$(1):\ -7.17 + 3.86\, Q_0/(1000\, C(t)) \tag{V.4.3}$$

$$(2):\ -7.60 + 3.94\, Q_0/(1000\, C(t)) - 0.00043\, (Q_0/(1000\, C(t)))^2 \tag{V.4.4}$$

$$(3):\ -4.023\, [1 - e^{-0.001}\, \{(Q_0/(1000\, C(t))) - 2.14\}] \tag{V.4.5}$$

Ham and Piepsz (1991) at t = 120

referenced to slope-intercept values with 0.85 correction (Chantler et al. 1969)

$$[2.602\, Q_0/(1000\, C(t))] - 0.273 \tag{V.4.6}$$

Ham and Piepsz (1992) at t = 120

referenced to slope-intercept values with 0.85 correction (Chantler et al. 1969)

$$[2.63\, Q_0/(1000\, C(t))] - 0.48 \tag{V.4.7}$$

Models Derived for Adults

Morgan et al. (1977) at t = 180

referenced to biexponential values

$$-23.92 + 2.78\, [Q_0/(1000\, C(t))] - 0.0111\, [Q_0/(1000\, C(t))]^2 \tag{V.4.8}$$

Chatterton (1978) at t = 180

referenced to biexponential values

$$-246 + 85.2 \ln[Q_0/(1000\, C(t))] \tag{V.4.9}$$

Constable et al. (1979) at t=180

referenced to slope-intercept values with 0.93 correction (Chantler et al. 1969)

$$24.5\sqrt{[Q_0/(1000\,C(t))-6.2]}-67 \tag{V.4.10}$$

Jacobsson (1983) at t=240

referenced to slope-intercept values with correction (Brochner-Mortensen 1972)
GFR is calculated in three stages:

1. approximate clearance value is calculated using the true distribution volume, V:

$$Cl=[1/\{(t/V)+0.0016\}]\ \ln[Q_0/(V\,C(t))]$$

where V is assumed to be 0.2 times body weight

2. modifying factor is calculated:

$$m=0.991-0.00122\ Cl$$

3. corrected clearance value is calculated using $V'=V/m$:

$$GFR=[1/\{(t/V')+0.0016\}]\ \ln[Q_0/(V'\,C(t))] \tag{V.4.11}$$

Russell et al. (1985) at t=190

referenced to biexponential analysis

$$A\ln(Q_0/C(t))+B \tag{V.4.12}$$

where
$A=-0.278t+119.1+2405/t$
$B=2.866t-1222.9-16820/t$

Tauxe (1986) at t=180

referenced to biexponential analysis

$$167.0\,[1-e^{-0.0206\,[\{Q_0/(1000\,C(t))\}-11.47]}] \tag{V.4.13}$$

Fawdry and Gruenewald (1987) at t=180

referenced to slope-intercept values with no correction

$$31.94\sqrt{[Q_0/(1000\,C(t))+16.92]}-161.7 \tag{V.4.14}$$

Waller et al. (1987) at t=240

referenced to slope-intercept values with no correction

$$[Q_0/C(t)]\,(k\,e^{-kt}) \qquad (V.4.15)$$

where k is the clearance constant (min^{-1})

Watson (1992) Modification of Christensen and Groth (1986) at t=240

referenced to biexponential analysis

$$709.6265 - \sqrt{[72792 - 0.7604 \ln\{ECV/(Q_0/C(t))\}\,ECV]}/0.3802 \qquad (V.4.16)$$

where ECV = 8116.6 BSA–28.2

References

Brochner-Mortensen J (1972) A simple method for the determination of glomerular filtration rate. J Clin Lab Invest 30:271–274

Chantler C, Garnett ES, Parsons V, Veall N (1969) Glomerular filtration rate measurement in man by the single injection method using ^{51}Cr-EDTA. Clin Sci 37:169–180

Chatterton BE (1978) Limitations of the single sample tracer method for determining glomerular filtration rate. Br J Radiol 51:981–985

Christensen AB, Groth S (1986) Determination of ^{99m}Tc-DTPA clearance by a single plasma sample method. Clin Physiol 6:579–588

Constable AR, Hussein MM, Albrecht FD, Thompson FD, Philalithis PE, Joekes AM (1979) Single sample estimates of renal clearances. Br J Urol 51:84–87

Fawdry RM, Gruenewald SM (1987) Three-hour volume of distribution method: an accurate simplified method of glomerular filtration rate measurement. J Nucl Med 28:510–513

Groth S (1984) Calculation of ^{51}Cr-EDTA clearance in children from the activity in one plasma sample by transformation of the biexponential plasma time-activity curve into a monoexponential with identical integral area below the time-activity curve. Clin Physiol 4:61–74

Groth S, Aasted M (1984) Chromium-51-EDTA clearance determined by one plasma sample in children. Clin Physiol 4:75–83

Ham HR, Piepsz A (1991) Estimation of glomerular filtration rate in infants and in children using a single-plasma sample method. J Nucl Med 32:1294–1297

Ham HR, Piepsz A (1992) Reply to Peters 1992 (letter). J Nucl Med 33:174

Jacobsson L (1983) A method for calculation of renal clearance based on a single plasma sample. Clin Physiol 3:297–305

Morgan WD, Birks LJ, Sivyer A, Ghose RR (1977) An efficient technique for the simultaneous estimation of GFR and ERPF, involving a single injection and two blood samples. Int J Nucl Med Biol 4:79–83

Russell CD, Bischoff PG, Kontzen FN, Rowell KL, Yester MV, Lloyd LK, Tauxe WN (1985) Measurement of glomerular filtration rate. Single injection plasma clearance method without urine collection. J Nucl Med 26:1243–1247

Tauxe WN (1986) Determination of glomerular filtration rate by single plasma sampling technique following injection of radioiodinated diatrizoate. J Nucl Med 27:45–50

Tauxe WN, Bagachi A, Tepe PG, Krishnaiah PR (1987) Single-sample method for the estimation of glomerular filtration rate in children. J Nucl Med 28:366–371

Waller DG, Keast CM, Fleming JS, Ackery DM (1987) Measurement of glomerular filtration rate with ^{99m}Tc-DTPA: comparison of plasma clearance techniques. J Nucl Med 28:372–377

Watson WS (1992) A simple method of estimating glomerular filtration rate. Eur J Nucl Med 19:827

Typical Acquisition and Processing Protocols for Single Photon Investigations

Abbreviations

GP	general purpose
HE	high energy
LEGP	low energy general purpose
LEHR	low energy high resolution
LEHS	low energy high sensitivity
LEUHR	low energy ultra high resolution
ME	medium energy
UHE	ultra high energy

Typical acquisition protocols, for single photon investigations, are listed in Table V.5.1. Guidance, where published, is listed under reference. Where a zoom is applied, the magnitude depends on the pixel size required to match the system spatial resolution, with the constraint of maintaining a sufficiently large field of view. The pinhole collimator is usually used with a medium (m), 5 mm, aperture. The LEHS collimator is used on short frame duration dynamic studies, except when there is an immediate subsequent acquisition which demands a higher spatial resolution collimator. If dual heads are available for whole body studies, the acquisition duration for $^{99}Tc^{m}$ should not be doubled because the image is probably already limited by spatial resolution rather than by count density. For other radionuclides, the acquisition duration should be doubled.

During processing, threshold and saturation are assumed to be applied in each study. Typical processing protocols are listed in Table V.5.2.

Table V.5.1. Acquisition protocols

	Collimator	Matrix	Zoom	Acquisition type	Termination conditions, frame rate, whole-body speed	Reference
Abdomen					Up to 40 min	
$^{99}Tc^{m}$	LEHR	128	No	ST		
Other radionuclides	GP	64	No	ST		
Bile acid retention	None	64	No	S	600 s	
Bone (perfusion)	LEGP	64	No	D	1–3 s frames for 30–90 s	Donohoe et al. (1996) Kelty et al. (1997) Hahn et al. (2001)
Bone (blood pool)	LEGP	128	No	S	180–300 s 3×10^5 cts	Donohoe et al. (1996) Kelty et al. (1997) Hahn et al. (2001)
Bone (blood pool)	LEGP	1024, 256	No	W	30 cm min^{-1} 10 min total	Kelty et al. (1997) Hahn et al. (2001)
Bone (late phase)	LEHR	256	No	S	$5–15\times10^5$ cts first image Same time on other images	Donohoe et al. (1996) Kelty et al. (1997) Hahn et al. (2001)
Bone (late phase)	LEHR	1024, 256	No	W	8–10 cm min^{-1} 20 min total $2.5–3.5\times10^6$ cts	Donohoe et al. (1996) Kelty et al. (1997) Hahn et al. (2001)
Bone (magnified)	Pinhole m	256	No	S	5–15 min $0.75–1.5\times10^5$ cts	Donohoe et al. (1996) Hahn et al. (2001)
Bone (SPET)	LEUHR	128	No	ST	Up to 40 min	Hahn et al. (2001)
Bone (^{201}Tl early phase)	LEHS	64	No	D	10 s frames for 5 min 60 s frames to 1 h	
Bone (^{201}Tl late phase)	LEGP	128	No	S	4×10^5 cts 10 min	
Bone marrow	LEHR	256	No	S	$3–4\times10^5$ cts 10 min	
Bone marrow	LEHR	1024, 256	No	W	10 cm min^{-1} 20 min total $2.5–3.5\times10^6$ cts	
Brain (perfusion)	LEGP	64	No	D	1 s frames for 1 min	
Brain (blood pool)	LEGP	128	No	S	5 min	
Brain (late phase)	LEGP	128	Yes	S	4×10^5 cts	

Table V.5.1. (Continued)

	Collimator	Matrix	Zoom	Acquisition type	Termination conditions, frame rate, whole-body speed	Reference
Brain (shunt perfusion)	LEGP	64	No	D	10 s frames for 10 min	
Brain (shunt late phase)	LEGP	128	No	S	60 s	
Brain (cisterns $^{99}Tc^{m}$)	LEHR	128	Yes	S	$2–5\times10^{5}$ cts 10 min	
Brain (cisterns ^{111}In)	ME	128	Yes	S	$2–3\times10^{5}$ cts 10 min	
Cerebral perfusion	LEHR fanbeam	128	Yes	ST	5×10^{6} cts 40 min	Tatsch et al. (2002)
Breast (MIBI)	LEHR	128	Yes	S	5 min	
Breast (SPET)	LEHR	128	No	ST	Up to 60 min	
Cardiac (FPRNA)	LEHS	32	No	D	0.025 s frames for up to 50 s	Port (1999) DePuey and Garcia (2001)
Cardiac (L to R shunt)	LEHS	32	No	D	0.1 s frames for 12 s	Port (1999) DePuey and Garcia (2001)
Cardiac (R to L shunt)	LEGP	512, 128	No	W	20 min	
Cardiac (ERNA)						Wittry et al. (1997)
Rest	LEHR LEHR	64	Yes	G	2×10^{4} counts cm^{-2} $3–7\times10^{6}$ cts	Port (1999) DePuey and Garcia (2001)
Exercise	LEGP				>2.5 min	
Dacroscintigraphy (early)	Pinhole s	128	No	D	10 s frames for 5 min	
Dacroscintigraphy (late)	Pinhole s	256	No	S	120 s	
^{52}Fe utilisation	UHE	128	No	S	5 min	
Gallium	ME	128	No	S	10–20 min $0.25–2\times10^{6}$ cts	Bartold et al. (1997) Seabold et al. (1997b)
Gallium	ME	1024, 256	No	W	>450 cts cm^{-1} >6–8 cm min^{-1} 25–35 min total $>1.5–2\times10^{6}$ cts	Bartold et al. (1997) Seabold et al. (1997b)
Gastric emptying ($^{99}Tc^{m}$)	LEGP	64	No	S	30–60 s over 90 min	Donohue et al. (1999)
Gastric emptying (^{111}In)	ME	64	No	S	30–60 s over 90 min	Donohue et al. (1999)
Gastroesophageal reflux	LEGP	64	No	D	30 s frames for 3 min	
GI transit (early phase)	LEHS	64	No	D	60 s frames for 60 min	
GI transit (late phase)	LEHS	64	No	S	60 s	

Table V.5.1. (Continued)

	Collimator	Matrix	Zoom	Acquisition type	Termination conditions, frame rate, whole-body speed	Reference
GI bleed (early phase)	LEGP	128	No	D	10–60 s frames for 60–90 min	Ford et al. (1999)
Haemangioma (early)	LEHS	64	No	D	1 s frames for 1 min	
Haemangioma (late)	LEGP	128	No	S	5 min	
Hepatobiliary (early)	LEGP	128	No	D	60 s frames for 30–60 min	Balon et al. (1997)
Kaposi sarcoma (^{201}Tl)	LEGP	128	No	S	5×10^5 cts	
Kaposi sarcoma (^{99}Tc^{m})	LEGP	128	No	S	5×10^5 cts	
Kaposi sarcoma (^{67}Ga)	ME	128	No	S	5×10^5 cts	
Leucocytes (^{99}Tc^{m})	LEHR	256	No	S	5×10^5 cts	Datz et al. (1997)
					5 min	
Leucocytes (^{111}In)	ME	128	No	S	5×10^5 cts or 600–900 s	Seabold et al. (1997a)
Leucocytes (^{111}In)	ME	1024, 256	No	W	5–6 cm min^{-1}	Seabold et al. (1997a)
					25–35 min	
Levine shunt	LEGP	128	No	S	60 s	
Liver (perfusion)	LEHS	64	No	D	1 s frames for 1 min	Royal et al. (1998)
Liver (blood pool)	LEHS	128	No	S	$1–2\times10^6$ cts	Royal et al. (1998)
Liver / spleen	LEGP	128	No	S	Anterior $5–10\times10^5$ cts,	
					then same time	Royal et al. (1998)
Lung (ventilation/perfusion)	LEGP	128	No	S	$2–5\times10^5$ cts	Parker et al. (1996)
	ME ^{81}Kr^{m}				5 min	
Lung (clearance)	LEHS	64	No	D	60 s frames for 70 min	
Lung (intrapulmonary shunt)	LEGP	64	No	D	0.5 s frames for 1 min	
	LEGP	128	No	S	10 min	
Lung (hepatopulmonary)	LEGP	128	No	S	100 s	
Lymph (early phase)	LEHS	64	No	D	30 s frames for 30 min	
Lymph (late phase)	LEGP	128	No	S	5×10^5 cts	
					10 min	
Lymph (late phase)	LEGP	512, 128	No	W	20 min	
Meckle's diverticulum	LEGP	128	No	D	30–60 s frames for 30–60 min	Ford et al. (1999)
MIBG ^{123}I	LEGP	128	No	S	5×10^5 cts	
					10 min	
MIBG ^{123}I	LEGP	512, 128	No	W	20 min	
MIBG ^{131}I	HE	128	No	S	10 min	
Milk scan (early phase)	LEHS	64	No	D	1 s frames for 5 min	

Table V.5.1. (Continued)

	Collimator	Matrix	Zoom	Acquisition type	Termination conditions, frame rate, whole-body speed	Reference
Milk scan (late phase)	LEGP	128	No	S	300 s	
Myocardial perfusion SPET						Strauss et al. (1998)
$^{99}Tc^{m}$	LEHR	64	Yes	ST	20–30 min	Port (1999)
^{201}Tl	LEGP			ST		DePuey and Garcia (2001)
Gated	LEGP			GST		
Oesophageal transit	LEHS	64	No	D	0.5 s frames for 2 min	
Pentetreotide ^{111}In	ME	128	No	S	5×10^{5} cts	
					10 min	
Pentetreotide ^{111}In	ME	512, 128	No	W	20 min	
Parathyroid (MIBI early)	Pinhole m	256	No	S	10 min	Greenspan et al. (1998)
Parathyroid (MIBI late)	LEHR	128	No	S	10 min	Greenspan et al. (1998)
Platelet ($^{99}Tc^{m}$ early)	LEHS	64	No	D	60 s frames for 20 min	
Platelet ($^{99}Tc^{m}$ late)	LEGP	128	No	S	5 min	
Renal (static)	LEHR	128	Yes	S	2.5–5×10^{5} cts	Mandell et al. (1997b)
					10 min	Prigent et al. (1999)
						Piepsz et al. (2001)
Renal (static) SPET	LEHR	128	No	ST	20–30 min	Mandell et al. (1997b)
Renal (cystography)	LEGP	128	No	D	5 s frames	Gordon et al. (2001a)
						Mandell et al. (1997c)
Renogram (child)	LEGP	64	Yes	D	10–30 s frames for 20–40 min	Gordon et al. (2001b)
						Mandell et al. (1997a)
Renogram (native)	LEGP	128	No	D	2 s frames for 1 min then	
					10–20 s frames for 20–30 min	O'Reilly et al. (1996)
						O'Reilly (2003)
						Prigent et al. (1999)
Renogram (transplant)	LEGP	128	No	D	1 s frames for 1 min then	
					10–20 s frames for 20–30 min	Dubovsky et al. (1999)
Renogram (late)	LEGP	128	No	S	5 min	
Reflux (ureteric)	LEGP	128	No	D	1 s frames for 5 min	
Sialoscintigraphy (early)	LEHR	128	Yes	D	20 s frames for 20 min	
Sialoscintigraphy (late)	LEHR	128	Yes	S	2 min	
Testicular (early)	LEGP	64	Yes	D	1 s frames for 2 min	
Testicular (late)	LEGP	64	Yes	S	7.5×10^{5} cts	
					10 min	

Table V.5.1. (Continued)

	Collimator	Matrix	Zoom	Acquisition type	Termination conditions, frame rate, whole-body speed	Reference
Thyroid ($^{99}Tc^{m}$ or ^{123}I)	Pinhole m	256	No	S	$0.5–1.0\times10^{5}$ cts 5 min	Becker et al. (1996a,b)
Thyroid ($^{99}Tc^{m}$)	Pinhole m	256	No	S	$1–2\times10^{5}$ cts 10 min	Becker et al. (1996a,b)
Thyroid (perchlorate)	Pinhole m	128	No	S	3×10^{4} cts 10 min	
Venography (early)	LEHS	64	No	D	2 s frames for 2 min	
Venography (late)	LEGP	128	No	S	4×10^{5} cts 5 min	
Whole body (^{131}I)	HE	128	No	S	1.5×10^{5} cts 10 min	
Whole body (^{123}I)	LEGP	128	No	S	1.5×10^{5} cts 10 min	
Whole body (^{123}I)	LEGP	512, 128	No	W	20 min	
Whole body (^{201}Tl)	LEGP	128	No	S	5×10^{5} cts 10 min	
Whole body (^{201}Tl)	LEGP	512, 128	No	W	20 min	

Acquisition type: *D* dynamic, *G* gated, *S* static, *ST* SPET, *W* whole body

Table V.5.2. Processing protocols

Bile acid retention	Total counts
Bone (late phase)	ROIs, ratio of counts
Brain (perfusion)	ROIs, activity–time curves
Cardiac (first pass)	ROIs, activity–time curves, ejection fraction
Cardiac (R to L shunt)	ROIs, activity–time curves, ratio of areas under curves
Cardiac (blood pool)	ROIs, activity–time curves, wall motion, various parameters, parametric images
Gallium lung	ROIs, ratio of counts, lung index
Gastric emptying	ROIs, activity–time curves, emptying indices
Gastroesophageal reflux	ROIs, activity–time curves
GI transit (early phase)	ROIs, activity–time curves
Hepatobiliary (early)	ROIs, activity–time curves, ejection fraction
Lacrimal ducts (early)	ROIs, activity–time curves
Liver (dynamic)	ROIs, activity–time curves
Lung (ventilation/perfusion)	Display intensity matching of images
Lung (clearance)	ROIs, activity–time curves, clearance indices
Milk scan (early phase)	ROIs, activity–time curves
Oesophageal transit	ROIs, activity–time curves, condensed images
Renal (static)	ROIs, geometric mean of counts, relative function
Renogram (transplant)	ROIs, activity–time curves, perfusion indices
Renogram	ROIs, activity–time curves, filtration and excretion indices
Reflux (ureteric)	ROIs, activity–time curves, condensed images
Salivary glands (early)	ROIs, activity–time curves
Thyroid ($^{99}Tc^{m}$ or ^{123}I)	ROIs, comparison with counts from phantom
Thyroid (perchlorate)	ROIs, comparison with counts from phantom
Whole body (^{131}I)	ROIs, comparison with counts from phantom
Whole body (^{123}I)	ROIs, comparison with counts from phantom

References

Balon HR, Fink-Bennett DM, Brill DR, Fig LM, Freitas JE, Krishnamurthy GT, Klingensmith WC 3rd, Royal HD (1997) Procedure guideline for hepatobiliary scintigraphy. Society of Nuclear Medicine. J Nucl Med 38:1654–1657

Bartold SP, Donohoe KJ, Fletcher JW, Haynie TP, Henkin RE, Silberstein EB, Royal HD, van den Abbeele A (1997) Procedure guideline for gallium scintigraphy in the evaluation of malignant disease. Society of Nuclear Medicine. J Nucl Med 38:990–994

Becker D, Charles ND, Dworkin H, Hurley J, McDougall IR, Price D, Royal H, Sarkar s (1996a) Procedure guideline for thyroid scintigraphy: 1.0. Society of Nuclear Medicine. J Nucl Med 37: 1264–1266

Becker D, Charkes ND, Dworkin H, Hurley J, McDougall IR, Price D, Royal H, Sarkar s (1996b) Procedure guideline for thyroid uptake measurement: 1.0. Society of Nuclear Medicine. J Nucl Med 37:1266–1268

Datz FL, Seabold JE, Brown ML, Forstrom LA, Greenspan BS, McAfee JG, Palestro CJ, Schauwecker DS, Royal HD (1997) Procedure guideline for technetium-99m-HMPAO-labeled leukocyte scintigraphy for suspected infection/inflammation. Society of Nuclear Medicine. J Nucl Med 38: 987–990

DePuey EG, Garcia EV (2001) Updated imaging guidelines for nuclear cardiology procedures, part 1. J Nucl Cardiol 8: G1-G58

Donohoe KJ, Henkin RE, Royal HD, Brown ML, Collier BD, O'Mara RE, Carretta RF (1996) Procedure guideline for bone scintigraphy: 1.0. Society of Nuclear Medicine. J Nucl Med 37: 1903–1906

Donohoe KJ, Maurer AH, Ziessman HA, Urbain JL, Royal HD (1999) Procedure guideline for gastric emptying and motility. Society of Nuclear Medicine. J Nucl Med 40: 1236–1239

Dubovsky EV, Russell CD, Biscof-Delaloye A, Bubeck B, Chaiwatanarat T, Hilson AJW, Rutland M, Oei HY, Sfakianakis GN, Taylor A Jr (1999) Report of the radionuclides in nephrourology committee for evaluation of transplanted kidney (review of techniques). Semin Nucl Med XXIX: 175–188

Ford PV, Bartold SP, Fink-Bennett DM, Jolles PR, Lull RJ, Maurer AH, Seabold JE (1999) Procedure guideline for gastrointestinal bleeding and Meckel's diverticulum scintigraphy. Society of Nuclear Medicine. J Nucl Med 40: 1226–1232

Gordon I, Colarinha P, Fettich J, Fischer S, Frokier J, Hahn K, Kabasakal L, Mitjavila M, Olivier P, Piepsz A, Porn U, Sixt R, van Velzen J (2001a) Guidelines for indirect radionuclide cystography. Eur J Nucl Med 28: BP16–BP20

Gordon I, Colarinha P, Fettich J, Fischer S, Frokier J, Hahn K, Kabasakal L, Mitjavila M, Olivier P, Piepsz A, Porn U, Sixt R, van Velzen J (2001b) Paediatric Committee of the European Association of Nuclear Medicine. Guidelines for standard and diuretic renography in children. Eur J Nucl Med 28: BP21–BP30

Greenspan BS, Brown ML, Dillehay GL, McBiles M, Sandler MP, Seabold JE, Sisson JC (1998) Procedure guideline for parathyroid scintigraphy. Society of Nuclear Medicine. J Nucl Med 39: 1111–1114

Hahn K, Fischer S, Colarinha P, Gordon I, Mann M, Piepsz A, Olivier P, Sixt R, van Velzen J (2001) Paediatric Committee of the European Association of Nuclear Medicine. Guidelines for bone scintigraphy in children. Eur J Nucl Med 28: BP42–BP47

Kelty NL, Cao Z-J, Holder LE (1997) Technical considerations for optimal orthopedic imaging. Semin Nucl Med XXVII: 328–333

Khalkhali I, Diggles LE, Taillefer R, Vandestreek PR, Peller PJ, Abdel-Nabi HH (1999) Procedure guideline for breast scintigraphy. Society of Nuclear Medicine. J Nucl Med 40: 1233–1235

Mandell GA, Cooper JA, Leonard JC, Majd M, Miller JH, Parisi MT, Sfakianakis GN (1997a) Procedure guideline for diuretic renography in children. Society of Nuclear Medicine. J Nucl Med 38: 1647–1650

Mandell GA, Eggli DF, Gilday DL, Heyman S, Leonard JC, Miller JH, Nadel HR, Treves ST (1997b) Procedure guideline for renal cortical scintigraphy in children. Society of Nuclear Medicine. J Nucl Med 38: 1644–1646

Mandell GA, Eggli DF, Gilday DL, Heyman S, Leonard JC, Miller JH, Nadel HR, Treves ST (1997c) Procedure guideline for radionuclide cystography in children. Society of Nuclear Medicine. J Nucl Med 38: 1650–1654

O'Reilly P (2003) Consensus Committee of the Society of Radionuclides in Nephrourology. Standardization of the renogram technique for investigating the dilated upper urinary tract and assessing the results of surgery. BJU Int 91: 239–243

O'Reilly P, Aurell M, Britton K, Kletter K, Rosenthal L, Testa T (1996) Consensus on diuresis renography for investigating the dilated upper urinary tract. Radionuclides in Nephrourology Group. Consensus Committee on Diuresis Renography. J Nucl Med 37: 1872–1876

Parker JA, Coleman RE, Siegel BA, Sostman HD, McKusick KA, Royal HD. (1996) Procedure guideline for lung scintigraphy: 1.0. Society of Nuclear Medicine. J Nucl Med 37: 1906–1910

Piepsz A, Colarinha P, Gordon I, Hahn K, Olivier P, Roca I, Sixt R, van Velzen J (2001) Paediatric Committee of the European Association of Nuclear Medicine. Guidelines for 99mTc-DMSA scintigraphy in children. Eur J Nucl Med 28: BP37–BP41

Port SC (1999) Imaging guidelines for nuclear cardiology procedures, part 2. American Society of Nuclear Cardiology. J Nucl Cardiol 6: G53–G83

Prigent A, Cosgriff P, Gates GF, Granerus G, Fine EJ, Itoh K, Peters M, Piepsz A, Rehling M, Rutland M, Taylor A Jr (1999) Consensus report on quality control of quantitative measurements of renal function obtained from the renogram: international consensus committee from the scientific committee of radionuclides in nephrourology. Semin Nucl Med XXIX: 146–159

Royal HD, Brown ML, Drum DE, Nagle CE, Sylvester JM, Ziessman HA (1998) Procedure guideline for hepatic and splenic imaging. Society of Nuclear Medicine. J Nucl Med 39: 1114–1116

Seabold JE, Forstrom LA, Schauwecker DS, Brown ML, Datz FL, McAfee JG, Palestro CJ, Royal HD (1997a) Procedure guideline for indium-111-leukocyte scintigraphy for suspected infection/inflammation. Society of Nuclear Medicine. J Nucl Med 38: 997–1001

Seabold JE, Palestro CJ, Brown ML, Datz FL, Forstrom LA, Greenspan BS, McAfee JG, Schauwecker DS, Royal HD (1997b) Procedure guideline for gallium scintigraphy in inflammation. Society of Nuclear Medicine. J Nucl Med 38:994–997

Strauss HW, Miller DD, Wittry MD, Cerqueira MD, Garcia EV, Iskandrian AS, Schelbert HR, Wackers FJ (1998) Procedure guideline for myocardial perfusion imaging. Society of Nuclear Medicine. J Nucl Med 39:918–923

Tatsch K, Asenbaum S, Bartenstein P, Catafau A, Halldin C, Pilowsky LS, Pupi A (2002) European Association of Nuclear Medicine procedure guidelines for brain perfusion SPET using (99m)Tc-labelled radiopharmaceuticals. Eur J Nucl Med 29:BP36–BP42

Wittry MD, Juni JE, Royal HD, Heller GV, Port SC (1997) Procedure guideline for equilibrium radionuclide ventriculography. Society of Nuclear Medicine. J Nucl Med 38:1658–1661

Subject Index

A

B

V

W

X

Z

Printing: Mercedes-Druck, Berlin
Binding: Stein+Lehmann, Berlin